MW00913897

Handbook of Multilevel Analysis

Jan de Leeuw · Erik Meijer
Editors

Handbook of Multilevel Analysis

Foreword by Harvey Goldstein

Springer

Editors
Jan de Leeuw
Department of Statistics
University of California at Los Angeles
Los Angeles, CA 90095-1554
USA
deleeuw@stat.ucla.edu

Erik Meijer
RAND Corporation
1776 Main St, PO Box 2138
Santa Monica, CA 90407-2138
USA
meijer@rand.org

Egret® is a registered trademark of Cytel Software Corporation.

SAS® and all other SAS Institute Inc. product or service names are registered trademarks or trademarks of SAS Institute Inc. in the USA and other countries.

S-PLUS® is either a registered trademark or trademark of Insightful Corporation in the United States and/or other countries.

SPSS® is a registered trademark of SPSS Inc.

Stata® is a registered trademark of StataCorp LP.

ISBN: 978-0-387-73183-4 e-ISBN: 978-0-387-73186-5

Library of Congress Control Number: 2007938348

© 2008 Springer Science+Business Media, LLC
All rights reserved. This work may not be translated or copied in whole or in part without the written permission of the publisher (Springer Science+Business Media, LLC, 233 Spring Street, New York, NY 10013, USA), except for brief excerpts in connection with reviews or scholarly analysis. Use in connection with any form of information storage and retrieval, electronic adaptation, computer software, or by similar or dissimilar methodology now known or hereafter developed is forbidden.
The use in this publication of trade names, trademarks, service marks and similar terms, even if they are not identified as such, is not to be taken as an expression of opinion as to whether or not they are subject to proprietary rights.

Printed on acid-free paper.

9 8 7 6 5 4 3 2 1

springer.com

Foreword

Social and medical researchers have long been concerned about the need properly to model complex data structures, especially those where there is a hierarchical structure such as pupils nested within schools or measurements nested within individuals. Statisticians, especially those involved in survey sampling, recognise that failure to take account of such structures in standard models can lead to incorrect inferences. What has been less well appreciated is that a failure to properly model complex data structures makes it impossible to capture that complexity that exists in the real world. It is only in the last 20 years or so, when appropriate and efficient model-based methods have become available to deal with this issue, that we have come to appreciate the power that more complex models provide for describing the world and providing new insights. This book sets out to present some of the most recent developments in what has come to be known as multilevel modelling.

An introductory chapter by de Leeuw and Meijer gives a brief history and a standard exposition of the basic multilevel model involving random coefficients at level 2 and above, together with a discussion of some likelihood-based estimation procedures. This is followed by a chapter by Draper that outlines a Bayesian approach to modelling multilevel structures using the MCMC algorithm, with a clear exposition of the rationale for such an approach and well worked through examples. This is as good an introduction as any to Bayesian analysis and MCMC estimation. The next chapter by Snijders and Berkhof deals with the important issue of diagnostics for multilevel models. It takes the reader carefully through the various model assumptions and how they can be examined, for example, making use of model elaborations and residual analysis. There is also a useful section on smoothing models. Moerbeek, van Breukelen and Berger look at ways of optimally sampling units in multilevel models. It includes clear examples for Normal and generalised linear models with useful discussions of repeated measures and schooling designs. Raudenbush contributes a chapter where he looks at the inferential problems that can arise when, in a 2-level model, the number of level-1 units per level-2 unit is small. He gives some examples, such as matched pairs and cluster randomised trials and explains how these can be interpreted and there is a brief discussion of issues in generalised linear models. The chapter by Hedeker deals in detail with discrete responses, either ordered or nominal. It has a clear exposition with useful examples. Skrondal and Rabe-Hesketh discuss models for longitudinal repeated measures data, including those with serial dependency structures, for Normal and discrete responses. Well-motivated examples are used for the exposition. Rasbash and Browne show how cross-classified and multiple membership structures can be modelled. They provide examples and a convincing exposition of why researchers should be looking beyond mere hierarchies when analysing real-life data. Rodriguez looks at generalised linear models with particular reference to survival data and gives a detailed discussion of various estimation algorithms, together with a useful example. Longford provides a chapter on missing data, where he describes the use of the EM algorithm and random multiple imputation. Van der Leeden, Meijer and Busing, in a comprehensive account, take a careful look at bootstrap and jackknife procedures for studying bias and for obtaining valid standard errors and confidence intervals in multilevel models. Finally, the du Toits present an account of multilevel structural equation models with some useful examples and detailed derivations.

The book covers a great number of important topics and there is a useful amount of cross-referencing with a good number of worked examples. The amount of methodological activity now underway is very impressive, and as these become incorporated into software packages, they will hopefully persuade researchers to undertake data analysis that more closely reflects the structure of real-world data than traditional methods assume. Most of the developments discussed leave room for further work. As hardware becomes more powerful, certain options will become more attractive. This is especially the case with resampling methods such as the bootstrap, multiple imputation and MCMC and these do seem to be where we may expect the most interesting future developments. In particular, given what is happening more generally, we should expect MCMC methods to become more and more prevalent. Not only do they allow proper Bayesian inference, especially for small samples, as emphasised by Draper, they also have great potential because of the modularity of the algorithm steps. This is clearly demonstrated in the chapter by Rasbash and Browne, where, as they point out, certain kinds of data simply cannot be treated properly using maximum likelihood.

So, apart from the increasing adoption of MCMC methods, what might be useful future directions for research? Several of these areas are described in this volume. I would single out cross-classified and multiple membership models that move us on from the consideration of simple hierarchies. It is very rare in the real world to find structures that are purely hierarchical. In education, students will typically belong to a school hierarchy at the same time as a neighbourhood hierarchy, both of which may influence the outcome variable of interest. In addition students may move among neighbourhoods or schools so that assignment to a single higher-level unit may be misleading and lead to important biases. As Rasbash and Browne describe, many other areas of the biological and social sciences have these structures and this provides an exciting and fruitful challenge for multilevel techniques.

I would also emphasise missing data procedures with nonrandom missingness, since problems such as nonresponse in surveys are becoming acute in many places. As Longford suggests, the existence of additional or "auxiliary" information in surveys can be especially useful in allowing the application of existing missing data procedures to handle informative nonresponse.

We also need good diagnostic procedures to test the assumptions of our models and more work here would be very useful, for example in testing the validity of the standard assumption of multivariate Normality. It is particularly important that these procedures are brought within existing modelling packages so that their use is encouraged.

Likewise, another large area of interest is in latent variable models of all kinds, including complex ones such as latent growth trajectory models. The application of multilevel latent structure models with binary and ordered responses is an important area for psychometrics where much current activity under the heading of item response modelling often ignores the inherent hierarchical structures.

Despite the wide coverage of the topics that are dealt with, there are also areas that are not so well covered in this book, which is inevitable in a rapidly changing field.

Thus, measurement and misclassification errors, while mentioned briefly, are not treated in depth, yet we know that ignoring them can have profound effects on inferences. In educational and medical research, for example, they abound and are often correlated, and we need research on both how to estimate measurement error variances and covariances and misclassification probabilities and then how to incorporate these estimates into our models.

Multivariate models are not as well covered as I would wish, since they are becoming more extensively used. An interesting problem is where there are multiple responses at more than one level together. Such models have important applications to prediction problems, multi-process modelling and multiple imputation. An example of the first case is where we have both repeated measures data on individuals and subsequent individual level measures we wish to predict, as in growth studies. Likewise, in multi-process models we may wish to jointly model, say, pupil responses together with teacher or school level variates, for prediction or adjustment purposes, as well as moving us towards better causal understandings. For imputation procedures we often need jointly to model responses at several levels if these variables have missing values. Additionally, in all these cases our responses may be mixtures of continuous and discrete variables, and this presents an additional challenge.

Whilst all these methodological developments are exciting and important, the methodological community still has the task of communicating them to potential users. As with all new techniques this requires a combination of clear exposition together with suitable software tools. Many of the authors of chapters in this book have themselves provided such combinations, but more is needed as the methodology advances. Nevertheless, we do need to be careful that we are not promoting multilevel modelling as a kind of magic wand that can transmute bad data into good or turn a poor design into a highly efficient one. Sensitivity to assumptions and accessible ways of investigating those assumptions are things we need continually to emphasise.

Finally, the editors are to be congratulated on bringing together a distinguished group of authors all of whom have interesting things to say. This volume gives us an insight into much current research and will hopefully attract others into this important area of activity.

November 2007

Harvey Goldstein
Professor of Social Statistics
University of Bristol

Contents

List of Contributors

Martijn P. F. Berger
Maastricht University
Department of Methodology
and Statistics
PO Box 616
6200 MD Maastricht
THE NETHERLANDS
martijn.berger@stat.unimaas.nl

Johannes Berkhof
VU University Medical Center
Department of Clinical Epidemiology
& Biostatistics
P.O. Box 7057
1007 MB Amsterdam
THE NETHERLANDS
h.berkhof@vumc.nl

William J. Browne
University of Bristol
Department of Clinical Veterinary
Science
Langford House
Langford
North Somerset BS40 5DU
UNITED KINGDOM
William.Browne@bristol.ac.uk

Frank M. T. A. Busing
Leiden University
Department of Psychology
PO Box 9555
2300 RB Leiden
THE NETHERLANDS
busing@fsw.leidenuniv.nl

Jan de Leeuw
Department of Statistics
University of California
at Los Angeles
Los Angeles, CA 90095-1554
USA
deleeuw@stat.ucla.edu

David Draper
Department of Applied Mathematics
and Statistics
Baskin School of Engineering
University of California
1156 High Street
Santa Cruz, CA 95064
USA
draper@ams.ucsc.edu

Mathilda du Toit
Scientific Software International
7383 N. Lincoln Ave., Suite 100
Lincolnwood, IL 60712-1747
USA
mdutoit@aol.com

Stephen H. C. du Toit
Scientific Software International
7383 N. Lincoln Ave., Suite 100
Lincolnwood, IL 60712-1747
USA
sdutoit@ssicentral.com

Donald Hedeker
University of Illinois at Chicago
School of Public Health
Division of Epidemiology
and Biostatistics
1603 West Taylor Street, Room 955
Chicago, IL 60612-4336
USA
hedeker@uic.edu

Nicholas T. Longford
SNTL
99 Swallows Croft
Reading RG1 6EH
UNITED KINGDOM
ntl@sntl.co.uk

Erik Meijer
RAND Corporation
1776 Main St
PO Box 2138
Santa Monica, CA 90407-2138
USA
meijer@rand.org

Mirjam Moerbeek
Utrecht University
Department of Methodology
and Statistics
PO Box 80140
3508 TC Utrecht
THE NETHERLANDS
m.moerbeek@fss.uu.nl

Sophia Rabe-Hesketh
Graduate School of Education
& Graduate Group in Biostatistics
University of California
3659 Tolman Hall
Berkeley, CA 94720
USA
sophiarh@berkeley.edu

Jon Rasbash
University of Bristol
Graduate School of Education
Helen Wodehouse Building
35 Berkeley Square
Clifton
Bristol BS8 1JA
UNITED KINGDOM
j.rasbash@bristol.ac.uk

Stephen W. Raudenbush
University of Chicago
Department of Sociology
1126 East 59th Street
Room SSR 416
Chicago, IL 60637
USA
sraudenb@uchicago.edu

Germán Rodríguez
Office of Population Research
Wallace Hall
Princeton University
Princeton, NJ 08544-2091
USA
grodri@princeton.edu

Anders Skrondal
Department of Statistics
and Methodology Institute
Columbia House
London School of Economics
Houghton Street
London WC2A 2AE
UNITED KINGDOM
a.skrondal@lse.ac.uk

Tom A. B. Snijders
Department of Statistics
University of Oxford
Nuffield College
New Road
Oxford OX1 1NF
UNITED KINGDOM
tom.snijders@nuffield.ox.ac.uk

Gerard J. P. Van Breukelen
Maastricht University
Department of Methodology
and Statistics
PO Box 616
6200 MD Maastricht
THE NETHERLANDS
gerard.vbreukelen@stat.
unimaas.nl

Rien van der Leeden
Leiden University
Department of Psychology
PO Box 9555
2300 RB Leiden
THE NETHERLANDS
vanderleeden@fsw.leidenuniv.nl

1

Introduction to Multilevel Analysis

Jan de Leeuw[1] and Erik Meijer[2]

[1] Department of Statistics, University of California at Los Angeles
[2] University of Groningen, Faculty of Economics and RAND Corporation

1.1 History

A common assumption in much of classical statistics is that observations are independently and identically distributed (or i.i.d.). In regression analysis, using the linear model, we cannot insist on identical distributions, because observations differ in expected value, but we generally continue to insist on independence. In fact, we continue to assume that the stochastic parts of the model, i.e., the *errors* or *disturbance terms*, are still i.i.d.

In educational statistics, and in various areas of quantitative sociology, researchers early on began looking for statistical techniques that could incorporate both information about individuals and information about groups to which these individuals belonged. They realized that one of the most challenging aspects of their discipline was to integrate *micro* and *macro* information into a single model. In particular, in the applications educational statisticians had in mind, students are nested in classes, and classes are nested in schools. And perhaps schools are nested in districts, and so on. We have predictors for variables of all these *levels*, and the challenge is to combine all these predictors into an appropriate statistical analysis, more specifically a regression analysis.

Previously, these problems had been approached by either *aggregating* individual-level variables to the group level or *disaggregating* group-level variables to the individual level. It was clear that both these two strategies were unpleasantly ad hoc and could introduce serious biases. Trying to integrate the results of such analyses, for instance by using group-level variables in individual-level regressions, was known as *contextual analysis* [9] or *ecological regression* [42]. It resulted in much discussion about *cross-level inference* and the possibility, or even the unavoidability, of committing an *ecological fallacy* [104].

J. de Leeuw, E. Meijer (eds.), *Handbook of Multilevel Analysis*,
© Springer 2008

In *school effectiveness research*, which became popular in the 1970s following the epochal studies of Coleman et al. [22] and Jencks et al. [62], educational researchers realized early on that taking group structure into account could result in dependencies between the individual observations. Economists and biostatisticians involved in agriculture and breeding had realized this earlier and had designed *variance and covariance component models* for the Analysis of Variance. But in school effectiveness research a somewhat different paradigm developed, which looked at dependencies in a more specific way. The emphasis was on regression analysis and on data of two levels, let's say students and schools. Performing a regression analysis for each school separately was not satisfactory, because often samples within schools were small and regression coefficients were unstable. Also, these separate analyses ignored the fact that all the schools were part of the same school system and that, consequently, it was natural to suppose the regression coefficients would be similar. This similarity should be used, in some way or another, to improve stability of the regression coefficients by what became known as *borrowing strength.* Finally, in large scale studies there were thousands of schools and long lists of regression coefficients did not provide enough data reduction to be useful.

On the other hand, requiring the regression coefficients in all schools to be the same was generally seen as much too restrictive, because there were many reasons why regressions within schools could be different. In some schools, test scores were relatively important, while in others, socio-economic status was a much more dominant predictor. Schools clearly differed in both average and variance of school success. Of course, requiring regression coefficients to be constant did provide a large amount of data reduction, and a small sampling variance, but the feeling was that the resulting regression coefficients were biased and not meaningful.

Thus, some intermediate form of analysis was needed, which did not result in a single set of regression coefficients, but which also did not compute regression coefficients separately for each school. This led naturally to the idea of random coefficient models, but it left open the problem of combining predictors of different levels into a single technique. In the early 1980s, Burstein and others came up with the idea of using the first-stage regression coefficients from the separate within-school regressions as dependent variables in a second-stage regression on school-level predictors. But in this second stage, the standard regression models that assumed independent observations could no longer be used, mainly because they resulted in inefficient estimates of the regression coefficients and biased estimates of their standard errors. Clearly, the first-stage regression coefficients could have widely different standard errors, because predictors could have very different distributions in different schools. The size of the school, as well as the covariance of the predictors within schools, determined the dispersions of the within-school regression coefficients. Typical of this stage in educational research are Langbein [71], Burstein et al. [15], and Burstein [14]. Attempts were made to estimate the second-stage regression coefficients by weighted least squares techniques, or to adjust in some other way for the bias in the standard errors [11, 50, 118]. These attempts were not entirely successful, because at the time the statistical aspects of these two-stage techniques were somewhat baffling. A more extensive historical overview of contextual analysis and Burstein's *slopes-as-outcomes* research is in de Leeuw and Kreft [28] and Kreft and de Leeuw [67].

It became clear, in the mid-1980s, that the models the educational researchers were looking for had already been around for quite some time in other areas of statistics. Under different names, to be sure, and usually in a slightly different form. They were known either as *mixed linear models* [51] or, in a Bayesian context, as *hierarchical linear models* [72]. The realization that the problems of contextual analysis could be imbedded in this classical linear model framework gave rise to what we now call *multilevel analysis.* Thus, multilevel analysis can be defined as the marriage of contextual analysis and traditional statistical mixed model theory.

In rapid succession the basic articles by Mason et al. [81], Aitkin and Longford [2], de Leeuw and Kreft [28], Goldstein [44], and Raudenbush and Bryk [100] appeared. All these articles were subsequently transformed into successful textbooks [46, 67, 76, 101]. The two major research groups in educational statistics led, respectively, by Goldstein and by Raudenbush produced and maintained major software packages [97, 102]. These textbooks and software packages, together with subsequent textbooks, such as Snijders and Bosker [111] and Hox [59], solidified the definition and demarcation of the field of multilevel analysis.

1.2 Application Areas

We have seen that multilevel analysis, at least as we have defined it, started in the mid-1980s in educational measurement and sociology. But it became clear quite rapidly that once you have discovered ways to deal with hierarchical data structures, you see them everywhere. The notion of individuals, or any other type of objects, that are naturally nested in groups, with membership in the same group leading to a possible correlation between the individuals, turned out to be very compelling in many disciplines. It generalizes the notion of intraclass correlation to a regression context. Moreover, the notion of regressing regression coefficients, or using slopes-as-outcomes, is an appealing way to code interactions and to introduce a particular structure for the dependencies within groups.

Survey Data

Many surveys are not simple random samples from a relatively homogeneous population, but are obtained from nested sampling in heterogeneous subgroups. Larger units (e.g., states) are drawn first; within these larger units, smaller units (e.g., counties) are drawn next; and so forth. Large surveys typically contain multiple levels of nesting. Sometimes, all units from a certain level are included, as with stratification. See, e.g., Muthén and Satorra [84] for some examples of the complicated sampling schemes used in survey design. The reason for such a complicated nesting structure of surveys is, of course, that it is assumed that the units are different in some respect. It is then natural to model the heterogeneity between groups through multilevel models. See, e.g., Skinner et al. [109] for a book-length discussion of many aspects of the analysis of survey data.

Repeated Measures

In *repeated measures models* (including *growth study models*) we have measurements on a number of individuals that are replicated at a number of fixed time points. Usually there is only a single outcome variable, but the generalization to multivariate outcomes is fairly straightforward. In addition, it is not necessary that all individuals be measured at the same time points. There can be missing data, or each individual can be measured at different time points. The number of books and articles on the analysis of repeated measures is rapidly approaching infinity, but in the context of multilevel analysis, the key publications are Strenio et al. [116] and Jennrich and Schluchter [63]. Chapter 7 of this volume discusses models for longitudinal data. For an extensive treatment of these longitudinal models in the more general context of mixed linear models, we refer to Verbeke and Molenberghs [122].

A different type of "repeated measures" is obtained with *conjoint choice* or *stated preference* data. With such data, subjects are asked to choose between several hypothetical alternatives, e.g., different products or different modes of transport, defined by a description of their alternatives. When subjects are given more than one choice task, a multilevel structure is induced by the repeated choices of the same individual. The corresponding models for such data are usually more straightforward multilevel models than in the case of longitudinal data, where problems such as dynamic dependence, causing non-interchangeability of the observations, and attrition (selective dropout of the sample) often have to be faced. See, e.g., Rouwendal and Meijer [105] for a multilevel logistic regression (or mixed logit) analysis of stated preference data. Similar data are common in experimental psychology, where multiple experiments are performed with the same subjects.

Twin Studies

In school-based attainment studies we often deal with a fairly small number of rather large groups. But the opposite can also occur, either by the nature of the problem or by design. We can decide to use only a small number of students from each class. Or, in repeated measures studies, we can only have two measurements per individual (a "before" and "after", for instance, with a treatment in between). Another "small groups" example is the twin study, in which group size is typically two. See Chapter 5 for a discussion of this type of data.

Meta-Analysis

Data, including historical data, are now much more accessible than in the past. Many data sets are online or are included in some way or another with published research. This makes it attractive to use previous data sets studying the same scientific problem to get larger sample sizes and perhaps a larger population to generalize to. Such (quantitative) analysis of data or results from multiple previous studies is called *meta-analysis.* In Raudenbush and Bryk [99], multilevel techniques specifically adapted to meta-analysis were proposed. Compare also Raudenbush and Bryk [101, Chapter 7].

Multivariate Data

There is a clever way, used by Goldstein [46, Chapter 6], to fit general multivariate data into the multilevel framework. If we have n observations on m variables, we can think of these m observations as nested in n groups with m group members each. This amounts to thinking of the $n \times m$ data matrix as a long vector with nm elements and then building the model with the usual regression components and a suitable specification for the dispersion of the within-group disturbances. It is quite easy to incorporate missing data into this framework, because having data missing simply means having fewer observations in some of the groups. On the other hand, in standard multilevel models, parameters such as regression coefficients are the same for different observations within the same group, whereas in multivariate analysis, this is rarely the case. Thus, writing the latter as a multilevel model requires some care.

1.3 Chapter Outline

In this first chapter of the Handbook we follow the general outline of de Leeuw and Kreft [29]. After this introduction, we first discuss the *statistical models* used in multilevel analysis, then we discuss the *loss functions* used to measure badness-of-fit, then the *techniques* used to minimize the loss functions, and, finally, the *computer programs* written for these techniques. By using these various steps in the development of multilevel statistical methods, it is easy to discuss the contributions of various authors. It can be used, for instance, to show that the most influential techniques in the field carefully discuss (and implement) all these sequential steps in the framework. After a section on sampling weights, we give an empirical illustration, in which much of the theory discussed in this chapter will be applied. We close with a few final remarks and appendixes that discuss notation and other useful technical background.

1.4 Models

A statistical model is a functional relationship between random variables. The observed data are supposed to be a realization of these random variables, or of a measurable function of these random variables. In most cases, random variables are only partly specified because we merely assert that their distribution belongs to some parametric family. In that case, the model is also only partly specified, and one of the standard statistical chores is to estimate the values of the unknown parameters.

In this section we discuss the multilevel model in the linear case in which there are, at least initially, only two levels. Nonlinear and multivariate generalizations will be discussed in later chapters of this handbook. We also relate it to variance components and mixed models, which, as we have mentioned above, have been around much longer.

Notation is explained in detail in Appendix 1.A. Our main conventions are to underline random variables and to write vectors and matrices in boldface.

1.4.1 Mixed Models

The *mixed linear model* or MLM is written as

$$\underline{\boldsymbol{y}} = \boldsymbol{X\beta} + \boldsymbol{Z}\underline{\boldsymbol{\delta}} + \underline{\boldsymbol{\epsilon}}, \tag{1.1}$$

with $\boldsymbol{X}[n,r]$, $\boldsymbol{Z}[n,p]$, and

$$\begin{pmatrix}\underline{\boldsymbol{\epsilon}}\\ \underline{\boldsymbol{\delta}}\end{pmatrix} \sim \mathcal{N}\left(\begin{pmatrix}\boldsymbol{\emptyset}\\ \boldsymbol{\emptyset}\end{pmatrix}, \begin{pmatrix}\boldsymbol{\Sigma} & \boldsymbol{\emptyset}\\ \boldsymbol{\emptyset} & \boldsymbol{\Omega}\end{pmatrix}\right).$$

To simplify the notation, we suppose throughout this chapter that both $\boldsymbol{X}$ and $\boldsymbol{Z}$ have full column rank.

The regression part of the model has a component with fixed regression coefficients and a component with random regression coefficients. Clearly,

$$\underline{\boldsymbol{y}} \sim \mathcal{N}(\boldsymbol{X\beta}, \boldsymbol{V}),$$

with

$$\boldsymbol{V} \stackrel{\Delta}{=} \boldsymbol{Z}\boldsymbol{\Omega}\boldsymbol{Z}' + \boldsymbol{\Sigma}. \tag{1.2}$$

This illustrates the consequences of making regression coefficients random. We see that the effects of the predictors in $\boldsymbol{Z}$ are shifted from the expected values to the dispersions of the normal distribution. We also see that MLM is a linear regression model with a very specific dispersion structure for the residuals. The form of the dispersion matrix for the residuals in (1.2) is somewhat reminiscent of the common factor analysis model [63], and this similarity can be used in extending multilevel models to covariance structure and latent variable models (see Chapter 12).

It is convenient to parametrize both dispersion matrices $\boldsymbol{\Sigma}$ and $\boldsymbol{\Omega}$ using vectors of parameters $\boldsymbol{\sigma}$ and $\boldsymbol{\xi}$. From now on we actually assume that $\boldsymbol{\Sigma}$ is *scalar*, i.e., $\boldsymbol{\Sigma} = \sigma^2\boldsymbol{I}$. A scalar dispersion matrix means we assume the disturbances $\underline{\boldsymbol{\epsilon}}$ are *homoskedastic*. This guarantees that if there are no random effects, i.e., if $\underline{\boldsymbol{\delta}}$ is zero almost everywhere, then we recover the classical linear model. We also parametrize $\boldsymbol{\Omega}$ as a *linear structure*, i.e., a linear combination of known matrices $\boldsymbol{C}_g$. Thus,

$$\boldsymbol{\Omega} = \xi_1\boldsymbol{C}_1 + \cdots + \xi_G\boldsymbol{C}_G = \sum_{g=1}^{G} \xi_g\boldsymbol{C}_g, \tag{1.3}$$

and, consequently, $\boldsymbol{V}$ also has linear structure

$$\boldsymbol{V} = \xi_1\boldsymbol{Z}\boldsymbol{C}_1\boldsymbol{Z}' + \cdots + \xi_G\boldsymbol{Z}\boldsymbol{C}_G\boldsymbol{Z}' + \sigma^2\boldsymbol{I} = \sum_{g=1}^{G} \xi_g\boldsymbol{Z}\boldsymbol{C}_g\boldsymbol{Z}' + \sigma^2\boldsymbol{I}.$$

The leading example is obtained when $\boldsymbol{\Omega} = (\omega_{kl})$ is completely free, apart from symmetry requirements. Then

$$\boldsymbol{\Omega} = \omega_{11}(\boldsymbol{e}_1\boldsymbol{e}_1') + \omega_{21}(\boldsymbol{e}_2\boldsymbol{e}_1' + \boldsymbol{e}_1\boldsymbol{e}_2') + \cdots + \omega_{pp}(\boldsymbol{e}_p\boldsymbol{e}_p'),$$

with $\boldsymbol{e}_k$ the k-th unit vector, i.e., the k-th column of $\boldsymbol{I}$, $\{\xi_1, \ldots, \xi_G\} = \{\omega_{11}, \omega_{21}, \ldots, \omega_{pp}\}$, and $\{\boldsymbol{C}_1, \ldots, \boldsymbol{C}_G\} = \{\boldsymbol{e}_1\boldsymbol{e}_1', \boldsymbol{e}_2\boldsymbol{e}_1' + \boldsymbol{e}_1\boldsymbol{e}_2', \ldots, \boldsymbol{e}_p\boldsymbol{e}_p'\}$. Another typical example is a restricted version of this where ω_{kl} is a given constant (such as 0) for some values of (k, l). These two examples cover the vast majority of specifications used in multilevel analysis.

In some cases it is useful to write models in scalar notation. Scalar notation is, in a sense, more constructive because it is closer to actual implementation on a computer. Also, it is useful for those who do not speak matrix algebra. In this notation, (1.1) becomes, for example,

$$\underline{y}_i = \sum_{q=1}^{r} x_{iq}\beta_q + \sum_{s=1}^{p} z_{is}\underline{\delta}_s + \underline{\epsilon}_i,$$

or

$$\underline{y}_i = x_{i1}\beta_1 + \cdots + x_{ir}\beta_r + z_{i1}\underline{\delta}_1 + \cdots + z_{ip}\underline{\delta}_p + \underline{\epsilon}_i.$$

A *two-level MLM*, which explicitly takes the group structure into account, is given by

$$\underline{\boldsymbol{y}}_j = \boldsymbol{X}_j\boldsymbol{\beta} + \boldsymbol{Z}_j\underline{\boldsymbol{\delta}}_j + \underline{\boldsymbol{\epsilon}}_j, \tag{1.4a}$$

with $j = 1, \ldots, m$, and

$$\begin{pmatrix} \underline{\boldsymbol{\epsilon}}_j \\ \underline{\boldsymbol{\delta}}_j \end{pmatrix} \sim \mathcal{N}\left(\begin{pmatrix} \emptyset \\ \emptyset \end{pmatrix}, \begin{pmatrix} \boldsymbol{\Sigma}_j & \emptyset \\ \emptyset & \boldsymbol{\Omega}_j \end{pmatrix}\right). \tag{1.4b}$$

and, using $\perp$ for independence,

$$(\underline{\boldsymbol{\epsilon}}_j, \underline{\boldsymbol{\delta}}_j) \perp (\underline{\boldsymbol{\epsilon}}_\ell, \underline{\boldsymbol{\delta}}_\ell) \tag{1.4c}$$

for all $j \neq \ell$.

As before, we assume that $\boldsymbol{\Sigma}_j = \sigma_j^2\boldsymbol{I}$, while, in addition, we assume that $\boldsymbol{\Omega}_j = \boldsymbol{\Omega}$. Thus,

$$\underline{\boldsymbol{y}}_j \sim \mathcal{N}(\boldsymbol{X}_j\boldsymbol{\beta}, \boldsymbol{V}_j),$$

with

$$\boldsymbol{V}_j \stackrel{\Delta}{=} \boldsymbol{Z}_j\boldsymbol{\Omega}\boldsymbol{Z}_j' + \sigma_j^2\boldsymbol{I},$$

and the $\underline{\boldsymbol{y}}_j$ for different j are independent.

Observe that the assumption that the $\boldsymbol{X}_j$ and the $\boldsymbol{Z}_j$ have full column rank can be quite restrictive in this case, because we could be dealing with many small groups (as in Chapter 5).

In most applications of multilevel analysis, it is assumed that all σ_j^2 are the same, so $\sigma_j^2 = \sigma^2$ for all j. This is not always a realistic assumption and, therefore, most of our discussion will use separate variances. This has its drawbacks as well, because, obviously, the number of parameters increases with the number of groups in the sample. Thus, when the sample consists of, say, 1000 schools, we would estimate 1000 variance parameters, which is unattractive. Furthermore, consistent estimation of σ_j^2 requires group sizes to diverge to infinity, and therefore in a practical sense, good estimators of σ_j^2 would require moderate within-group sample sizes (e.g., $n_j = 30$). In applications with many small groups, this is obviously not the case.

We can view $\sigma_j^2 = \sigma^2$ as a no-between-groups variation specification and all σ_j^2 treated as separate parameters as a fixed effects specification. From this, it seems that it would be in the spirit of multilevel analysis to treat σ_j^2 as a random parameter, $\underline{\sigma}_j^2$, and use a specification like

$$\log \underline{\sigma}_j^2 = \boldsymbol{z}_{j,p+1}'\gamma_{p+1} + \underline{\delta}_{j,p+1},$$

with, say, $\underline{\delta}_{j,p+1} \sim \mathcal{N}(0, \omega_{p+1,p+1})$, which may be correlated with the other random terms. Such a specification is uncommon in multilevel analysis, but it would be particularly straightforward to incorporate in the Bayesian approach to multilevel analysis (Chapter 2). In the Bayesian approach, it is more common to use Gamma or inverse Gamma distributions for variance parameters though, but adaptation of this specification to such distributions is fairly easy.

We will not further discuss specification of $\underline{\sigma}_j^2$ as random parameters in this chapter, and treat σ_j^2 as separate parameters. For a specification with $\sigma_j^2 = \sigma^2$, most expressions are unaltered except for dropping the j subscript. However, there are some instances where the differences are a little bit more pronounced, e.g., in the derivatives of the loglikelihood functions. Then we will indicate how the expressions change. Thus, we cover both specifications.

1.4.2 Random Coefficient Models

The *random coefficient model* or RCM is the model with

$$\underline{\boldsymbol{y}} = \boldsymbol{X}\underline{\boldsymbol{\beta}} + \underline{\boldsymbol{\epsilon}},$$
$$\underline{\boldsymbol{\beta}} = \boldsymbol{\beta} + \underline{\boldsymbol{\delta}},$$

with

$$\begin{pmatrix} \underline{\boldsymbol{\epsilon}} \\ \underline{\boldsymbol{\delta}} \end{pmatrix} \sim \mathcal{N}\left(\begin{pmatrix} \boldsymbol{\emptyset} \\ \boldsymbol{\emptyset} \end{pmatrix}, \begin{pmatrix} \boldsymbol{\Sigma} & \boldsymbol{\emptyset} \\ \boldsymbol{\emptyset} & \boldsymbol{\Omega} \end{pmatrix} \right).$$

Obviously, in an RCM we have

$$\underline{\boldsymbol{y}} = \boldsymbol{X}\boldsymbol{\beta} + \boldsymbol{X}\underline{\boldsymbol{\delta}} + \underline{\boldsymbol{\epsilon}},$$

which shows that the RCM is an MLM in which $\boldsymbol{Z} = \boldsymbol{X}$.

The RCM in this form is not very useful, because without additional assumptions, it is not identified. We give it in this form here to introduce the notion of random coefficients and to prepare for the multilevel RCM.

The two-level RCM that has been studied most extensively looks like

$$\underline{\boldsymbol{y}}_j = \boldsymbol{X}_j\underline{\boldsymbol{\beta}}_j + \underline{\boldsymbol{\epsilon}}_j, \tag{1.5a}$$
$$\underline{\boldsymbol{\beta}}_j = \boldsymbol{\beta} + \underline{\boldsymbol{\delta}}_j, \tag{1.5b}$$

with the same distributional assumptions as above for the two-level MLM. Observe that the fixed part of $\underline{\boldsymbol{\beta}}_j$ is assumed to be the same for all groups. This is necessary for identification of the model.

In this form the random coefficient model has been discussed in the econometric literature, starting from Swamy [117]. It has also become more popular in statistics as one form of the *varying coefficient model*, although this term is mostly used for models with (partly) systematic or deterministic variation of the coefficients, such as a deterministic function of time or some other explanatory variable [54, 61].

The fact that we are dealing with a two-level model here is perhaps clearer if we use scalar notation. This gives

$$\underline{y}_{ij} = x_{ij1}\underline{\beta}_{j1} + \cdots + x_{ijp}\underline{\beta}_{jp} + \underline{\epsilon}_{ij},$$
$$\underline{\beta}_{js} = \beta_{js} + \underline{\delta}_{js}.$$

An important subclass of the RCM is the *random intercept model* or RIM. It is the same as RCM, except for the fact that we assume that all regression coefficients that are not intercepts have no random component. Thus, all slopes are fixed. For a two-level RIM, we consequently have, with some obvious modifications of the notation,

$$\underline{\boldsymbol{y}}_j = \underline{\mu}_j \mathbf{1}_{n_j} + \boldsymbol{X}_j \boldsymbol{\beta} + \underline{\boldsymbol{\epsilon}}_j,$$
$$\underline{\mu}_j = \mu + \underline{\delta}_j.$$

There is an extensive discussion of RIMs, with many applications, in Longford [76]. The econometric panel data literature also discusses this model extensively; see, e.g., Chamberlain [18], Wooldridge [126, Chapter 10], Arellano [4, Chapter 3], or Hsiao [60, Chapter 3]. Observe that for a RIM,

$$\boldsymbol{V}_j = \omega^2 \boldsymbol{E} + \sigma_j^2 \boldsymbol{I},$$

where $\boldsymbol{E}$ has all its elements equal to $+1$. This is the well-known intraclass covariance structure, with intraclass correlation

$$\rho_j^2 = \frac{\omega^2}{\omega^2 + \sigma_j^2}.$$

1.4.3 Slopes-as-Outcomes Models

We are now getting close to what is usually called multilevel analysis. The *slopes-as-outcomes model* or SOM is the model with

$$\underline{\boldsymbol{y}} = \boldsymbol{X}\underline{\boldsymbol{\beta}} + \underline{\boldsymbol{\epsilon}},$$
$$\underline{\boldsymbol{\beta}} = \boldsymbol{Z}\gamma + \underline{\boldsymbol{\delta}},$$

with $\boldsymbol{X}[n,p]$, $\boldsymbol{Z}[p,r]$, and

$$\begin{pmatrix} \underline{\boldsymbol{\epsilon}} \\ \underline{\boldsymbol{\delta}} \end{pmatrix} \sim \mathcal{N}\left(\begin{pmatrix} \boldsymbol{\emptyset} \\ \boldsymbol{\emptyset} \end{pmatrix}, \begin{pmatrix} \boldsymbol{\Sigma} & \boldsymbol{\emptyset} \\ \boldsymbol{\emptyset} & \boldsymbol{\Omega} \end{pmatrix} \right).$$

The characteristic that is unique to this model, compared to others discussed here, is that the random coefficients $\underline{\boldsymbol{\beta}}$ are themselves dependent variables in a second regression equation. Of course, in a SOM we have

$$\underline{\boldsymbol{y}} = \boldsymbol{X}\boldsymbol{Z}\gamma + \boldsymbol{X}\underline{\boldsymbol{\delta}} + \underline{\boldsymbol{\epsilon}},$$

which shows that the SOM is an MLM in which the fixed regressors are $\boldsymbol{X} = \boldsymbol{X}\boldsymbol{Z}$ and the random regressors are $\boldsymbol{X}$.

The two-level SOM is

$$\underline{\boldsymbol{y}}_j = \boldsymbol{X}_j\underline{\boldsymbol{\beta}}_j + \underline{\boldsymbol{\epsilon}}_j, \tag{1.6a}$$

$$\underline{\boldsymbol{\beta}}_j = \boldsymbol{Z}_j\gamma + \underline{\boldsymbol{\delta}}_j, \tag{1.6b}$$

again with the same distributional assumptions. Here $\boldsymbol{X}_j[n_j, p]$ and $\boldsymbol{Z}_j[p, r]$. It is possible, in principle, to have different numbers of predictors in the different $\boldsymbol{X}_j$, but we will ignore this possibility. The regression equations (1.6b) for the random coefficients imply that differences between the regression coefficients of different groups are partly explained by observed characteristics of the groups. These equations are often of great substantive interest.

By substituting the second-level equations (1.6b) in the first-level equations (1.6a) and by stacking the resulting m equations, we find

$$\underline{\boldsymbol{y}} = \boldsymbol{U}\gamma + \boldsymbol{X}\underline{\boldsymbol{\delta}} + \underline{\boldsymbol{\epsilon}},$$

with

$$\boldsymbol{U} \triangleq \begin{pmatrix} \boldsymbol{X}_1\boldsymbol{Z}_1 \\ \vdots \\ \boldsymbol{X}_m\boldsymbol{Z}_m \end{pmatrix} \tag{1.7}$$

and with the remaining terms stacked in the same way, except $\boldsymbol{X}$, which has the direct sum form

$$\boldsymbol{X} = \bigoplus_{j=1}^{m} \boldsymbol{X}_j = \begin{pmatrix} \boldsymbol{X}_1 & & \emptyset \\ & \ddots & \\ \emptyset & & \boldsymbol{X}_m \end{pmatrix}.$$

Again, this shows that the two-level SOM is just an MLM with some special structure. We analyze this structure in more detail below.

In the first place, the dispersion matrix of $\underline{\boldsymbol{y}}$ has block-diagonal or direct-sum structure:

$$\underline{\boldsymbol{y}} \sim \mathcal{N}\Big(\boldsymbol{U}\gamma, \bigoplus_{j=1}^{m} \boldsymbol{V}_j\Big),$$

with

$$\boldsymbol{V}_j \triangleq \boldsymbol{X}_j\boldsymbol{\Omega}_j\boldsymbol{X}_j' + \sigma_j^2\boldsymbol{I}.$$

Second, the design matrix $\boldsymbol{U}$ in the fixed part has the structure (1.7). In fact, there usually is even more structure than that. In the two-level SOM, we often have

$$\boldsymbol{Z}_j = \bigoplus_{s=1}^{p} \boldsymbol{z}_j'; \tag{1.8}$$

i.e., $\boldsymbol{Z}_j$ is the direct sum of p row vectors, all equal to a vector $\boldsymbol{z}_j'$ with q elements. The vector $\boldsymbol{z}_j$ describes group j in terms of q second-level variables. More elaborately,

$$\boldsymbol{Z}_j = \begin{pmatrix} \boldsymbol{z}_j' & \emptyset & \emptyset & \cdots & \emptyset \\ \emptyset & \boldsymbol{z}_j' & \emptyset & \cdots & \emptyset \\ \emptyset & \emptyset & \boldsymbol{z}_j' & \cdots & \emptyset \\ \vdots & \vdots & \vdots & \ddots & \vdots \\ \emptyset & \emptyset & \emptyset & \cdots & \boldsymbol{z}_j' \end{pmatrix}.$$

This is easily generalized to direct sums of different vectors, even if they have different numbers of elements. It follows that, if we partition γ accordingly into p subvectors of length q, we have

$$E(\underline{\beta}_{js}) = \boldsymbol{z}_{js}' \gamma_s.$$

Also

$$\boldsymbol{U}_j = \boldsymbol{X}_j \boldsymbol{Z}_j = \begin{bmatrix} \boldsymbol{x}_{j1} \boldsymbol{z}_{j1}' & \boldsymbol{x}_{j2} \boldsymbol{z}_{j2}' & \cdots & \boldsymbol{x}_{jp} \boldsymbol{z}_{jp}' \end{bmatrix},$$

where $\boldsymbol{x}_{js}$ is the s-th column of $\boldsymbol{X}_j$. Thus, $\boldsymbol{U}$ is a block-matrix, consisting of m by p blocks, and each block is of rank 1. Consequently, we say the $\boldsymbol{U}$ is a *block-rank-one matrix.*

From the point of view of interpretation, each column of a block-rank-one matrix is the product of a first-level predictor from $\boldsymbol{X}$ and a second-level predictor from $\boldsymbol{Z}$. Because generally both $\boldsymbol{X}$ and $\boldsymbol{Z}$ include an intercept, i.e., a column with all elements equal to 1, this means that the columns of $\boldsymbol{X}$ and $\boldsymbol{Z}$ themselves also occur in $\boldsymbol{U}$, with $\boldsymbol{Z}$ disaggregated. Thus, SOM models have predictors with fixed regression coefficients that are *interactions*, and much of the classical literature on interaction in the linear model, such as Cox [23] and Aiken and West [1], applies to these models as well.

There is one additional consequence of the structure (1.8). We can write

$$[\boldsymbol{U}\gamma]_{ij} = \sum_{s=1}^{p} x_{ijs} \boldsymbol{z}_j' \gamma_s = \sum_{s=1}^{p} \sum_{v=1}^{q} x_{ijs} \gamma_{sv} z_{jv}.$$

Now define the *balanced case* of SOM, in which all $\boldsymbol{X}_j$ are the same. This seems very far fetched if we are thinking of students in classes, but it is actually quite natural for repeated measures. There $\boldsymbol{X}$ could be a basis of growth functions, such as polynomials or exponentials. If measurements are made at the same time points, then indeed all $\boldsymbol{X}_j$ are the same. Other situations in which this may happen are medical or biological experiments, in which dosages of drugs or other treatment variables could be the same, or psychological experiments, in which the stimuli presented to all participants are the same.

In the balanced case, we can rearrange SOM as

$$\underline{Y} = Z\Gamma X' + \underline{\Delta} X' + \underline{E},$$

where the (j,i)-th element of $\underline{Y}$ is $\underline{y}_{ij}$, the j-th row of Z is z_j', the j-th row of $\underline{\Delta}$ is $\underline{\delta}_j'$, the s-th column of Γ is γ_s, and the meaning of the other symbols follows. Thus, the rows are independent. This shows that SOM in this case is a random coefficient version of the classical growth curve model of Potthoff and Roy [91]. Conversely, SOM can be seen as a far-reaching generalization of these classical fixed-effect growth models.

1.4.4 Multilevel Models

Most of the classical multilevel literature, with its origins in education and sociology, deals with the SOM. But in more recent literature, multilevel analysis can refer to more general *Hierarchical Linear Models* or HLMs, of which the two-level MLM (1.4) and the two-level RCM (1.5) are examples. A good example of this more general use, which we also follow throughout the Handbook, is the discussion in Gelman [40].

1.4.5 Generalizations

We shall be very brief about the various generalizations of the multilevel model, because most of these are discussed extensively in the subsequent chapters of this Handbook.

Heteroskedasticity and Conditional Intragroup Dependence

Heteroskedasticity is the phenomenon that residual variances are different for different units. More specifically, it usually means that the variance of the residual depends in some way on the explanatory variables. Heteroskedasticity is a frequently occurring phenomenon in cross-sectional data analysis (and some forms of time series analysis, in particular financial time series). Therefore, we may expect that heteroskedasticity will also be prevalent in many multilevel data analyses. This is indeed the case. In fact, heteroskedasticity is an explicit part of most multilevel models. For example, in the model that we focus on, the covariance matrix of the dependent variables for the j-th group, $\underline{y}_j$, is $V_j = X_j \Omega X_j' + \sigma_j^2 I$. Clearly, this depends on X_j, so if X_j contains more than just the constant and the corresponding elements of Ω are not restricted to zero, this induces heteroskedasticity. Furthermore, allowing different residual variances σ_j^2 is also a form of heteroskedasticity.

However, in this specification, the residual variances within the same group are the same, i.e., $\text{Var}(\underline{\epsilon}_{ij}) = \sigma_j^2$, which is the same for all i. Thus, there is heteroskedasticity between groups, but not within groups. This may be unrealistic in many applications. In such cases, one may want to specify an extended model that explicitly includes within-groups heteroskedasticity. Such a model, and how it can be used to detect heteroskedasticity and thus misspecification of the random part of the model, is described in Chapter 3.

Another widespread phenomenon is lack of independence of observations. Again, this is one of the features of a typical multilevel model: It is assumed that observations within groups are dependent. This gives rise to the well-known intraclass correlation. As we have seen, this is modeled in a typical multilevel model through the random coefficients and, more specifically, through the random terms $\underline{\boldsymbol{\delta}}_j$ in our model specification. However, again this feature does not extend to conditional within-groups comparisons. The units are assumed conditionally independent within their groups, reflected in the diagonality of the covariance matrix of $\underline{\boldsymbol{\epsilon}}_j$. This assumption may also not always be realistic. The leading example in which it is likely to be violated is in longitudinal (or *panel*) data, where the within-groups observations are different observations of the same subject (or object) over time. In such data, residuals often show considerable autocorrelation; i.e., there is a high correlation between residuals that are not far apart. This phenomenon, and how it can be modeled, is discussed extensively in Chapter 7. A similar situation is encountered with *spatial* data, such as data on geographic regions. Then there tends to be spatial autocorrelation; i.e., neighboring regions are "more similar" than regions further apart. See, e.g., Anselin [3] for an overview of modeling spatial autocorrelation. This type of model was integrated in a multilevel model with random coefficients by Elhorst and Zeilstra [37].

More Levels and Different Dependence Structures

Slopes-as-outcomes models can be generalized quite easily to more than two levels. One problem is, however, that matrix notation does not work any more. Switching to scalar notation, we indicate how to generalize by giving the multi-level model for student i_1 in class i_2 in school i_3, and so on. For a model with L levels, it is

$$\underline{\beta}^{(v)}_{i_v,\ldots,i_{v+L-1}} = \sum_{i_{v+L}=1}^{p^{(v)}_{L+1,\ldots,v+L-1}} x^{(v)}_{i_v,\ldots,i_{v+L}} \underline{\beta}^{(v+1)}_{i_{v+1},\ldots,i_{v+L}} + \underline{\epsilon}^{(v)}_{i_v,\ldots,i_{v+L-1}},$$

where superscripts in parentheses indicate the level of the variable. In order to complete the model, we have to assume something about the boundary cases. For level $v = 1$, $\underline{\beta}_{i_1,\ldots,i_L}$ is what we previously wrote as $\underline{y}_{ij}$ for a two-level model, i.e., the value of the outcome for student ij. For the highest level $(L+1)$, the random coefficients are set to fixed constants, because otherwise we would have to go on making further specifications. Although the notation becomes somewhat unwieldy, the idea is simple enough.

Other types of different dependence structures are cross-classifications and multiple membership classifications. In the former, an observation is nested in two or more higher-level units, but these higher-level units are not nested within each other. An example is a sample of individuals who are nested within the primary schools and secondary schools that they attended, but not all students from a primary school necessarily attended the same secondary school or vice versa. Multiple membership classifications occur when observations are nested within multiple higher-level units of the same type. For example, patients can be treated by several nurses. These two types of dependency structure are discussed at length in Chapter 8. The notation that is used in that chapter can also be applied to "ordinary" (i.e., nested) multiple-level models, somewhat reducing the unwieldiness mentioned above.

Nonlinear Mixed Models

Nonlinear mixed models come in two flavors. And of course, these nonlinear generalizations specialize in the obvious way to random coefficient and slopes-as-outcomes models.

First, we have *nonlinear mixed models* in which the linear combinations of the predictors are replaced by nonlinear parametric functions, both for the fixed part and the random part. An obvious variation, to reduce the complexity, is to use a nonlinear combination of linear combinations. These nonlinear mixed models are usually fitted with typical nonlinear regression techniques; i.e., we linearize the model around the current estimate and then use linear multilevel techniques. For details we refer to Pinheiro and Bates [89]. Detection and nonparametric modeling of nonlinearities in the fixed part of the model is discussed in more detail in Chapter 3.

Second, we have *generalized linear mixed models.* In the same way as the generalized linear model extends the linear model, the generalized linear mixed model extends the mixed linear model. The basic trick is (in the two-level case) to condition on the random effects and to assume a generalized linear model for the conditional distribution of the outcomes. Then the full model is obtained by multiplying the conditional density by the marginal density of the random effects and integrating. This is, of course, easier said than done, because the high-dimensional integrals that are involved cannot be evaluated in closed form. Thus, sophisticated approximations and algorithms are needed. These are discussed in many of the subsequent chapters, in particular Chapters 2, 5, and 9.

The leading case of applications of nonlinear models is the modeling of nominal and ordinal categorical dependent variables. Several competing specifications exist, and each has its advantages and disadvantages. These are discussed and compared in detail in Chapter 6.

Multivariate Models, Endogeneity, Measurement Errors, and Latent Variables

In this chapter, we focus on models with one dependent variable, called $\underline{y}$, and explanatory variables (generically called x and z) that are assumed to be fixed constants. Instead of the latter, we can also assume that the explanatory variables are strictly *exogenous* random variables and then do our analysis conditionally on their realizations. This does not change the treatment, the results, or the notation.

In fact, most of the multilevel literature is based on a similar setup, so in that sense this chapter reflects the mainstream of multilevel analysis. In many practical situations, however, this setup is not sufficient, or even clearly incorrect, and extensions or modifications are needed. Here, we briefly mention a few such topics that are somewhat related.

Of these, multiple dependent variables are often most easily accommodated. In most situations, one can simply estimate the models for each of these dependent variables separately. If the different equations do not share any parameters and the dependent variable of one equation does not enter another as explanatory variable, this should be sufficient. Also, as mentioned earlier, multivariate models can be viewed as univariate models with an additional level and thus be estimated within a relatively standard multilevel modeling setup.

Endogeneity is the situation where (at least) one of the explanatory variables in a regression equation is a random variable that is correlated with the error term in the equation of interest. Statistically, this leads to biased and inconsistent estimators. Substantively, this is often the result of one or more unobserved variables that influence both the explanatory variable and the dependent variable in the equation. If it is only considered a statistical nuisance, consistent estimators can usually be obtained by using some form of instrumental variables method [e.g., 126], which has been developed for multilevel analysis by Kim and Frees [65]. In many cases, however, it is of some substantive interest to model the dependence more extensively. Examples of such models are especially abundant in longitudinal situations. Chapter 7 discusses these in detail.

A special source of endogeneity that occurs frequently in the social sciences is measurement error in an explanatory variable. Almost all psychological test scores can be considered as, at best, imperfect measures of some concept that one tries to measure. A notorious example from economics is income. Let us assume that true (log) consumption $\underline{c}^*$ of a household depends on true (log) household income $\underline{y}^*$ through a simple linear regression equation, but the measurements $\underline{c}$ and $\underline{y}$ of consumption and income are only crude estimates. In formulas,

$$\underline{c}^* = \beta_1 + \beta_2 \underline{y}^* + \underline{\epsilon},$$
$$\underline{c} = \underline{c}^* + \underline{v},$$
$$\underline{y} = \underline{y}^* + \underline{w},$$

where we assume that the error terms $\underline{\epsilon}$, $\underline{v}$, and $\underline{w}$ are all mutually independent and independent of $\underline{y}^*$, and we have omitted the indices denoting the observations. We can write the model in terms of the observed variables as

$$\underline{c} = \beta_1 + \beta_2 \underline{y} + \underline{u},$$

where $\underline{u} = \underline{\epsilon} + \underline{v} - \beta_2 \underline{w}$. Because $\underline{w}$ is part of both the explanatory variable $\underline{y}$ and the error term $\underline{u}$, these two are correlated and thus we have the endogeneity problem. An extensive general treatment of measurement error, its statistical consequences, and how to obtain suitable estimators, is given by Wansbeek and Meijer [123]. Goldstein [46, Chapter 13] discusses the handling of measurement errors in multilevel models.

Models that include measurement errors explicitly are a subset of *latent variable models*. Latent variable models typically specify a relationship between substantive concepts, the *structural model*, and a relationship between these concepts and the observed variables (the indicators), which is the *measurement model*. The concepts may be fairly concrete, like income above, but may also be highly abstract theoretical concepts, like personality traits. Most latent variable models are members of the class of *structural equation models*. Because of the flexibility in selecting (multiple) observed variables to analyze and the flexibility in defining latent variables, structural equation models encompass a huge class of models. In particular, multivariate models, endogeneity, measurement errors, and latent variables can all be combined into a single structural equation model. Structural equation models for multilevel data are described extensively in Chapter 12.

Nonnormality

It is customary to specify normal distributions for the random terms in a multilevel model. A normality assumption for error terms can typically be defended by arguing that the error term captures many small unobserved influences, and a central limit theorem then implies that it should be approximately normally distributed. However, normality of random coefficients is often not at all logical. Empirically, in effectiveness studies of schools, hospitals, etc., we might find that many perform "average", whereas there are a few that perform exceptionally well or exceptionally poor. Such a pattern would suggest a distribution with heavy tails or a mixture distribution. Moreover, the normal distribution has positive density for both positive and negative values, whereas in many cases, theory or common sense (which often coincide) says that a coefficient should have a specific sign. For example, in economics, a higher price should decrease (indirect) utility, and in education, higher intelligence should lead to higher scores on school tests.

In economics, marketing, and transportation, the lognormal distribution has been proposed as a convenient alternative distribution for random coefficients in discrete choice models, perhaps after changing the sign of the explanatory variable. Meijer and Rouwendal [83] discuss this literature and compare normal, lognormal, and Gamma distributions, as well as a nonparametric alternative. In their travel preference data, lognormal and Gamma clearly outperform normal and nonparametric, on the basis of fit and interpretability. Chapter 7 further discusses the nonparametric maximum likelihood estimator.

For *linear* multilevel models, it is fairly straightforward that all the usual estimators are still consistent if the random terms are nonnormally distributed [121]. The standard errors of the fixed coefficients are still correct under nonnormality, but standard errors of the variance parameters must be adjusted. This can be done by using a robust covariance matrix, which will be discussed in Section 1.6.3 below, or by using resampling techniques specifically developed for multilevel data (see Chapter 11).

Estimators of *nonlinear* multilevel analysis models are inconsistent if the distribution of the random coefficients is misspecified. Robust covariance matrices and resampling can give asymptotically correct variability estimators, but it may be questionable whether these are useful if it is unclear whether the estimators of the model parameters are meaningful under gross misspecification of the distributions.

An interesting logical consequence of the line of reasoning that leads to nonnormal distributions is that it also suggests that in cases where the coefficient should have a specific sign, the functional form of the level-2 model should also change. For example, if a level-1 random coefficient $\underline{\beta}$ should be positive, then a specification $\underline{\beta} = \boldsymbol{z}'\boldsymbol{\gamma} + \underline{\delta}$, even with nonnormal $\underline{\delta}$, may be problematic, and a specification

$$\log \underline{\beta} = \boldsymbol{z}'\boldsymbol{\gamma} + \underline{\delta}$$

may make more sense, where now there is nothing wrong with a normal $\underline{\delta}$, because it induces a lognormal $\underline{\beta}$. Remarkably, with this specification, although both level-1 and level-2 submodels are linear in parameters, the combined model is not.

1.5 Loss Functions

Loss functions are used in statistics to measure the badness-of-fit of the model and the given data. In most circumstances, they measure the distance between the observed and the expected values of appropriately chosen statistics such as the means, the dispersions, or the distribution functions. It is quite common in the multilevel literature to concentrate exclusively on the likelihood function or, in a Bayesian context, the posterior density function. We will pay more attention than usual to least squares loss functions, both for historical and didactic reasons.

1.5.1 Least Squares

A general least squares loss function for the multilevel problem (in particular, the SOM) is of the form

$$\rho(\gamma) = \sum_{j=1}^{m} (\boldsymbol{y}_j - \boldsymbol{X}_j \boldsymbol{Z}_j \gamma)' \boldsymbol{A}_j^{-1} (\boldsymbol{y}_j - \boldsymbol{X}_j \boldsymbol{Z}_j \gamma), \tag{1.9}$$

where the weight matrices $\boldsymbol{A}_j$ are supposed to be known (not estimated).

There is a simple trick that can be used to simplify the computations, and to give additional insight into the structure of the loss function. Define the regression coefficients

$$\boldsymbol{b}_j = (\boldsymbol{X}_j' \boldsymbol{A}_j^{-1} \boldsymbol{X}_j)^{-1} \boldsymbol{X}_j' \boldsymbol{A}_j^{-1} \boldsymbol{y}_j$$

and the residuals

$$\boldsymbol{r}_j = \boldsymbol{y}_j - \boldsymbol{X}_j \boldsymbol{b}_j \, .$$

Then $\boldsymbol{y}_j = \boldsymbol{X}_j \boldsymbol{b}_j + \boldsymbol{r}_j$, and $\boldsymbol{X}_j' \boldsymbol{A}_j^{-1} \boldsymbol{r}_j = \boldsymbol{\emptyset}$. Now, for group j,

$$\rho_j(\gamma) = (\boldsymbol{b}_j - \boldsymbol{Z}_j \gamma)' \boldsymbol{X}_j' \boldsymbol{A}_j^{-1} \boldsymbol{X}_j (\boldsymbol{b}_j - \boldsymbol{Z}_j \gamma) + \boldsymbol{r}_j' \boldsymbol{A}_j^{-1} \boldsymbol{r}_j. \tag{1.10}$$

This expression of the loss function is considerably more convenient than (1.9), because it involves smaller vectors and matrices.

If we choose $\boldsymbol{A}_j$ of the form $\boldsymbol{V}_j = \boldsymbol{X}_j \boldsymbol{\Omega} \boldsymbol{X}_j' + \sigma_j^2 \boldsymbol{I}$, again with $\boldsymbol{\Omega}$ and σ_j^2 assumed known, then we can simplify the loss function some more, using the matrix results in Appendix 1.C. Let $\boldsymbol{P}_j \triangleq \boldsymbol{X}_j (\boldsymbol{X}_j' \boldsymbol{X}_j)^{-1} \boldsymbol{X}_j'$, and $\boldsymbol{Q}_j \triangleq \boldsymbol{I} - \boldsymbol{P}_j$. We will also write, in the sequel,

$$\boldsymbol{W}_j \triangleq \boldsymbol{\Omega} + \sigma_j^2 (\boldsymbol{X}_j' \boldsymbol{X}_j)^{-1}.$$

Observe that if $\underline{\boldsymbol{b}}_j \triangleq (\boldsymbol{X}_j' \boldsymbol{X}_j)^{-1} \boldsymbol{X}_j' \underline{\boldsymbol{y}}_j$, then $\boldsymbol{W}_j$ is the dispersion of $\underline{\boldsymbol{b}}_j$. Accordingly, from now on we redefine $\boldsymbol{b}_j \triangleq (\boldsymbol{X}_j' \boldsymbol{X}_j)^{-1} \boldsymbol{X}_j' \boldsymbol{y}_j$ and $\boldsymbol{r}_j \triangleq \boldsymbol{y}_j - \boldsymbol{X}_j \boldsymbol{b}_j$, regardless of the definition of $\boldsymbol{A}_j$.

From Theorem 1.2 in the appendix,

$$\boldsymbol{V}_j^{-1} = \boldsymbol{X}_j(\boldsymbol{X}_j'\boldsymbol{X}_j)^{-1}\boldsymbol{W}_j^{-1}(\boldsymbol{X}_j'\boldsymbol{X}_j)^{-1}\boldsymbol{X}_j' + \sigma_j^{-2}\boldsymbol{Q}_j, \tag{1.11}$$

and thus

$$\boldsymbol{r}_j'\boldsymbol{V}_j^{-1}\boldsymbol{r}_j = \sigma_j^{-2}\boldsymbol{r}_j'\boldsymbol{r}_j = (n_j - p)s_j^2/\sigma_j^2$$

and

$$\boldsymbol{X}_j'\boldsymbol{V}_j^{-1}\boldsymbol{X}_j = \boldsymbol{W}_j^{-1}.$$

Hence,

$$\rho_j(\boldsymbol{\gamma}) = (\boldsymbol{b}_j - \boldsymbol{Z}_j\boldsymbol{\gamma})'\boldsymbol{W}_j^{-1}(\boldsymbol{b}_j - \boldsymbol{Z}_j\boldsymbol{\gamma}) + (n_j - p)s_j^2/\sigma_j^2. \tag{1.12}$$

Computing least squares loss in this way is even more efficient than using (1.10).

1.5.2 Full Information Maximum Likelihood (FIML)

The least squares approach supposes that the weight matrix is known, but, of course, in a more general case the weight function will depend on some unknown parameters that have to be estimated from the same data as the regression coefficients. In that case, we need a loss function that not only measures how close the fitted regression coefficients are to their expected values, but also measures, at the same time, how well the fitted dispersion matrices correspond with the dispersion of the residuals. For this we use the log-likelihood.

As is well known, the method of maximum likelihood has a special position in statistics, especially in applied statistics. Maximum likelihood estimators are introduced as if they are by definition optimal, in all situations. Another peculiarity of the literature is that maximum likelihood methods are introduced by assuming a specific probability model, which is often quite obviously false in the situations one has in mind. In our context, this means that typically it is assumed that the disturbances, and thus the observed $\boldsymbol{y}$, are realizations of jointly normal random variables. Of course, such an assumption is highly debatable in many educational research situations, and quite absurd in others.

Consequently, we take a somewhat different position. Least squares estimates are obtained by minimizing a given loss function. Afterward, we derive their properties and we discover that they behave nicely in some situations. We approach multinormal maximum likelihood in a similar way. The estimates are defined as those values of $\boldsymbol{\gamma}$, $\boldsymbol{\Omega}$, and $\{\sigma_j^2\}$ that minimize the loss function

$$\mathcal{L}^F(\boldsymbol{\gamma}, \boldsymbol{\Omega}, \{\sigma_j^2\}) \stackrel{\Delta}{=} \log|\boldsymbol{V}| + (\boldsymbol{y} - \boldsymbol{U}\boldsymbol{\gamma})'\boldsymbol{V}^{-1}(\boldsymbol{y} - \boldsymbol{U}\boldsymbol{\gamma}). \tag{1.13}$$

This loss function, which is the negative logarithm of the likelihood function (except for irrelevant constants), is often called the *deviance*. The important fact here is not that we assume multivariate normality but that (1.13) defines quite a natural loss function. It measures closeness of $\boldsymbol{y}$ to $\boldsymbol{U}\gamma$ by weighted least squares, and it measures at the same time closeness of $\boldsymbol{R}(\gamma) \triangleq (\boldsymbol{y} - \boldsymbol{U}\gamma)(\boldsymbol{y} - \boldsymbol{U}\gamma)'$ to $\boldsymbol{V}$.

This last property may not be immediately apparent from the form of (1.13). It follows from the inequality $\log|\boldsymbol{A}| + \operatorname{tr} \boldsymbol{A}^{-1}\boldsymbol{B} \geq \log|\boldsymbol{B}| + m$, which is true for all pairs of positive definite matrices of order m. We have equality if and only if $\boldsymbol{A} = \boldsymbol{B}$. Thus, in our context, $\log|\boldsymbol{V}| + \operatorname{tr} \boldsymbol{V}^{-1}\boldsymbol{R}(\gamma)$ measures the distance between $\boldsymbol{V}$ and the residuals $\boldsymbol{R}(\gamma)$. We want to make residuals small, and we want the dispersion to be maximally similar to the dispersion of the residuals. Moreover, we want to combine these two objectives in a single loss function.

To find simpler expressions for the inverse and the determinant in (1.13), we use the matrix results in Appendix 1.C, in the same way as they were used in Section 1.5.1. From Theorem 1.1 in the appendix,

$$\log|\boldsymbol{V}_j| = (n_j - p)\log\sigma_j^2 + \log|\boldsymbol{X}_j'\boldsymbol{X}_j| + \log|\boldsymbol{W}_j|.$$

If we combine this with result (1.12), we find for group j, ignoring terms that do not depend on the parameters,

$$\mathcal{L}_j^F(\gamma, \boldsymbol{\Omega}, \sigma_j^2) = (n_j - p)\big(\log\sigma_j^2 + s_j^2/\sigma_j^2\big) + \log|\boldsymbol{W}_j| \\ + (\boldsymbol{b}_j - \boldsymbol{Z}_j\gamma)'\boldsymbol{W}_j^{-1}(\boldsymbol{b}_j - \boldsymbol{Z}_j\gamma).$$

To distinguish the resulting estimators explicitly from the REML estimators below, these ML estimators are called *full information maximum likelihood* (FIML) in this chapter.

1.5.3 Residual Maximum Likelihood (REML)

In the simplest possible linear model $\underline{y}_i = \mu + \underline{\epsilon}_i$, with $\underline{\epsilon}_i \overset{iid}{\sim} \mathcal{N}(0, \sigma^2)$, the maximum likelihood estimator of μ is the mean and that of σ^2 is the sum of squares around the mean, divided by the number of observations n. This estimate of the variance is biased and, as a consequence, the sample variance is usually defined by dividing the sum of squares by $n-1$. The same reasoning, adjusting for bias, in the linear regression model leads to dividing the residual sum of squares by $n - s$, where s is the number of predictors.

We can also arrive at these bias adjustments in a slightly different way, which allows us to continue to use the log-likelihood. Suppose we compute the likelihood of the deviations of the mean, or in the more general case the likelihood of the observed regression residuals. These residuals have a singular multivariate normal distribution, and the maximum likelihood estimate of the variance turns out to be precisely the bias-adjusted estimate. Thus, in these simple cases, *residual maximum likelihood* (REML; also frequently called *restricted maximum likelihood*) estimates can actually be computed from full information maximum likelihood estimates by a simple multiplicative bias adjustment.

In multilevel models, or more generally in MLMs, bias adjustment is not that easy, but we can continue to use the same reasoning as in the simpler cases and then expect to get an estimator with smaller bias. Let us start with the MLM $\underline{\boldsymbol{y}} = \boldsymbol{U}\gamma + \boldsymbol{X}\underline{\boldsymbol{\delta}} + \underline{\epsilon}$. Suppose $\boldsymbol{U}$ is $n \times s$ and of full column rank. Also suppose $\boldsymbol{K}$ is any orthonormal basis for the orthogonal complement of the column space of $\boldsymbol{U}$; i.e., $\boldsymbol{K}$ is an $n \times (n-s)$ matrix with $\boldsymbol{K}'\boldsymbol{K} = \boldsymbol{I}$ and $\boldsymbol{K}'\boldsymbol{U} = \boldsymbol{\emptyset}$. Then define the residuals $\underline{\boldsymbol{r}} \stackrel{\Delta}{=} \boldsymbol{K}'\underline{\boldsymbol{y}} \sim \mathcal{N}(\boldsymbol{\emptyset}, \boldsymbol{K}'\boldsymbol{V}\boldsymbol{K})$. Thus, the negative loglikelihood or deviance of a realization of $\underline{\boldsymbol{r}}$ is, ignoring the usual constants,

$$\mathcal{L}^R(\boldsymbol{\Omega}, \{\sigma_j^2\}) = \log|\boldsymbol{K}'\boldsymbol{V}\boldsymbol{K}| + \boldsymbol{r}'(\boldsymbol{K}'\boldsymbol{V}\boldsymbol{K})^{-1}\boldsymbol{r}.$$

Observe that this is no longer a function of γ. Thus, we cannot compute maximum likelihood estimates of the fixed regression coefficients by minimizing this loss function.

Now use Theorem 1.3 from Appendix 1.C, which shows that

$$\boldsymbol{r}'(\boldsymbol{K}'\boldsymbol{V}\boldsymbol{K})^{-1}\boldsymbol{r} = \min_{\gamma}(\boldsymbol{y} - \boldsymbol{U}\gamma)'\boldsymbol{V}^{-1}(\boldsymbol{y} - \boldsymbol{U}\gamma).$$

Harville [52] shows that

$$\log|\boldsymbol{K}'\boldsymbol{V}\boldsymbol{K}| = \log|\boldsymbol{V}| + \log|\boldsymbol{U}'\boldsymbol{V}^{-1}\boldsymbol{U}| - \log|\boldsymbol{U}'\boldsymbol{U}|$$

and, consequently, except for irrelevant constants,

$$\mathcal{L}^R(\boldsymbol{\Omega}, \{\sigma_j^2\}) = \log|\boldsymbol{U}'\boldsymbol{V}^{-1}\boldsymbol{U}| + \min_{\gamma} \mathcal{L}^F(\gamma, \boldsymbol{\Omega}, \{\sigma_j^2\}).$$

It follows that the loss functions for FIML and REML only differ by the term $\log|\boldsymbol{U}'\boldsymbol{V}^{-1}\boldsymbol{U}|$, which can be thought of as a bias correction. In SOM, we can use

$$\boldsymbol{U}'\boldsymbol{V}^{-1}\boldsymbol{U} = \sum_{j=1}^{m} \boldsymbol{Z}_j'\boldsymbol{W}_j^{-1}\boldsymbol{Z}_j,$$

and, if (1.8) applies, then

$$\boldsymbol{U}'\boldsymbol{V}^{-1}\boldsymbol{U} = \sum_{j=1}^{m} \boldsymbol{W}_j^{-1} \otimes \boldsymbol{z}_j\boldsymbol{z}_j'.$$

1.5.4 Bayesian Multilevel Analysis

In the Bayesian approach to multilevel analysis, the parameters are treated as random variables, so in our notation they would be written as $\underline{\gamma}$, $\underline{\boldsymbol{\Omega}}$, and

$\{\sigma_j^2\}$, jointly denoted as $\underline{\boldsymbol{\theta}}$. Then a *prior distribution* for $\underline{\boldsymbol{\theta}}$ is specified, which is completely known. The parameters of this prior distribution are called *hyperparameters* and their values reflect the state of knowledge about $\underline{\boldsymbol{\theta}}$. In the absence of prior knowledge, this typically means that variances of the parameters are chosen to be infinite or at least very large. Given the specification of the prior distribution, the *posterior distribution* of $\underline{\boldsymbol{\theta}}$, given the observed sample, is found by application of Bayes' theorem:

$$p(\boldsymbol{\theta} \mid \boldsymbol{y}) = \frac{f(\boldsymbol{y} \mid \boldsymbol{\theta})\pi(\boldsymbol{\theta})}{f(\boldsymbol{y})} = Cf(\boldsymbol{y} \mid \boldsymbol{\theta})\pi(\boldsymbol{\theta}),$$

where $p(\boldsymbol{\theta} \mid \boldsymbol{y})$ is the posterior density, $\pi(\boldsymbol{\theta})$ is the specified prior density, $f(\boldsymbol{y} \mid \boldsymbol{\theta})$ is the conditional normal density that we have been using all along (which is equal to the likelihood function), and C is a normalizing constant that does not depend on $\boldsymbol{\theta}$. An explicit expression for C is rarely needed. The posterior density contains all information about $\underline{\boldsymbol{\theta}}$; all inferences about $\underline{\boldsymbol{\theta}}$ are derived from it. It combines the prior information and the information contained in the sample in a sound (and optimal) way.

From this description, it appears that the Bayesian approach does not fit into our framework of specifying a loss function and then optimizing it. However, in the Bayesian approach, it is common to use the posterior mode or posterior mean as an "estimator" and to compute intervals that contain $100(1-\alpha)\%$ (e.g., 95%) of the probability mass, which act as a kind of "confidence interval". The posterior mean $\hat{\mu}_g$ of parameter g is the argument for which the loss function $E[(\underline{\theta}_g - \mu_g)^2]$, where the expectation is taken over the posterior distribution, attains its minimum, whereas the posterior mode $\hat{\boldsymbol{\theta}}_M$ is by definition the value for which the posterior density $p(\boldsymbol{\theta} \mid \boldsymbol{y})$ attains its maximum or, equivalently, the loss function $-p(\boldsymbol{\theta} \mid \boldsymbol{y})$ attains its minimum. Both are very natural loss functions and, thus, in this way the Bayesian approach neatly fits within our framework. An important advantage of the Bayesian "confidence intervals", especially for the variance parameters, is that they may be asymmetric, reflecting a nonnormal posterior distribution. This is often more realistic for the variance parameters in small to moderate samples.

An important reason for the increasing popularity of the Bayesian approach is that it is able to deal with nonlinear models in a fairly straightforward way, using *Markov chain Monte Carlo* (MCMC) techniques. This gives good results where non-Bayesian approaches often have great difficulty in obtaining good estimators. Chapter 2 is an extensive discussion of the Bayesian approach, and in several other chapters, especially those dealing with nonlinear models, it is also discussed, applied, and compared to likelihood-based approaches. Therefore, we will not discuss it in more detail in this chapter.

1.5.5 Missing Data

It is implicit in the discussion thus far that we have assumed that there are no missing data. In practice, the fact that there are missing data is a widespread phenomenon and often a problem. We can distinguish between *unit nonresponse*, in which no information is available for a targeted observation, and *item nonresponse*, where information is available for some variables but not for others. If we assume that unit nonresponse is not related to any of the random variables ($\underline{\boldsymbol{\delta}}$, $\underline{\epsilon}$) of interest for the missing unit, we can simply proceed by analyzing the observed data set. If it is suspected that unit nonresponse leads to distortions, weighting can be applied (and is often applied) to let the sample distribution of some key variables match the (assumed known) population distribution. See Section 1.8 below for a discussion of sampling weights in multilevel models.

With *item nonresponse*, the simplest and most frequently applied solution is to simply omit all observations for which one or more variables are missing (*listwise deletion*). Although widely used, it is generally considered a bad method. It omits useful information and thus gives inefficient estimators. Even more importantly, it may easily lead to biases in the analyses, if the missing data patterns are related to the variables of interest. Chapter 10 extensively discusses how missing data can be treated in a sound and systematic way.

1.6 Techniques and Algorithms

If we have a loss function, then the obvious associated technique to estimate parameters is to minimize the loss function. Of course, for nonlinear optimization problems there are many different minimization methods. Some are general-purpose optimization methods that can be applied to any multivariate function, and some take the properties of the loss function explicitly into account.

1.6.1 Ordinary and Weighted Least Squares

As we have see in a previous section, the SOM model can be expressed in two steps, as in

$$\underline{\boldsymbol{y}}_j = \boldsymbol{X}_j \underline{\boldsymbol{\beta}}_j + \underline{\epsilon}_j, \tag{1.14a}$$

$$\underline{\boldsymbol{\beta}}_j = \boldsymbol{Z}_j \gamma + \underline{\boldsymbol{\delta}}_j, \tag{1.14b}$$

or in a single-step, as in

$$\underline{\boldsymbol{y}}_j = \boldsymbol{X}_j \boldsymbol{Z}_j \gamma + \boldsymbol{X}_j \underline{\boldsymbol{\delta}}_j + \underline{\epsilon}_j. \tag{1.15}$$

The one-step (1.15) and the two-step (1.14) specifications of the multilevel model suggest two different ordinary least squares methods for fitting the model. This was already discussed in detail by Boyd and Iversen [11]. We follow the treatment of de Leeuw and Kreft [28].

The two-step method first estimates the $\boldsymbol{\beta}_j$ by

$$\boldsymbol{b}_j = (\boldsymbol{X}_j'\boldsymbol{X}_j)^{-1}\boldsymbol{X}_j'\boldsymbol{y}_j, \tag{1.16}$$

and then γ by

$$\hat{\gamma} = \left(\sum_{j=1}^{m} \boldsymbol{Z}_j'\boldsymbol{Z}_j\right)^{-1} \sum_{j=1}^{m} \boldsymbol{Z}_j'\boldsymbol{b}_j. \tag{1.17}$$

Within the framework of Section 1.5.1, this is obtained by choosing $\boldsymbol{A}_j = \boldsymbol{X}_j\boldsymbol{X}_j' + \boldsymbol{Q}_j$, so that $\boldsymbol{A}_j^{-1} = \boldsymbol{X}_j(\boldsymbol{X}_j'\boldsymbol{X}_j)^{-2}\boldsymbol{X}_j' + \boldsymbol{Q}_j$.

The one-step method estimates γ directly from (1.15) as

$$\hat{\gamma} = \left(\sum_{j=1}^{m} \boldsymbol{Z}_j'\boldsymbol{X}_j'\boldsymbol{X}_j\boldsymbol{Z}_j\right)^{-1} \sum_{j=1}^{m} \boldsymbol{Z}_j'\boldsymbol{X}_j'\boldsymbol{y}_j.$$

By using (1.16), we see immediately, however, that the one-step method can also be written as

$$\hat{\gamma} = \left(\sum_{j=1}^{m} \boldsymbol{Z}_j'\boldsymbol{X}_j'\boldsymbol{X}_j\boldsymbol{Z}_j\right)^{-1} \sum_{j=1}^{m} \boldsymbol{Z}_j'\boldsymbol{X}_j'\boldsymbol{X}_j\boldsymbol{b}_j. \tag{1.18}$$

Thus, the one-step estimate can be computed in two steps as well. Within the framework of Section 1.5.1, the one-step estimate is obtained by choosing $\boldsymbol{A}_j = \boldsymbol{I}$.

Both methods provide unbiased estimators of γ, they are non-iterative, and they are easy to implement. An expression for their dispersion matrices is easily obtained by using $\text{Cov}(\underline{\boldsymbol{b}}_j) = \boldsymbol{W}_j$, which was obtained above. Hence, the dispersion matrix of the two-step estimator is

$$\left(\sum_{j=1}^{m} \boldsymbol{Z}_j'\boldsymbol{Z}_j\right)^{-1}\left(\sum_{j=1}^{m} \boldsymbol{Z}_j'\boldsymbol{W}_j\boldsymbol{Z}_j\right)\left(\sum_{j=1}^{m} \boldsymbol{Z}_j'\boldsymbol{Z}_j\right)^{-1}$$

and the dispersion matrix of the one-step estimator is

$$\left(\sum_{j=1}^{m} \boldsymbol{Z}_j'\boldsymbol{X}_j'\boldsymbol{X}_j\boldsymbol{Z}_j\right)^{-1}\left(\sum_{j=1}^{m} \boldsymbol{Z}_j'\boldsymbol{X}_j'\boldsymbol{X}_j\boldsymbol{W}_j\boldsymbol{X}_j'\boldsymbol{X}_j\boldsymbol{Z}_j\right)\left(\sum_{j=1}^{m} \boldsymbol{Z}_j'\boldsymbol{X}_j'\boldsymbol{X}_j\boldsymbol{Z}_j\right)^{-1}.$$

Despite their virtues, these least squares estimators have fallen into disgrace in the mainstream multilevel world, because they are neither BLUE nor BLUP

[43, 103]. This is somewhat supported by the simulations reported (for a three-level model) in Cheong et al. [21], where especially for level-1 covariates efficiencies of ML estimators are substantially higher (up to 55%). The one-step OLS estimator still enjoys a great popularity in economics, though.

The next candidate that comes to mind applies if both $\boldsymbol{\Omega}$ and $\{\sigma_j^2\}$ are known. We can then compute the WLS estimate

$$\hat{\gamma} = \left(\sum_{j=1}^{m} \boldsymbol{Z}_j' \boldsymbol{X}_j' \boldsymbol{V}_j^{-1} \boldsymbol{X}_j \boldsymbol{Z}_j\right)^{-1} \sum_{j=1}^{m} \boldsymbol{Z}_j' \boldsymbol{X}_j' \boldsymbol{V}_j^{-1} \boldsymbol{y}_j. \tag{1.19}$$

As we have seen, this can be simplified to

$$\hat{\gamma} = \left(\sum_{j=1}^{m} \boldsymbol{Z}_j' \boldsymbol{W}_j^{-1} \boldsymbol{Z}_j\right)^{-1} \sum_{j=1}^{m} \boldsymbol{Z}_j' \boldsymbol{W}_j^{-1} \boldsymbol{b}_j. \tag{1.20}$$

Within the framework of Section 1.5.1, the WLS estimate is obtained by choosing $\boldsymbol{A}_j = \boldsymbol{V}_j$. The dispersion matrix of the WLS estimator is obtained analogously to the ones above, and in this case it simplifies to

$$\left(\sum_{j=1}^{m} \boldsymbol{Z}_j' \boldsymbol{W}_j^{-1} \boldsymbol{Z}_j\right)^{-1}.$$

The formal similarity of (1.17), (1.18), and (1.20) is clear. They can all be thought of as two-step methods, which first compute the $\boldsymbol{b}_j$ and then do a weighted regression of the $\boldsymbol{b}_j$ on the $\boldsymbol{Z}_j$. Of course, (1.20) is mostly useless by itself, because we do not know what σ_j^2 and $\boldsymbol{\Omega}$ are, but we can insert consistent estimators of these instead. A method to compute consistent estimators of the elements of the variance parameters from the OLS residuals is discussed in de Leeuw and Kreft [28], and is also discussed below. The resulting method for estimating $\boldsymbol{\gamma}$ is fully efficient and non-iterative.

For WLS estimators with estimators of the variance parameters inserted, the exact covariance matrix generally cannot be computed. However, it follows from standard large sample theory (Slutsky's theorem; see, e.g., Ferguson [38] or Wansbeek and Meijer [123, pp. 369–370]) that if the estimators of $\boldsymbol{\Omega}$ and σ_j^2 are consistent, then the asymptotic distribution of the WLS estimator of $\boldsymbol{\gamma}$ is the same as the (asymptotic) distribution of the hypothetical estimator (1.20) that uses the true values of $\boldsymbol{\Omega}$ and σ_j^2 in the weight matrix, so we can still use the covariance matrix given above, especially with larger sample sizes.

The BLUE and the BLUP

Consider the model $\underline{\boldsymbol{y}} \sim \mathcal{N}(\boldsymbol{U}\boldsymbol{\gamma}, \boldsymbol{V})$. A linear estimator of the form $\underline{\hat{\gamma}} = \boldsymbol{L}'\underline{\boldsymbol{y}}$ is unbiased if $\boldsymbol{L}'\boldsymbol{U} = \boldsymbol{I}$, and it has dispersion $\boldsymbol{L}'\boldsymbol{V}\boldsymbol{L}$. The dispersion matrix is minimized, in the Löwner [77] ordering of matrices (i.e., $\boldsymbol{A} \geq \boldsymbol{B}$ if $\boldsymbol{A} - \boldsymbol{B}$ is positive semidefinite), by choosing $\boldsymbol{L} = \boldsymbol{V}^{-1}\boldsymbol{U}(\boldsymbol{U}'\boldsymbol{V}^{-1}\boldsymbol{U})^{-1}$. Thus,

$$\underline{\hat{\gamma}} = (\boldsymbol{U}'\boldsymbol{V}^{-1}\boldsymbol{U})^{-1}\boldsymbol{U}'\boldsymbol{V}^{-1}\underline{\boldsymbol{y}}$$

is the *best linear unbiased estimator* or BLUE. In the SOM,

$$\boldsymbol{U}'\boldsymbol{V}^{-1}\boldsymbol{U} = \sum_{j=1}^{m} \boldsymbol{Z}_j'\boldsymbol{W}_j^{-1}\boldsymbol{Z}_j$$

and

$$\boldsymbol{U}'\boldsymbol{V}^{-1}\underline{\boldsymbol{y}} = \sum_{j=1}^{m} \boldsymbol{Z}_j'\boldsymbol{W}_j^{-1}\underline{\boldsymbol{b}}_j.$$

Thus, the BLUE is given by (1.20).

We can also look at estimates of the error components. Of course, this means we are estimating random variables and, consequently, the *best linear unbiased predictor* or BLUP is a more appropriate term than the BLUE. To find the BLUP, we minimize the mean squared prediction error

$$\text{MSPE} \triangleq E\left[(\boldsymbol{L}'\underline{\boldsymbol{y}} + \boldsymbol{a} - \underline{\boldsymbol{\delta}})(\boldsymbol{L}'\underline{\boldsymbol{y}} + \boldsymbol{a} - \underline{\boldsymbol{\delta}})'\right] \tag{1.21a}$$

over $\boldsymbol{L}$ and $\boldsymbol{a}$ on the condition that

$$E(\boldsymbol{L}'\underline{\boldsymbol{y}} + \boldsymbol{a} - \underline{\boldsymbol{\delta}}) = \boldsymbol{\emptyset}. \tag{1.21b}$$

From (1.21b) we obtain $\boldsymbol{a} = -\boldsymbol{L}'\boldsymbol{U}\boldsymbol{\gamma}$, which means that the mean squared prediction error (1.21a) is

$$\begin{aligned}\text{MSPE} &= \boldsymbol{L}'\boldsymbol{V}\boldsymbol{L} - \boldsymbol{L}'\boldsymbol{X}\boldsymbol{\Omega} - \boldsymbol{\Omega}\boldsymbol{X}'\boldsymbol{L} + \boldsymbol{\Omega}\\ &= (\boldsymbol{V}\boldsymbol{L} - \boldsymbol{X}\boldsymbol{\Omega})'\boldsymbol{V}^{-1}(\boldsymbol{V}\boldsymbol{L} - \boldsymbol{X}\boldsymbol{\Omega}) + \boldsymbol{\Omega} - \boldsymbol{\Omega}\boldsymbol{Z}'\boldsymbol{V}^{-1}\boldsymbol{Z}\boldsymbol{\Omega}\\ &\geq \boldsymbol{\Omega} - \boldsymbol{\Omega}\boldsymbol{Z}'\boldsymbol{V}^{-1}\boldsymbol{Z}\boldsymbol{\Omega},\end{aligned}$$

with equality if $\boldsymbol{L} = \boldsymbol{V}^{-1}\boldsymbol{X}\boldsymbol{\Omega}$, i.e., if

$$\hat{\boldsymbol{\delta}} = \boldsymbol{\Omega}\boldsymbol{X}'\boldsymbol{V}^{-1}(\boldsymbol{y} - \boldsymbol{U}\boldsymbol{\gamma}).$$

In the SOM, using (1.11),

$$\hat{\boldsymbol{\delta}}_j = \boldsymbol{\Omega}\boldsymbol{W}_j^{-1}(\boldsymbol{b}_j - \boldsymbol{Z}_j\boldsymbol{\gamma}),$$

and thus

$$\hat{\boldsymbol{\beta}}_j = \boldsymbol{Z}_j\boldsymbol{\gamma} + \hat{\boldsymbol{\delta}}_j = \boldsymbol{\Omega}\boldsymbol{W}_j^{-1}\boldsymbol{b}_j + (\boldsymbol{I} - \boldsymbol{\Omega}\boldsymbol{W}_j^{-1})\boldsymbol{Z}_j\boldsymbol{\gamma}. \tag{1.22}$$

Thus, the BLUP of the random effects is a matrix weighted average [19] of the least squares estimates $\boldsymbol{b}_j$ and the expected values $\boldsymbol{Z}_j\boldsymbol{\gamma}$. The within-group least squares estimates are shrunken toward the overall model-based estimate $\boldsymbol{Z}_j\boldsymbol{\gamma}$ of the regression coefficients. This shrinking, which is common in BLUP and related empirical Bayes procedures, is also the basis for the discussion of *borrowing strength*, which has played a major role in the multilevel literature [cf. 13, 101].

Of course, (1.22) contains unknown parameters, and in order to use it in practice, we substitute whatever estimates we have for these unknown parameters.

Estimating the Variance Parameters

As we have seen, for the WLS estimator of $\boldsymbol{\gamma}$ and the BLUP of the random effects, we need consistent estimators of σ_j^2 and $\boldsymbol{\Omega}$. Moreover, estimating these parameters is often one of the main goals of a multilevel analysis and the focus on the random effects is perhaps the most salient difference between multilevel analysis and ordinary regression analysis.

A simple unbiased estimator of σ_j^2 is, of course, the within-groups residual variance $\underline{s}_j^2$. Given the assumptions above,

$$(n_j - p)\underline{s}_j^2/\sigma_j^2 \sim \chi^2_{n_j-p},$$

so that in addition to $E(\underline{s}_j^2) = \sigma_j^2$, we also have $\mathrm{Var}(\underline{s}_j^2) = 2(\sigma_j^2)^2/(n_j - p)$. Furthermore, $\underline{s}_j^2$ is independent of $\underline{\boldsymbol{b}}_j$. However, the variance, chi-square distribution, and independence result depend critically on the normality assumption. If all σ_j^2 are assumed equal, then its natural unbiased estimator is

$$\underline{s}^2 \triangleq \frac{1}{n-p} \sum_{j=1}^{m} (n_j - p)\underline{s}_j^2,$$

where n is total sample size. Under the model assumptions,

$$(n - p)\underline{s}^2/\sigma^2 \sim \chi^2_{n-p},$$

so that $E(\underline{s}^2) = \sigma^2$ and $\mathrm{Var}(\underline{s}^2) = 2(\sigma^2)^2/(n - p)$. Note that consistency of $\underline{s}_j^2$ requires $n_j \to \infty$. This is a little problematic because in some standard asymptotic theory for multilevel analysis (e.g., Longford [76, p. 252]; Verbeke and Lesaffre [120, Lemma 3]), it is assumed that the group sizes are bounded. However, close scrutiny of their theories reveals that the general asymptotic theory should still be valid under a hypothetical sequence such that $m \to \infty$, $n_j \to \infty$, and $n_j/m \to 0$. Maybe even weaker assumptions suffice. Of course, with (many) small groups, $n_j \to \infty$ may not be a useful assumption anyway. On the other hand, consistency of $\underline{s}^2$ only requires $n \to \infty$, which is obviously much weaker. However, the latter also requires the much stronger assumption that all residual variances are equal.

Observing that $\boldsymbol{\Omega} = \mathrm{Cov}(\underline{\boldsymbol{\beta}}_j) = E\big[(\underline{\boldsymbol{\beta}}_j - \boldsymbol{Z}_j\boldsymbol{\gamma})(\underline{\boldsymbol{\beta}}_j - \boldsymbol{Z}_j\boldsymbol{\gamma})'\big]$, a simple estimator of $\boldsymbol{\Omega}$ is obtained by inserting the least squares estimators of $\underline{\boldsymbol{\beta}}_j$ and $\boldsymbol{\gamma}$ in this expression:

$$\underline{\hat{\boldsymbol{\Omega}}} = \frac{1}{m}\sum_{j=1}^{m}(\underline{\boldsymbol{b}}_j - \boldsymbol{Z}_j\underline{\hat{\boldsymbol{\gamma}}})(\underline{\boldsymbol{b}}_j - \boldsymbol{Z}_j\underline{\hat{\boldsymbol{\gamma}}})',$$

or perhaps with $m-1$ instead of m in the denominator, and where $\underline{\hat{\boldsymbol{\gamma}}}$ is the one-step or two-step OLS estimator. Such an estimator is used in the MLA program [16] as "least squares estimator" of $\boldsymbol{\Omega}$ and as starting value for the iterations for obtaining the ML estimators. However, this estimator is biased for two reasons: The variability of $\underline{\hat{\boldsymbol{\gamma}}}$ is not taken into account and the covariance matrix of $\underline{\boldsymbol{b}}_j$ is not $\boldsymbol{\Omega}$, but $\boldsymbol{W}_j$. The first cause of bias vanishes as $m \to \infty$ and the second vanishes as $n_j \to \infty$, so it is only a reasonably good estimator if sample sizes at both levels are large. We can compute its exact expectation and exact variances of its elements, but we will not do that here. In addition to its simplicity, however, it has the virtue that it is guaranteed to be positive (semi)definite. This may prevent numerical problems when used as a starting value in an iterative procedure. Kovačević and Rai [66] propose a similar estimator, with $\boldsymbol{Z}_j\hat{\boldsymbol{\gamma}}$ replaced by the sample average of the $\boldsymbol{b}_j$'s, as a "conservative approximation".

Based on earlier formulas of Swamy [117], de Leeuw and Kreft [28] derive an unbiased estimator of $\boldsymbol{\Omega}$. The estimator of $\boldsymbol{\Omega}$ is derived elementwise. Thus, we look at its (k,l)-th element ω_{kl} and define an unbiased estimator of this element. By doing this for all distinct elements of $\boldsymbol{\Omega}$, we obtain an unbiased estimator of $\boldsymbol{\Omega}$.

Consider the k-th element of $\underline{\boldsymbol{\beta}}_j$, $\underline{\beta}_{jk}$. According to the model assumptions,

$$\underline{\beta}_{jk} = \boldsymbol{z}_{jk}'\boldsymbol{\gamma}_k + \underline{\delta}_{jk},$$

where $\boldsymbol{\gamma}_k$ is a subvector of $\boldsymbol{\gamma}$. The corresponding subvector of the two-step OLS estimator $\underline{\hat{\boldsymbol{\gamma}}}$ is $\underline{\hat{\boldsymbol{\gamma}}}_k$. Let $\boldsymbol{Z}_k$ be the $m \times q_k$ matrix with j-th row $\boldsymbol{z}_{jk}'$, where q_k is the number of elements of $\boldsymbol{z}_{jk}$, i.e., the number of explanatory variables for the k-th random coefficient. Correspondingly, let $\underline{\boldsymbol{b}}_k$ be the vector of length m with $\underline{b}_{jk}$ as its j-th element. Then it follows straightforwardly from the derivation of $\underline{\hat{\boldsymbol{\gamma}}}$ and the structure of $\boldsymbol{Z}_j$ that

$$\underline{\hat{\boldsymbol{\gamma}}}_k = (\boldsymbol{Z}_k'\boldsymbol{Z}_k)^{-1}\boldsymbol{Z}_k'\underline{\boldsymbol{b}}_k.$$

Let $\underline{\hat{\boldsymbol{t}}}_k$ be the vector of length m with $\underline{\hat{t}}_{jk} = \underline{b}_{jk} - \boldsymbol{z}_{jk}'\underline{\hat{\boldsymbol{\gamma}}}_k$ as its j-th element. Then we have

$$\underline{\hat{\boldsymbol{t}}}_k = \boldsymbol{Q}_k\underline{\boldsymbol{b}}_k = \boldsymbol{Q}_k(\underline{\boldsymbol{b}}_k - \boldsymbol{Z}_k\boldsymbol{\gamma}_k) = \boldsymbol{Q}_k\underline{\boldsymbol{t}}_k,$$

where $\boldsymbol{Q}_k = \boldsymbol{I}_m - \boldsymbol{Z}_k(\boldsymbol{Z}_k'\boldsymbol{Z}_k)^{-1}\boldsymbol{Z}_k'$ and $\underline{\boldsymbol{t}}_k$ is implicitly defined. Note that $E(\underline{\boldsymbol{b}}_k) = \boldsymbol{Z}_k\boldsymbol{\gamma}_k$ and

$$\text{Cov}(\underline{b}_k, \underline{b}_l') = \bigoplus_{j=1}^{m} (\boldsymbol{W}_j)_{kl} = \text{diag}[(\boldsymbol{W}_j)_{kl}] = \omega_{kl}\boldsymbol{I}_m + \boldsymbol{\Sigma}\boldsymbol{\nabla}_{kl},$$

where $\boldsymbol{\Sigma}$ is the diagonal matrix with j-th diagonal element equal to σ_j^2 and $\boldsymbol{\nabla}_{kl}$ is the diagonal matrix with j-th diagonal element equal to $[(\boldsymbol{X}_j'\boldsymbol{X}_j)^{-1}]_{kl}$. It follows that $E(\hat{\underline{t}}_k) = \boldsymbol{\emptyset}$ and

$$E(\hat{\underline{t}}_k\hat{\underline{t}}_l') = \text{Cov}(\hat{\underline{t}}_k, \hat{\underline{t}}_l') = \omega_{kl}\boldsymbol{Q}_k\boldsymbol{Q}_l + \boldsymbol{Q}_k\boldsymbol{\Sigma}\boldsymbol{\nabla}_{kl}\boldsymbol{Q}_l.$$

It is now natural to define the estimator

$$\hat{\underline{\omega}}_{kl} \triangleq \frac{\text{tr}\left[\hat{\underline{t}}_k\hat{\underline{t}}_l' - \boldsymbol{Q}_k\hat{\underline{\boldsymbol{\Sigma}}}\boldsymbol{\nabla}_{kl}\boldsymbol{Q}_l\right]}{\text{tr}(\boldsymbol{Q}_k\boldsymbol{Q}_l)} = \frac{1}{m^*}\left[\hat{\underline{t}}_l'\hat{\underline{t}}_k - \text{tr}(\hat{\underline{\boldsymbol{\Sigma}}}\boldsymbol{\nabla}_{kl}\boldsymbol{Q}_l\boldsymbol{Q}_k)\right],$$

where $m^* = \text{tr}(\boldsymbol{Q}_k\boldsymbol{Q}_l)$ and $\hat{\underline{\boldsymbol{\Sigma}}}$ is the diagonal matrix with j-th diagonal element equal to $\underline{s}_j^2$. This estimator of ω_{kl} is optimal in the least squares sense and it is evidently unbiased. However, unbiasedness in this context is not necessarily good, because it can easily lead to negative variance estimates.

Noticing that $\hat{\underline{\omega}}_{kl}$ is a quadratic function of the data, its variance can be found by using standard results about the expectations of quadratic forms in normally distributed random variables. The resulting expression is

$$\begin{aligned}\text{Var}(\hat{\underline{\omega}}_{kl}) = \frac{1}{(m^*)^2}\Bigg\{\sum_{i=1}^{m}\sum_{j=1}^{m} &[(\boldsymbol{W}_i)_{ll}(\boldsymbol{W}_j)_{kk}(\boldsymbol{Q}_l\boldsymbol{Q}_k)_{ij}^2 \\ &+ (\boldsymbol{W}_i)_{kl}(\boldsymbol{W}_j)_{kl}(\boldsymbol{Q}_l\boldsymbol{Q}_k)_{ij}(\boldsymbol{Q}_l\boldsymbol{Q}_k)_{ji}] \\ &+ \sum_{j=1}^{m}\frac{2(\sigma_j^2)^2}{n_j - p}[(\boldsymbol{X}_j'\boldsymbol{X}_j)^{-1}]_{kl}^2(\boldsymbol{Q}_l\boldsymbol{Q}_k)_{jj}^2\Bigg\}.\end{aligned}$$

An estimator of this variance is obtained by inserting the estimators $\underline{s}_j^2$ for σ_j^2 and $\hat{\underline{\boldsymbol{\Omega}}}$ for $\boldsymbol{\Omega}$ (the latter in $\boldsymbol{W}_j$) in this formula.

A somewhat related but slightly different method for estimating the variance parameters uses the same ideas as the WLS estimator above, but reverses the roles of the fixed coefficients and the variance parameters. In particular, assume that γ is known and that $\boldsymbol{\Omega}$ is written in the linear form (1.3). Then

$$\begin{aligned}E\left[(\underline{\boldsymbol{y}} - \boldsymbol{U}\gamma)(\underline{\boldsymbol{y}} - \boldsymbol{U}\gamma)'\right] &= \boldsymbol{V} \\ &= \bigoplus_{j=1}^{m}(\boldsymbol{X}_j\boldsymbol{\Omega}\boldsymbol{X}_j' + \sigma_j^2\boldsymbol{I}_{n_j}) \\ &= \sum_{g=1}^{G}\left(\bigoplus_{j=1}^{m}\boldsymbol{X}_j\boldsymbol{C}_g\boldsymbol{X}_j'\right)\xi_g + \sum_{j=1}^{m}(\boldsymbol{e}_j\boldsymbol{e}_j' \otimes \boldsymbol{I}_{n_j})\sigma_j^2,\end{aligned}$$

where e_j is the j-th column of I_m, and if all residual variances are equal, the last summation reduces to $\sigma^2 I_n$. Clearly, this expectation is linear in the parameters $\{\xi_g\}$ and $\{\sigma_j^2\}$.

Now, let $U^*[n, G+m]$ be the matrix with g-th column equal to

$$U_g^* \triangleq \text{vec}\left(\bigoplus_{j=1}^{m} X_j C_g X_j'\right)$$

and $(G+j)$-th column equal to

$$U_{G+j}^* \triangleq \text{vec}(e_j e_j' \otimes I_{n_j}).$$

Furthermore, let $\gamma^*[G+m]$ be the vector with g-th element ξ_g $(g = 1, \ldots, G)$ and $(G+j)$-th element σ_j^2 $(j = 1, \ldots, m)$. If all σ_j^2 are equal, U^* has $G+1$ columns, the last one being vec I_n, and γ^* has $G+1$ elements, the last one being σ^2. The rest of the discussion is unaltered. Finally, let

$$y^* \triangleq \text{vec}\big[(y - U\gamma)(y - U\gamma)'\big]. \tag{1.23}$$

Then $E\,\underline{y}^* = U^*\gamma^*$, which suggests that the variance parameters in γ^* can be jointly estimated by a least squares method. Although an OLS method would be computationally much easier, a WLS method is typically used, for reasons that become clear in Section 1.6.2 below. From the characteristics of the normal distribution, it follows that the dispersion matrix of $\underline{y}^*$ is $2N_n(V \otimes V)$ (e.g., Magnus and Neudecker [79, Lemma 9]), where $N_n[n^2, n^2]$ is a symmetric idempotent matrix of rank $n(n+1)/2$, which projects a column vector of order n^2 onto the space of vec's of symmetric matrices. It is therefore called the *symmetrization matrix* by Wansbeek and Meijer [123, p. 361]. Thus, the dispersion matrix of $\underline{y}^*$ is singular, the reason being that $\underline{y}^*$ contains duplicated elements. We can remove the duplicated elements and then compute the nonsingular dispersion matrix and use it in a WLS procedure. Due to the structure of the problem, this is equivalent to computing the estimate

$$\hat{\gamma}^* = \big((U^*)'(V^*)^{-1}(U^*)\big)^{-1}(U^*)'(V^*)^{-1}(y^*), \tag{1.24}$$

where $V^* = 2(V \otimes V)$. From the derivation, it follows immediately that

$$\text{Cov}(\underline{\hat{\gamma}}^*) = \big((U^*)'(V^*)^{-1}(U^*)\big)^{-1},$$

where the symmetrization matrix drops out because of the structure of the matrices involved.

It appears that (1.24) suffers from a few problems. The first is that the right-hand side contains unknown parameters: not only γ, but also the very parameters that the left-hand side estimates, through its dependence on V^*.

Thus, as before, we have to insert (preliminary) estimators of these. This leads to the following typical estimation procedure: (1) compute the 1-step or 2-step OLS estimate of γ; (2) use this to compute an estimate of $\boldsymbol{y}^*$ and compute a preliminary estimate of γ^* from (1.24) with $\boldsymbol{V}^* = \boldsymbol{I}$; (3) use this to compute an estimate $\tilde{\boldsymbol{V}}$ of $\boldsymbol{V}$ and compute the WLS estimator of γ from (1.20); (4) use this to compute an improved estimate of $\boldsymbol{y}^*$ and compute the WLS estimate of γ^* from (1.24) with $\boldsymbol{V}^* = 2(\tilde{\boldsymbol{V}} \otimes \tilde{\boldsymbol{V}})$. Variations, e.g., using the estimators of de Leeuw and Kreft [28] as preliminary estimators, are possible, but as it is presented here, it suggests further iterating steps (3) and (4). Indeed, this is typically done and leads to the IGLS algorithm discussed in Section 1.6.2 below.

The second problem with direct application of (1.24) is that it is a computational disaster. The matrix $\boldsymbol{V}^*$ is of order $n^2 \times n^2$, so if $n = 20,000$ as in the application reported below, then we would have to store and invert a 400 million $\times$ 400 million matrix. Fortunately, however, the problem has so much structure that this is not necessary: $\boldsymbol{V}^* = 2(\boldsymbol{V} \otimes \boldsymbol{V})$, which reduces the problem to $n \times n$, but the direct sum form of $\boldsymbol{V}$ reduces this further to $n_j \times n_j$. Then, reductions like the ones used above to arrive at (1.20) as a more convenient version of (1.19) further simplify the computations. Efficient computational procedures are discussed in Goldstein and Rasbash [47].

A variant of (1.24) is obtained by recognizing that the WLS estimator $\underline{\hat{\gamma}}$ that is inserted in the computation of $\boldsymbol{y}^*$ is not equal to γ, but is an unbiased estimator with variance $(\boldsymbol{U}'\boldsymbol{V}^{-1}\boldsymbol{U})^{-1}$, ignoring variance due to estimation error in the preliminary estimate of $\boldsymbol{V}$. More specifically, by writing

$$\boldsymbol{y} - \boldsymbol{U}\hat{\gamma} = \left[\boldsymbol{I} - \boldsymbol{U}(\boldsymbol{U}'\boldsymbol{V}^{-1}\boldsymbol{U})^{-1}\boldsymbol{U}'\boldsymbol{V}^{-1}\right]\boldsymbol{y},$$

it follows that

$$\begin{aligned} &E\left[(\underline{\boldsymbol{y}} - \boldsymbol{U}\underline{\hat{\gamma}})(\underline{\boldsymbol{y}} - \boldsymbol{U}\underline{\hat{\gamma}})'\right] \\ &\quad = \left[\boldsymbol{I} - \boldsymbol{U}(\boldsymbol{U}'\boldsymbol{V}^{-1}\boldsymbol{U})^{-1}\boldsymbol{U}'\boldsymbol{V}^{-1}\right]\boldsymbol{V}\left[\boldsymbol{I} - \boldsymbol{U}(\boldsymbol{U}'\boldsymbol{V}^{-1}\boldsymbol{U})^{-1}\boldsymbol{U}'\boldsymbol{V}^{-1}\right]' \\ &\quad = \boldsymbol{V} - \boldsymbol{U}(\boldsymbol{U}'\boldsymbol{V}^{-1}\boldsymbol{U})^{-1}\boldsymbol{U}', \end{aligned}$$

or

$$E\left[(\underline{\boldsymbol{y}} - \boldsymbol{U}\underline{\hat{\gamma}})(\underline{\boldsymbol{y}} - \boldsymbol{U}\underline{\hat{\gamma}})' + \boldsymbol{U}(\boldsymbol{U}'\boldsymbol{V}^{-1}\boldsymbol{U})^{-1}\boldsymbol{U}'\right] = \boldsymbol{V}.$$

This suggests replacing (1.23) by

$$\boldsymbol{y}^* \stackrel{\Delta}{=} \operatorname{vec}\left[(\underline{\boldsymbol{y}} - \boldsymbol{U}\underline{\hat{\gamma}})(\underline{\boldsymbol{y}} - \boldsymbol{U}\underline{\hat{\gamma}})' + \boldsymbol{U}(\boldsymbol{U}'\tilde{\boldsymbol{V}}^{-1}\boldsymbol{U})^{-1}\boldsymbol{U}'\right] \tag{1.25}$$

and then proceeding with the estimation process as described above. The term $\boldsymbol{U}(\boldsymbol{U}'\tilde{\boldsymbol{V}}^{-1}\boldsymbol{U})^{-1}\boldsymbol{U}'$ can be viewed as a bias correction. The resulting estimator is again consistent with the same expression for the asymptotic covariance matrix, but is generally less biased in finite samples. The iteration procedure described above with this estimator leads to RIGLS estimators, which are also discussed in Section 1.6.2 below.

1.6.2 Maximum Likelihood

Except for some special cases, explicit closed-form expressions for the maximum likelihood estimators are not available. The loglikelihood function has to be optimized by using some kind of numerical algorithm. This section discusses several of the available algorithms. We can distinguish, on the one hand, generic numerical optimization techniques that can be used for any well-behaved function and, on the other hand, algorithms that are more specific to the problem at hand.

Let $f(\boldsymbol{\theta})$ be a loss function of a parameter vector $\boldsymbol{\theta}$. We want to find the value $\hat{\boldsymbol{\theta}}$ of $\boldsymbol{\theta}$ that minimizes $f(\boldsymbol{\theta})$. Throughout, we assume that $f(\boldsymbol{\theta})$ is well behaved, i.e., that it is continuous and has continuous first and second partial derivatives, is locally Lipschitz, etc. The loss functions for FIML and REML satisfy these and other regularity conditions except in pathological situations where the sample data have no variation or predictor matrices are not of full rank. Thus, we assume these away.

For a short introduction to generic numerical optimization, we refer to Appendix 1.B. The (modified) Newton-Raphson method mentioned there is described for multilevel models by Jennrich and Schluchter [63] and Lindstrom and Bates [73] and it is used in the BMDP5V program [107] for repeated measures models and the nlme package [90] for multilevel analysis in R. The BFGS method is implemented in most general-purpose optimization functions and is used in the MLA program for multilevel analysis [16]. From the discussion in Appendix 1.B, it is clear that we typically need at least first partial derivatives of the loss function, and for Newton-Raphson also the second partial derivatives. We will give their formulas for the FIML and REML loss functions below.

Derivatives of FIML

Computing the partial derivatives of the loglikelihood function with respect to the parameters is a straightforward, albeit tedious, application of (matrix) calculus as developed by, e.g., Magnus and Neudecker [80]. Here we only give the results, the derivations are available from us upon request. Throughout, we will assume that $\boldsymbol{\Omega}$ is parametrized as in (1.3). The first partial derivatives are

$$\frac{\partial \mathcal{L}^F}{\partial \gamma} = -2 \sum_{j=1}^{m} \boldsymbol{Z}_j' \boldsymbol{W}_j^{-1} \boldsymbol{t}_j, \tag{1.26a}$$

$$\frac{\partial \mathcal{L}^F}{\partial \sigma_j^2} = -(n_j - p) \left(\frac{s_j^2 - \sigma_j^2}{(\sigma_j^2)^2} \right) - \operatorname{tr}\left[\boldsymbol{T}_j (\boldsymbol{X}_j' \boldsymbol{X}_j)^{-1} \right], \tag{1.26b}$$

$$\frac{\partial \mathcal{L}^F}{\partial \xi_g} = -\sum_{j=1}^{m} \operatorname{tr}(\boldsymbol{T}_j \boldsymbol{C}_g), \tag{1.26c}$$

where

$$\boldsymbol{t}_j \stackrel{\Delta}{=} \boldsymbol{b}_j - \boldsymbol{Z}_j \boldsymbol{\gamma},$$
$$\boldsymbol{T}_j \stackrel{\Delta}{=} \boldsymbol{W}_j^{-1}(\boldsymbol{t}_j \boldsymbol{t}_j' - \boldsymbol{W}_j)\boldsymbol{W}_j^{-1}.$$

It is easy to check that the expected values of these partials (when viewed as functions of random variables) are zero, as they should be. It follows immediately from (1.26a) that after convergence (first partials are zero), (1.20) holds. Thus, the FIML estimator of γ is a WLS estimator based on the FIML estimates of the variance parameters.

The second partial derivatives with respect to the parameters are

$$\frac{\partial^2 \mathcal{L}^F}{\partial \boldsymbol{\gamma}\, \partial \boldsymbol{\gamma}'} = 2\sum_{j=1}^{m} \boldsymbol{Z}_j' \boldsymbol{W}_j^{-1} \boldsymbol{Z}_j,$$
$$\frac{\partial^2 \mathcal{L}^F}{\partial \boldsymbol{\gamma}\, \partial \sigma_j^2} = 2\boldsymbol{Z}_j' \boldsymbol{W}_j^{-1} (\boldsymbol{X}_j' \boldsymbol{X}_j)^{-1} \boldsymbol{W}_j^{-1} \boldsymbol{t}_j,$$
$$\frac{\partial^2 \mathcal{L}^F}{\partial \boldsymbol{\gamma}\, \partial \xi_g} = 2\sum_{j=1}^{m} \boldsymbol{Z}_j' \boldsymbol{W}_j^{-1} \boldsymbol{C}_g \boldsymbol{W}_j^{-1} \boldsymbol{t}_j,$$
$$\frac{\partial^2 \mathcal{L}^F}{\partial \sigma_j^2\, \partial \sigma_j^2} = (n_j - p)\left(\frac{2s_j^2 - \sigma_j^2}{(\sigma_j^2)^3}\right) + \operatorname{tr}\left[\boldsymbol{\Upsilon}_j (\boldsymbol{X}_j' \boldsymbol{X}_j)^{-1}\right],$$
$$\frac{\partial^2 \mathcal{L}^F}{\partial \sigma_j^2\, \partial \sigma_k^2} = 0 \qquad \text{for } k \neq j,$$
$$\frac{\partial^2 \mathcal{L}^F}{\partial \sigma_j^2\, \partial \xi_g} = \operatorname{tr}(\boldsymbol{\Upsilon}_j \boldsymbol{C}_g),$$
$$\frac{\partial^2 \mathcal{L}^F}{\partial \xi_g\, \partial \xi_h} = \sum_{j=1}^{m} \operatorname{tr}(\boldsymbol{T}_j \boldsymbol{C}_h \boldsymbol{W}_j^{-1} \boldsymbol{C}_g + \boldsymbol{W}_j^{-1} \boldsymbol{C}_h \boldsymbol{T}_j \boldsymbol{C}_g + \boldsymbol{W}_j^{-1} \boldsymbol{C}_h \boldsymbol{W}_j^{-1} \boldsymbol{C}_g),$$

where

$$\boldsymbol{\Upsilon}_j \stackrel{\Delta}{=} \boldsymbol{T}_j (\boldsymbol{X}_j' \boldsymbol{X}_j)^{-1} \boldsymbol{W}_j^{-1} + \boldsymbol{W}_j^{-1} (\boldsymbol{X}_j' \boldsymbol{X}_j)^{-1} \boldsymbol{T}_j + \boldsymbol{W}_j^{-1} (\boldsymbol{X}_j' \boldsymbol{X}_j)^{-1} \boldsymbol{W}_j^{-1}.$$

As mentioned above, often it is assumed that all residual variances are the same: $\sigma_j^2 = \sigma^2$. This leads to fairly trivial changes in these formulas: Every explicit or implicit occurrence of σ_j^2 on the right-hand side is replaced by σ^2, and the derivatives with respect to σ^2 are simply the sums over all groups of the derivatives with respect to σ_j^2 as given here:

$$\frac{\partial \mathcal{L}^F}{\partial \sigma^2} = -\sum_{j=1}^{m} \left\{ (n_j - p) \left(\frac{s_j^2 - \sigma^2}{(\sigma^2)^2} \right) + \operatorname{tr}\left[\boldsymbol{T}_j (\boldsymbol{X}_j' \boldsymbol{X}_j)^{-1} \right] \right\}, \tag{1.27}$$

$$\frac{\partial^2 \mathcal{L}^F}{\partial \gamma \, \partial \sigma^2} = 2 \sum_{j=1}^{m} \boldsymbol{Z}_j' \boldsymbol{W}_j^{-1} (\boldsymbol{X}_j' \boldsymbol{X}_j)^{-1} \boldsymbol{W}_j^{-1} \boldsymbol{t}_j,$$

$$\frac{\partial^2 \mathcal{L}^F}{\partial \sigma^2 \, \partial \sigma^2} = \sum_{j=1}^{m} \left\{ (n_j - p) \left(\frac{2s_j^2 - \sigma^2}{(\sigma^2)^3} \right) + \operatorname{tr}\left[\boldsymbol{\Upsilon}_j (\boldsymbol{X}_j' \boldsymbol{X}_j)^{-1} \right] \right\},$$

$$\frac{\partial^2 \mathcal{L}^F}{\partial \sigma^2 \, \partial \xi_g} = \sum_{j=1}^{m} \operatorname{tr}(\boldsymbol{\Upsilon}_j \boldsymbol{C}_g).$$

The derivatives can now be used in a standard numerical optimization algorithm to obtain the FIML estimates.

Derivatives of REML

The first partial derivatives of the REML loss function with respect to the parameters are

$$\frac{\partial \mathcal{L}^R}{\partial \sigma_j^2} = -(n_j - p) \left(\frac{s_j^2 - \sigma_j^2}{(\sigma_j^2)^2} \right) - \operatorname{tr}\left[\boldsymbol{\Delta}_j (\boldsymbol{X}_j' \boldsymbol{X}_j)^{-1} \right], \tag{1.28a}$$

$$\frac{\partial \mathcal{L}^R}{\partial \xi_g} = -\sum_{j=1}^{m} \operatorname{tr}(\boldsymbol{\Delta}_j \boldsymbol{C}_g), \tag{1.28b}$$

where

$$\boldsymbol{\Delta}_j \triangleq \boldsymbol{W}_j^{-1} (\hat{\boldsymbol{t}}_j \hat{\boldsymbol{t}}_j' - \boldsymbol{W}_j + \boldsymbol{Z}_j \boldsymbol{A} \boldsymbol{Z}_j') \boldsymbol{W}_j^{-1},$$

$$\hat{\boldsymbol{t}}_j \triangleq \boldsymbol{b}_j - \boldsymbol{Z}_j \hat{\gamma},$$

$$\hat{\gamma} \triangleq \boldsymbol{A} \sum_{j=1}^{m} \boldsymbol{Z}_j' \boldsymbol{W}_j^{-1} \boldsymbol{b}_j,$$

$$\boldsymbol{A} \triangleq \left(\sum_{j=1}^{m} \boldsymbol{Z}_j' \boldsymbol{W}_j^{-1} \boldsymbol{Z}_j \right)^{-1}.$$

Note that there are no derivatives with respect to γ, because $\mathcal{L}^R$ is not a function of γ. We use $\hat{\gamma}$ as a shorthand, but it is not a parameter, it is a function of the data and the variance parameters. Of course, after convergence, this same definition is used to obtain a WLS estimate of γ, but in deriving statistical properties of the REML estimators, we must treat $\hat{\gamma}$ as a function and not as a mathematical variable.

The second partial derivatives of the REML loss function with respect to the parameters are

$$\frac{\partial^2 \mathcal{L}^R}{\partial \sigma_j^2 \, \partial \sigma_j^2} = (n_j - p)\left(\frac{2s_j^2 - \sigma_j^2}{(\sigma_j^2)^3}\right) + \operatorname{tr}\big[\boldsymbol{\Theta}_j(\boldsymbol{X}_j'\boldsymbol{X}_j)^{-1}\big]$$
$$- 2\hat{\boldsymbol{u}}_j'\boldsymbol{A}\hat{\boldsymbol{u}}_j - \operatorname{tr}(\boldsymbol{\Lambda}_j\boldsymbol{A}\boldsymbol{\Lambda}_j\boldsymbol{A}),$$
$$\frac{\partial^2 \mathcal{L}^R}{\partial \sigma_j^2 \, \partial \sigma_k^2} = -2\hat{\boldsymbol{u}}_j'\boldsymbol{A}\hat{\boldsymbol{u}}_k - \operatorname{tr}(\boldsymbol{\Lambda}_j\boldsymbol{A}\boldsymbol{\Lambda}_k\boldsymbol{A}) \qquad \text{for } k \neq j,$$
$$\frac{\partial^2 \mathcal{L}^R}{\partial \sigma_j^2 \, \partial \xi_g} = \operatorname{tr}(\boldsymbol{\Theta}_j\boldsymbol{C}_g) - 2\hat{\boldsymbol{u}}_j'\boldsymbol{A}\hat{\boldsymbol{\tau}}_g - \operatorname{tr}(\boldsymbol{\Lambda}_j\boldsymbol{A}\boldsymbol{\Xi}_g\boldsymbol{A}),$$
$$\frac{\partial^2 \mathcal{L}^R}{\partial \xi_g \, \partial \xi_h} = \sum_{j=1}^{m} \operatorname{tr}(\boldsymbol{\Delta}_j\boldsymbol{C}_h\boldsymbol{W}_j^{-1}\boldsymbol{C}_g + \boldsymbol{W}_j^{-1}\boldsymbol{C}_h\boldsymbol{\Delta}_j\boldsymbol{C}_g + \boldsymbol{W}_j^{-1}\boldsymbol{C}_h\boldsymbol{W}_j^{-1}\boldsymbol{C}_g)$$
$$- 2\hat{\boldsymbol{\tau}}_h'\boldsymbol{A}\hat{\boldsymbol{\tau}}_g - \operatorname{tr}(\boldsymbol{\Xi}_h\boldsymbol{A}\boldsymbol{\Xi}_g\boldsymbol{A}),$$

where

$$\boldsymbol{\Theta}_j \triangleq \boldsymbol{\Delta}_j(\boldsymbol{X}_j'\boldsymbol{X}_j)^{-1}\boldsymbol{W}_j^{-1} + \boldsymbol{W}_j^{-1}(\boldsymbol{X}_j'\boldsymbol{X}_j)^{-1}\boldsymbol{\Delta}_j + \boldsymbol{W}_j^{-1}(\boldsymbol{X}_j'\boldsymbol{X}_j)^{-1}\boldsymbol{W}_j^{-1},$$
$$\boldsymbol{\Lambda}_j \triangleq \boldsymbol{Z}_j'\boldsymbol{W}_j^{-1}(\boldsymbol{X}_j'\boldsymbol{X}_j)^{-1}\boldsymbol{W}_j^{-1}\boldsymbol{Z}_j,$$
$$\hat{\boldsymbol{u}}_j \triangleq \boldsymbol{Z}_j'\boldsymbol{W}_j^{-1}(\boldsymbol{X}_j'\boldsymbol{X}_j)^{-1}\boldsymbol{W}_j^{-1}\hat{\boldsymbol{t}}_j,$$
$$\boldsymbol{\Xi}_g \triangleq \sum_{j=1}^{m} \boldsymbol{Z}_j'\boldsymbol{W}_j^{-1}\boldsymbol{C}_g\boldsymbol{W}_j^{-1}\boldsymbol{Z}_j,$$
$$\hat{\boldsymbol{\tau}}_g \triangleq \sum_{j=1}^{m} \boldsymbol{Z}_j'\boldsymbol{W}_j^{-1}\boldsymbol{C}_g\boldsymbol{W}_j^{-1}\hat{\boldsymbol{t}}_j.$$

When all σ_j^2 are equal, the first partial derivative with respect to σ^2 becomes

$$\frac{\partial \mathcal{L}^R}{\partial \sigma^2} = -\sum_{j=1}^{m}\left\{(n_j - p)\left(\frac{s_j^2 - \sigma^2}{(\sigma^2)^2}\right) + \operatorname{tr}\big[\boldsymbol{\Delta}_j(\boldsymbol{X}_j'\boldsymbol{X}_j)^{-1}\big]\right\} \tag{1.29}$$

and the second partial derivatives involving σ^2 become

$$\frac{\partial^2 \mathcal{L}^R}{\partial \sigma^2 \, \partial \sigma^2} = \sum_{j=1}^{m}\left\{(n_j - p)\left(\frac{2s_j^2 - \sigma^2}{(\sigma^2)^3}\right) + \operatorname{tr}\big[\boldsymbol{\Theta}_j(\boldsymbol{X}_j'\boldsymbol{X}_j)^{-1}\big]\right\}$$
$$- 2\hat{\boldsymbol{u}}'\boldsymbol{A}\hat{\boldsymbol{u}} - \operatorname{tr}(\boldsymbol{\Lambda}\boldsymbol{A}\boldsymbol{\Lambda}\boldsymbol{A}),$$
$$\frac{\partial^2 \mathcal{L}^R}{\partial \sigma^2 \, \partial \xi_g} = \sum_{j=1}^{m} \operatorname{tr}(\boldsymbol{\Theta}_j\boldsymbol{C}_g) - 2\hat{\boldsymbol{u}}'\boldsymbol{A}\hat{\boldsymbol{\tau}}_g - \operatorname{tr}(\boldsymbol{\Lambda}\boldsymbol{A}\boldsymbol{\Xi}_g\boldsymbol{A}),$$

where

$$\hat{\boldsymbol{u}} \triangleq \sum_{j=1}^{m} \hat{\boldsymbol{u}}_j\,,$$

$$\boldsymbol{\Lambda} \triangleq \sum_{j=1}^{m} \boldsymbol{\Lambda}_j.$$

Standard Errors

For the standard errors, we need the expectations of the second derivatives instead of the second derivatives themselves. This simplifies the formulas considerably, because many terms have expectation zero and thus drop out. In particular, using $E(\underline{\boldsymbol{t}}_j) = \boldsymbol{\emptyset}$, we obtain

$$E\left(\frac{\partial^2 \underline{\mathcal{L}}^F}{\partial \gamma \, \partial \gamma'}\right) = 2 \sum_{j=1}^{m} \boldsymbol{Z}_j' \boldsymbol{W}_j^{-1} \boldsymbol{Z}_j \, ,$$

$$E\left(\frac{\partial^2 \underline{\mathcal{L}}^F}{\partial \gamma \, \partial \sigma_j^2}\right) = \boldsymbol{\emptyset},$$

$$E\left(\frac{\partial^2 \underline{\mathcal{L}}^F}{\partial \gamma \, \partial \xi_g}\right) = \boldsymbol{\emptyset}.$$

Hence, the matrix of expectations of the second derivatives of the FIML loss function is a block-diagonal matrix with a diagonal block for the fixed coefficients and a diagonal block for the variance parameters.

For the latter part, we observe that $E(\underline{\boldsymbol{T}}_j) = \boldsymbol{\emptyset}$ implies that

$$E(\underline{\boldsymbol{\Upsilon}}_j) = \boldsymbol{W}_j^{-1} (\boldsymbol{X}_j' \boldsymbol{X}_j)^{-1} \boldsymbol{W}_j^{-1}.$$

Consequently,

$$E\left(\frac{\partial^2 \underline{\mathcal{L}}^F}{\partial \sigma_j^2 \, \partial \sigma_j^2}\right) = \frac{n_j - p}{(\sigma_j^2)^2} + \operatorname{tr}\left[\boldsymbol{W}_j^{-1} (\boldsymbol{X}_j' \boldsymbol{X}_j)^{-1} \boldsymbol{W}_j^{-1} (\boldsymbol{X}_j' \boldsymbol{X}_j)^{-1}\right],$$

$$E\left(\frac{\partial^2 \underline{\mathcal{L}}^F}{\partial \sigma_j^2 \, \partial \sigma_k^2}\right) = 0 \qquad \text{for } k \neq j,$$

$$E\left(\frac{\partial^2 \underline{\mathcal{L}}^F}{\partial \sigma_j^2 \, \partial \xi_g}\right) = \operatorname{tr}\left[\boldsymbol{W}_j^{-1} (\boldsymbol{X}_j' \boldsymbol{X}_j)^{-1} \boldsymbol{W}_j^{-1} \boldsymbol{C}_g\right],$$

$$E\left(\frac{\partial^2 \underline{\mathcal{L}}^F}{\partial \xi_g \, \partial \xi_h}\right) = \sum_{j=1}^{m} \operatorname{tr}(\boldsymbol{W}_j^{-1} \boldsymbol{C}_h \boldsymbol{W}_j^{-1} \boldsymbol{C}_g).$$

When all σ_j^2 are the same, the first three of these are replaced by

$$E\left(\frac{\partial^2 \underline{\mathcal{L}}^F}{\partial \sigma^2 \, \partial \sigma^2}\right) = \sum_{j=1}^{m} \left\{ \frac{n_j - p}{(\sigma^2)^2} + \operatorname{tr}\left[\boldsymbol{W}_j^{-1} (\boldsymbol{X}_j' \boldsymbol{X}_j)^{-1} \boldsymbol{W}_j^{-1} (\boldsymbol{X}_j' \boldsymbol{X}_j)^{-1}\right] \right\},$$

$$E\left(\frac{\partial^2 \underline{\mathcal{L}}^F}{\partial \sigma^2 \, \partial \xi_g}\right) = \sum_{j=1}^{m} \operatorname{tr}\left[\boldsymbol{W}_j^{-1} (\boldsymbol{X}_j' \boldsymbol{X}_j)^{-1} \boldsymbol{W}_j^{-1} \boldsymbol{C}_g\right].$$

The information matrix $\mathcal{I}^F$ is defined as

$$\mathcal{I}^F \triangleq E\left(-\frac{\partial^2 \underline{\ell}^F}{\partial\boldsymbol{\theta}\,\partial\boldsymbol{\theta}'}\right),$$

where $\underline{\ell}^F$ is the FIML loglikelihood function viewed as a random variable and $\boldsymbol{\theta}$ is the parameter vector. Up till now, we have ignored some constants that do not affect the estimators, but we need to be a little more precise for the standard errors. In fact, $\underline{\mathcal{L}}_j^F = -2(\underline{\ell}_j^F - K_j)$, where K_j is a constant that does not depend on the parameters. Hence, it follows that

$$\mathcal{I}^F = \tfrac{1}{2}\, E\left(\frac{\partial^2 \underline{\mathcal{L}}^F}{\partial\boldsymbol{\theta}\,\partial\boldsymbol{\theta}'}\right),$$

so we have to divide the formulas that have just been given by 2. Standard maximum likelihood theory tells us that the standard errors of the estimators are the square roots of the diagonal elements of $(\mathcal{I}^F)^{-1}$. In particular, the submatrix of $\mathcal{I}^F$ corresponding to γ is

$$\mathcal{I}^F_{\gamma\gamma} = \sum_{j=1}^{m} \boldsymbol{Z}_j'\mathbf{W}_j^{-1}\boldsymbol{Z}_j.$$

Because of the block-diagonal structure of $\mathcal{I}^F$, it follows that the standard errors of $\underline{\hat{\gamma}}$ are the square roots of the elements of

$$(\mathcal{I}^F_{\gamma\gamma})^{-1} = \left(\sum_{j=1}^{m} \boldsymbol{Z}_j'\mathbf{W}_j^{-1}\boldsymbol{Z}_j\right)^{-1},$$

which corroborates the results obtained earlier for the WLS estimator.

Analogously, for the REML estimators, the expressions are

$$\begin{aligned}
E\left(\frac{\partial^2 \underline{\mathcal{L}}^R}{\partial\sigma_j^2\,\partial\sigma_j^2}\right) &= \frac{n_j - p}{(\sigma_j^2)^2} + \operatorname{tr}\left[\boldsymbol{W}_j^{-1}(\boldsymbol{X}_j'\boldsymbol{X}_j)^{-1}\boldsymbol{W}_j^{-1}(\boldsymbol{X}_j'\boldsymbol{X}_j)^{-1}\right] \\
&\quad - \operatorname{tr}(\boldsymbol{\Lambda}_j\boldsymbol{A}\boldsymbol{\Lambda}_j\boldsymbol{A}), \\
E\left(\frac{\partial^2 \underline{\mathcal{L}}^R}{\partial\sigma_j^2\,\partial\sigma_k^2}\right) &= -\operatorname{tr}(\boldsymbol{\Lambda}_j\boldsymbol{A}\boldsymbol{\Lambda}_k\boldsymbol{A}) \qquad \text{for } k \neq j, \\
E\left(\frac{\partial^2 \underline{\mathcal{L}}^R}{\partial\sigma_j^2\,\partial\xi_g}\right) &= \operatorname{tr}\left[\boldsymbol{W}_j^{-1}(\boldsymbol{X}_j'\boldsymbol{X}_j)^{-1}\boldsymbol{W}_j^{-1}\boldsymbol{C}_g\right] - \operatorname{tr}(\boldsymbol{\Lambda}_j\boldsymbol{A}\boldsymbol{\Xi}_g\boldsymbol{A}), \\
E\left(\frac{\partial^2 \underline{\mathcal{L}}^R}{\partial\xi_g\,\partial\xi_h}\right) &= \sum_{j=1}^{m}\operatorname{tr}(\boldsymbol{W}_j^{-1}\boldsymbol{C}_h\boldsymbol{W}_j^{-1}\boldsymbol{C}_g) - \operatorname{tr}(\boldsymbol{\Xi}_h\boldsymbol{A}\boldsymbol{\Xi}_g\boldsymbol{A}).
\end{aligned}$$

When all σ_j^2 are the same, the first three of these are replaced by

$$E\left(\frac{\partial^2\mathcal{L}^R}{\partial\sigma^2\,\partial\sigma^2}\right) = \sum_{j=1}^{m}\left\{\frac{n_j - p}{(\sigma^2)^2} + \operatorname{tr}\left[\boldsymbol{W}_j^{-1}(\boldsymbol{X}_j'\boldsymbol{X}_j)^{-1}\boldsymbol{W}_j^{-1}(\boldsymbol{X}_j'\boldsymbol{X}_j)^{-1}\right]\right\}$$
$$- \operatorname{tr}(\boldsymbol{\Lambda A \Lambda A}),$$
$$E\left(\frac{\partial^2\mathcal{L}^R}{\partial\sigma^2\,\partial\xi_g}\right) = \sum_{j=1}^{m}\operatorname{tr}\left[\boldsymbol{W}_j^{-1}(\boldsymbol{X}_j'\boldsymbol{X}_j)^{-1}\boldsymbol{W}_j^{-1}\boldsymbol{C}_g\right] - \operatorname{tr}(\boldsymbol{\Lambda A \Xi}_g\boldsymbol{A}).$$

The information matrix $\mathcal{I}^R$ is again obtained by dividing the formulas for the expectations of the second derivatives by 2. Standard errors are the square roots of the diagonal elements of the inverse of the information matrix.

As indicated above, after convergence, we use the expression for $\hat{\gamma}$ used in the expressions for the REML derivatives as an estimator of γ. It is immediately clear that this is a WLS estimator with $\boldsymbol{W}_j$ based on the REML estimators for the variance parameters. Hence, the standard error formulas given for WLS above apply directly to this estimator.

Scoring

We have seen above that expressions for the second derivatives of the ML loss functions are rather unwieldy, whereas the expressions for their expectations are much simpler. In fact, because the asymptotic covariance matrix of the estimators is a positive constant times the inverse of the matrix of expected second derivatives, the matrix of expected second derivatives must be a positive definite matrix. Furthermore, in large samples, the exact second derivatives should be close to the expected second derivatives. Combining these statistical observations with the general theory of numerical optimization suggests that a convenient alternative to the Newton-Raphson algorithm would be to replace the Hessian by its expectation. Because the expected Hessian is guaranteed to be positive definite, this does not need to be checked and modifications of it are not necessary. Thus, an easier expression is used, which is computationally less demanding, and the block-diagonality of the expected Hessian reduces the computational burden in computing the inverse as well.

The resulting algorithm, which is specific to loglikelihood functions (but certainly not to multilevel models), is called *Method of Scoring*, *Fisher scoring*, or simply *Scoring*. It was proposed for multilevel models by Longford [74] and implemented in the VARCL program [75]. It tends to be very fast and stable.

Iteratively Reweighted Least Squares

In (1.20), we have seen a simple, yet statistically efficient estimator of the fixed coefficients γ, given knowledge of the variance parameters. In practice, this means that consistent estimators of the variance parameters are plugged in. Conversely, in (1.24), combined with either (1.23) or (1.25), we have given a (conceptually) simple and statistically efficient estimator of the variance parameters γ^*, given γ and a preliminary estimate of the variance parameters. As noted there, this suggests an iterative algorithm, in which these two steps are alternated.

This algorithm was introduced for multilevel models by Goldstein [44] using (1.23) to compute $\boldsymbol{y}^*$ and by Goldstein [45] using (1.25) to compute $\boldsymbol{y}^*$. In the former case, the algorithm is called *iterative generalized least squares* (IGLS), whereas in the latter, it is called *restricted iterative generalized least squares* (RIGLS). Similar procedures, also known as *iterative reweighted least squares* (IRLS), are used in many branches of statistics. For example, the standard estimation method for generalized linear models is IRLS [82] and it can be used to compute estimators based on "robust" loss functions, which are less sensitive to outliers [48]. An overview, relating IGLS to various numerical optimization algorithms, is given by del Pino [32]. From these sources, it is known that IGLS produces maximum likelihood estimators.

The equivalence of IGLS to FIML was shown explicitly for the multilevel model by Goldstein [44]. Goldstein [45] showed that RIGLS gives REML estimators. Paralleling his proofs, we can see here, as we have noted above, that setting (1.26a) to zero is equivalent to the IGLS/RIGLS condition (1.20). Furthermore, it is easy to show that (1.24) combined with (1.23) and (1.20) implies that (1.26b) and (1.26c) are zero. Thus, after convergence of the IGLS algorithm, the first partial derivatives of the FIML loglikelihood are zero and, thus (assuming regularity), the IGLS estimates must be equal to the FIML estimates. Analogously, it is equally easy to show that (1.24) combined with (1.25) and (1.20) implies that (1.28a) and (1.28b) are zero and, thus, that after convergence, the RIGLS estimates are equal to the REML estimates.

EM Algorithm

The *EM algorithm* is an iterative method for optimizing functions of the form $f(\boldsymbol{\theta}) = \log \int g(\boldsymbol{\theta}, \boldsymbol{z})\, \mathrm{d}\boldsymbol{z}$ with respect to $\boldsymbol{\theta}$. It was presented in its full generality by Dempster et al. [33]. Typically, $f(\boldsymbol{\theta})$ is a loglikelihood function and $\log g(\boldsymbol{\theta}, \boldsymbol{z})$ the *complete-data loglikelihood* function, i.e., the loglikelihood function that would have been obtained if the realization of the random variables $\underline{\boldsymbol{z}}$ would have been observed. Thus, both are also implicitly functions of the observed data $\boldsymbol{y}$. Maximization of $f(\boldsymbol{\theta})$ proceeds by iteratively maximizing the expectation of the complete-data loglikelihood. That is, in each iteration, the function

$$Q(\boldsymbol{\theta} \mid \boldsymbol{\theta}^{(i)}) \triangleq E\left[\log g(\boldsymbol{\theta}, \underline{\boldsymbol{z}}) \mid \boldsymbol{y}, \boldsymbol{\theta}^{(i)}\right]$$

is maximized, where the expectation is taken over the conditional distribution of $\underline{z}$ given the observed data $\boldsymbol{y}$ and the value $\boldsymbol{\theta}^{(i)}$ of the parameter vector after the previous iteration. Appendix 1.D explains in more detail why this works.

For the multilevel model, $\underline{z}$ consists of the random effects $\{\underline{\boldsymbol{\delta}}_j\}$, and $\boldsymbol{\theta}$ and $\boldsymbol{y}$ have their usual meaning. As derived in Appendix 1.D, when applied to the FIML loglikelihood, this means that in the expectation step, the following quantities are computed:

$$\begin{aligned}\boldsymbol{\mu}_j^{(i)} &= \boldsymbol{\Omega}\boldsymbol{W}_j^{-1}(\boldsymbol{b}_j - \boldsymbol{Z}_j\gamma),\\ \boldsymbol{\Sigma}_j^{(i)} &= \sigma_j^2\boldsymbol{\Omega}\boldsymbol{W}_j^{-1}(\boldsymbol{X}_j'\boldsymbol{X}_j)^{-1},\end{aligned}$$

where the right-hand sides are evaluated in $\boldsymbol{\theta}^{(i)}$. If $\boldsymbol{\Omega}$ is completely free (apart from the requirements of symmetry and positive definiteness, of course), the maximization step leads to the updates

$$\begin{aligned}\boldsymbol{\Omega}^{(i+1)} &= \frac{1}{m}\sum_{j=1}^{m}(\boldsymbol{\Sigma}_j^{(i)} + \boldsymbol{\mu}_j^{(i)}\boldsymbol{\mu}_j^{(i)\prime}),\\ \gamma^{(i+1)} &= \left(\sum_{j=1}^{m}\boldsymbol{Z}_j'\boldsymbol{X}_j'\boldsymbol{X}_j\boldsymbol{Z}_j\right)^{-1}\sum_{j=1}^{m}\boldsymbol{Z}_j'\boldsymbol{X}_j'\boldsymbol{X}_j(\boldsymbol{b}_j - \boldsymbol{\mu}_j^{(i)}),\\ (\sigma_j^2)^{(i+1)} &= \frac{1}{n_j}\left[(n_j - p)s_j^2 + \mathrm{tr}(\boldsymbol{X}_j'\boldsymbol{X}_j\boldsymbol{\Lambda}_j^{(i)})\right],\end{aligned}$$

or, instead of the latter,

$$(\sigma^2)^{(i+1)} = \frac{1}{n}\sum_{j=1}^{m}\left[(n_j - p)s_j^2 + \mathrm{tr}(\boldsymbol{X}_j'\boldsymbol{X}_j\boldsymbol{\Lambda}_j^{(i)})\right],$$

where

$$\boldsymbol{\Lambda}_j^{(i)} \triangleq \boldsymbol{\Sigma}_j^{(i)} + (\boldsymbol{b}_j - \boldsymbol{\mu}_j^{(i)} - \boldsymbol{Z}_j\gamma^{(i+1)})(\boldsymbol{b}_j - \boldsymbol{\mu}_j^{(i)} - \boldsymbol{Z}_j\gamma^{(i+1)})'.$$

If $\boldsymbol{\Omega}$ is restricted, typically by (1.3) with $G < p(p+1)/2$ parameters, the update of the variance parameters $\boldsymbol{\xi}$ is a bit more complicated; see Appendix 1.D.

A great advantage of the EM algorithm is that the loglikelihood is improved in each iteration, i.e., the algorithm is monotonic. Furthermore, the computations in each iteration are often very simple, much simpler than with other numerical optimization algorithms. Another strength of the EM algorithm is that it is able to deal with missing data in a very natural way (see Chapter 10). A drawback of EM is that it tends to converge very slowly. Formally, it converges linearly, whereas, for example, Newton-Raphson converges quadratically when in the neighborhood of the optimum. On the other hand, when far from the optimum, the EM algorithm shows more stable convergence in the direction of the optimum. For this reason, the nlme package [90] uses

EM for the initial iterations and switches to Newton-Raphson later on in the algorithm. An incomplete list of other multilevel packages that use EM, either as an option or for specific tasks, is BMDP-5V [107], MLA [16], and especially HLM [102], which popularized the algorithm for multilevel analysis. The EM algorithm is described for multilevel analysis and especially its special case of repeated measures models in Dempster et al. [34], Laird and Ware [70], Jennrich and Schluchter [63], Laird et al. [69], Lindstrom and Bates [73], and Raudenbush and Bryk [101, Chapter 14].

Further Numerical and Computational Issues

As we have seen, most formulas for computing estimates for multilevel models can be expressed in different ways. Some of these are clearly computationally inefficient, whereas others use the structure of the problem in better ways. This pertains to usage of memory, sizes of inverses needed, and other ways to compute the same expressions. Given the sizes of typical multilevel datasets and the ways in which computations can be done inefficiently, implementing an estimator for a multilevel model for general use needs considerable fine-tuning.

In many cases, we have presented results using $\boldsymbol{Z}_j$, $\boldsymbol{W}_j$, $\boldsymbol{b}_j$, and a few other matrices and vectors. These are of smaller sizes than $\boldsymbol{U}_j$, $\boldsymbol{V}_j$, and $\boldsymbol{y}_j$, so that this already improves the computations considerably. Longford [74] gives further computational formulas, such that the amount of storage needed is further reduced (but dimensions of inverses do not become smaller).

However, our formulas still use expressions like $\boldsymbol{b}_j = (\boldsymbol{X}_j'\boldsymbol{X}_j)^{-1}\boldsymbol{X}_j'\boldsymbol{y}_j$. Actually computing an estimator in this way is generally considered undesirable, because it exacerbates any numerical problems that may exist. A good way to compute a least squares estimator is to use the QR decomposition. Pinheiro and Bates [89] discuss these issues at length and present detailed analyses in which the multilevel loglikelihood is transformed in a way that makes computations fast, numerically stable, and memory efficient. We do not present these here, but recommend their book to interested readers.

1.6.3 Robust Covariance Matrix Estimation

We have seen above that the two-step OLS estimator of γ is

$$\hat{\underline{\gamma}} = \left(\sum_{j=1}^{m} \boldsymbol{Z}_j'\boldsymbol{Z}_j\right)^{-1} \sum_{j=1}^{m} \boldsymbol{Z}_j'\underline{\boldsymbol{b}}_j = \boldsymbol{A}\sum_{j=1}^{m} \boldsymbol{Z}_j'\underline{\boldsymbol{b}}_j,$$

with $\boldsymbol{A}$ implicitly defined. Its covariance matrix is

$$\boldsymbol{C} \triangleq \mathrm{Cov}(\hat{\underline{\gamma}}) = \boldsymbol{A}\left(\sum_{j=1}^{m} \boldsymbol{Z}_j' \,\mathrm{Cov}(\underline{\boldsymbol{b}}_j)\boldsymbol{Z}_j\right)\boldsymbol{A}.$$

If $m \to \infty$, $\underline{\hat{\gamma}}$ is a consistent estimator of γ, and instead of using the model-based estimator of $\boldsymbol{C}$ presented earlier, $\boldsymbol{C}$ can be straightforwardly estimated by the cluster-robust covariance matrix [e.g., 98]

$$\underline{\hat{C}}_{\mathrm{cr}} = \boldsymbol{A}\left(\sum_{j=1}^{m} \boldsymbol{Z}_j' \underline{\hat{\boldsymbol{t}}}_j \underline{\hat{\boldsymbol{t}}}_j' \boldsymbol{Z}_j\right)\boldsymbol{A},$$

where $\underline{\hat{\boldsymbol{t}}}_j = \underline{\boldsymbol{b}}_j - \boldsymbol{Z}_j\underline{\hat{\gamma}}$. When m is large, this is an accurate estimator, but in moderately large samples, it tends to be biased because the variability in estimation of γ is not taken into account. That is, the difference between $\underline{\hat{\boldsymbol{t}}}_j$ and $\underline{\boldsymbol{t}}_j \stackrel{\Delta}{=} \underline{\boldsymbol{b}}_j - \boldsymbol{Z}_j\gamma$ is ignored. Inspired by similar problems with the (Eicker-Huber-)White heteroskedasticity-consistent covariance matrix, and fairly successful corrections thereof [25, pp. 552–556], corrections to the cluster-robust covariance matrix can be computed, which take the form of multiplication by a certain factor, e.g.,

$$\frac{m}{m-1}\,\frac{n-1}{n-r},$$

where n is total sample size and r is the number of elements of γ. Cameron and Trivedi [17, p. 834] mention this correction in the context of the one-step OLS estimator.

Analogously, abusing the same notation for different estimators, the one-step OLS estimator is

$$\underline{\hat{\gamma}} = \left(\sum_{j=1}^{m} \boldsymbol{U}_j'\boldsymbol{U}_j\right)^{-1} \sum_{j=1}^{m} \boldsymbol{U}_j'\underline{\boldsymbol{y}}_j = \boldsymbol{A}\sum_{j=1}^{m} \boldsymbol{U}_j'\underline{\boldsymbol{y}}_j.$$

Thus, we can estimate its covariance matrix by the cluster-robust covariance estimator

$$\underline{\hat{C}}_{\mathrm{cr}} = \boldsymbol{A}\left(\sum_{j=1}^{m} \boldsymbol{U}_j' \underline{\hat{\boldsymbol{r}}}_j \underline{\hat{\boldsymbol{r}}}_j' \boldsymbol{U}_j\right)\boldsymbol{A},$$

[e.g., 126, p. 152], where $\underline{\hat{\boldsymbol{r}}}_j = \underline{\boldsymbol{y}}_j - \boldsymbol{U}_j\underline{\hat{\gamma}}$. As observed above, the one-step OLS estimator can also be written as

$$\underline{\hat{\gamma}} = \boldsymbol{A}\sum_{j=1}^{m} \boldsymbol{Z}_j'\boldsymbol{X}_j'\boldsymbol{X}_j\underline{\boldsymbol{b}}_j,$$

where $\boldsymbol{A}$ is now written as

$$\boldsymbol{A} = \left(\sum_{j=1}^{m} \boldsymbol{Z}_j'\boldsymbol{X}_j'\boldsymbol{X}_j\boldsymbol{Z}_j\right)^{-1}.$$

Hence, the cluster-robust covariance estimator can be rewritten as

$$\underline{\hat{\boldsymbol{C}}}_{\mathrm{cr}} = \boldsymbol{A}\left(\sum_{j=1}^{m} \boldsymbol{Z}_j'\boldsymbol{X}_j'\boldsymbol{X}_j\underline{\hat{\boldsymbol{t}}}_j\underline{\hat{\boldsymbol{t}}}_j'\boldsymbol{X}_j'\boldsymbol{X}_j\boldsymbol{Z}_j\right)\boldsymbol{A},$$

where it is now natural to use the one-step estimator of the coefficient vector γ in the definition of $\underline{\hat{\boldsymbol{t}}}_j$.

In the same way, a straightforward cluster-robust covariance matrix of the WLS estimator $\underline{\hat{\gamma}}$ is found to be

$$\underline{\hat{\boldsymbol{C}}}_{\mathrm{cr}} = \underline{\hat{\boldsymbol{A}}}\left(\sum_{j=1}^{m} \boldsymbol{U}_j'\underline{\hat{\boldsymbol{V}}}_j^{-1}\underline{\hat{\boldsymbol{r}}}_j\underline{\hat{\boldsymbol{r}}}_j'\underline{\hat{\boldsymbol{V}}}_j^{-1}\boldsymbol{U}_j\right)\underline{\hat{\boldsymbol{A}}},$$

where now the WLS estimator of γ is used in the definition of $\underline{\hat{\boldsymbol{r}}}_j$,

$$\underline{\hat{\boldsymbol{V}}}_j = \boldsymbol{X}_j\underline{\hat{\boldsymbol{\Omega}}}\boldsymbol{X}_j' + \underline{\hat{\sigma}}_j^2\boldsymbol{I}_{n_j},$$

$$\underline{\hat{\boldsymbol{A}}} = \left(\sum_{j=1}^{m} \boldsymbol{U}_j'\underline{\hat{\boldsymbol{V}}}_j^{-1}\boldsymbol{U}_j\right)^{-1},$$

or, equivalently,

$$\underline{\hat{\boldsymbol{C}}}_{\mathrm{cr}} = \underline{\hat{\boldsymbol{A}}}\left(\sum_{j=1}^{m} \boldsymbol{Z}_j'\underline{\hat{\boldsymbol{W}}}_j^{-1}\underline{\hat{\boldsymbol{t}}}_j\underline{\hat{\boldsymbol{t}}}_j'\underline{\hat{\boldsymbol{W}}}_j^{-1}\boldsymbol{Z}_j\right)\underline{\hat{\boldsymbol{A}}},$$

with

$$\underline{\hat{\boldsymbol{W}}}_j = \underline{\hat{\boldsymbol{\Omega}}} + \underline{\hat{\sigma}}_j^2(\boldsymbol{X}_j'\boldsymbol{X}_j)^{-1},$$

$$\underline{\hat{\boldsymbol{A}}} = \left(\sum_{j=1}^{m} \boldsymbol{Z}_j'\underline{\hat{\boldsymbol{W}}}_j^{-1}\boldsymbol{Z}_j\right)^{-1},$$

and the WLS estimator of γ is used in the definition of $\underline{\hat{\boldsymbol{t}}}_j$. Note that for the asymptotic results, it does not matter which estimators of $\boldsymbol{\Omega}$ and σ_j^2 are used, as long as they are consistent. Of course, in finite samples, it does matter and we would expect that more precise estimators of $\boldsymbol{\Omega}$ and σ_j^2 result in better estimators of γ and $\boldsymbol{C}$.

Robust Covariance Matrices for ML Estimators

A robust covariance estimator for the FIML estimator of γ is immediately obtained from the one for the WLS estimator given above. The same applies to the two-step ML ("REML") estimator obtained as a WLS estimator that uses the REML estimates of the variance parameters in computing the weight matrix.

It is also possible to compute a robust covariance matrix for the variance parameters. However, because no closed-form expression for the estimators of the variance parameters exists, this requires a bit more asymptotic statistical theory. The basic idea starts from the first-order condition for ML estimators

$$\sum_{j=1}^{m} \frac{\partial \underline{\mathcal{L}}_j}{\partial \boldsymbol{\theta}}(\hat{\underline{\boldsymbol{\theta}}}) = \boldsymbol{\emptyset}.$$

Then a first-order Taylor series expansion of this, around the true value $\boldsymbol{\theta}_0$, is taken, giving

$$\sum_{j=1}^{m} \left\{ \frac{\partial \underline{\mathcal{L}}_j}{\partial \boldsymbol{\theta}}(\boldsymbol{\theta}_0) + \frac{\partial^2 \underline{\mathcal{L}}_j}{\partial \boldsymbol{\theta}\, \partial \boldsymbol{\theta}'}(\boldsymbol{\theta}_0)\, (\hat{\underline{\boldsymbol{\theta}}} - \boldsymbol{\theta}_0) + O_p \|\hat{\underline{\boldsymbol{\theta}}} - \boldsymbol{\theta}_0\|^2 \right\} = \boldsymbol{\emptyset}.$$

Under suitable regularity conditions, a form of the central limit theorem implies that

$$\frac{1}{\sqrt{m}} \sum_{j=1}^{m} \frac{\partial \underline{\mathcal{L}}_j}{\partial \boldsymbol{\theta}}(\boldsymbol{\theta}_0) \stackrel{\mathcal{L}}{\Longrightarrow} \mathcal{N}(\boldsymbol{\emptyset}, \boldsymbol{\Psi})$$

from some finite positive definite matrix $\boldsymbol{\Psi}$, and a form of the law of large numbers implies that

$$\frac{1}{m} \sum_{j=1}^{m} \frac{\partial^2 \underline{\mathcal{L}}_j}{\partial \boldsymbol{\theta}\, \partial \boldsymbol{\theta}'}(\boldsymbol{\theta}_0) \stackrel{\mathcal{P}}{\Longrightarrow} \boldsymbol{\mathcal{H}}$$

for some finite positive definite matrix $\boldsymbol{\mathcal{H}}$. Combining results, we obtain

$$\sqrt{m}(\hat{\underline{\boldsymbol{\theta}}} - \boldsymbol{\theta}_0) = -\boldsymbol{\mathcal{H}}^{-1} \frac{1}{\sqrt{m}} \sum_{j=1}^{m} \frac{\partial \underline{\mathcal{L}}_j}{\partial \boldsymbol{\theta}}(\boldsymbol{\theta}_0) + o_p(1) \stackrel{\mathcal{L}}{\Longrightarrow} \mathcal{N}(\boldsymbol{\emptyset}, \boldsymbol{\mathcal{H}}^{-1} \boldsymbol{\Psi} \boldsymbol{\mathcal{H}}^{-1}).$$

Obviously, consistent estimators of $\boldsymbol{\mathcal{H}}$ and $\boldsymbol{\Psi}$ are

$$\hat{\underline{\boldsymbol{\mathcal{H}}}} = \frac{1}{m} \sum_{j=1}^{m} \frac{\partial^2 \underline{\mathcal{L}}_j}{\partial \boldsymbol{\theta}\, \partial \boldsymbol{\theta}'}(\hat{\underline{\boldsymbol{\theta}}}),$$

$$\hat{\underline{\boldsymbol{\Psi}}} = \frac{1}{m} \sum_{j=1}^{m} \frac{\partial \underline{\mathcal{L}}_j}{\partial \boldsymbol{\theta}}(\hat{\underline{\boldsymbol{\theta}}})\, \frac{\partial \underline{\mathcal{L}}_j}{\partial \boldsymbol{\theta}'}(\hat{\underline{\boldsymbol{\theta}}}).$$

For computing a robust covariance matrix for $\hat{\underline{\boldsymbol{\theta}}}$, all factors of m drop out and we obtain

$$\hat{\underline{\boldsymbol{C}}}_{\mathrm{cr}} = \left(\sum_{j=1}^{m} \frac{\partial^2 \underline{\mathcal{L}}_j}{\partial \boldsymbol{\theta}\, \partial \boldsymbol{\theta}'}(\hat{\underline{\boldsymbol{\theta}}}) \right)^{-1} \left(\sum_{j=1}^{m} \frac{\partial \underline{\mathcal{L}}_j}{\partial \boldsymbol{\theta}}(\hat{\underline{\boldsymbol{\theta}}})\, \frac{\partial \underline{\mathcal{L}}_j}{\partial \boldsymbol{\theta}'}(\hat{\underline{\boldsymbol{\theta}}}) \right) \left(\sum_{j=1}^{m} \frac{\partial^2 \underline{\mathcal{L}}_j}{\partial \boldsymbol{\theta}\, \partial \boldsymbol{\theta}'}(\hat{\underline{\boldsymbol{\theta}}}) \right)^{-1}. \quad (1.30)$$

The theory underlying the robust covariance matrices for ML estimators in a multilevel model is derived in detail, with appropriate regularity conditions, in Verbeke and Lesaffre [120, 121].

From (1.26), (1.27), (1.28), and (1.29), it follows that this theory should work for the FIML and REML estimators of γ and ξ_g, and for the corresponding estimators of σ^2 if all residual variances are assumed to be the same. However, if separate residual variances σ_j^2 are estimated, the corresponding first-order conditions do not satisfy the central limit theorem as presented here, because they have only one term. In that case, assuming that $n_j \to \infty$, it is still possible to derive some kind of robust variance estimators for the variance estimators $\hat{\underline{\sigma}}_j^2$, using within-groups asymptotics along the lines of Browne [12], but this tends to require large within-group sample sizes, so this may not work well in practice.

Note that when all the model assumptions are met, we have the well-known result (correcting for our scaling of the loglikelihood)

$$\tfrac{1}{2}\boldsymbol{\mathcal{H}} = \tfrac{1}{4}\boldsymbol{\Psi} = \lim_{m\to\infty} \frac{1}{m}\boldsymbol{\mathcal{I}},$$

which leads to the standard (model-based) covariance matrix presented earlier.

Robust Versus Model-Based Covariance Matrices

With a few exceptions, the model-based covariance matrices are only correct if the complete model is correctly specified ("true"). The robust covariance matrices are consistent under a wider range of assumptions, including fairly general forms of misspecification of the random part of the model, such as intragroup dependence and heteroskedasticity. So if the main interest of the analyses is the fixed part of the model (i.e., γ), a cluster-robust covariance matrix may be preferred.

On the other hand, if the random part of the model is the main focus of interest, i.e., modeling/explaining between-group variation is important, then an estimator of the covariance matrix of the fixed part that is robust to misspecification of the random part is only of secondary interest. If the random part is (severely) misspecified, the primary aim of the analysis is not met. This is even more salient for robust covariance matrices of the variance parameters themselves. If the model is misspecified, it is generally unclear what is estimated, and thus it is questionable whether a robust covariance matrix is of any use [39].

There is, however, a leading example where the random part is misspecified, but the estimators are still consistent estimators of meaningful parameters. This is the case when the model is correctly specified, except for the distribution of the random variables. If these are nonnormally distributed, the model-based covariance matrices for the estimators of γ are still correct, but standard model-based covariance matrices of the variance parameters are incorrect. But $\boldsymbol{\Omega}$ and σ^2 are still meaningful parameters and their estimators are consistent. So then using a robust covariance matrix is clearly useful [120, 121].

The robust covariance matrices are typically far less precise if the model is (approximately) correctly specified and the sample size is small to moderate. Therefore, in not-too-large samples, the model-based covariance matrices will typically be preferred if the analyst believes that the random part of the model is reasonably well specified. Maas and Hox [78] performed a simulation study to investigate these issues for REML estimators and concluded that the model-based standard errors of the estimator of σ^2 performed well under nonnormality, while the robust standard errors are often too large. However, both model-based and robust standard errors of level-2 variance parameters did not perform very well at small sample size, although the robust ones were clearly better than the model-based ones. They conclude that at least 100 groups are needed for reliable robust standard errors. As a general strategy, they recommend comparing the robust standard errors with the model-based ones to diagnose possible misspecification of the model.

An alternative way for robust statistical inference under possible misspecification is to use resampling methods. Moreover, the bootstrap in particular has the additional potential advantage that it can generate asymmetric confidence intervals, thereby reflecting nonnormal finite-sample distributions of especially the level-2 variance parameters. However, confidence intervals based on resampling methods tend to perform less than satisfactory as well with small or moderate level-2 sample sizes. See Chapter 11 for a detailed description of resampling methods for multilevel models and their empirical properties.

1.7 Software

We will be brief about software here, if only because details about software are likely to be quickly outdated. An overview of the history of the development of software for multilevel analysis, and the state of affairs ca. 2000 is given in de Leeuw and Kreft [30]. The overview is still broadly valid, except that the details have changed and there are some additions.

As mentioned earlier in this chapter, the software packages have largely been developed by the same authors who pioneered the development of multilevel analysis as a statistical method and who have written successful textbooks about multilevel modeling. And, for that matter, are contributors to this Handbook.

Two software packages dominate the market for dedicated multilevel analysis software. These are HLM [102] and MLwiN [97]. These packages offer a broad range of linear and nonlinear specifications of multilevel models and have user-friendly graphical user interfaces. There are some differences in the algorithms used, but these are not particularly interesting for the average user. There are also some differences in the more advanced options or less frequently used model specifications, so users with specific desires may prefer one over the other for this reason.

Originally, VARCL [75] was also one of the major packages, but development of this package has been terminated. There are many packages that focus on more specific multilevel models, options, or other aspects. These tend to be research software, with fewer options and less user-friendly interfaces, and development of these progresses faster if the authors are working on new directions in their research that requires additions to the programs. Examples of these are MLA [16], which focuses on resampling methods (see Chapter 11) and PINT [10], which focuses on power calculations (see Chapter 4). The MIXFOO suite [55, 56, 57, etc.] also belongs in this category, although taken as a whole, it is a fairly comprehensive multilevel package.

The BUGS program and its variants, most notably WinBUGS [113], are programs for Bayesian data analysis. They offer extensive possibilities for Bayesian multilevel analysis and are particularly useful for estimating nonlinear multilevel models. See also Chapter 2.

Many general-purpose (or almost-all-encompassing) statistical packages now have multilevel options as well. Important examples are SAS® [106], which has PROC MIXED and PROC NLMIXED, SPSS® [114], which has MIXED and several other procedures that can be used for multilevel analyses, Stata® [115], which has many "survey", "cluster", and "panel" programs and options, and the extensive `gllamm` program [95], and R [93], for which the lme4 and nlme packages are available [7, 90].

A relatively recent development is the incorporation of multilevel facilities in programs for structural equation modeling, such as LISREL [35, 64], EQS [8], and Mplus [85]. The possibilities of these programs are somewhat different from the standard multilevel programs. They often have less extensive options for estimating nonlinear models and models with three or more levels, but are better equipped for estimating multivariate models and models with latent variables and measurement errors, i.e., multilevel structural equation models (see Chapter 12). Thus, they complement traditional multilevel packages.

Throughout this Handbook, other software packages (perhaps less well known or more specialized) are mentioned where appropriate and useful.

1.8 Sampling Weights

Surveys are often nonrepresentative of the population of interest, in the sense that persons (or, more generally, units) with certain characteristics are more prevalent in the data than in the population. There are essentially two reasons for this: deliberate oversampling of certain groups and different nonresponse rates. An example of the former is the oversampling of relatively small groups, like minorities, to obtain more reliable information about these groups. An example of the latter is the tendency to obtain an overrepresentation of women in a study that was designed to be neutral, which may happen because women tend to be more often at home than men.

Agencies that collect such surveys typically provide *sampling weights* with the data set. The idea is that applying these sampling weights in the analysis corrects for the nonrepresentativeness of the data by giving underrepresented groups more weight and overrepresented groups less weight. For example, assume that we are interested in the mean height of adults in a country of interest. Assume further that we have a sample of 1000 adults, 600 of which are women, whereas in the population 50% of adults is female. Height is expected to be related to sex, so if we simply computed the sample average, we would likely obtain an underestimate of our parameter of interest. However, if we give women a weight of $w_i = 5/6$ and men a weight of $w_i = 5/4$, then the weighted average

$$\begin{aligned} \bar{h}_w &\triangleq \frac{\sum_{i=1}^{1000} w_i h_i}{\sum_{i=1}^{1000} w_i} \\ &= \frac{600 \cdot (5/6) \cdot \bar{h}_f + 400 \cdot (5/4) \cdot \bar{h}_m}{600 \cdot (5/6) + 400 \cdot (5/4)} \\ &= 0.5\bar{h}_f + 0.5\bar{h}_m \end{aligned} \tag{1.31}$$

is clearly (the realization of) an unbiased estimator of average height in the population, where h_i is the height of the i-th observation in the sample and $\bar{h}_f$ and $\bar{h}_m$ are the average heights of females and males in the sample, respectively. (Note that apparently some software packages define weights as the reciprocals of the definition we use here, so check your manuals.)

For regression models, there is some discussion in the literature about whether weights should be applied, even if the sample is nonrepresentative and weights are available. In fact, if the standard regression model $\underline{y}_i = \boldsymbol{x}_i'\boldsymbol{\beta} + \underline{\epsilon}_i$, with $\underline{\epsilon}_i$ i.i.d., holds and the nonrepresentativeness is possibly related to $\boldsymbol{x}$ but not to $\underline{\epsilon}$, then OLS is still the most efficient estimator, and all statistical inference is correct. However, in many circumstances, it is quite likely that the error term represents the influence of a large number of variables that each have a fairly small effect, most of which are unknown and/or unobserved, but some of which may be somehow related to the probabilities of being included in the sample. In such cases, OLS would be biased, whereas a weighted analysis would still give an unbiased estimator.

An important special case where a weighted analysis gives simple consistent estimators and an unweighted analysis does not is in the analysis of so-called *choice-based samples* or, more generally, *endogenously stratified* samples. In this case, samples are drawn from strata defined by the dependent variable. An example is a sample consisting of 500 bus passengers sampled on board bus lines and 500 car drivers sampled along the road, and the dependent variable is mode choice. Another important example is a medical study in which a sample of people having a rare disease is drawn from hospital records and a similar-sized sample of people not having the disease is drawn from the general public, and the dependent variable in the study is whether or not one has the disease.

These issues are extensively discussed in Cameron and Trivedi [17, pp. 817–829] and Wooldridge [124, 125, 127], who also give detailed derivations and explanations, showing why unweighted analyses are sometimes inconsistent and under different circumstances consistent and efficient. For the remainder of this section, we assume that a weighted analysis is desired.

For multilevel analysis, an additional complication is how to deal with units at different levels. To continue our example, assume that we have a two-level sample, where level-1 is individuals and level-2 is counties. Perhaps heights are correlated within counties because of environmental factors, different socio-economic composition, different ethnic composition or more specifically family relations, and therefore a multilevel approach is desired, but still females are overrepresented. Furthermore, let us assume that we know the population percentages of males and females in each county (not necessarily 50%). Then a straightforward adaptation of (1.31) gives an estimate of the within-county mean height:

$$\bar{h}_{wj} \triangleq \frac{\sum_{i=1}^{n_j} w_{i|j} h_{ij}}{\sum_{i=1}^{n_j} w_{i|j}}$$

in obvious notation. If each county had the same population size (or height was unrelated to population size) and the sample of counties is representative of all counties in whatever way this is defined, a simple average of the county averages gives an unbiased estimate of the parameter of interest. More generally, however, we also have a county weight w_j, and the overall weighted mean is computed as

$$\bar{h}_{w\cdot} \triangleq \frac{\sum_{j=1}^{m} w_j \bar{h}_{wj}}{\sum_{j=1}^{m} w_j}.$$

Determining the value of w_j depends on the sampling scheme and the resulting representativeness at the county level. For example, if the counties are a simple random sample of all counties in the country, then counties with small population size are overrepresented given that we are interested in the mean height of individuals. It is easy to see then that w_j should be proportional to county population size N_j. Often, however, sampling at county level is done proportional to size, so that w_j should be the same for each county.

When a survey data set is given, it typically contains an individual weight w_{ij} and the clusters are defined by the researcher. Then the multilevel weights can be computed as

$$
\begin{aligned}
w_j &\stackrel{\Delta}{=} \sum_{i=1}^{n_j} w_{ij}, \\
w_{i|j} &\stackrel{\Delta}{=} w_{ij}/w_j.
\end{aligned}
$$

See, however, Potthoff et al. [92], Pfeffermann et al. [88], Grilli and Pratesi [49], Asparouhov [6], and Rabe-Hesketh and Skrondal [94] for a discussion of different definitions of weights and empirical studies of their properties. Chantala et al. [20] provide software that computes appropriate multilevel sampling weights for usage in several software packages.

Let us now assume that we have a set of weights, and we would like to compute the weighted version of the within-groups OLS estimate $\boldsymbol{b}_j$. The formula for the latter can be written as

$$
\boldsymbol{b}_j \stackrel{\Delta}{=} (\boldsymbol{X}_j'\boldsymbol{X}_j)^{-1}\boldsymbol{X}_j'\boldsymbol{y}_j = \left(\frac{1}{n_j}\sum_{i=1}^{n_j}\boldsymbol{x}_{ij}\boldsymbol{x}_{ij}'\right)^{-1}\left(\frac{1}{n_j}\sum_{i=1}^{n_j}\boldsymbol{x}_{ij}y_{ij}\right).
$$

Clearly, each of the two factors contains some kind of average, so that the analogy with average height mentioned above gives the following weighted estimate:

$$
\begin{aligned}
\boldsymbol{b}_{wj} &\stackrel{\Delta}{=} \left(\frac{\sum_{i=1}^{n_j} w_{i|j}\boldsymbol{x}_{ij}\boldsymbol{x}_{ij}'}{\sum_{i=1}^{n_j} w_{i|j}}\right)^{-1}\left(\frac{\sum_{i=1}^{n_j} w_{i|j}\boldsymbol{x}_{ij}y_{ij}}{\sum_{i=1}^{n_j} w_{i|j}}\right) \\
&= \left(\sum_{i=1}^{n_j} w_{i|j}\boldsymbol{x}_{ij}\boldsymbol{x}_{ij}'\right)^{-1}\left(\sum_{i=1}^{n_j} w_{i|j}\boldsymbol{x}_{ij}y_{ij}\right) \\
&= (\boldsymbol{X}_j'\boldsymbol{\mathcal{W}}_j\boldsymbol{X}_j)^{-1}\boldsymbol{X}_j'\boldsymbol{\mathcal{W}}_j\boldsymbol{y}_j,
\end{aligned}
$$

where $\boldsymbol{\mathcal{W}}_j$ (not to be confused with $\boldsymbol{W}_j$) is the diagonal matrix with elements $w_{i|j}$ on its diagonal. A corresponding suitable estimator of σ_j^2 is obtained by a properly scaled version of the weighted sum of squared residuals. For the *unbiased* estimator, the denominator in this is a bit more complicated than in the unweighted case. The resulting formula is

$$
s_{wj}^2 \stackrel{\Delta}{=} (\boldsymbol{y}_j - \boldsymbol{X}_j\boldsymbol{b}_{wj})'\boldsymbol{\mathcal{W}}_j(\boldsymbol{y}_j - \boldsymbol{X}_j\boldsymbol{b}_{wj})/(n_j^* - p^*),
$$

where

$$
\begin{aligned}
n_j^* &\stackrel{\Delta}{=} \sum_{i=1}^{n_j} w_{i|j} = \operatorname{tr}\boldsymbol{\mathcal{W}}_j\,, \\
p^* &\stackrel{\Delta}{=} \operatorname{tr}\left[(\boldsymbol{X}_j'\boldsymbol{\mathcal{W}}_j\boldsymbol{X}_j)^{-1}(\boldsymbol{X}_j'\boldsymbol{\mathcal{W}}_j^2\boldsymbol{X}_j)\right].
\end{aligned}
$$

Then, paraphrasing our earlier discussion and simplifying somewhat, for estimating $\boldsymbol{\gamma}$, least squares loss functions incorporating sampling weights can be defined as

$$\rho_w(\gamma) \triangleq \sum_{j=1}^{m} w_j(\boldsymbol{b}_{wj} - \boldsymbol{Z}_j\gamma)'\boldsymbol{B}_j^{-1}(\boldsymbol{b}_{wj} - \boldsymbol{Z}_j\gamma),$$

leading to the estimators

$$\underline{\hat{\gamma}}_{w,\boldsymbol{B}} \triangleq \left(\sum_{j=1}^{m} w_j \boldsymbol{Z}_j'\boldsymbol{B}_j^{-1}\boldsymbol{Z}_j\right)^{-1} \sum_{j=1}^{m} w_j \boldsymbol{Z}_j'\boldsymbol{B}_j^{-1}\underline{\boldsymbol{b}}_{wj}.$$

Because

$$\boldsymbol{W}_{wj} \triangleq \mathrm{Cov}(\underline{\boldsymbol{b}}_{wj}) = \boldsymbol{\Omega} + \sigma_j^2(\boldsymbol{X}_j'\boldsymbol{\mathcal{W}}_j\boldsymbol{X}_j)^{-1}(\boldsymbol{X}_j'\boldsymbol{\mathcal{W}}_j^2\boldsymbol{X}_j)(\boldsymbol{X}_j'\boldsymbol{\mathcal{W}}_j\boldsymbol{X}_j)^{-1},$$

the covariance matrices of these least squares estimators are

$$\left(\sum_{j=1}^{m} w_j \boldsymbol{Z}_j'\boldsymbol{B}_j^{-1}\boldsymbol{Z}_j\right)^{-1} \left(\sum_{j=1}^{m} w_j^2 \boldsymbol{Z}_j'\boldsymbol{B}_j^{-1}\boldsymbol{W}_{wj}\boldsymbol{B}_j^{-1}\boldsymbol{Z}_j\right) \left(\sum_{j=1}^{m} w_j \boldsymbol{Z}_j'\boldsymbol{B}_j^{-1}\boldsymbol{Z}_j\right)^{-1}.$$

The estimators corresponding to the 1-step and 2-step OLS estimators are obtained by choosing $\boldsymbol{B}_j = (\boldsymbol{X}_j'\boldsymbol{\mathcal{W}}_j\boldsymbol{X}_j)^{-1}$ and $\boldsymbol{B}_j = \boldsymbol{I}$, respectively. The most logical analog of the WLS estimator seems to be the one based on $\boldsymbol{B}_j = \boldsymbol{W}_{wj}$, but the optimality properties of the unweighted version do not hold and the covariance matrix does not simplify considerably. A different WLS estimator for data with sampling weights,

$$\begin{aligned}
\underline{\hat{\gamma}}_{w,\mathrm{KR}} &\triangleq \left(\sum_{j=1}^{m} w_j \boldsymbol{U}_j'\boldsymbol{V}_j^{-1}\boldsymbol{U}_j\right)^{-1} \sum_{j=1}^{m} w_j \boldsymbol{U}_j'\boldsymbol{V}_j^{-1}\underline{\boldsymbol{y}}_j \\
&= \left(\sum_{j=1}^{m} w_j \boldsymbol{Z}_j'\boldsymbol{W}_j^{-1}\boldsymbol{Z}_j\right)^{-1} \sum_{j=1}^{m} w_j \boldsymbol{Z}_j'\boldsymbol{W}_j^{-1}\underline{\boldsymbol{b}}_j,
\end{aligned}$$

using the unweighted within-groups estimates $\boldsymbol{b}_j$ and $\boldsymbol{W}_j$, was proposed by Kovačević and Rai [66]. This also does not have the optimality properties of the WLS estimator without sampling weights.

Generally, we need an estimate of $\boldsymbol{\Omega}$ as well. The estimators discussed earlier can be adapted relatively straightforwardly, but we omit this here, with the exception of a general treatment of ML with sampling weights.

The loglikelihood function for a two-level model that is not necessarily linear can be written as

$$\mathcal{L} = \sum_{j=1}^{m} \log \int \exp(\mathcal{L}_{j|\boldsymbol{\delta}_j})\, f_{\boldsymbol{\delta}}(\boldsymbol{\delta}_j)\, \mathrm{d}\boldsymbol{\delta}_j,$$

where we have suppressed the dependence on the parameter vector $\boldsymbol{\theta}$. The function $f_{\boldsymbol{\delta}}(\cdot)$ is the density function of $\underline{\boldsymbol{\delta}}_j$ and $\mathcal{L}_{j|\boldsymbol{\delta}_j}$ is the loglikelihood of the j-th group conditional on $\boldsymbol{\delta}_j$. Thus,

$$\mathcal{L}_{j|\boldsymbol{\delta}_j} = \sum_{i=1}^{n_j} \log f_{y\,|\,\boldsymbol{\delta}}(y_{ij} \mid \boldsymbol{\delta}_j)$$

in obvious notation. From this form, the adaptation for sampling weights is straightforward, leading to

$$\mathcal{L}_{w,j|\boldsymbol{\delta}_j} \triangleq \sum_{i=1}^{n_j} w_{i|j} \log f_{y\,|\,\boldsymbol{\delta}}(y_{ij} \mid \boldsymbol{\delta}_j),$$

$$\mathcal{L}_w \triangleq \sum_{j=1}^{m} w_j \log \int \exp(\mathcal{L}_{w,j|\boldsymbol{\delta}_j})\, f_{\boldsymbol{\delta}}(\boldsymbol{\delta}_j)\, \mathrm{d}\boldsymbol{\delta}_j = \sum_{j=1}^{m} w_j \mathcal{L}_{wj},$$

with $\mathcal{L}_{wj}$ implicitly defined. Thus, the first-order condition for the ML estimator with sampling weights is

$$\sum_{j=1}^{m} w_j \frac{\partial \mathcal{L}_{wj}}{\partial \boldsymbol{\theta}} = \boldsymbol{\emptyset}, \tag{1.32}$$

so that, adapting (1.30), the covariance estimate for the resulting estimator $\hat{\underline{\boldsymbol{\theta}}}$ becomes

$$\left(\sum_{j=1}^{m} w_j \frac{\partial^2 \mathcal{L}_{wj}}{\partial \boldsymbol{\theta}\, \partial \boldsymbol{\theta}'}(\hat{\boldsymbol{\theta}}) \right)^{-1} \left(\sum_{j=1}^{m} w_j^2 \frac{\partial \mathcal{L}_{wj}}{\partial \boldsymbol{\theta}}(\hat{\boldsymbol{\theta}})\, \frac{\partial \mathcal{L}_{wj}}{\partial \boldsymbol{\theta}'}(\hat{\boldsymbol{\theta}}) \right) \left(\sum_{j=1}^{m} w_j \frac{\partial^2 \mathcal{L}_{wj}}{\partial \boldsymbol{\theta}\, \partial \boldsymbol{\theta}'}(\hat{\boldsymbol{\theta}}) \right)^{-1}.$$

Unlike the covariance matrix without sampling weights, this formula does not simplify considerably even if all model assumptions are met. Thus, this illustrates that the resulting estimators are not proper ML estimators and the weighted loglikelihood function is not a proper loglikelihood. The estimators can, however, be viewed as generalized estimating equation (GEE) estimators based on the estimating equations (1.32) and, under weak regularity conditions, have desirable statistical properties (consistency, asymptotic normality). From this theory, it also follows that it is immaterial whether the weights are predetermined (by the sampling scheme) or estimated afterward (because of differential nonresponse), in which case they would be random variables. The estimating equations are still valid, unless the nonresponse is related to the dependent variable of interest ("nonignorable"), in which case analyzing the data becomes much more complicated and perhaps consistent estimators do not exist.

Of course, the formulas for the ML estimators with sampling weights simplify considerably for the linear multilevel model. This is straightforward and we do not give the expresssions here.

More extensive discussions of how to treat sampling weights in survey data in general and with multilevel models in particular can be found in Skinner [108], Pfeffermann [87], Pfeffermann et al. [88], and Asparouhov [5, 6].

1.9 A School Effects Example

In this section, we apply some of the techniques discussed in this chapter by analyzing the well-known NELS-88 data. These have been used to illustrate multilevel techniques by several authors and, of course, they have been used in substantive research as well.

The part of the NELS data that we use contains information about the score on a mathematics test, which will be our dependent variable, and the amount of time spent on homework, which will be our level-1 explanatory variable, and the student-teacher ratio of the school, which will be our level-2 explanatory variable. The math test score is a continuous variable having a sample average of 51, with a range of 27–71. Homework is coded from 0 = "None" to 7 = "10 or more hours per week". This is a slightly nonlinear transformation of the hours, reflecting expected diminishing returns from additional hours of homework. Both the average and the median of this variable are 2. The student-teacher ratio varies from 10 to 30, with mean and median approximately equal to 17. The data set consists of 21,580 students in 1003 schools, so the average number of observations per school is about 22. The number of observations per school varies from 1 to 67.

Kreft and de Leeuw [67] have previously analyzed this data set with multilevel analysis. We base our analyses on the model they describe in their Chapter 4. However, whereas their goal is to discuss different model specifications and the choice between them, we focus on comparing results for the same model obtained with different estimators.

In line with the description in this chapter, we start by computing the within-school regressions. This immediately illustrates a drawback of our focus on two-step estimators: In 10 schools, the within-groups regression coefficients $\boldsymbol{b}_j$ and/or the within groups residual variance s_j^2 cannot be computed because the sample size is too small ($n_j \leq p = 2$) or because $\boldsymbol{X}_j$ is not of full column rank, which is presumably also due to small sample size. Thus, we drop these 10 schools and proceed with the 993 remaining schools, leaving us with 21,558 observations. We do not expect that this seriously affects the results, and this is confirmed by the closeness of our results with the corresponding ones in Kreft and de Leeuw [67]. However, this also indicates that models that use different within-groups residual variances (σ_j^2) will not reliably estimate these parameters for schools with small numbers of observations.

After these disclaimers, we report the within-schools results for the first 30 successfully analyzed schools in Table 1.1. It shows considerable variation both in the regression coefficients and in the residual variances. This is corroborated by summary statistics for the whole sample: The within-groups intercept varies from 34 to 72, with mean and median approximately equal to 48, and the regression coefficient for homework varies from −12 to +15, with mean and median equal to 1.3, but more than 75% are positive. Finally, the

Table 1.1 Within-school statistics for the first 30 successfully analyzed schools: school identifier, number of pupils, student-teacher ratio, regression coefficients, and residual variance.

			Regression coefficient		Residual
School ID	Observations	S-t ratio	Constant	Homework	variance
1249	24	21	54.0969	−0.5760	66.6295
1755	14	16	45.9339	0.3330	60.6991
1806	15	25	45.8242	3.0579	70.4722
1846	36	28	45.3300	1.5674	62.4661
2114	19	13	57.5974	−0.6658	83.7773
2335	19	11	60.0461	0.5249	16.1703
2666	20	14	43.0026	3.1134	69.1364
2759	17	10	57.3730	−2.8981	86.0793
2861	21	17	52.5275	2.6298	73.8099
2888	20	30	53.5131	0.4496	71.1451
2988	23	22	51.0928	0.5839	99.0531
6043	10	23	57.0538	0.5509	54.7340
6044	24	23	55.4732	0.1090	65.2169
6053	44	18	51.6696	2.0880	75.1713
6091	8	22	47.7969	−0.3928	108.5720
6185	3	19	47.9300	0.7850	41.3438
6327	8	23	63.8000	−8.6350	25.8185
6358	10	28	60.6133	0.5409	16.9813
6375	4	20	57.5608	0.4358	21.8832
6420	7	25	53.0421	0.2061	70.3876
6442	11	12	48.8171	0.1168	101.7292
6467	5	19	41.0639	6.9128	11.8384
6518	21	29	60.2006	0.9153	64.9436
6631	5	20	68.5750	−7.4025	40.5725
6641	29	15	50.2446	1.5950	70.2012
6656	4	16	37.7940	3.9710	10.9923
6738	3	26	54.9100	−6.0000	10.7648
6868	18	13	52.3958	0.9523	63.7598
7000	24	13	41.6905	1.2020	72.8585
7011	20	24	45.9697	1.6501	62.8256

residual variance varies from 5 to 180, with mean and median approximately equal to 71. It is the goal of the second step of the analysis to model at least some of the variation in the regression coefficients.

Of course, a negative coefficient for time spent on homework does not make sense substantively. Rather, in addition to the possibility of sheer random fluctuation, this points to a possible endogeneity problem, caused by students who have more problems with mathematics spending more time on their homework. That is, it may be the result of a partial reversal of causality. For the analysis here, we will ignore this possibility, given that we are primarily interested in differences between estimators.

We proceed by computing the one-step and two-step OLS estimates of the regression coefficients γ. These are reported in the first two columns of Table 1.2. The estimates are in the first panel, model-based standard errors (computed using the de Leeuw and Kreft [28] estimate of $\boldsymbol{\Omega}$) in the second panel, and robust standard errors in the third panel. Unlike a similar comparison for different data in de Leeuw and Kreft [28], we see some important differences between these estimates. The estimated main effect of the student-teacher ratio is twice as large for the two-step estimator, whereas the main effect of homework is less than half as large and the interaction term is also considerably less important, even statistically insignificant.

By using the within-groups and two-step OLS estimates, we can estimate $\boldsymbol{\Omega}$ by the method of de Leeuw and Kreft [28] discussed above. The estimate is denoted by DLK in Table 1.3. Fortunately, this is positive definite, so we do not encounter the problems faced by de Leeuw and Kreft for their example. Thus, we can use this estimate to compute the WLS estimates of γ. They are given in the third column of Table 1.2. They are very similar to the two-step estimates. As mentioned above, the estimate of $\boldsymbol{\Omega}$ is also be used in computing the model-based standard errors of the one-step and two-step OLS and WLS estimates, which are given in the second panel of Table 1.2. The third panel contains standard errors obtained from the cluster-robust covariance matrices.

Table 1.2 Estimates of fixed regression coefficients for the NELS-88 data and their standard errors.

	OLS (1-step)	OLS (2-step)	WLS (DLK)	FIML (1 σ)	REML (1 σ)	FIML (sep. σ's)	REML (sep. σ's)
				Estimates			
Constant	49.1477	52.1147	52.1062	51.4428	51.4434	51.9983	51.9988
S-t ratio	−0.1113	−0.2217	−0.2290	−0.2006	−0.2006	−0.2242	−0.2242
Homework	2.8520	1.2834	1.2785	1.5272	1.5272	1.3557	1.3561
hw × ratio	−0.0522	−0.0003	0.0058	−0.0030	−0.0030	0.0028	0.0028
				Model-based standard errors			
Constant	0.7857	0.7303	0.6913	0.7003	0.7011	0.7307	0.7314
S-t ratio	0.0428	0.0398	0.0378	0.0382	0.0382	0.0399	0.0400
Homework	0.2642	0.2362	0.1875	0.1823	0.1825	0.1781	0.1783
hw × ratio	0.0142	0.0127	0.0103	0.0100	0.0100	0.0098	0.0099
				Robust standard errors			
Constant	0.8176	0.8287	0.8862	0.8077	0.8049	0.8751	0.8639
S-t ratio	0.0433	0.0437	0.0469	0.0428	0.0427	0.0462	0.0457
Homework	0.2166	0.2225	0.1922	0.1828	0.1782	0.1961	0.1767
hw × ratio	0.0117	0.0118	0.0105	0.0098	0.0097	0.0105	0.0097

Next, we compute ML estimates. There are four of them: FIML and REML, each with a common variance parameter σ^2 or with separate variances σ_j^2. The results for the fixed coefficients are listed in the last four columns of Table 1.2. As argued before, these REML results are better called "WLS based on REML estimates of the variance parameters", but for convenience we call them REML here, and similarly WLS based on the DLK variance parameter estimates will be simply called WLS. The model-based standard errors for the ML estimators are obtained from the information matrix, whereas the robust standard errors are obtained from the cluster-robust covariance matrices described above. An exception is formed by the robust standard errors accompanying FIML with separate residual variances. These have been computed by formulas based on a combination of within-groups and between-groups asymptotics, as briefly mentioned but not worked out above (details are available upon request). This is intended to avoid the problems with the cluster-based estimator of the variance of the first derivatives of the loglikelihood, because its σ part is based on only 1 independent observation. However, the within-groups asymptotics involve sample fourth-order moments, which are highly inaccurate for the many small within-groups sample sizes. Nevertheless, the numerical results are similar to the ones for the other ML estimators, and also very similar to the two-step OLS and WLS results.

Note that the robust s.e.'s of the REML estimator are simply the WLS formulas, and thus are not affected by this problem. Given that the FIML estimators of γ are also WLS estimators, based on the FIML estimates of the variance parameters, we could have done the same for FIML. On the other hand, these WLS-based variance estimates essentially ignore any variability in the estimators of the variance parameters, which is also only asymptotically warranted.

The DLK and ML estimates of the elements of the level-2 covariance matrix $\boldsymbol{\Omega}$ are given in Table 1.3. The ML estimates using a single residual variance parameter are very similar to the DLK estimates (which are, incidentally, based on separate residual variances). The standard errors are a bit smaller, reflecting the higher precision of ML. When separate residual variances are estimated with ML, the estimates of $\boldsymbol{\Omega}$ are noticeably larger.

For both ML estimators with a single residual variance parameter, the estimate of σ^2 is 71.74 with a model-based standard error of 0.72 and a robust standard error of 0.85. The value of 71 corresponds closely with the average of the within-groups residual variance estimates.

For FIML with separate variances, the estimates of the residual variances vary from 8 to 161, with mean and median again approximately equal to 71. Similarly, for REML with separate variances, the estimates of the residual variances vary from 8 to 157, with mean and median also approximately equal to 71. This range is slightly narrower than the range of the within-groups

Table 1.3 Estimates of level-2 variance parameters for the NELS-88 data and their standard errors.

	DLK	FIML (1 σ)	REML (1 σ)	FIML (sep. σ's)	REML (sep. σ's)
			Estimates		
Constant, constant	23.9283	23.2633	23.3326	27.8982	27.9745
Homework, constant	−0.9319	−0.9105	−0.9197	−1.6088	−1.6197
Homework, homework	0.8678	0.5190	0.5243	0.6828	0.6878
			Model-based standard errors		
Constant, constant	1.8298	1.5125	1.5172	1.6826	1.6826
Homework, constant	0.6159	0.3138	0.3149	0.3296	0.3296
Homework, homework	0.3691	0.0993	0.0998	0.0971	0.0971
			Robust standard errors		
Constant, constant	—	1.5646	1.5591	1.7509	1.7509
Homework, constant	—	0.2983	0.2931	0.3197	0.3197
Homework, homework	—	0.1048	0.1047	0.1262	0.1262

Note: Robust standard errors are not available for the DLK [28] estimator.

estimates of the residual variances, but otherwise seems to confirm that the residual variances are not equal.

We can compute a likelihood ratio test statistic comparing the model with a common residual variance with the model with separate variances. For both FIML and REML, its value is approximately 1500, with 992 degrees of freedom, which gives a hugely significant p-value of approximately 2.2×10^{-23}. Even though the chi-square approximation is possibly inaccurate with such a large number of degrees of freedom and such small within-groups sample sizes, it clearly points in the direction of heterogeneous variances.

This leaves us with the conclusion that a model with a common variance is likely misspecified and a model with separate variances cannot be estimated reliably. Thus, this is a case in point for a more genuine multilevel approach in which the residual variance is modeled with a systematic part and a random residual, as suggested earlier.

Fortunately, however, the estimates and standard errors of the fixed coefficients, and to a lesser degree also the results for the level-2 covariance matrix, appear fairly insensitive to the specification of the level-1 random part. Thus, substantive conclusions would also be largely unaffected by this issue.

Clearly, this single empirical example is only an illustration and cannot be viewed as representative of all multilevel analyses. Many more examples, showing various issues in model specification and estimation, are discussed in detail in the textbooks [46, 59, 67, 76, 89, 101, 110, 111], the program manuals, and many empirical articles cited here and in the mentioned textbooks. Finally, the remaining chapters of this Handbook contain many empirical applications as well, although for more complicated models.

1.10 Final Remarks

In this final section, we would like to briefly mention a few topics that have not been addressed in the previous sections. The first is *hypothesis tests.* Of course, this is one of the main topics of statistics (and typically the one that gives statistics its bad reputation among students in the social sciences). However, there is almost nothing that is specific to multilevel analysis. Thus, the general theory of hypothesis testing as presented in, e.g., Cameron and Trivedi [17, Chapter 7], and in particular, the well-known Wald, likelihood ratio, and Lagrange multiplier tests, can be directly applied. The only thing worth mentioning is that the REML loglikelihood cannot be used to test hypotheses concerning γ, i.e., exclusion of certain variables from the fixed part of the model, because when viewed as a proper loglikelihood, it does not contain γ.

More generally, model fit is an important subject. In addition to formal hypothesis tests, this typically involves certain more descriptive indexes of model fit, like R^2 in linear regression. Several such indexes have been proposed for multilevel analysis, but these tend to have serious drawbacks. Sometimes it is not guaranteed that the fit index improves as variables (or, more generally, parameters) are added to the model, whereas other fit indexes do not have a clear intuitive interpretation. Thus, the literature does not seem to have converged on this topic. See, e.g., Snijders and Bosker [111, Chapter 7], Hox [59, Section 4.4], Spiegelhalter et al. [112], Xu [128], and Gelman and Pardoe [41] for some proposed indexes and their properties. A systematic approach to diagnosing model (mis)specification, directed at various directions of misspecification, is given in Chapter 3 of this volume.

An important issue in multilevel model specification is *centering.* In social science data, variables typically do not have a natural zero point, and even if there is a natural zero, it may still not be an important baseline value. Therefore, in regression analysis and other multivariate statistical analysis methods, variables are often centered, so that the zero point is the sample average, which *is* an important baseline value. This tends to ease the interpretation of the parameters, especially the intercept, and it sometimes has some computational advantages as well. This practice has also been advocated for multilevel analysis, but the consequences for multilevel analysis are not as innocuous as for ordinary linear regression analysis. Moreover, in multilevel analysis, there are two possibilities for centering the data. The first is *grand mean centering,* i.e., the sample average of all observations is subtracted, and the second is *within-groups centering,* where the sample average of only the observations within the same group is subtracted. Generally, grand mean centering does not change the model and is thus innocuous, but within-groups centering implicitly changes the model that is estimated, unless the sample averages of all level-1 predictor variables are included as level-2 predictors. For an extensive analysis, see Kreft et al. [68], Van Landeghem et al. [119], de Leeuw [27], and the references therein.

We close by noting that the quality of every data analysis crucially depends on the quality of the data. Most issues in data quality are not specific to multilevel analysis and are thus not discussed here. One important aspect, however, is the *sampling design.* Because a multilevel data set has observations at different levels, deciding on issues like sample size and randomization becomes more complicated than with single-level data. This subject is treated in detail in Chapter 4 in this volume.

Appendix

1.A Notational Conventions

This appendix describes the notation used in this chapter. The notation throughout this Handbook has been made as consistent as possible, so that this appendix also serves as a reference for the other chapters. However, the reader may occasionally discern slight differences in notation between the chapters.

1.A.1 Existing Notation

We used the most common books on mixed, random coefficient, and multilevel models to find a compromise notation [24, 46, 67, 76, 89, 101, 111]. There is a substantial agreement on notation in these books, although there are of course many differences of detail.

1.A.2 Matrices and Vectors

Matrices are boldface capitals; vectors are lowercase bold. In general, we use Greek symbols for unknowns and unobservables, such as parameters or latent variables (disturbances, variance components).

As another convention, we write $\boldsymbol{X}[n,r]$ for "$\boldsymbol{X}$ is an $n \times r$ matrix" and $\boldsymbol{y}[n]$ for "$\boldsymbol{y}$ is an n-element vector". Also, $\boldsymbol{X} = (x_{ij})$ is used to define a matrix in terms of its elements.

Two special matrix symbols we use are $\oplus$ for the *direct sum* and $\otimes$ for the *direct* (or *Kronecker*) *product.* If $\boldsymbol{A}_1, \ldots, \boldsymbol{A}_p$ are matrices, with $\boldsymbol{A}_s[n_s, m_s]$, then the direct sum is the $\sum_{s=1}^p n_s \times \sum_{s=1}^p m_s$ matrix

$$\bigoplus_{s=1}^{p} \boldsymbol{A}_s = \boldsymbol{A}_1 \oplus \cdots \oplus \boldsymbol{A}_p = \begin{pmatrix} \boldsymbol{A}_1 & \boldsymbol{\emptyset} & \boldsymbol{\emptyset} & \dots & \boldsymbol{\emptyset} \\ \boldsymbol{\emptyset} & \boldsymbol{A}_2 & \boldsymbol{\emptyset} & \cdots & \boldsymbol{\emptyset} \\ \boldsymbol{\emptyset} & \boldsymbol{\emptyset} & \boldsymbol{A}_3 & \cdots & \boldsymbol{\emptyset} \\ \vdots & \vdots & \vdots & \ddots & \vdots \\ \boldsymbol{\emptyset} & \boldsymbol{\emptyset} & \boldsymbol{\emptyset} & \cdots & \boldsymbol{A}_p \end{pmatrix},$$

where $\boldsymbol{\emptyset}$ denotes a (sub-)matrix with all elements equal to zero. The direct product is a $\prod_{s=1}^{p} n_s \times \prod_{s=1}^{p} m_s$ matrix, which we can best define recursively starting with two matrices $\boldsymbol{A}$ and $\boldsymbol{B}$. If $\boldsymbol{A}$ is $n \times m$, then

$$\boldsymbol{A} \otimes \boldsymbol{B} = \begin{pmatrix} a_{11}\boldsymbol{B} & a_{12}\boldsymbol{B} & a_{13}\boldsymbol{B} & \cdots & a_{1m}\boldsymbol{B} \\ a_{21}\boldsymbol{B} & a_{22}\boldsymbol{B} & a_{23}\boldsymbol{B} & \cdots & a_{2m}\boldsymbol{B} \\ a_{31}\boldsymbol{B} & a_{32}\boldsymbol{B} & a_{33}\boldsymbol{B} & \cdots & a_{3m}\boldsymbol{B} \\ \vdots & \vdots & \vdots & \ddots & \vdots \\ a_{n1}\boldsymbol{B} & a_{n2}\boldsymbol{B} & a_{n3}\boldsymbol{B} & \cdots & a_{nm}\boldsymbol{B} \end{pmatrix}$$

and, by recursion,

$$\bigotimes_{s=1}^{p} \boldsymbol{A}_s = \boldsymbol{A}_1 \otimes (\boldsymbol{A}_2 \otimes \cdots \otimes \boldsymbol{A}_p).$$

Superscripted delta is the *Kronecker delta*, i.e.,

$$\delta^{st} = \begin{cases} 1 & \text{if } i = j \\ 0 & \text{if } i \neq j. \end{cases}$$

The identity matrix is $\boldsymbol{I}$, a vector with all elements equal to 1 is $\mathbf{1}$. The matrix $\boldsymbol{E}$ has all elements equal to 1. The size of these matrices and vectors will often be clear from the context. If we need to be explicit, we can always write, for instance, $\boldsymbol{E}[n, m]$, but we also use the forms $\boldsymbol{I}_n$ and $\mathbf{1}_n$. Unit vectors $\boldsymbol{e}_i$ have all elements equal to zero, except for element i, which is equal to 1. Thus, $\mathbf{1}$ is the sum of the $\boldsymbol{e}_i$.

1.A.3 Special Symbols

We use the following special symbols:

- $\stackrel{\Delta}{=}$ is defined as
- $\sim$ is distributed as
- $\mathcal{N}$ normal distribution
- $\stackrel{\mathcal{L}}{\Longrightarrow}$ convergence in law (distribution)
- $\stackrel{a.d.}{=}$ has the same asymptotic distribution
- $\stackrel{\mathcal{P}}{\Longrightarrow}$ convergence in probability
- $\stackrel{iid}{\sim}$ i.i.d. with given distribution

1.A.4 Underlining Random Variables

A non-standard part of our notation is that we *underline random variables* [28]. Thus, vector or matrix random variables are both underlined and bold.

The advantage of distinguishing between random variables and fixed known or unknown constants in the context of mixed models is clear. We use constants (the design matrix, unknown parameters) and random variables (the outcome variables, of which we observe a realization, and the random effects, which we do not observe at all). We also estimate parameters. Estimates are fixed values, realization of estimators, which are random variables. Underlining gives us an extra alphabet, it also gives us a method to indicate how constants and random variables are related, because we can use y for a realization of $\underline{y}$. The advantages of underlining, known as the *Dutch Convention* or *Van Dantzig Convention*, are discussed in more detail in Hemelrijk [58].

As a simple example, the classical linear model is

$$\underline{\boldsymbol{y}} = \boldsymbol{X}\boldsymbol{\beta} + \underline{\boldsymbol{\epsilon}},$$

with

$$\underline{\boldsymbol{\epsilon}} \sim \mathcal{N}(\boldsymbol{\emptyset}, \sigma^2 \boldsymbol{I}).$$

Thus,

$$\underline{\boldsymbol{y}} \sim \mathcal{N}(\boldsymbol{X}\boldsymbol{\beta}, \sigma^2 \boldsymbol{I}).$$

We observe $\boldsymbol{y}$ and $\boldsymbol{X}$, and we compute

$$\hat{\boldsymbol{\beta}} = (\boldsymbol{X}'\boldsymbol{X})^{-1}\boldsymbol{X}'\boldsymbol{y}, \tag{1.33}$$

which is a realization of a random variable $\underline{\hat{\boldsymbol{\beta}}}$, satisfying

$$\underline{\hat{\boldsymbol{\beta}}} \sim \mathcal{N}(\boldsymbol{\beta}, \sigma^2(\boldsymbol{X}'\boldsymbol{X})^{-1}).$$

It obviously makes sense to write $E(\underline{\hat{\boldsymbol{\beta}}}) = \boldsymbol{\beta}$, and it does not make sense to write $E(\hat{\boldsymbol{\beta}}) = \boldsymbol{\beta}$.

Equation (1.33) also illustrates the convention of writing the estimate of a parameter by putting a hat on the parameter symbol. We also use this convention for "estimating" a random component, for instance,

$$\hat{\boldsymbol{\epsilon}} = \boldsymbol{y} - \boldsymbol{X}\hat{\boldsymbol{\beta}}.$$

For conditional expectations, we can both have $E(\underline{x} \mid y)$ and $E(\underline{x} \mid \underline{y})$, because we can condition on both a random variable and its realization. The first expression defines a deterministic function of y, the second a function of $\underline{y}$, i.e., a random variable.

It is important to emphasize some basic consequences of our conventions. Anything we actually compute cannot be underlined, because we only compute with realizations, not with random variables. Anything that is underlined is by definition part of a statistical model, because it implies a framework of replication or a degree of belief. In Bayesian models, there will be more underlining than in empirical Bayes models, and empirical Bayes models have more underlining than classical frequentist models. Ultimately, of course, even fully Bayesian models will have fixed hyperparameters, because otherwise the specification of the model will never stop.

1.B Generic Numerical Optimization

The most common starting point for numerical optimization of a generic well-behaved function is a second-order Taylor series expansion around a point $\boldsymbol{\theta}_1$:

$$f(\boldsymbol{\theta}) = f_1 + \boldsymbol{g}_1'(\boldsymbol{\theta} - \boldsymbol{\theta}_1) + \tfrac{1}{2}(\boldsymbol{\theta} - \boldsymbol{\theta}_1)'\boldsymbol{H}_1(\boldsymbol{\theta} - \boldsymbol{\theta}_1) + o\|\boldsymbol{\theta} - \boldsymbol{\theta}_1\|^2,$$

where f_1, $\boldsymbol{g}_1$, and $\boldsymbol{H}_1$ are the function $f(\cdot)$, its gradient $\boldsymbol{g}(\cdot)$ (vector of first partial derivatives with respect to $\boldsymbol{\theta}$), and its Hessian $\boldsymbol{H}(\cdot)$ (matrix of second partial derivatives with respect to $\boldsymbol{\theta}$), respectively, all evaluated in $\boldsymbol{\theta}_1$.

Thus, if we ignore the approximation error reflected by the last term, we find that the function is minimized for

$$\hat{\boldsymbol{\theta}} = \boldsymbol{\theta}_1 - \boldsymbol{H}_1^{-1}\boldsymbol{g}_1,$$

provided that $\boldsymbol{H}_1$ is positive definite. Of course, in practice the approximation error is not zero, so that this does not minimize the loss function immediately. But we can assert that we have come closer and repeat the process, leading to the algorithm

$$\boldsymbol{\theta}_{i+1} = \boldsymbol{\theta}_i - \boldsymbol{H}_i^{-1}\boldsymbol{g}_i,$$

where i denotes the iteration number. This algorithm defines the well-known *Newton-Raphson* method, also known simply as Newton's method. In practice, two modifications are often necessary to ensure that this algorithm works well. The first is that the *search direction* $-\boldsymbol{H}_i^{-1}\boldsymbol{g}_i$ is only guaranteed to point in the direction of smaller function values if $\boldsymbol{H}_i$ is positive definite. Hence, if the loss function is not globally convex, $\boldsymbol{H}_i$ may have to be modified in some iterations to ensure that it is positive definite. This is typically done by adding a positive multiple of the identity matrix until all eigenvalues are positive. The second modification that is often used is to insert a *step size* α_i, with which the search direction is multiplied, so that the algorithm becomes

$$\boldsymbol{\theta}_{i+1} = \boldsymbol{\theta}_i - \alpha_i\boldsymbol{H}_i^{-1}\boldsymbol{g}_i, \tag{1.34}$$

where it is understood that $\boldsymbol{H}_i$ may be the modified version to make it positive definite. Even though it is guaranteed that the search direction points toward smaller function values, the unmodified update may "overshoot" if the function decreases slowly in the neighborhood of the current point, but then increases sharply. Therefore, the factor α_i is chosen such that the function value in the next point is smaller than in the current point. A value of α_i that ensures this always exists if $\boldsymbol{H}_i$ is positive definite and $\boldsymbol{g}_i$ is nonzero. Typically, one would start with $\alpha_i = 1$, halving step size until such a point is reached. The (modified) Newton-Raphson method is implemented in most general-purpose optimization functions.

There exist many alternative generic numerical optimization methods, most of which use the same form (1.34) of an iteration, but with $\boldsymbol{H}_i^{-1}$ replaced by another positive (semi)definite matrix. The reason for this is that it is often computationally demanding to compute $\boldsymbol{H}_i^{-1}$, and places a larger burden on the researcher and/or programmer, because the second derivatives have to be computed and programmed. In principle, these methods converge more slowly, because in the neighborhood of the minimum, the loss function is closely approximated by a quadratic function, so that Newton-Raphson converges very fast. In contrast, the *steepest descent* method, which simply replaces $\boldsymbol{H}_i^{-1}$ by the identity matrix, tends to converge extremely slowly. In many cases, however, the better alternative methods are not noticeably worse (in terms of speed and accuracy) than Newton-Raphson. A good and popular method is the BFGS method, which replaces $\boldsymbol{H}_i^{-1}$ by the matrix $\boldsymbol{G}_i$. The latter matrix is computed using the update formula

$$\boldsymbol{G}_{i+1} = (\boldsymbol{I} - \rho_i \triangle\boldsymbol{\theta}_i \triangle\boldsymbol{g}_i')\boldsymbol{G}_i(\boldsymbol{I} - \rho_i \triangle\boldsymbol{g}_i \triangle\boldsymbol{\theta}_i') + \rho_i \triangle\boldsymbol{\theta}_i \triangle\boldsymbol{\theta}_i',$$

where $\triangle\boldsymbol{\theta}_i = \boldsymbol{\theta}_{i+1} - \boldsymbol{\theta}_i$, $\triangle\boldsymbol{g}_i = \boldsymbol{g}_{i+1} - \boldsymbol{g}_i$, and $\rho_i = 1/\triangle\boldsymbol{g}_i'\triangle\boldsymbol{\theta}_i$. Clearly, if $\boldsymbol{G}_i$ is positive semidefinite, then $\boldsymbol{G}_{i+1}$ is also positive semidefinite. Moreover, it can be proved that if $\boldsymbol{G}_i$ is positive definite, then $\boldsymbol{G}_{i+1}$ is also positive definite. Typically, the starting value $\boldsymbol{G}_0$ is the identity matrix, which is clearly positive definite, or an informed guess of $\boldsymbol{H}^{-1}$. When BFGS is applied to a (convex) quadratic function of an n-element vector $\boldsymbol{\theta}$, and the step size is chosen to minimize the function along the line defined by the update formula, the global minimum is attained in n iterations and $\boldsymbol{G}_{n+1} = \boldsymbol{H}^{-1}$ (which is a constant matrix). Therefore, unless the number of parameters is large, BFGS tends to converge quickly in the neighborhood of the minimum, where the loss function is approximately quadratic. The BFGS method is also implemented in most general-purpose optimization functions.

An extensive treatment of many generic numerical optimization procedures, including Newton-Raphson and BFGS, with derivations of their properties, can be found in Nocedal and Wright [86].

1.C Some Matrix Expressions

Here we collect some convenient results to deal with two-level linear models. The first two results have been known for a long time [26, 36, 117]. Proofs of the first three results are given, for example, in de Leeuw and Liu [31]. Many additional useful matrix results are provided by Wansbeek and Meijer [123, appendix A] and Harville [53].

Theorem 1.1 If $\boldsymbol{A} = \boldsymbol{B} + \boldsymbol{TCT}'$ with $\boldsymbol{A}$ and $\boldsymbol{B}$ positive definite, then

$$\log|\boldsymbol{A}| = \log|\boldsymbol{B}| + \log|\boldsymbol{C}| + \log|\boldsymbol{C}^{-1} + \boldsymbol{T}'\boldsymbol{B}^{-1}\boldsymbol{T}|.$$

If, in addition, $\boldsymbol{T}$ is of full column rank, then

$$\log|\boldsymbol{A}| = \log|\boldsymbol{B}| + \log|\boldsymbol{T}'\boldsymbol{B}^{-1}\boldsymbol{T}| + \log|\boldsymbol{C} + (\boldsymbol{T}'\boldsymbol{B}^{-1}\boldsymbol{T})^{-1}|.$$

Theorem 1.2 If $\boldsymbol{A} = \boldsymbol{B} + \boldsymbol{TCT}'$ with $\boldsymbol{A}$ and $\boldsymbol{B}$ positive definite, then

$$\boldsymbol{A}^{-1} = \boldsymbol{B}^{-1} - \boldsymbol{B}^{-1}\boldsymbol{T}(\boldsymbol{C}^{-1} + \boldsymbol{T}'\boldsymbol{C}^{-1}\boldsymbol{T})^{-1}\boldsymbol{T}'\boldsymbol{B}^{-1}.$$

If, in addition, $\boldsymbol{T}$ is of full column rank, then

$$\begin{aligned}\boldsymbol{A}^{-1} = \boldsymbol{T}(\boldsymbol{T}'\boldsymbol{T})^{-1}(\boldsymbol{C} + (\boldsymbol{T}'\boldsymbol{B}^{-1}\boldsymbol{T})^{-1})^{-1}(\boldsymbol{T}'\boldsymbol{T})^{-1}\boldsymbol{T}' \\ + \{\boldsymbol{B}^{-1} - \boldsymbol{B}^{-1}\boldsymbol{T}(\boldsymbol{T}'\boldsymbol{B}^{-1}\boldsymbol{T})^{-1}\boldsymbol{T}'\boldsymbol{B}^{-1}\}.\end{aligned}$$

Theorem 1.3 If $\boldsymbol{A} = \boldsymbol{B} + \boldsymbol{TCT}'$ with $\boldsymbol{A}$ and $\boldsymbol{B}$ positive definite,then

$$\boldsymbol{y}'\boldsymbol{A}^{-1}\boldsymbol{y} = \min_{\boldsymbol{x}}\{(\boldsymbol{y} - \boldsymbol{Tx})'\boldsymbol{B}^{-1}(\boldsymbol{y} - \boldsymbol{Tx}) + \boldsymbol{x}'\boldsymbol{C}^{-1}\boldsymbol{x}\}.$$

The fourth result was proved by de Hoog et al. [26] by letting $\boldsymbol{C}^{-1} \to \boldsymbol{\emptyset}$ on both sides of Theorem 1.2.

Theorem 1.4 If $\boldsymbol{B}$ is positive definite and $\boldsymbol{T}$ is of full column-rank, then

$$\boldsymbol{B}^{-1} - \boldsymbol{B}^{-1}\boldsymbol{T}(\boldsymbol{T}'\boldsymbol{B}^{-1}\boldsymbol{T})^{-1}\boldsymbol{T}'\boldsymbol{B}^{-1} = (\boldsymbol{QBQ})^{+},$$

where $\boldsymbol{Q} = \boldsymbol{I} - \boldsymbol{T}(\boldsymbol{T}'\boldsymbol{T})^{-1}\boldsymbol{T}'$ and superscript + denotes the Moore-Penrose inverse.

1.D The EM Algorithm

The EM algorithm of Dempster et al. [33] is a general method to optimize functions of the form $f(\boldsymbol{\theta}) = \log\int g(\boldsymbol{\theta}, \boldsymbol{z})\, d\boldsymbol{z}$ over $\boldsymbol{\theta}$, where $g(\boldsymbol{\theta}, \boldsymbol{z}) > 0$ for all $\boldsymbol{\theta}$ and $\boldsymbol{z}$ in the domain. It is usually presented in probabilistic terminology, but the reason why it works is the concavity of the logarithm, which is obviously not a probabilistic result.

Define $h(\boldsymbol{\theta}) \stackrel{\Delta}{=} \int g(\boldsymbol{\theta}, \boldsymbol{z})\, \mathrm{d}\boldsymbol{z}$ and $k(\boldsymbol{z} \mid \boldsymbol{\theta}) \stackrel{\Delta}{=} g(\boldsymbol{\theta}, \boldsymbol{z})/h(\boldsymbol{\theta})$. Then, by the concavity of the logarithm, it follows from Jensen's inequality [96, p. 58] that for all $\boldsymbol{\theta}$ and $\tilde{\boldsymbol{\theta}}$,

$$f(\boldsymbol{\theta}) \geq f(\tilde{\boldsymbol{\theta}}) + \int \log g(\boldsymbol{\theta}, \boldsymbol{z})\, k(\boldsymbol{z} \mid \tilde{\boldsymbol{\theta}})\, \mathrm{d}\boldsymbol{z} - \int \log g(\tilde{\boldsymbol{\theta}}, \boldsymbol{z})\, k(\boldsymbol{z} \mid \tilde{\boldsymbol{\theta}})\, \mathrm{d}\boldsymbol{z}, \quad (1.35)$$

with equality if and only if $g(\boldsymbol{\theta}, \boldsymbol{z}) = g(\tilde{\boldsymbol{\theta}}, \boldsymbol{z})$ almost everywhere.

In each iteration of the EM algorithm we take $\tilde{\boldsymbol{\theta}}$ to be our current best approximation to the optimum and improve it by maximizing the right-hand side of (1.35) over $\boldsymbol{\theta}$ for this given $\tilde{\boldsymbol{\theta}}$. In other words, we find $\boldsymbol{\theta}^{(i+1)}$ by maximizing

$$Q(\boldsymbol{\theta} \mid \boldsymbol{\theta}^{(i)}) \stackrel{\Delta}{=} \int \log g(\boldsymbol{\theta}, \boldsymbol{z})\, k(\boldsymbol{z} \mid \boldsymbol{\theta}^{(i)})\, \mathrm{d}\boldsymbol{z}$$

over $\boldsymbol{\theta}$. The algorithm is monotone, in the sense that $f(\boldsymbol{\theta}^{(i+1)}) > f(\boldsymbol{\theta}^{(i)})$ and in many cases this is enough to guarantee (linear) convergence to a local maximum of $f(\cdot)$.

In the probabilistic interpretation, $f(\boldsymbol{\theta})$ is a loglikelihood function and EM stands for expectation-maximization. The E-step computes $Q(\boldsymbol{\theta} \mid \boldsymbol{\theta}^{(i)})$, which is the conditional expectation of the *complete-data loglikelihood* $g(\boldsymbol{\theta}, \boldsymbol{z})$, given the observed data and the current parameter value $\boldsymbol{\theta}^{(i)}$, and the M-step maximizes the resulting function.

We can now apply the EM algorithm to the multilevel FIML loglikelihood. Here, $\underline{\boldsymbol{z}}$ consists of all the random effects $\underline{\boldsymbol{\delta}}_j$, and $\boldsymbol{\theta}$ is the usual parameter vector. The complete-data loglikelihood has the form

$$g(\boldsymbol{\theta}, \boldsymbol{\delta}) = \prod_{j=1}^{m} g_j(\boldsymbol{\theta}, \boldsymbol{\delta}_j),$$

where $g_j(\boldsymbol{\theta}, \boldsymbol{\delta}_j)$ is the joint density of $\underline{\boldsymbol{y}}_j$ and $\underline{\boldsymbol{\delta}}_j$. Using standard probability theory, we can write

$$\begin{aligned}
g_j(\boldsymbol{\theta}, \boldsymbol{\delta}_j) &= f_{\boldsymbol{\delta} \mid \boldsymbol{y}}(\boldsymbol{\delta}_j \mid \boldsymbol{y}_j) f_{\boldsymbol{y}}(\boldsymbol{y}_j), \\
h_j(\boldsymbol{\theta}) &\stackrel{\Delta}{=} \int g_j(\boldsymbol{\theta}, \boldsymbol{\delta}_j)\, \mathrm{d}\boldsymbol{\delta}_j = f_{\boldsymbol{y}}(\boldsymbol{y}_j), \\
k_j(\boldsymbol{\delta}_j \mid \boldsymbol{\theta}) &\stackrel{\Delta}{=} g_j(\boldsymbol{\theta}, \boldsymbol{\delta}_j)/h_j(\boldsymbol{\theta}) = f_{\boldsymbol{\delta} \mid \boldsymbol{y}}(\boldsymbol{\delta}_j \mid \boldsymbol{y}_j), \\
Q_j(\boldsymbol{\theta} \mid \boldsymbol{\theta}^{(i)}) &\stackrel{\Delta}{=} \int \log g_j(\boldsymbol{\theta}, \boldsymbol{\delta}_j)\, k_j(\boldsymbol{\delta}_j \mid \boldsymbol{\theta}^{(i)})\, \mathrm{d}\boldsymbol{\delta}_j\,, \\
Q(\boldsymbol{\theta} \mid \boldsymbol{\theta}^{(i)}) &= \sum_{j=1}^{m} Q_j(\boldsymbol{\theta} \mid \boldsymbol{\theta}^{(i)}).
\end{aligned}$$

The joint distribution of $\underline{\boldsymbol{y}}_j$ and $\underline{\boldsymbol{\delta}}_j$ is normal:

$$\begin{pmatrix} \underline{y}_j \\ \underline{\delta}_j \end{pmatrix} \sim \mathcal{N}\left(\begin{pmatrix} U_j\gamma \\ \emptyset \end{pmatrix}, \begin{pmatrix} V_j & X_j\Omega \\ \Omega X_j' & \Omega \end{pmatrix} \right),$$

from which we obtain the conditional distribution of $\underline{\delta}_j$ given y_j as

$$\underline{\delta}_j \mid y_j \sim \mathcal{N}(\mu_j, \Sigma_j),$$

with

$$\begin{aligned} \mu_j &= \Omega X_j' V_j^{-1}(y_j - U_j\gamma) = \Omega W_j^{-1}(b_j - Z_j\gamma), \\ \Sigma_j &= \Omega - \Omega X_j' V_j^{-1} X_j \Omega = \sigma_j^2 \Omega W_j^{-1}(X_j' X_j)^{-1}. \end{aligned}$$

By writing $g_j(\theta, \delta_j) = f_{y\,|\,\delta}(y_j \mid \delta_j) f_\delta(\delta_j)$, and observing that the marginal distribution of $\underline{\delta}_j$ is normal with mean zero and covariance matrix Ω, and the conditional distribution of $\underline{y}_j$ given δ_j is normal with mean $U_j\gamma + X_j\delta_j$ and covariance matrix $\sigma_j^2 I_{n_j}$, we obtain, after some simplification,

$$\begin{aligned} \log g_j(\theta, \delta_j) = &-\frac{n_j+p}{2}\log(2\pi) - \frac{n_j}{2}\log\sigma_j^2 - \frac{1}{2\sigma_j^2}(n_j - p)s_j^2 \\ &- \frac{1}{2\sigma_j^2}(b_j - Z_j\gamma)' X_j' X_j (b_j - Z_j\gamma) + \frac{1}{\sigma_j^2}(b_j - Z_j\gamma)' X_j' X_j \delta_j \\ &- \tfrac{1}{2}\log|\Omega| - \tfrac{1}{2}\operatorname{tr}\left[(\sigma_j^{-2} X_j' X_j + \Omega^{-1})\delta_j\delta_j'\right]. \end{aligned}$$

The function $Q_j(\theta \mid \theta^{(i)})$ is obtained by integrating the product of this with $k_j(\delta_j \mid \theta^{(i)})$. That is, it is obtained as the expectation of $\log g_j(\theta, \underline{\delta}_j)$ when viewed as a function of the random variable $\underline{\delta}_j$ that is normally distributed with mean $\mu_j^{(i)}$ and covariance matrix $\Sigma_j^{(i)}$, which are μ_j and Σ_j evaluated in $\theta^{(i)}$. For this distribution, we evidently have $E(\underline{\delta}_j) = \mu_j^{(i)}$ and $E(\underline{\delta}_j\underline{\delta}_j') = \Sigma_j^{(i)} + \mu_j^{(i)}\mu_j^{(i)\prime}$, so that, after some simplification, we obtain

$$\begin{aligned} Q_j(\theta \mid \theta^{(i)}) = &\left(-\frac{n_j+p}{2}\log(2\pi)\right) - \tfrac{1}{2}\left(\log|\Omega| + \operatorname{tr}\left[\Omega^{-1}(\Sigma_j^{(i)} + \mu_j^{(i)}\mu_j^{(i)\prime})\right]\right) \\ &- \frac{n_j}{2}\log\sigma_j^2 - \frac{1}{2\sigma_j^2}(n_j - p)s_j^2 - \frac{1}{2\sigma_j^2}\operatorname{tr}(X_j' X_j \Sigma_j^{(i)}) \\ &- \frac{1}{2\sigma_j^2}(b_j - \mu_j^{(i)} - Z_j\gamma)' X_j' X_j (b_j - \mu_j^{(i)} - Z_j\gamma). \end{aligned}$$

Consequently, the parameter values that optimize $Q(\theta \mid \theta^{(i)})$ are

$$\begin{aligned} \Omega^{(i+1)} &= \frac{1}{m}\sum_{j=1}^{m}(\Sigma_j^{(i)} + \mu_j^{(i)}\mu_j^{(i)\prime}), \\ \gamma^{(i+1)} &= \left(\sum_{j=1}^{m} Z_j' X_j' X_j Z_j\right)^{-1} \sum_{j=1}^{m} Z_j' X_j' X_j (b_j - \mu_j^{(i)}), \\ (\sigma_j^2)^{(i+1)} &= \frac{1}{n_j}\left[(n_j - p)s_j^2 + \operatorname{tr}(X_j' X_j \Lambda_j^{(i)})\right], \end{aligned}$$

or, instead of the latter,

$$(\sigma^2)^{(i+1)} = \frac{1}{n} \sum_{j=1}^{m} \left[(n_j - p)s_j^2 + \mathrm{tr}(\boldsymbol{X}_j' \boldsymbol{X}_j \boldsymbol{\Lambda}_j^{(i)}) \right],$$

where

$$\boldsymbol{\Lambda}_j^{(i)} \triangleq \boldsymbol{\Sigma}_j^{(i)} + (\boldsymbol{b}_j - \boldsymbol{\mu}_j^{(i)} - \boldsymbol{Z}_j \boldsymbol{\gamma}^{(i+1)})(\boldsymbol{b}_j - \boldsymbol{\mu}_j^{(i)} - \boldsymbol{Z}_j \boldsymbol{\gamma}^{(i+1)})'.$$

Note that when $\boldsymbol{\Omega}$ is not completely free (apart from the requirements of symmetry and positive definiteness, of course), then the M-step with respect to the parameters $\{\xi_g\}$ is nontrivial. We then need to minimize the function

$$F(\boldsymbol{\xi}) \triangleq \log |\boldsymbol{\Omega}| + \mathrm{tr}\big(\boldsymbol{\Omega}^{-1} \boldsymbol{S}^{(i)}\big)$$

with respect to $\boldsymbol{\xi} \triangleq (\xi_1, \ldots, \xi_G)'$, where

$$\boldsymbol{S}^{(i)} \triangleq \frac{1}{m} \sum_{j=1}^{m} (\boldsymbol{\Sigma}_j^{(i)} + \boldsymbol{\mu}_j^{(i)} \boldsymbol{\mu}_j^{(i)\prime}).$$

Assuming (1.3), the first-order conditions are

$$\mathrm{tr}[\boldsymbol{\Omega}^{-1}(\boldsymbol{S}^{(i)} - \boldsymbol{\Omega})\boldsymbol{\Omega}^{-1}\boldsymbol{C}_g] = 0.$$

Letting $\boldsymbol{C}^*$ be the matrix with g-th column equal to $\mathrm{vec}(\boldsymbol{C}_g)$, these can be jointly written as

$$\boldsymbol{C}^{*\prime}(\boldsymbol{\Omega}^{-1} \otimes \boldsymbol{\Omega}^{-1})(\mathrm{vec}\, \boldsymbol{S}^{(i)} - \boldsymbol{C}^* \boldsymbol{\xi}) = \boldsymbol{\emptyset},$$

which is a nonlinear equation that does not generally have a closed-form solution. However, it strongly suggests that one or more IGLS iterations of the form

$$\boldsymbol{\xi}^{(i+1,k+1)} = \left[\boldsymbol{C}^{*\prime}(\boldsymbol{\Omega}^{-1} \otimes \boldsymbol{\Omega}^{-1})\boldsymbol{C}^*\right]^{-1} \boldsymbol{C}^{*\prime}(\boldsymbol{\Omega}^{-1} \otimes \boldsymbol{\Omega}^{-1})\, \mathrm{vec}\, \boldsymbol{S}^{(i)},$$

where in the right-hand side $\mathrm{vec}\, \boldsymbol{S}^{(i)}$ is held fixed throughout these subiterations, but $\boldsymbol{\Omega}$ is the value from the previous (k-th) subiteration, should also increase the loglikelihood, so that full optimization in this step is not necessary.

References

1. L. S. Aiken and S. G. West. *Multiple Regression: Testing and Interpreting Interaction.* Sage Publications, Newbury Park, CA, 1991.

2. M. Aitkin and N. Longford. Statistical modelling issues in school effectiveness studies. *Journal of the Royal Statistical Society, Series A*, 149:1–43, 1986. (with discussion)
3. L. Anselin. Spatial econometrics. In B. H. Baltagi, editor, *A Companion to Theoretical Econometrics*, pages 310–330. Blackwell, Malden, MA, 2001.
4. M. Arellano. *Panel Data Econometrics.* Oxford University Press, Oxford, UK, 2003.
5. T. Asparouhov. Sampling weights in latent variable modeling. *Structural Equation Modeling*, 12:411–434, 2005.
6. T. Asparouhov. General multi-level modeling with sampling weights. *Communications in Statistics—Theory & Methods*, 35:439–460, 2006.
7. D. Bates and D. Sarkar. *The lme4 Package*, 2006. URL `http://cran.r-project.org`
8. P. M. Bentler. *EQS 6 Structural Equations Program Manual.* Multivariate Software, Encino, CA, 2006.
9. H. M. Blalock. Contextual effects models: Theoretical and methodological issues. *Annual Review of Sociology*, 10:353–372, 1984.
10. R. J. Bosker, T. A. B. Snijders, and H. Guldemond. *PINT: Estimating Standard Errors of Regression Coefficients in Hierarchical Linear Models for Power Calculations. User's Manual Version 1.6.* University of Twente, Enschede, The Netherlands, 1999.
11. L. H. Boyd and G. R. Iversen. *Contextual Analysis: Concepts and Statistical Techniques.* Wadsworth, Belmont, CA, 1979.
12. M. W. Browne. Asymptotically distribution-free methods for the analysis of covariance structures. *British Journal of Mathematical and Statistical Psychology*, 37:62–83, 1984.
13. A. S. Bryk and S. W. Raudenbush. *Hierachical Linear Models: Applications and Data Analysis Methods.* Sage, Newbury Park, CA, 1992.
14. L. Burstein. The analysis of multilevel data in educational research and evaluation. *Review of Research in Education*, 8:158–233, 1980.
15. L. Burstein, R. L. Linn, and F. J. Capell. Analyzing multilevel data in the presence of heterogeneous within-class regressions. *Journal of Educational Statistics*, 3:347–383, 1978.
16. F. M. T. A. Busing, E. Meijer, and R. Van der Leeden. *MLA: Software for MultiLevel Analysis of Data with Two Levels. User's Guide for Version 4.1.* Leiden University, Department of Psychology, Leiden, 2005.
17. A. C. Cameron and P. K. Trivedi. *Microeconometrics: Methods and Applications.* Cambridge University Press, Cambridge, UK, 2005.
18. G. Chamberlain. Panel data. In Z. Griliches and M. D. Intriligator, editors, *Handbook of Econometrics*, volume 2, pages 1247–1318. North-Holland, Amsterdam, 1984.
19. G. Chamberlain and E. E. Leamer. Matrix weighted averages and posterior bounds. *Journal of the Royal Statistical Society, Series B*, 38:73–84, 1976.
20. K. Chantala, D. Blanchette, and C. M. Suchindran. Software to compute sampling weights for multilevel analysis, 2006. URL `http://www.cpc.unc.edu/restools/data_analysis/ml_sampling_weights`

21. Y. F. Cheong, R. P. Fotiu, and S. W. Raudenbush. Efficiency and robustness of alternative estimators for two- and three-level models: The case of NAEP. *Journal of Educational and Behavioral Statistics*, 26:411–429, 2001.
22. J. S. Coleman, E. Q. Campbell, C. J. Hobson, J. McPartland, A. M. Mood, F. D. Weinfeld, and R. L. York. *Equality of Educational Opportunity*. U.S. Government Printing Office, Washington, DC, 1966.
23. D. R. Cox. Interaction. *International Statistical Review*, 52:1–31, 1984.
24. M. Davidian and D. M. Giltinan. *Nonlinear Models for Repeated Mesurement Data*. Chapman & Hall, London, 1995.
25. R. Davidson and J. G. MacKinnon. *Estimation and Inference in Econometrics*. Oxford University Press, Oxford, UK, 1993.
26. F. R. de Hoog, T. P. Speed, and E. R. Williams. On a matrix identity associated with generalized least squares. *Linear Algebra and its Applications*, 127:449–456, 1990.
27. J. de Leeuw. Centering in multilevel analysis. In B. S. Everitt and D. C. Howell, editors, *Encyclopedia of Statistics in Behavioral Science*, volume 1, pages 247–249. Wiley, New York, 2005.
28. J. de Leeuw and I. G. G. Kreft. Random coefficient models for multilevel analysis. *Journal of Educational Statistics*, 11:57–85, 1986.
29. J. de Leeuw and I. G. G. Kreft. Questioning multilevel models. *Journal of Educational and Behavioral Statistics*, 20:171–190, 1995.
30. J. de Leeuw and I. G. G. Kreft. Software for multilevel analysis. In A. H. Leyland and H. Goldstein, editors, *Multilevel Modelling of Health Statistics*, pages 187–204. Wiley, Chichester, 2001.
31. J. de Leeuw and G. Liu. Augmentation algorithms for mixed model analysis. Preprint 115, UCLA Statistics, Los Angeles, CA, 1993.
32. G. del Pino. The unifying role of iterative generalized least squares in statistical algorithms. *Statistical Science*, 4:394–408, 1989. (with discussion)
33. A. P. Dempster, N. M. Laird, and D. B. Rubin. Maximum likelihood from incomplete data via the *EM* algorithm. *Journal of the Royal Statistical Society, Series B*, 39:1–38, 1977. (with discussion)
34. A. P. Dempster, D. B. Rubin, and R. K. Tsutakawa. Estimation in covariance components models. *Journal of the American Statistical Association*, 76:341–353, 1981.
35. M. du Toit and S. H. C. du Toit. *Interactive LISREL: User's Guide*. Scientific Software International, Chicago, 2002.
36. W. J. Duncan. Some devices for the solution of large sets of simultaneous linear equations (with an appendix on the reciprocation of partitioned matrices). *The London, Edinburgh and Dublin Philosophical Magazine and Journal of Science, 7th Series*, 35:660–670, 1944.
37. J. P. Elhorst and A. S. Zeilstra. Labour force participation rates at the regional and national levels of the European Union: An integrated analysis. *Papers in Regional Science*, forthcoming.
38. T. S. Ferguson. *A Course in Large Sample Theory*. Chapman & Hall, London, 1996.

39. D. A. Freedman. On the so-called "Huber sandwich estimator" and "robust standard errors". Unpublished manuscript, 2006.
40. A. Gelman. Multilevel (hierarchical) modeling: What it can and cannot do. *Technometrics*, 48:432–435, 2006.
41. A. Gelman and I. Pardoe. Bayesian measures of explained variance and pooling in multilevel (hierarchical) models. *Technometrics*, 48:241–251, 2006.
42. A. Gelman, D. K. Park, S. Anselobehere, P. N. Price, and L. C. Minnete. Models, assumptions and model checking in ecological regressions. *Journal of the Royal Statistical Society, Series A*, 164:101–118, 2001.
43. A. S. Goldberger. Best linear unbiased prediction in the generalized linear regression model. *Journal of the American Statistical Association*, 57:369–375, 1962.
44. H. Goldstein. Multilevel mixed linear model analysis using iterative generalized least squares. *Biometrika*, 73:43–56, 1986.
45. H. Goldstein. Restricted unbiased iterative generalized least-squares estimation. *Biometrika*, 76:622–623, 1989.
46. H. Goldstein. *Multilevel Statistical Models*, 3rd edition. Edward Arnold, London, 2003.
47. H. Goldstein and J. Rasbash. Efficient computational procedures for the estimation of parameters in multilevel models based on iterative generalised least squares. *Computational Statistics & Data Analysis*, 13:63–71, 1992.
48. P. J. Green. Iteratively reweighted least squares for maximum likelihood estimation, and some robust and resistant alternatives. *Journal of the Royal Statistical Society, Series B*, 46:149–192, 1984. (with discussion)
49. L. Grilli and M. Pratesi. Weighted estimation in multilevel ordinal and binary models in the presence of informative sampling designs. *Survey Methodology*, 30:93–103, 2004.
50. E. A. Hanushek. Efficient estimates for regressing regression coefficients. *American Statistician*, 28:66–67, 1974.
51. H. O. Hartley and J. N. K. Rao. Maximum likelihood estimation for the mixed analysis of variance model. *Biometrika*, 54:93–108, 1967.
52. D. A. Harville. Baysian inference for variance components using only error contrasts. *Biometrika*, 61:383–385, 1974.
53. D. A. Harville. *Matrix Algebra From a Statistician's Perspective.* Springer, New York, 1997.
54. T. Hastie and R. Tibshirani. Varying-coefficient models. *Journal of the Royal Statistical Society, Series B*, 55:757–796, 1993. (with discussion)
55. D. Hedeker. MIXNO: A computer program for mixed-effects nominal logistic regression. *Journal of Statistical Software*, 4(5):1–92, 1999.
56. D. Hedeker and R. D. Gibbons. MIXOR: A computer program for mixed-effects ordinal regression analysis. *Computer Methods and Programs in Biomedicine*, 49:157–176, 1996.
57. D. Hedeker and R. D. Gibbons. MIXREG: A computer program for mixed-effects regression analysis with autocorrelated errors. *Computer Methods and Programs in Biomedicine*, 49:229–252, 1997.

58. J. Hemelrijk. Underlining random variables. *Statistica Neerlandica*, 20:1–7, 1966.
59. J. J. Hox. *Multilevel Analysis: Techniques and Applications.* Erlbaum, Mahwah, NJ, 2002.
60. C. Hsiao. *Analysis of Panel Data*, 2nd edition. Cambridge University Press, Cambridge, UK, 2003.
61. J. Z. Huang, C. O. Wu, and L. Zhou. Varying-coefficient models and basis function approximations for the analysis of repeated measurements. *Biometrika*, 89:111–128, 2002.
62. C. Jencks, M. Smith, H. Acland, M. J. Bane, D. Cohen, H. Gintis, B. Heyns, and S. Michelson. *Inequality: A Reassessment of the Effect of Family and Schooling in America.* Basic Books, New York, 1972.
63. R. I. Jennrich and M. D. Schluchter. Unbalanced repeated-measures models with structured covariance matrices. *Biometrics*, 42:805–820, 1986.
64. K. G. Jöreskog, D. Sörbom, S. H. C. du Toit, and M. du Toit. *LISREL 8: New Statistical Features.* Scientific Software International, Chicago, 2001. (3rd printing with revisions)
65. J. Kim and E. W. Frees. Multilevel modeling with correlated effects. *Psychometrika*, forthcoming.
66. M. S. Kovačević and S. N. Rai. A pseudo maximum likelihood approach to multilevel modelling of survey data. *Communications in Statistics—Theory and Methods*, 32:103–121, 2003.
67. I. G. G. Kreft and J. de Leeuw. *Introducing Multilevel Modeling.* Sage, London, 1998.
68. I. G. G. Kreft, J. de Leeuw, and L. S. Aiken. The effect of different forms of centering in hierarchical linear models. *Multivariate Behavioral Research*, 30: 1–21, 1995.
69. N. M. Laird, N. Lange, and D. Stram. Maximum likelihood computations with repeated measures: Application of the EM algorithm. *Journal of the American Statistical Association*, 82:97–105, 1987.
70. N. M. Laird and J. H. Ware. Random-effects models for longitudinal data. *Biometrics*, 38:963–974, 1982.
71. L. I. Langbein. Schools or students: Aggregation problems in the study of student achievement. *Evaluation Studies Review Annual*, 2:270–298, 1977.
72. D. V. Lindley and A. F. M. Smith. Bayes estimates for the linear model. *Journal of the Royal Statistical Society, Series B*, 34:1–41, 1972.
73. M. J. Lindstrom and D. M. Bates. Newton-Raphson and EM algorithms for linear mixed-effects models for repeated-measures data. *Journal of the American Statistical Association*, 83:1014–1022, 1988.
74. N. T. Longford. A fast scoring algorithm for maximum likelihood estimation in unbalanced mixed models with nested random effects. *Biometrika*, 74:817–827, 1987.
75. N. T. Longford. *VARCL. Software for Variance Component Analysis of Data with Nested Random Effects (Maximum Likelihood).* Educational Testing Service, Princeton, NJ, 1990.

76. N. T. Longford. *Random Coefficient Models.* Oxford University Press, Oxford, UK, 1993.
77. K. Löwner. Über monotone Matrixfunktionen. *Mathematische Zeitschrift*, 38: 177–216, 1934.
78. C. J. M. Maas and J. J. Hox. The influence of violations of assumptions on multilevel parameter estimates and their standard errors. *Computational Statistics & Data Analysis*, 46:427–440, 2004.
79. J. R. Magnus and H. Neudecker. Symmetry, 0–1 matrices and Jacobians: A review. *Econometric Theory*, 2:157–190, 1986.
80. J. R. Magnus and H. Neudecker. *Matrix Differential Calculus with Applications in Statistics and Econometrics.* Wiley, Chichester, 1988.
81. W. M. Mason, G. Y. Wong, and B. Entwisle. Contextual analysis through the multilevel linear model. *Sociological Methodology*, 14:72–103, 1983.
82. P. McCullagh and J. A. Nelder. *Generalized Linear Models*, 2nd edition. Chapman & Hall, London, 1989.
83. E. Meijer and J. Rouwendal. Measuring welfare effects in models with random coefficients. *Journal of Applied Econometrics*, 21:227–244, 2006.
84. B. O. Muthén and A. Satorra. Complex sample data in structural equation modeling. *Sociological Methodology*, 25:267–316, 1995.
85. L. K. Muthén and B. O. Muthén. *Mplus User's Guide*, 4th edition. Muthén & Muthén, Los Angeles, 1998–2006.
86. J. Nocedal and S. J. Wright. *Numerical Optimization.* Springer, New York, 1999.
87. D. Pfeffermann. The role of sampling weights when modeling survey data. *International Statistical Review*, 61:317–337, 1993.
88. D. Pfeffermann, C. J. Skinner, D. J. Holmes, H. Goldstein, and J. Rasbash. Weighting for unequal selection probabilities in multilevel models. *Journal of the Royal Statistical Society, Series B*, 60:23–56, 1998. (with discussion)
89. J. C. Pinheiro and D. M. Bates. *Mixed-Effects Models in S and S-PLUS.* Springer, New York, 2000.
90. J. C. Pinheiro, D. M. Bates, S. DebRoy, and D. Sarkar. *The nlme Package*, 2006. URL `http://cran.r-project.org`
91. R. F. Potthoff and S. N. Roy. A generalized multivariate analysis of variance model useful especially for growth curve problems. *Biometrika*, 51:313–326, 1964.
92. R. F. Potthoff, M. A. Woodbury, and K. G. Manton. "Equivalent sample size" and "equivalent degrees of freedom" refinements for inference using survey weights under superpopulation models. *Journal of the American Statistical Association*, 87:383–396, 1992.
93. R Development Core Team. *R: A language and environment for statistical computing.* R Foundation for Statistical Computing, Vienna, Austria, 2006. URL `http://www.r-project.org`
94. S. Rabe-Hesketh and A. Skrondal. Multilevel modelling of complex survey data. *Journal of the Royal Statistical Society, Series A*, 169:805–827, 2006.

95. S. Rabe-Hesketh, A. Skrondal, and A. Pickles. GLLAMM manual. Working Paper 160, U.C. Berkeley Division of Biostatistics, Berkeley, CA, 2004. (Downloadable from `http://www.bepress.com/ucbbiostat/paper160/`)
96. C. R. Rao. *Linear Statistical Inference and its Applications*, 2nd edition. Wiley, New York, 1973.
97. J. Rasbash, F. Steele, W. J. Browne, and B. Prosser. *A User's Guide to MLwiN. Version 2.0.* Centre for Multilevel Modelling, University of Bristol, Bristol, UK, 2005.
98. S. W. Raudenbush. Reexamining, reaffirming, and improving application of hierarchical models. *Journal of Educational and Behavioral Statistics*, 20:210–220, 1995.
99. S. W. Raudenbush and A. S. Bryk. Empirical Bayes meta-analysis. *Journal of Educational Statistics*, 10:75–98, 1985.
100. S. W. Raudenbush and A. S. Bryk. A hierarchical model for studying school effects. *Sociology of Education*, 59:1–17, 1986.
101. S. W. Raudenbush and A. S. Bryk. *Hierarchical Linear Models: Applications and Data Analysis Methods*, 2nd edition. Sage, Thousand Oaks, CA, 2002.
102. S. W. Raudenbush, A. S. Bryk, Y. F. Cheong, and R. Congdon. *HLM 6: Hierarchical Linear and Nonlinear Modeling.* Scientific Software International, Chicago, 2004.
103. G. K. Robinson. That BLUP is a good thing: the estimation of random effects. *Statistical Science*, 6:15–51, 1991. (with discussion)
104. W. S. Robinson. Ecological correlations and the behavior of individuals. *Sociological Review*, 15:351–357, 1950.
105. J. Rouwendal and E. Meijer. Preferences for housing, jobs, and commuting: A mixed logit analysis. *Journal of Regional Science*, 41:475–505, 2001.
106. SAS/Stat. *SAS/Stat User's Guide, version 9.1.* SAS Institute, Cary, NC, 2004.
107. M. D. Schluchter. BMDP5V – Unbalanced repeated measures models with structured covariance matrices. Technical Report 86, BMDP Statistical Software, Los Angeles, 1988.
108. C. J. Skinner. Domain means, regression and multivariate analysis. In C. J. Skinner, D. Holt, and T. M. F. Smith, editors, *Analysis of Complex Surveys*, pages 59–87. Wiley, New York, 1989.
109. C. J. Skinner, D. Holt, and T. M. F. Smith, editors. *Analysis of Complex Surveys.* Wiley, New York, 1989.
110. A. Skrondal and S. Rabe-Hesketh. *Generalized Latent Variable Modeling: Multilevel, Longitudinal, and Structural Equation Models.* Chapman & Hall/CRC, Boca Raton, FL, 2004.
111. T. A. B. Snijders and R. J. Bosker. *Multilevel Analysis: An Introduction to Basic and Advanced Multilevel Modeling.* Sage, London, 1999.
112. D. J. Spiegelhalter, N. G. Best, B. P. Carlin, and A. van der Linde. Bayesian measures of model complexity and fit. *Journal of the Royal Statistical Society, Series B*, 64:583–639, 2002. (with discussion)
113. D. J. Spiegelhalter, A. Thomas, N. G. Best, and D. Lunn. *WinBUGS User Manual, Version 1.4.* MRC Biostatistics Unit, Cambridge, UK, 2003.
114. SPSS. *SPSS Advanced Models™ 15.0 Manual.* SPSS, Chicago, 2006.

115. StataCorp. *Stata Statistical Software: Release 9.* Stata Corporation, College Station, TX, 2005.
116. J. L. F. Strenio, H. I. Weisberg, and A. S. Bryk. Empirical Bayes estimation of individual growth curve parameters and their relationship to covariates. *Biometrics*, 39:71–86, 1983.
117. P. A. V. B. Swamy. *Statistical Inference in a Random Coefficient Model.* Springer, New York, 1971.
118. R. L. Tate and Y. Wongbundhit. Random versus nonrandom coefficient models for multilevel analysis. *Journal of Educational Statistics*, 8:103–120, 1983.
119. G. Van Landeghem, P. Onghena, and J. Van Damme. The effect of different forms of centering in hierarchical linear models re-examined. Technical Report 2001-04, Catholic University of Leuven, University Centre for Statistics, Leuven, Belgium, 2001.
120. G. Verbeke and E. Lesaffre. Large sample properties of the maximum likelihood estimators in linear mixed models with misspecified random-effects distributions. Technical Report 1996.1, Catholic University of Leuven, Biostatistical Centre for Clinical Trials, Leuven, 1996.
121. G. Verbeke and E. Lesaffre. The effect of misspecifying the random-effects distribution in linear mixed models for longitudinal data. *Computational Statistics & Data Analysis*, 23:541–556, 1997.
122. G. Verbeke and G. Molenberghs. *Linear Mixed Models for Longitudinal Data.* Springer, New York, 2000.
123. T. Wansbeek and E. Meijer. *Measurement Error and Latent Variables in Econometrics.* North-Holland, Amsterdam, 2000.
124. J. M. Wooldridge. Asymptotic properties of weighted M-estimators for variable probability samples. *Econometrica*, 67:1385–1406, 1999.
125. J. M. Wooldridge. Asymptotic properties of weighted M-estimators for standard stratified samples. *Econometric Theory*, 17:451–470, 2001.
126. J. M. Wooldridge. *Econometric Analysis of Cross Section and Panel Data.* MIT Press, Cambridge, MA, 2002.
127. J. M. Wooldridge. Inverse probability weighted M-estimators for sample selection, attrition, and stratification. *Portuguese Economic Journal*, 1:117–139, 2002.
128. R. Xu. Measuring explained variation in linear mixed effects models. *Statistics in Medicine*, 22:3527–3541, 2003.

2

Bayesian Multilevel Analysis and MCMC

David Draper

Department of Applied Mathematics and Statistics, Baskin School of Engineering, University of California Santa Cruz

2.1 Introduction

Multilevel models have gained wide acceptance over the past 20 years in many fields, including education and medicine [e.g., 26, 43, 45], as an important methodology for dealing appropriately with nested or clustered data. The idea of conducting an experiment in such a way that the levels of one factor are nested inside those of another goes back all the way to the initial development, in the 1920s, of the analysis of variance (ANOVA; [34]), so there's nothing new in working with nested data; the novelty in recent decades is in the methods for fitting multilevel models, the ability to work with data possessing many levels of nesting and multiple predictor variables at any or all levels, and an increased flexibility in distributional assumptions. The earliest designs featured one-way ANOVA models such as[1]

$$
\begin{aligned}
y_{ij} = \mu + \alpha_j^T + a_{ij}^S, \quad j = 1, \dots, J, \quad i = 1, \dots, n_j, \\
\sum_{j=1}^J n_j = N, \quad \sum_{j=1}^J \alpha_j^T = 0, \quad a_{ij}^S \stackrel{iid}{\sim} \mathcal{N}(0, \sigma_S^2),
\end{aligned}
\tag{2.1}
$$

in which the subject factor S (indexed by i), treated as *random*, is nested within the treatment factor T (indexed by j), treated as *fixed*. Under the normality assumption in (2.1) such models required little for the (frequentist) estimation of the parameters μ, σ_S^2, and the α_j^T beyond minor extensions of the least squares methods known since the time of Legendre [51] and Gauss [36]. Regarding the treatment factor as random, however, by changing the α_j^T to $a_j^T \stackrel{iid}{\sim} \mathcal{N}(0, \sigma_T^2)$ (with the a_j^T and a_{ij}^S mutually independent), created substantial new difficulties in model fitting—indeed, as late as the 1950s, one

[1] Note that random variables are not underlined in this chapter.

J. de Leeuw, E. Meijer (eds.), *Handbook of Multilevel Analysis*,
© Springer 2008

of the leading estimation methods [e.g., 65] was based on *unbiased* estimates of the *variance components* σ_T^2 and σ_S^2, the former of which can easily, and embarrassingly, go negative when σ_T^2 is small. Fisher [33] had much earlier pioneered the use of *maximum likelihood* estimation, but before the widespread use of fast computers this approach was impractical in *random-effects* and *mixed* models such as

$$
\begin{aligned}
y_{ij} &= \beta_0 + \beta_1 (x_{ij} - \bar{x}) + a_j^T + a_{ij}^S, \quad j = 1, \ldots, J, \quad i = 1, \ldots, n_j, \\
&\sum_{j=1}^{J} n_j = N, \quad a_j^T \stackrel{iid}{\sim} \mathcal{N}(0, \sigma_T^2), \quad a_{ij}^S \stackrel{iid}{\sim} \mathcal{N}(0, \sigma_S^2)
\end{aligned}
\tag{2.2}
$$

(where the x_{ij} are fixed known values of a predictor variable and $\bar{x}$ is the sample mean of this variable), because the likelihood equations in such models can only be solved iteratively. Multilevel modeling entered a new phase in the 1980s, with the development of computer programs such as `ML3`, `VARCL`, and `HLM` using likelihood-based estimation approaches based on iterative generalized least squares [42], Fisher scoring [52], and the EM algorithm [e.g., 15], respectively. In particular, the latest versions of `MLwiN` (the successor to `ML3`; [60]) and `HLM` [66] have worldwide user bases in the social and biomedical sciences numbering in the thousands, and likelihood-based fitting of at least some multilevel models is also now obtainable in more general-purpose statistical packages such as `SAS` [64] and `Stata` [71].

However, the use of the likelihood function alone in multilevel modeling can lead to the following technical problems:

- *Maximum-likelihood estimates* (MLEs) and their *(estimated asymptotic) standard errors* (SEs) can readily be found by iterative means for the parameters in Gaussian multilevel models such as (2.2), but interval estimates of those parameters can be problematic when J, the number of level-2 units, is small. For example, simple "95%" intervals of the form $\hat{\sigma}_T^2 \pm 1.96\, \widehat{\text{se}}(\hat{\sigma}_T^2)$ (based on the large-sample Gaussian repeated-sampling distribution of $\hat{\sigma}_T^2$) can go negative and can have actual coverage levels substantially below 95%, and other methods based only on $\hat{\sigma}_T^2$ and $\widehat{\text{se}}(\hat{\sigma}_T^2)$ (which are the default outputs of packages such as `MLwiN` and `HLM`) are not guaranteed to do much better, in part because (with small sample sizes) the MLE of σ_T^2 can be 0 even when the true value of σ_T^2 is well away from 0 [e.g., 12].
- The situation becomes even more difficult when the outcome variable y in the multilevel model is dichotomous rather than Gaussian, as in *random-effects logistic regression* (RELR) models such as

$$
\begin{aligned}
(y_{ij} \mid p_{ij}) &\stackrel{\text{indep}}{\sim} \text{Bernoulli}(p_{ij}), \quad \text{where} \\
\text{logit}(p_{ij}) &= \beta_0 + \beta_1 (x_{ij} - \bar{x}) + u_j, \quad u_j \stackrel{iid}{\sim} \mathcal{N}(0, \sigma_u^2).
\end{aligned}
\tag{2.3}
$$

Here the likelihood methods that work with Gaussian outcomes fail; the likelihood function itself cannot even be evaluated without integrating out the random effects u_j from (2.3). Available software such as MLwiN fits RELR models via *quasi-likelihood* methods [7]; this approach to fitting nonlinear models such as (2.3) proceeds by linearizing the second line of the model via Taylor series expansion, yielding *marginal* and *penalized* quasi-likelihood (MQL and PQL) estimates according to the form of the expansion used. These are not full likelihood methods and would be better termed *likelihood-based* techniques. Browne and Draper [12] have shown that the actual coverage of nominal 95% interval estimates with this approach in RELR models can be far less than 95% when the intervals are based only on MQL and PQL point estimates and their (estimated asymptotic) SEs; see Section 2.3.3 below. *Calibration* results of this kind for other methods which attempt to more accurately approximate the actual likelihood function [e.g., 1, 50, 53, 57, 61] are sparse and do not yet fully cover the spectrum of models in routine use, and user-friendly software for many of these methods is still hard to come by.

This chapter concerns the *Bayesian* approach to fitting multilevel models, which (a) attempts to remedy the above problems (though not without introducing some new challenges of its own) and (b) additionally provides a mechanism for the formal incorporation of any *prior information* which may be available about the parameters of the multilevel model of interest external to the current data set. A computing revolution based on *Markov chain Monte Carlo* (MCMC) methods, and the availability of much faster (personal) computers, have together made the Bayesian fitting of multilevel models increasingly easier since the early 1990s. In this chapter I (1) describe the basic outline of a Bayesian analysis (multilevel or not), in the context of a case study, (2) motivate the need for simulation-based computing methods, (3) describe MCMC methods in general and their particular application to multilevel modeling, (4) discuss MCMC diagnostic methods (to ensure accuracy of the computations), and (5) present an MCMC solution to the multilevel modeling case study.

2.1.1 A Case Study

In the spring of 1993 a survey was taken of bicycle and other traffic in the vicinity of the University of California, Berkeley, campus [37]. Ten city blocks were selected at random in each of the six cells of a 2×3 table that cross-tabulates presence or absence of a bike route on a street against whether the street was residential, fairly busy, or busy. This street classification was made before the data were gathered. Each block was observed for one hour at the same time and day of the week on a randomly chosen day, and a record was

Table 2.1 Raw data from the Berkeley traffic survey. Entries are of the form p/n, where p is the proportion of bicycle traffic (PBT) and n is the number of vehicles in each block. Data from two (No, Residential) blocks are missing, and one bicycle was added to the starred block in the (No, Fairly Busy) cell to avoid a zero PBT value.

Bike			Street Type			
Route?	Residential		Fairly Busy		Busy	
	.216/ 74	.091/ 99	.216/ 37	.078/ 450	.037/1605	.033/1550
	.172/ 58	.186/ 70	.068/ 456	.311/ 61	.035/1656	.105/ 562
Yes	.156/122	.260/ 77	.174/ 218	.065/ 722	.115/ 460	.044/1562
	.173/104	.132/129	.066/ 664	.091/ 481	.042/1626	.034/1766
	.114/308	.462/119	.382/ 76	.038/ 480	.130/ 547	.077/ 815
	.096/125	.053/ 19	.018/ 567	.033/1301	.006/1256	.007/1255
	.125/ 16	.083/ 48	.010/ 504	.023/ 615	.004/1602	.005/1774
No	.041/217	.095/ 74	.048/1221	.021/ 715	.015/1309	.024/2559
	.237/ 38	.049/162	.011/ 91*	.041/1140	.013/2377	.024/3176
			.034/1510	.029/1118	.007/1932	.011/2343

kept of the numbers of bicycles and other vehicles traveling in the sampled blocks. The data for two of the residential blocks without a bike route were lost. The study was observational—for instance, no attempt was made to assign bike routes to streets at random to see what that would do to vehicular traffic in Berkeley—but interest nevertheless focuses on the "effects" of (a) having or not having a bike route and (b) street type on the proportion of bicycle traffic (PBT).

Table 2.1 presents the raw data from this study, and Table 2.2 offers summaries of the means and standard deviations (SDs) of the block-level PBT values on the raw and logit (log-odds) scales. It's clear from these tables that

- the street type classifications are fairly accurate as to volume of traffic, although there is overlap; for instance, 8 of the 18 residential streets were busier during the chosen observation periods than 4 of the 20 fairly busy streets;
- street type and bike route both have strong effects on PBT in the intuitively reasonable directions (e.g., it's 16 times more likely that a vehicle will be a bicycle on residential streets with bike routes than busy streets without them), although there is substantial (unexplained) between-block variation within cells of the 2×3 table;
- there is a strong relationship between the cell means and SDs on the raw PBT scale, and this is substantially diminished when the log-odds of the

Table 2.2 Summaries of the Berkeley traffic survey data. Entries are means and (SDs) of the block-level PBT values, on the raw (top table) and logit (bottom table) scales.

Bike Route?	Residential	Street Type Fairly Busy	Busy	Total
Yes	.196 (.105)	.149 (.112)	.065 (.038)	.137 (.106)
No	.097 (.063)	.026 (.014)	.012 (.007)	.041 (.049)
Total	.152 (.100)	.087 (.103)	.038 (.038)	.091 (.096)

Bike Route?	Residential	Street Type Fairly Busy	Busy	Total
Yes	−1.50 (.598)	−2.01 (.917)	−2.81 (.598)	−2.11 (.884)
No	−2.38 (.639)	−3.71 (.541)	−4.63 (.634)	−3.66 (1.08)
Total	−1.89 (.746)	−2.86 (1.14)	−3.72 (1.11)	−2.85 (1.25)

PBT values are considered, suggesting either additive modeling on the logit scale or multiplicative modeling on the raw scale;[2] and

- there is a fairly strong interaction between street type and bike route (e.g., on the logit scale the effect of having or not having a bike route is about half as large for residential streets as it is for busier streets).

These data have a multilevel (or *hierarchical*) character: bike route R and street type T are fully crossed, city block B is nested in $R \times T$, and vehicle V is nested in B (and therefore also in $R \times T$). It's natural in this study to treat R and T as fixed factors (at 2 and 3 levels, respectively) and to regard B and V as random. Letting y_{ijkl} be 1 if vehicle i observed in block j of bike route status k and street type l is a bicycle and 0 otherwise, one possible model for these data is

$$\begin{aligned} (y_{ijkl} \mid p_{ijkl}) &\stackrel{\text{indep}}{\sim} \text{Bernoulli}(p_{ijkl}), \quad \text{where} \\ \text{logit}(p_{ijkl}) &= \mu + \alpha_k^R + \alpha_l^T + \alpha_{kl}^{RT} + a_{jkl}^B, \quad a_{jkl}^B \stackrel{iid}{\sim} \mathcal{N}(0, \sigma_B^2), \end{aligned} \tag{2.4}$$

with appropriate side conditions on the fixed effects such as $\sum_{k=1}^K \alpha_k^R = 0$. The normal distribution for the random effects a_{jkl}^B in (2.4) and the choice of the logistic link function in this RELR model are both conventional assumptions, not automatically motivated by the real-world details of this case study, and

[2] 40 of the 58 PBT values are less than 0.1, and for p close to 0, $\text{logit}(p) \stackrel{\Delta}{=} \log(p/(1-p)) \doteq \log(p)$; thus the log transform, which is routinely used to produce approximate additivity of multiplicative (raw-scale) treatment effects, and the logit have almost the same effect here.

would require checking (see Chapter 3 of this volume for multilevel diagnostic methods).

An alternative formulation equivalent to (2.4) would define y_{jkl} to be the number of bicycles among the n_{jkl} vehicles in block j with $R \times T$ status (k, l) and would then take

$$\begin{aligned} (y_{jkl} \mid p_{jkl}) &\overset{\text{indep}}{\sim} \text{Binomial}(n_{jkl}, p_{jkl}), \quad \text{where} \\ \text{logit}(p_{jkl}) &= \mu + \alpha_k^R + \alpha_l^T + \alpha_{kl}^{RT} + a_{jkl}^B, \quad a_{jkl}^B \overset{iid}{\sim} \mathcal{N}(0, \sigma_B^2), \end{aligned} \tag{2.5}$$

with the analysis conditional on the observed n_{jkl}. One final class of models for these data begins as in the first line of (2.5) but makes distributional assumptions about the p_{jkl} on the raw scale, e.g., by replacing the second line of (2.5) by

$$p_{jkl} \sim \text{Beta}(\alpha_{jkl}, \beta_{jkl}) \tag{2.6}$$

and then linking the α_{jkl} and β_{jkl} values for different (k, l) to the levels of the R and T factors.[3] Exact small-sample likelihood inferences for functions of the α_{jkl} and β_{jkl} such as the mean of p_{jkl}, $\alpha_{jkl}/(\alpha_{jkl} + \beta_{jkl})$, would be difficult in this model, but (as will be seen below) such inferences are straightforward with Bayesian fitting via MCMC methods.

2.1.2 Prior, Likelihood, Posterior, and Predictive Distributions

To motivate the ingredients of a Bayesian analysis (not necessarily of multilevel data), consider the first residential city block with a bike route in Table 2.1, where $s = y_{111} = 16$ of the $n = n_{111} = 74$ vehicles observed were bicycles (with the data gathered, say, on a Tuesday afternoon from 3 to 4 pm), and suppose that these were the only data available. For ease of notation in this section let the individual indicators $(y_{1,111}, \ldots, y_{74,111})$ of bicycle-or-not be denoted $(b_1, \ldots, b_n)$. In the predictivist approach to Bayesian statistics that makes the most sense to me [23], I'm encouraged to consider the binary observables b_i before the data have arrived and to quantify my uncertainty about them by means of a joint (predictive) probability distribution, $p(b_1, \ldots, b_n)$. I notice that my predictive uncertainty is the same for (say) b_{17} as it is for (say) b_{31}, which is another way of saying that my $p(b_1, \ldots, b_n)$ would be unchanged under any permutation of the indices $i = 1, \ldots, n$; de Finetti called this a judgment of *exchangeability*[4] of (my predictive distribution for)

[3] Equation (2.5) accomplishes something similar by means of what might be called the *logit-normal* distribution, which models the behavior of p when logit(p) is assumed to follow a $\mathcal{N}(\mu, \sigma^2)$ distribution. The logit-normal family exhibits a range of shapes similar to that of the Beta family.

[4] See Draper et al. [28] for an exploration of how data are used to make such judgments in practice in more complicated situations.

the b_i. de Finetti [22] showed that if I'm willing to regard $(b_1, \ldots, b_n)$ as the beginning of an indefinitely long sequence of exchangeable binary observables, this judgment is functionally equivalent to assuming the hierarchical model

$$\begin{aligned} \theta &\sim p(\theta) \\ (b_i \mid \theta) &\stackrel{iid}{\sim} \text{Bernoulli}(\theta) \end{aligned} \tag{2.7}$$

(for $i = 1, \ldots, n$), or equivalently the model

$$\begin{aligned} \theta &\sim p(\theta) \\ (s \mid \theta) &\sim \text{Binomial}(n, \theta), \end{aligned} \tag{2.8}$$

where in both models n is treated as fixed and known. Here

- θ is interpretable both (a) as the marginal probability $\Pr(b_i = 1)$ that any vehicle in the indefinitely long sequence is a bicycle and (b) as the long-run average of the b_i, which could also be thought of in this case study as the underlying PBT value for this city block (previously denoted p_{111}) during other periods (e.g., Tuesday afternoons from 3–4 pm) judged similar to the day on which the data were gathered; and
- logically θ is a fixed (unknown) constant, but to use model (2.7) or (2.8) it's necessary to regard it as a random quantity possessing a probability distribution $p(\theta)$. This is my *prior* distribution for θ, and represents an opportunity to quantitatively summarize what (if anything) I know about θ external to the present data set.

Notice that in de Finetti's formulation θ is not the primitive construct; prediction of future observables is the fundamental operation, and θ arises as a quantity which makes this prediction easier, by rendering the b_i conditionally IID[5] given θ.

Once the b_i are observed my state of knowledge about θ will change. Denoting the data vector by $\boldsymbol{b} = (b_1, \ldots, b_n)$, it can be shown [e.g., 2] that—to avoid internal inconsistencies in my probability assessments—this new state of knowledge must be given by the (conditional) *posterior* distribution $p(\theta \mid \boldsymbol{b})$ for θ given $\boldsymbol{b}$, and that passing from the prior to posterior states of knowledge must be accomplished via *Bayes' Theorem*:[6]

[5] Exchangeability and IID are not the same thing. IID implies exchangeability, and exchangeable random variables do have identical marginal distributions, but they're not independent: If you didn't know anything about θ, the knowledge of how some of the b_i turn out would help you to predict the other b_i, whereas if you somehow knew the exact value of θ in (2.7) or (2.8), the b_i become conditionally independent given this knowledge, because information about any of the b_i (given θ) would be irrelevant in predicting any of the other b_i.

[6] Here, $p(\cdot)$ denotes a probability density or probability mass function, i.e., in this chapter the same symbol is used for a distribution, e.g., $p(\theta) = \text{Beta}(\alpha, \beta)$, and its density function, e.g., $p(\theta) = c\,\theta^{\alpha-1}\,(1-\theta)^{\beta-1}$.

$$p(\theta \mid \boldsymbol{b}) = \frac{p(\theta)\, p(\boldsymbol{b} \mid \theta)}{p(\boldsymbol{b})}. \tag{2.9}$$

After the data have arrived the left side of (2.9) is a probability distribution for θ with $\boldsymbol{b}$ a known quantity, so the same must be true of the right side. This means that (i) the $p(\boldsymbol{b})$ term in the denominator is just a constant[7] and (ii) $p(\boldsymbol{b} \mid \theta)$, which before the data were gathered would be recognizable as the joint sampling distribution $p(b_1, \ldots, b_n \mid \theta)$ for the b_i given θ, needs to be interpreted, after the data are known, as a function of θ for fixed $\boldsymbol{b}$. Fisher [33] called this the *likelihood function* $l(\theta \mid \boldsymbol{b})$; more precisely he noticed that this function is only determined up to a constant multiple and defined $l(\theta \mid \boldsymbol{b}) = c\, p(\boldsymbol{b} \mid \theta)$ (here and below I'll use $c > 0$ as a generic positive constant). In Bayesian work it's often useful to choose this constant so that the likelihood integrates to 1; call the result the *likelihood distribution* for θ given $\boldsymbol{b}$. Then (2.9) can be rewritten as

$$p(\theta \mid \boldsymbol{b}) = c\, p(\theta)\, l(\theta \mid \boldsymbol{b}). \tag{2.10}$$

This provides a prescription for calculating a posterior distribution when the parameter θ in (2.10) is univariate: multiply the prior and likelihood distributions pointwise (in θ) and normalize the product to integrate to 1.

The first step in applying (2.10) in the case of the model (2.7) is to compute the likelihood distribution, which is obtained by writing out the joint sampling distribution $p(b_1, \ldots, b_n \mid \theta)$ for the Bernoulli model and reinterpreting it as a function of θ for fixed $\boldsymbol{b}$. Here, because the b_i are conditionally IID given θ, this is just the product of the marginal Bernoulli sampling distributions

$$p(b_i \mid \theta) = \begin{cases} \theta & \text{if } b_i = 1 \\ 1 - \theta & \text{if } b_i = 0, \end{cases}$$

and, since this can be written $p(b_i \mid \theta) = \theta^{b_i} (1 - \theta)^{1 - b_i}$, the result is

$$p(b_1, \ldots, b_n \mid \theta) = \prod_{i=1}^{n} p(b_i \mid \theta) = \prod_{i=1}^{n} \theta^{b_i} (1 - \theta)^{1 - b_i} = \theta^s (1 - \theta)^{n - s}, \tag{2.11}$$

where $s = \sum_{i=1}^{n} b_i$ counts the number of bicycles among the n vehicles. Thus in this case[8] $l(\theta \mid \boldsymbol{b}) = c\, \theta^s (1 - \theta)^{n - s}$, with c chosen to make l a density in θ for fixed s. This is recognizable as the Beta$(s + 1, n - s + 1)$ distribution. It's worth noting that the likelihood here depends on the data vector $\boldsymbol{b}$ only through s; according to Fisher's [33] definition, this makes s a *sufficient statistic* for

[7] In fact, it's a *normalizing* constant, determined by the condition that for all possible data vectors $\boldsymbol{b}$, $\int_0^1 p(\theta \mid \boldsymbol{b})\, d\theta = 1$.

[8] The same result is immediate from (2.8): by definition the Binomial sampling distribution is $p(s \mid \theta) = c\, \theta^s (1 - \theta)^{n - s}$ with $c = n! / \big(s!\, (n - s)!\big)$.

θ in the Bernoulli/Binomial model,[9] and this additionally implies that the posterior for θ given $\boldsymbol{b}$ also depends only on s: $p(\theta \mid \boldsymbol{b}) = p(\theta \mid s)$.

What should I take for my prior distribution $p(\theta)$? As long ago as in the work of Laplace [49] it was observed that in this problem a computational simplification arises from assuming that the prior has the same Beta form as the likelihood: if $p(\theta) = c\,\theta^{\alpha-1}(1-\theta)^{\beta-1}$ for some $\alpha, \beta > 0$ then

$$\begin{aligned} p(\theta \mid \boldsymbol{b}) &= c\left[c\,\theta^{\alpha-1}(1-\theta)^{\beta-1}\right]\left[c\,\theta^{s}(1-\theta)^{n-s}\right] \\ &= c\,\theta^{\alpha+s-1}(1-\theta)^{\beta+n-s-1} \\ &= \text{Beta}(\alpha+s, \beta+n-s). \end{aligned}$$

The Beta prior is said to be *conjugate* to the Bernoulli/Binomial likelihood,[10] and this choice of a *conjugate prior* leads to a simple updating rule:

$$\left.\begin{aligned} \theta &\sim \text{Beta}(\alpha_0, \beta_0) \\ (b_i \mid \theta) &\stackrel{iid}{\sim} \text{Bernoulli}(\theta), \\ i &= 1, \ldots, n \end{aligned}\right\} \Rightarrow (\theta \mid \boldsymbol{b}) = (\theta \mid s) \sim \text{Beta}(\alpha_0+s, \beta_0+n-s). \quad (2.12)$$

It's important to note that this line of reasoning has only demonstrated that the Beta distribution is computationally convenient, not necessarily that it's scientifically compelling (by which I mean an accurate reflection of my prior information), although the Beta family does exhibit a wide variety of (unimodal and U-shaped[11]) behaviors as α and β range freely over $(0, \infty)$.

The choice of a conjugate prior brings with it interpretational as well as computational advantages. For example, the mean of the Beta(α_0, β_0) distribution is $\alpha_0/(\alpha_0+\beta_0)$; from this, having used a Beta prior, it's possible to write the posterior mean $E(\theta \mid \boldsymbol{b})$ as a weighted average of the prior mean $E(\theta)$ and the data mean $\bar{b} = s/n$:

---

9 Fisher was interested in dimensionality reduction, and it appealed to him that (conditional on the "truth" of model (2.7)) you don't have to carry around the full n-dimensional data vector $\boldsymbol{b}$ to draw inferences about θ; the one-dimensional summary s is enough. In fact he would have called s a *minimal* sufficient statistic, meaning that all other sufficient statistics are of dimensionality at least as large as that of s. For example, $(\sum_{i=1}^{k} b_i, \sum_{i=k+1}^{n} b_i)$ for any $k = 1, \ldots, n-1$ is sufficient but not minimal sufficient here.

10 Informally, a prior $p(\theta)$ is conjugate to a likelihood $l(\theta \mid \boldsymbol{y})$ if the resulting posterior $p(\theta \mid \boldsymbol{y})$ has the same distributional form as $p(\theta)$; see Bernardo and Smith [2] for a formal definition.

11 Other shapes can be achieved by using Beta distributions as building blocks; in fact, Diaconis and Ylvisaker [24] have shown that *all possible* prior distributions for parameters of models that can be expressed as members of the *exponential family* can be approximated arbitrarily closely by *mixtures* of conjugate priors (see, e.g., Bernardo and Smith [2] for a thorough discussion of the exponential family).

$$
\begin{aligned}
E(\theta \mid \boldsymbol{b}) &= \frac{\alpha_0 + s}{\alpha_0 + \beta_0 + n} \\
&= \left(\frac{\alpha_0}{\alpha_0 + \beta_0} \right) \left(\frac{\alpha_0 + \beta_0}{\alpha_0 + \beta_0 + n} \right) + \left(\frac{s}{n} \right) \left(\frac{n}{\alpha_0 + \beta_0 + n} \right) \\
&= E(\theta) \left(\frac{\alpha_0 + \beta_0}{\alpha_0 + \beta_0 + n} \right) + \bar{b} \left(\frac{n}{\alpha_0 + \beta_0 + n} \right).
\end{aligned}
$$

Thus the data mean s/n receives n votes and the prior mean gets $\alpha_0 + \beta_0$ votes in the posterior compromise between data and prior information, and since the data sample size is n it's natural to refer to $n_0 = \alpha_0 + \beta_0$ as the *prior sample size*: as far as the prior-to-posterior updating is concerned it's as if the prior information were equivalent to a *prior data set* consisting of α_0 1s and β_0 0s which is merged with the *current data set* consisting of s 1s and $(n - s)$ 0s to yield the *posterior data set*.[12]

Consider two seemingly rather different sets of prior information/beliefs in this problem:

- In the first set, before the data arrive I'd be quite surprised if θ, the proportion of bicycle traffic in the residential city block with a bike route at issue here, were less than 5% or greater than 50%.
- In the second set, before $\boldsymbol{b}$ is observed I wish to express comparative ignorance about θ across the entire range of its possible values from 0 to 1.

One way to make the first set of prior information/beliefs operational within the conjugate Beta family is to take the phrase "quite surprised" to mean, e.g., $\Pr(0.05 \le \theta \le 0.5) = 0.9$, and to split the remaining 10% of prior probability equally between the two tails, leading to the two equations

$$
\begin{aligned}
\Pr(\theta < 0.05) &= \int_0^{0.05} \text{Beta}(\theta; \alpha_0, \beta_0) \, \mathrm{d}\theta = 0.05, \\
\Pr(\theta > 0.5) &= \int_{0.5}^{1} \text{Beta}(\theta; \alpha_0, \beta_0) \, \mathrm{d}\theta = 0.05,
\end{aligned}
\tag{2.13}
$$

where $\text{Beta}(\theta; \alpha, \beta) = c\,\theta^{\alpha-1}(1-\theta)^{\beta-1}$ is the Beta density[13] with *hyperparameters* α and β. The equations in (2.13) may be solved numerically in a package such as `Maple` [74] or `R` [58] to yield $(\alpha_0, \beta_0) \doteq (2.0, 6.4)$. With this specification (a) the prior mean for θ is $2.0/(2.0 + 6.4) \doteq 0.24$, (b) its prior

[12] This idea provides a direct bridge between Bayesian and frequentist analyses of the same data: if I conduct the Bayesian analysis described here and instead you feed the posterior data set based on my prior into the likelihood machinery of Section 2.1.3 below, you and I will draw the same conclusions.

[13] The normalizing constant is $c = \Gamma(\alpha + \beta) / \big(\Gamma(\alpha)\, \Gamma(\beta)\big)$.

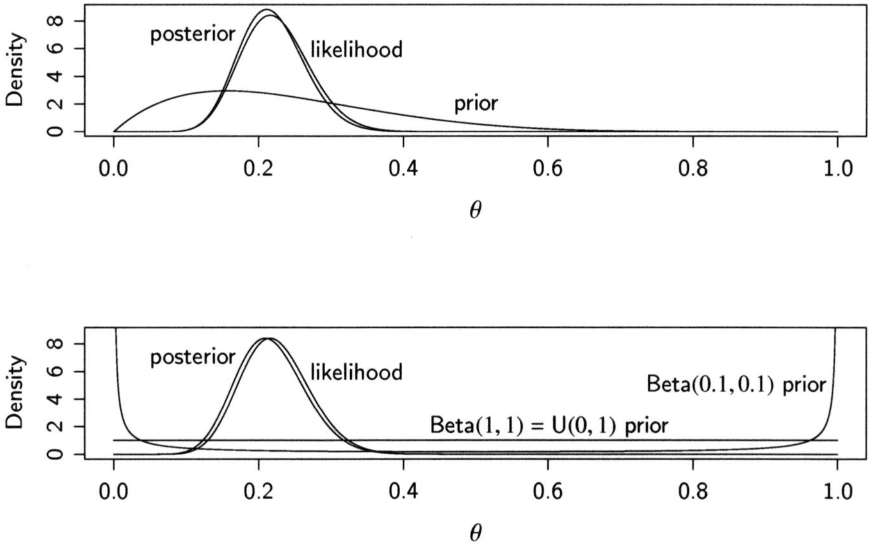

Fig. 2.1 Prior-to-posterior updating with two prior specifications in the single-city-block data set.

standard deviation[14] (SD) is about 0.14, and (c) the corresponding prior data set has $2.0 + 6.4 \doteq 8$ observations worth of data in it.

As for the second specification above in the Beta family, complete prior ignorance would correspond to a prior sample size of $n_0 = 0$, which would be obtained by letting both α_0 and β_0 tend to 0. The result is an *improper* prior which cannot be normalized to integrate to 1 (because its integral is infinite). However, any positive small choice of α_0 and β_0, e.g., $\alpha_0 = \beta_0 = 0.1$ or 0.5 or 1.0, will yield a proper prior with a small prior sample size, and all such choices should lead to similar posterior distributions because the data sample size ($n = 74$) is so much larger than the resulting n_0. The choice $\alpha_0 = \beta_0 = 1$ yields the familiar Uniform $U(0, 1)$ distribution, with prior mean 0.5 and SD $1/\sqrt{12} \doteq 0.29$.

Figure 2.1 illustrates prior-to-posterior updating with the two prior specifications examined above. The top panel plots the prior, likelihood, and posterior distributions with the first specification (Beta(2.0, 6.4)); the bottom panel plots the Beta(0.1, 0.1) and Beta(1, 1) $= U(0, 1)$ prior distributions, the likelihood (which is also the posterior with the $U(0, 1)$ prior), and the posterior with the Beta(0.1, 0.1) prior. For technical reasons the Beta(0.1, 0.1) distribution has regrettable asymptotic behavior near 0 and 1, but this does not affect the posterior because the likelihood is so close to zero in those regions that the spikes are irrelevant. It's clear from this figure and the form of (2.10) that *any prior that is locally (nearly) uniform in the region in which*

[14] The variance of the Beta(α, β) distribution is $\alpha\beta / \big((\alpha + \beta)^2 (\alpha + \beta + 1)\big)$.

the likelihood is appreciable will have negligible effect on the posterior.[15] The terms *noninformative*, *diffuse*, and *flat* tend to be used interchangeably to describe the second type of prior specification examined here (the meaning of the terms diffuse and flat is motivated by plots such as Fig. 2.1), but—given that all choices of prior specification embody one prior information base or another—"noninformative" seems a less satisfactory term.

It's also evident from Fig. 2.1 that a piece of prior information like that embodied in the first specification has little effect in this problem with a data sample size as large as $n = 74$. For example, the posterior mean, SD, and central 95% interval[16] for θ are 0.218, 0.045, and $(0.137, 0.313)$, respectively, under the Beta$(2.0, 6.4)$ prior; the corresponding values from the Beta$(0.1, 0.1)$ prior are 0.217, 0.048, and $(0.132, 0.317)$. Notice that the data sample size is large enough here that the likelihood and posterior distributions are fairly close to Gaussian (by the Central Limit Theorem (CLT)): an approximate 95% central interval for θ under the Beta$(2.0, 6.4)$ prior using this normal approximation would run from $0.218 - (1.96)(0.045) \doteq 0.124$ to $0.218 + (1.96)(0.045) \doteq 0.310$.

The language of this section has emphasized that the Bayesian approach to uncertainty quantification is *personal*, or *subjective*: my prior is mine, and may differ from yours, because you and I have different knowledge bases or we invoke different types of judgment to bring that knowledge to bear on the issue at hand. People have sometimes tried to argue in the past that personal judgments have no valid part to play in science, a position which would cast doubt on the relevance of Bayesian inference in scientific reasoning. But in situations of realistic complexity, particularly in the modeling of observational data, it's equally true—under all forms of statistical inference in current use, not just Bayesian—that my *likelihood* is mine, and may differ from yours: you and I may legitimately disagree in our judgments about what is appropriate to assume about the *structure* of the model[17] (consider, for example, the range of possibilities mentioned in (2.4) to (2.6) above for the full Berkeley traffic survey dataset), and it's not always possible to definitively settle these differences with data-driven model diagnostics. Personal judgment cannot be eradicated from complex statistical work in science; the laudable-sounding goal of "objectivity," as the word is generally used,[18] is unattainable in actual scientific practice. In light of this, attention should evidently focus—in all statistical inference, Bayesian or not[19]—on the *stability* or *robustness* of the

---

[15] This is the basis of what Edwards et al. [31] called the *stable estimation principle.*

[16] This can also readily be obtained numerically in `Maple` or `R`.

[17] See Draper [25] for a Bayesian approach to the quantification of structural model uncertainty.

[18] In Bayesian language saying that a probability assessment is objective just means that many people would agree with it, at least approximately.

[19] The Bayesian approach highlights the need/opportunity to quantify prior information about parameters conditional on model structure, and any such choice

mapping from assumptions to conclusions. Figure 2.1 is a simple example of one such robustness investigation: if I'm trying to quantify relatively diffuse prior beliefs, I need to convince myself (and you, if I want you to be a willing consumer of my conclusions) that a variety of plausible attempts at diffuse prior specification all lead to essentially the same findings, as is true in this case.[20]

This section began with a predictive motivation of Bayesian inference; a good way to end it is to examine how one would construct the *posterior predictive distribution* $p(b_{n+1} \mid \boldsymbol{b})$ for the next observable b_{n+1}, given the n binary indicators seen so far and assuming the Bayesian model (2.12). This predictive distribution initially seems a bit difficult to compute formally, although intuition says that (a) in this simple subset of the full case study it has to be a Bernoulli distribution and (b) the posterior predictive mean $E(b_{n+1} \mid \boldsymbol{b})$ cannot be anything other than the current posterior mean $(\alpha_0 + s)/(\alpha_0 + \beta_0 + n)$ of θ. The formal reasoning proceeds as follows.

(1) It's hard to say what I know about b_{n+1} by itself, but I know quite a lot about $(b_{n+1} \mid \theta)$, so it would help to introduce θ into the calculation. By the law of total probability

$$p(b_{n+1} \mid \boldsymbol{b}) = \int_0^1 p(b_{n+1}, \theta \mid \boldsymbol{b}) \, \mathrm{d}\theta.$$

(2) Now I want to move the θ to the other side of the conditioning bar. By the definition of conditional probability

$$\int_0^1 p(b_{n+1}, \theta \mid \boldsymbol{b}) \, \mathrm{d}\theta = \int_0^1 p(b_{n+1} \mid \theta, \boldsymbol{b}) \, p(\theta \mid \boldsymbol{b}) \, \mathrm{d}\theta. \tag{2.14}$$

(3) $p(\theta \mid \boldsymbol{b})$ in (2.14) is recognizable as the posterior distribution for θ given the data seen so far, namely Beta$(\alpha_0 + s, \beta_0 + n - s)$.

(4) (and this step is crucial) Given θ, there's no useful information in $\boldsymbol{b} = (b_1, \ldots, b_n)$ for predicting b_{n+1} (informally, "the past and the future are conditionally independent given the truth"), so $p(b_{n+1} \mid \theta, \boldsymbol{b}) = p(b_{n+1} \mid \theta)$, and this last expression is just the sampling distribution for observation $n + 1$, which under model (2.12) is Bernoulli(θ).

---

should be justified, but—given that likelihood analyses correspond to Bayesian answers with a particular form of diffuse prior—it would seem that the imperative to think about (and justify) priors is not unique to the Bayesian paradigm.

[20] In a careful analysis I should also plausibly vary my 0.9 translation of the phrase "quite surprised" and the θ values 0.05 and 0.5 in (2.13), for instance by increasing and decreasing each of these specifications by (say) 10% to see what happens, but you can see from Fig. 2.1 that all such variations would lead to essentially the same posterior here.

(5) Therefore
$$p(b_{n+1} \mid \boldsymbol{b}) = \int_0^1 p(b_{n+1} \mid \theta)\, p(\theta \mid \boldsymbol{b})\, \mathrm{d}\theta.$$
In other words, the posterior predictive distribution for b_{n+1} given $\boldsymbol{b}$ is a weighted average, or *mixture*, of Bernoulli(θ) sampling distributions, with the mixing weights given by the current posterior distribution $p(\theta \mid \boldsymbol{b})$ for θ given $\boldsymbol{b}$.

With $\alpha^* = \alpha_0 + s$ and $\beta^* = \beta_0 + n - s$, calculation in this example reveals that

$$\begin{aligned} p(b_{n+1} \mid \boldsymbol{b}) &= \int_0^1 \theta^{b_{n+1}} (1-\theta)^{1-b_{n+1}} \frac{\Gamma(\alpha^*+\beta^*)}{\Gamma(\alpha^*)\,\Gamma(\beta^*)} \theta^{\alpha^*-1} (1-\theta)^{\beta^*-1}\, \mathrm{d}\theta \\ &= \frac{\Gamma(\alpha^*+\beta^*)}{\Gamma(\alpha^*)\,\Gamma(\beta^*)} \int_0^1 \theta^{(\alpha^*+b_{n+1})-1} (1-\theta)^{(\beta^*-b_{n+1}+1)-1}\, \mathrm{d}\theta \qquad (2.15) \\ &= \left[\frac{\Gamma(\alpha^*+b_{n+1})}{\Gamma(\alpha^*)}\right] \left[\frac{\Gamma(\beta^*-b_{n+1}+1)}{\Gamma(\beta^*)}\right] \left[\frac{\Gamma(\alpha^*+\beta^*)}{\Gamma(\alpha^*+\beta^*+1)}\right]. \end{aligned}$$

Recalling that for any real number x, $\Gamma(x+1)/\Gamma(x) = x$, (2.15) agrees with intuition: for example, $\Pr(b_{n+1} = 1 \mid \boldsymbol{b}) = E(b_{n+1} \mid \boldsymbol{b}) = \alpha^*/(\alpha^*+\beta^*)$. With any of the prior distributions examined above, I predict that the next vehicle in this block on the sampled day of the week and time of day will be a bicycle with probability about 0.22.

2.1.3 A Comparison with Likelihood Inference

A likelihood inferential[21] analysis of the single-city-block data would begin by computing the MLE for θ, which may be found by maximizing either the likelihood function or its logarithm (the latter tends to be mathematically easier and more numerically stable to work with, since likelihood functions like (2.11) are typically products of a (possibly large) number of values not far from zero). Here $\log l(\theta \mid s) = \log c + s \log \theta + (n-s)\log(1-\theta)$, a concave function with a single maximum at the value of θ for which

$$\frac{\partial}{\partial\theta} \log l(\theta \mid s) = \frac{s}{\theta} - \frac{n-s}{1-\theta} = 0, \quad \text{namely} \quad \theta = \hat{\theta}_{\mathrm{MLE}} = \frac{s}{n} = \bar{b}. \qquad (2.16)$$

Fisher [33] showed that the estimated asymptotic variance of the MLE (in repeated sampling) is given by

$$\widehat{\mathrm{Var}}(\hat{\theta}_{\mathrm{MLE}}) = \hat{\mathcal{I}}^{-1},$$

[21] Prediction is often more difficult with the repeated-sampling approach to probability, especially in small-sample non-Gaussian situations; this is a distinct advantage for the Bayesian approach.

where $\hat{\mathcal{I}}$ is the *observed information* content of the sample:

$$\hat{\mathcal{I}} = -\left[\frac{\partial^2}{\partial\theta^2}\log l(\theta \mid s)\right]_{\theta=\hat{\theta}_{\text{MLE}}}.$$

Here the sampling variance of the MLE reduces to the familiar expression

$$\widehat{\text{Var}}(\hat{\theta}_{\text{MLE}}) = \left[\frac{s}{\theta^2} + \frac{n-s}{(1-\theta)^2}\right]^{-1}_{\theta=s/n} = \frac{\hat{\theta}_{\text{MLE}}(1-\hat{\theta}_{\text{MLE}})}{n}.$$

In this example $\hat{\theta}_{\text{MLE}} \doteq 0.216$ with estimated asymptotic standard error $\widehat{\text{se}}(\hat{\theta}_{\text{MLE}}) = \sqrt{\widehat{\text{Var}}(\hat{\theta}_{\text{MLE}})} \doteq 0.048$, and an approximate 95% confidence interval for θ based on the CLT would run from $0.216 - (1.96)(0.048) \doteq 0.122$ to $0.216 + (1.96)(0.048) \doteq 0.310$. These results are similar to those from the Bayesian analyses above with both prior specifications, which is typical of situations with fairly large n and relatively diffuse prior information. Note, however, that the interpretation of the results from the two approaches differs:

- In the (frequentist) likelihood approach θ is fixed but unknown and $\bar{b}$ is random, with the analysis based on imagining what would happen if the random sampling of the observed vehicles in the chosen city block were hypothetically repeated, and appealing to the fact that across these repetitions $(\bar{b} - \theta) \overset{\cdot}{\sim} \mathcal{N}(0, .048^2)$; whereas
- In the Bayesian approach $\bar{b}$ is fixed at its observed value and θ is treated as random, as a means of quantifying uncertainty about it: $(\theta - \bar{b} \mid \bar{b}) \overset{\cdot}{\sim} \mathcal{N}(0, .048^2)$.

This means among other things that, while it's not legitimate with the frequentist approach to say that $\Pr_f(.12 \le \theta \le .31) \doteq .95$, which is what many users of confidence intervals would like them to mean, the corresponding statement $\Pr_B(.12 \le \theta \le .31 \mid \boldsymbol{b}, \text{diffuse prior information}) \doteq .95$ is a natural consequence of the Bayesian approach. In the case of diffuse prior information this justifies the fairly common practice of computing inferential summaries in a frequentist way and then interpreting them in Bayesian language.

2.2 The Need for Simulation-Based Bayesian Computation

The example above illustrates two approaches to Bayesian computation in a situation where the parameter of interest is one-dimensional:

- *conjugate analysis*—showing that some prior family of distributions is conjugate to the likelihood in the model under investigation, and finding a member of that conjugate family which (at least approximately) expresses the relevant prior information; and

- *asymptotic analysis*—appealing to the fact that when n is large (a) the likelihood and posterior distributions will be similar because the prior sample size n_0 will be negligible in relation to n, and (b) both these distributions will be close to normal by the CLT, so that

$$(\theta \mid \boldsymbol{y}) \overset{\cdot}{\sim} \mathcal{N}(\hat{\theta}_{\text{MLE}}, \hat{\mathcal{I}}^{-1}). \tag{2.17}$$

In (2.17) $\boldsymbol{y}$ is a generic data vector of length n and θ is a generic one-dimensional *regular*[22] parameter.

Asymptotic analysis extends directly to situations where the unknown $\boldsymbol{\theta}$ is a vector of dimension (say) $k > 1$, with three main differences: finding the MLE often then involves solving (perhaps iteratively) a system of k equations like (2.16) in the k unknowns $\theta_1, \ldots, \theta_k$; the normal distribution in (2.17) is multivariate; and the analogue of observed information $\hat{\boldsymbol{\mathcal{I}}}$ then becomes the negative Hessian (matrix of second partial derivatives) of the log likelihood evaluated at $\hat{\boldsymbol{\theta}}_{\text{MLE}}$. Conjugate analysis also has a direct extension to cases with $k > 1$: Bayes' Theorem (2.9) is still valid when $\boldsymbol{\theta}$ is a vector. However, it's far easier to find a conjugate family when $k = 1$ than in problems of higher dimension, and a new set of Bayesian challenges arises when $k > 1$: interest often focuses on the *marginal* posterior distributions of individual components of $\boldsymbol{\theta}$, and these require calculating $(k-1)$-dimensional integrals of the form

$$p(\theta_j \mid \boldsymbol{y}) = \int \cdots \int p(\theta_1, \ldots, \theta_k \mid \boldsymbol{y}) \, \mathrm{d}\theta_1 \ldots \mathrm{d}\theta_{j-1} \, \mathrm{d}\theta_{j+1} \ldots \mathrm{d}\theta_k \,. \tag{2.18}$$

Moreover, four other kinds of high-dimensional integrals also arise when k is large: (a) with a generic data vector $\boldsymbol{y} = (y_1, \ldots, y_n)$ in place of $\boldsymbol{b}$ in Bayes' Theorem (2.9), the normalizing constant in that equation is

$$c^{-1} = p(\boldsymbol{y}) = \int p(\boldsymbol{y}, \boldsymbol{\theta}) \, \mathrm{d}\boldsymbol{\theta} = \int p(\boldsymbol{y} \mid \boldsymbol{\theta}) \, p(\boldsymbol{\theta}) \, \mathrm{d}\boldsymbol{\theta}, \tag{2.19}$$

and this is a k-dimensional integral; (b) as was noted at the end of Section 2.1.2, the predictive distribution for the next observation y_{n+1} is

$$p(y_{n+1} \mid \boldsymbol{y}) = \int p(y_{n+1} \mid \boldsymbol{\theta}) \, p(\boldsymbol{\theta} \mid \boldsymbol{y}) \, \mathrm{d}\boldsymbol{\theta}, \tag{2.20}$$

---

[22] In most Bayesian work there are three main types of parameters: *location*, *scale*, and *range-restricting*. Location and scale parameters typically pin down the center and spread of a sampling distribution and are regular in the sense of this footnote; as an example of a range-restricting parameter, consider basing the likelihood function on the Uniform$(0, \theta)$ sampling distribution for unknown $\theta > 0$. Range-restricting parameters are irregular in the sense of this footnote because a different type of asymptotics than (2.17) typically applies to them; in (2.17) the asymptotic posterior variance $\hat{\mathcal{I}}^{-1}$ typically goes down as the amount of data increases at a $1/n$ rate, whereas with range-restricting parameters this rate is typically $1/n^2$. See Bernardo and Smith [2] and Draper [27] for more details.

which is another k-dimensional integral; (c) it's often useful to summarize a marginal distribution like $p(\theta_j \mid \boldsymbol{y})$ with a few of its low-order moments, such as its mean

$$E(\theta_j \mid \boldsymbol{y}) = \int \theta_j \, p(\theta_j \mid \boldsymbol{y}) \, \mathrm{d}\theta_j, \tag{2.21}$$

and you can see from (2.18) that this also involves calculating a k-dimensional integral; and (d) inference about a function of the parameters, such as the coefficient of variation σ/μ in the $\mathcal{N}(\mu, \sigma^2)$ sampling model for positive data distributed well away from 0, also requires complicated manipulations with high-dimensional integrals. Accurate numerical evaluation of integrals of this type for large k has been the central technical challenge of Bayesian statistical work for the past two and a half centuries.[23]

The conjugate and asymptotic approaches to Bayesian computation are useful as far as they go, but conjugate priors are rarely (if ever) available for the complicated likelihoods arising in multilevel models, and asymptotic analysis can be highly misleading when the sample sizes are small. Consider, for instance, one of the simplest multilevel settings, a *variance components* model, arising (for example) in the measurement of the quality of hospital care: I choose a random sample of J hospitals in (say) California in (say) January 2007 and a random sample of n_j patients in the chosen hospitals (a *single-stage cluster sample*), and initially I fit the model

$$\begin{gathered} y_{ij} = \beta_0 + a_j^H + a_{ij}^P, \quad j = 1, \ldots, J, \quad i = 1, \ldots, n_j, \\ \sum_{j=1}^{J} n_j = N, \quad a_j^H \overset{iid}{\sim} \mathcal{N}(0, \sigma_H^2), \quad a_{ij}^P \overset{iid}{\sim} \mathcal{N}(0, \sigma_P^2) \end{gathered} \tag{2.22}$$

(with the a_j^H and a_{ij}^P mutually conditionally independent given the parameters) as a way of quantifying how much of the variation in the quality of care scores y_{ij} is within and between hospitals.[24] The parameter vector $\boldsymbol{\theta}$ in this model has three components: the intercept β_0 and the variance components σ_H^2 and σ_P^2, measuring variability at the hospital and patient levels, respectively, with both the hospital factor H and the patient factor P treated as random because interest focuses on the populations of hospitals and patients in California in January 2007 from which the cluster sample was drawn.

---

23 More than 200 years ago Laplace [48] developed an approach, based on a clever use of Taylor series, to approximating integrals of the form (2.18)–(2.21) which can work well when n is large; his method was ignored/forgotten for a long time until it was independently reinvented under the name *saddlepoint approximations* [e.g., 20]. See Raudenbush et al. [61] for an application of *Laplace approximations* to multilevel models.

24 Equation (2.22) is also sometimes called a *random-intercepts regression* model, because it's like a regression with no predictor variables in which the intercept $\beta_0 + a_j^H$ is allowed to vary randomly from hospital to hospital.

Conjugate analysis of variance-components models is impossible: there is no conjugate prior for the parameters of model (2.22). The success of asymptotic analysis would depend on the sizes J and N of the hospital- and patient-level samples and the values of the variance components: if J were on the order of (say) 50 (or more) and N were in the hundreds (or more), and if additionally both σ_H^2 and σ_P^2 were well away from zero, large-sample normal approximations for the marginal posterior distributions of all three parameters (given fairly diffuse priors) could well be adequate. However, it's important to note that the usual intuitions about sample size require some modification in multilevel modeling: data sets in this quality of care example with $N = 1{,}000$ could still be "small samples" as far as the accuracy of (at least one of) the asymptotic approximations is concerned. A large value of N will typically translate into approximately normal marginal posteriors for β_0 and σ_P^2, but the behavior of the marginal posterior for σ_H^2 depends on J, N, and the *intraclass* (or *intracluster*) *correlation*

$$\rho = \frac{\sigma_H^2}{\sigma_H^2 + \sigma_P^2},$$

which is just the ordinary correlation between any two patients y_{ij} and $y_{i'j}$ ($i \neq i'$) in the same hospital (a measure of the degree to which patients in any given hospital receive care of similar quality). It's intuitively evident that the *effective sample size* of the sampling plan as far as σ_H^2 is concerned will be much closer to J than to N if ρ is large,[25] and if J is small the marginal posterior for σ_H^2 can be far from normal.

The bottom line from all of this is that the Bayesian approach to multilevel modeling was severely restricted as long as asymptotic analysis was the only computational way forward. This situation changed suddenly in the early 1990s, with the introduction (to the discipline of statistics, at least) of a new class of simulation-based computational tools, *Markov Chain Monte Carlo* (MCMC) methods.

2.3 Markov Chain Monte Carlo (MCMC) Methods

The evolution of ideas toward the current set of MCMC methods began with the efforts of two mathematicians, Nick Metropolis and Stanislav Ulam, near the end of World War II, in their work on the project that led to the development of the atomic bomb. For reasons unrelated to those of Bayesian

[25] For example, in the limit as $\rho \to 1$, having chosen 50 patients from each of 20 hospitals is the same as having chosen only 1 patient in each hospital, so a sample of 1,000 patients produces only 20 independent observations for learning about σ_H^2.

statistics they needed accurate approximations to integrals like the right-hand side of (2.18), and (through their work on the bomb, and parallel efforts in England at about the same time by Alan Turing and others to break the German Enigma codes) they could see that high-speed computers were about to become a reality. Metropolis and Ulam [55], in a visionary paper that laid the groundwork for an explosion of new scientific activity decades later, made two fundamental observations:

- Anything you want to know about a probability distribution $p(\boldsymbol{\theta})$ of essentially arbitrary complexity can be learned to arbitrary accuracy by *sampling* a large enough number of random draws from it; and
- If performed correctly, it's not necessary for the *validity* of this approach that the draws from $p(\boldsymbol{\theta})$ be made in an IID fashion.

They called this technique the *Monte Carlo method*, a reference to the European principality of the same name famous for its gambling casinos.

Suppose, for example, that you're interested in a k-dimensional posterior distribution $p(\boldsymbol{\theta} \mid \boldsymbol{y})$ which can't be worked with (easily) in closed form. Three types of things of direct interest to you about $p(\boldsymbol{\theta} \mid \boldsymbol{y})$ would be

- the marginal means $\mu_j = E(\theta_j \mid \boldsymbol{y})$ and the marginal standard deviations $\sigma_j = \sqrt{\text{Var}(\theta_j \mid \boldsymbol{y})}$ of the components of $\boldsymbol{\theta}$,
- the shapes of these marginal distributions (basically you'd like to be able to trace out the entire density curves), and
- one or more of the quantiles of the marginal distributions (e.g., to construct a 95% central posterior interval for θ_j you need to know its 2.5% and 97.5% quantiles, and sometimes the posterior median (the 50th percentile) is of interest too).

Suppose you could take an arbitrarily large random sample from $p(\boldsymbol{\theta} \mid \boldsymbol{y})$, say $\boldsymbol{\theta}_1^*, \dots, \boldsymbol{\theta}_m^*$, where each $\boldsymbol{\theta}_i^*$ is a vector of sampled values of $(\theta_1, \dots, \theta_k)$. Imagine collecting these vectors together into an $m \times k$ matrix or table $\{\theta_{ij}^*\}$, with individual sampled vectors as rows and components of $\boldsymbol{\theta}$ as columns; call this the *Monte Carlo (MC) data set.* Then each of the above three aspects of $p(\boldsymbol{\theta} \mid \boldsymbol{y})$ can be estimated from this data set, in straightforward fashion:

- $\hat{\mu}_j = \hat{E}(\theta_j \mid \boldsymbol{y}) = \bar{\theta}_j^* = \frac{1}{m} \sum_{i=1}^m \theta_{ij}^*$, and

$$\hat{\sigma}_j = \sqrt{\widehat{\text{Var}}(\theta_j \mid \boldsymbol{y})} = \sqrt{\frac{1}{m-1} \sum_{i=1}^m \left(\theta_{ij}^* - \bar{\theta}_j^*\right)^2};$$

- the marginal posterior density of $(\theta_j \mid \boldsymbol{y})$ can be estimated by a histogram or kernel density estimate based on the values in column j of the MC data set; and

- percentiles for θ_j can be estimated by counting how many of the θ^*_{ij} values fall below a series of specified points. For example, to find an estimate of the 2.5% quantile you solve the equation

$$\hat{F}_{\theta_j | \boldsymbol{y}}(q) = \frac{1}{m} \sum_{i=1}^{m} I(\theta^*_{ij} \le q) = 0.025 \tag{2.23}$$

for q, where $I(\mathcal{A})$ is the indicator function (1 if $\mathcal{A}$ is true, 0 otherwise).[26]

This simple idea beautifully solves the *marginalization* problem—and the low-order moments problem—posed by having to calculate integrals like (2.18) and (2.21): to learn anything you want about θ_j you just use simple descriptive methods on the values in column j of the MC data set, ignoring all other columns. Moreover,

- if you're interested in the relationship between two of the parameters in the posterior, for instance as summarized by their correlation, you can just compute the sample correlation coefficient based on the relevant columns in the MC data set;
- if there's some function of the parameters that interests you, such as $\eta = f(\theta_2, \theta_5, \theta_{10}) = (\theta_2 + \theta_{10})/\sqrt{\theta_5}$, all you have to do to learn about it is to *monitor* η by creating a new column in the MC data set with values $\eta^*_i = (\theta^*_{i,2} + \theta^*_{i,10})/\sqrt{\theta^*_{i,5}}$ and then apply the usual descriptive summaries to η^*; and

[26] Notice how literally all of these estimates do their job, e.g., $\bar{\theta}^*_j = \frac{1}{m} \sum_{i=1}^{m} \theta^*_{ij}$ is an estimate of $\int \theta_j \, p(\theta_j \mid \boldsymbol{y}) \, d\theta_j$—the integral is asking us to compute a weighted average of θ_j values with weights given by $p(\theta_j \mid \boldsymbol{y})$, which is exactly what $\bar{\theta}^*_j$ does when the rows of the MC data set are random draws from $p(\boldsymbol{\theta} \mid \boldsymbol{y})$. More formally, if $(\theta_j \mid \boldsymbol{y})$ is a real-valued random variable with density $p_{\theta_j|\boldsymbol{y}}(q)$ (in a change of notation) and cumulative distribution function (CDF) $F_{\theta_j|\boldsymbol{y}}(q) = \Pr(\theta_j \le q \mid \boldsymbol{y})$, so that $p_{\theta_j|\boldsymbol{y}}(q) = \mathrm{d}F_{\theta_j|\boldsymbol{y}}(q)/\,\mathrm{d}q$ and $p_{\theta_j|\boldsymbol{y}}(q)\,\mathrm{d}q = \mathrm{d}F_{\theta_j|\boldsymbol{y}}(q)$, the posterior mean is

$$\mu_j = \int q \, p_{\theta_j|\boldsymbol{y}}(q) \, \mathrm{d}q = \int q \, \mathrm{d}F_{\theta_j|\boldsymbol{y}}(q),$$

and it's reasonable to estimate this with the *empirical CDF* $\hat{F}_{\theta_j|\boldsymbol{y}}(q)$ used to compute quantiles in (2.23):

$$\begin{aligned} \hat{\mu}_j &= \int q \, \mathrm{d}\hat{F}_{\theta_j|\boldsymbol{y}}(q) = \int q \, \mathrm{d}\left[\frac{1}{m} \sum_{i=1}^{m} I(\theta^*_{ij} \le q) \right] \\ &= \frac{1}{m} \sum_{i=1}^{m} \int q \, \mathrm{d}I(\theta^*_{ij} \le q) = \frac{1}{m} \sum_{i=1}^{m} \theta^*_{ij} \, , \end{aligned}$$

from basic properties of step functions and the (Dirac) delta function; see, e.g., Butkov [16].

- this approach also solves the problem of calculating the predictive distribution $p(y_{n+1} \mid \boldsymbol{y})$ for future data, as follows. As noted above, the integral in (2.20) expresses the predictive distribution as a mixture (in $\boldsymbol{\theta}$) of the sampling distributions $p(y_{n+1} \mid \boldsymbol{\theta})$ weighted by the posterior distribution $p(\boldsymbol{\theta} \mid \boldsymbol{y})$. This indicates that to sample a draw from $p(y_{n+1} \mid \boldsymbol{y})$ you just sample a $\boldsymbol{\theta}^*$ from $p(\boldsymbol{\theta} \mid \boldsymbol{y})$ and then sample a y_{n+1} from $p(y_{n+1} \mid \boldsymbol{\theta}^*)$. In this way the predictive distribution can simply be monitored as a new column in the MC data set.

In fact, straightforward use of the Monte Carlo method solves all of the difficult integration problems mentioned in Section 2.2 except the calculation of the normalization constant c that makes $p(\boldsymbol{\theta})$ integrate to 1, and this problem will disappear with the introduction of the *Markov chain* Monte Carlo methods examined below.

There's just one question: what do we have to assume about the nature of the random sampling of the $\boldsymbol{\theta}^*$ values from $p(\boldsymbol{\theta} \mid \boldsymbol{y})$ for this idea to work? Basic repeated-sampling theory based on the (weak) Law of Large Numbers (see, e.g., Bickel and Doksum [5] for details) shows that figuring out how to draw the $\boldsymbol{\theta}_i^*$ in an IID fashion would be sufficient: with IID sampling the above *Monte Carlo estimates* of the true summaries of $p(\boldsymbol{\theta} \mid \boldsymbol{y})$ are *consistent*, meaning that they can be made arbitrarily close to the truth with arbitrarily high probability as $m \to \infty$. The problem, of course, is that it can be extremely difficult to figure out how to make IID draws in an efficient manner from a high-dimensional distribution. Metropolis and Ulam [55] sketched a possible solution to this problem by noting that IID sampling is not necessary: *dependent* draws from $p(\boldsymbol{\theta} \mid \boldsymbol{y})$ will also work if the dependence takes a particular form. Think of iteration number i in the Monte Carlo sampling process as a discrete index of time t, so that the columns of the MC data set can be viewed as *time series*. IID draws from $p(\boldsymbol{\theta} \mid \boldsymbol{y})$ correspond to *white noise*: a time series with zero *autocorrelations* at all *lags* (time intervals)[27] $k \neq 0$. However, it can be shown [e.g., 32] that all of the above descriptive summaries are still consistent as long as the columns of the MC data set form *stationary* time series, in the sense that the joint distributions of any blocks $\{\theta^*_{n_1+t,j}, \theta^*_{n_2+t,j}, \ldots, \theta^*_{n_r+t,j}\}$ of MC data in any given column j are invariant under time shifts (i.e., these distributions should be independent of t). It might not seem readily apparent how relaxing the IID assumption in this way constitutes progress, but Metropolis again helped to provide the breakthrough a few years after his paper with Ulam, this time working with a different set of colleagues in 1953 [54].

[27] The autocorrelation ρ_k of a stationary time series θ_t^* at lag k (see, e.g., Chatfield [18]) is γ_k / γ_0, where $\gamma_k = \text{Cov}(\theta_t^*, \theta_{t+k}^*)$, the covariance of the series with itself k iterations in the future (or past).

2.3.1 IID Monte Carlo Sampling

Consider first how to implement the Monte Carlo method with IID sampling, and for simplicity in this section assume that θ is real-valued. If $\bar{\theta}^* = \frac{1}{m}\sum_{t=1}^{m}\theta_t^*$ is based on an IID sample of size m from $p(\theta \mid \boldsymbol{y})$, you can use the frequentist fact that in repeated sampling $\text{Var}(\bar{\theta}^*) = \sigma^2/m$, where (as above) σ^2 is the variance of $p(\theta \mid \boldsymbol{y})$, to construct a *Monte Carlo standard error* (MCSE) for $\bar{\theta}^*$:

$$\widehat{\text{se}}(\bar{\theta}^*) = \frac{\hat{\sigma}}{\sqrt{m}},$$

where $\hat{\sigma}$ is the sample SD of the θ^* values. This can be used, possibly after some preliminary experimentation, to decide on m, the *Monte Carlo sample size*, which will also be referred to below as the length of the *monitoring run.*

As an unrealistically simple first example, consider employing the Monte Carlo approach to estimate the posterior mean in the conjugate Beta-Bernoulli example of Section 2.1.2: I want to simulate draws from the Beta$(\alpha_0 + s, \beta_0 + n - s)$ distribution with $(\alpha_0, \beta_0, s, n) = (2.0, 6.4, 16, 74)$ (pretend here that you don't know the formulas for the mean and variance of this distribution). One of the most computationally efficient ways to generate random draws from a given density function is *rejection sampling*, which was first developed by von Neumann [73]. The idea is as follows. Suppose the target density $p(\theta \mid \boldsymbol{y})$ is difficult to sample from, but you can find an integrable *envelope function* $G(\theta \mid \boldsymbol{y})$ such that (a) G dominates p in the sense that $G(\theta \mid \boldsymbol{y}) \geq p(\theta \mid \boldsymbol{y}) \geq 0$ for all θ and (b) the density g obtained by normalizing G—later to be called the *proposal distribution*—is easy and fast to sample from. Then to get a random draw from p, make a draw θ^* from g instead and accept or reject it according to an *acceptance probability* $\alpha_R(\theta^* \mid \boldsymbol{y})$; if you reject the draw, repeat this process until you accept. von Neumann showed that the choice $\alpha_R(\theta^* \mid \boldsymbol{y}) = p(\theta^* \mid \boldsymbol{y})/G(\theta^* \mid \boldsymbol{y})$ correctly produces IID draws from p, and you can intuitively see that he's right by the following argument. Making a draw from the posterior distribution of interest is like choosing a point at random (in two dimensions) under the density curve $p(\theta \mid \boldsymbol{y})$ in such a way that all possible points are equally likely, and then writing down its θ value. If you instead draw from G so that all points under G are equally likely, to get correct draws from p you'll need to throw away any point that falls between p and G, and this can be accomplished by accepting each sampled point θ^* with probability $p(\theta^* \mid \boldsymbol{y})/G(\theta^* \mid \boldsymbol{y})$, as von Neumann said. A summary of this method[28] is as follows.

---

[28] After having absorbed the idea that Algorithm 2.1 works for univariate θ, notice that there's nothing about it that makes this restriction necessary: the algorithm is valid when $\boldsymbol{\theta}$ is a vector of length k for any $k \geq 1$.

Algorithm 2.1 (Rejection sampling) To make m draws at random from the density $p(\theta \mid \boldsymbol{y})$, select an integrable envelope function G—which when normalized to integrate to 1 is the proposal distribution g—such that $G(\theta \mid \boldsymbol{y}) \geq p(\theta \mid \boldsymbol{y}) \geq 0$ for all θ; define the acceptance probability $\alpha_R(\theta^* \mid \boldsymbol{y}) = p(\theta^* \mid \boldsymbol{y})/G(\theta^* \mid \boldsymbol{y})$; and

```
Initialize t ← 0
Repeat {
    Sample θ* ~ g(θ | y)
    Sample u ~ Uniform(0,1)
    If u ≤ α_R(θ* | y) then { θ_{t+1} ← θ*; t ← (t+1) }
}
until t = m.
```

Figure 2.2 demonstrates this method on the Beta$(18.0, 64.4)$ density arising in the Beta-Bernoulli case study examined earlier. Rejection sampling permits considerable flexibility in the choice of envelope function. Here, borrowing an idea from Gilks and Wild [41], I've noted that the relevant Beta density is concave on the log scale, meaning that it's easy to construct an envelope on that scale in a piecewise linear fashion, by choosing points on the log density and constructing tangents to the curve at those points. The simplest possible such envelope involves two line segments, one on either side of the mode. The optimal choice of the tangent points would maximize the marginal probability of acceptance of a draw in the rejection algorithm, which can be shown to be $\left[\int G(\theta)\, d\theta\right]^{-1}$; in other words, you should minimize the area under the (un-normalized) envelope function subject to the constraint that it dominates the target density $p(\theta \mid \boldsymbol{y})$. Here this optimum turns out to be attained by locating the two tangent points at about 0.17 and 0.26, as in Fig. 2.2; the resulting acceptance probability of about 0.75 could clearly be improved by adding more tangents. Piecewise linear envelope functions on the log scale are a good choice because the resulting envelope density on the raw scale is a piecewise set of scaled exponential distributions (see the bottom panel in Fig. 2.2), from which random samples can be taken quickly.

A preliminary sample of $m_0 = 500$ IID draws from the Beta$(18.0, 64.4)$ distribution using the above rejection sampling method yields $\bar{\theta}^* = 0.2197$ and $\hat{\sigma} = 0.04505$, meaning that the posterior mean has already been estimated with an MCSE of only $\hat{\sigma}/\sqrt{m_0} = 0.002$ even with just 500 draws. Suppose, however, that I wanted $\bar{\theta}^*$ to differ from the true posterior mean μ by no more than some (perhaps even smaller) tolerance T with Monte Carlo probability at least $1 - \epsilon$:

$$\Pr(|\bar{\theta}^* - \mu| \leq T) \geq 1 - \epsilon,$$

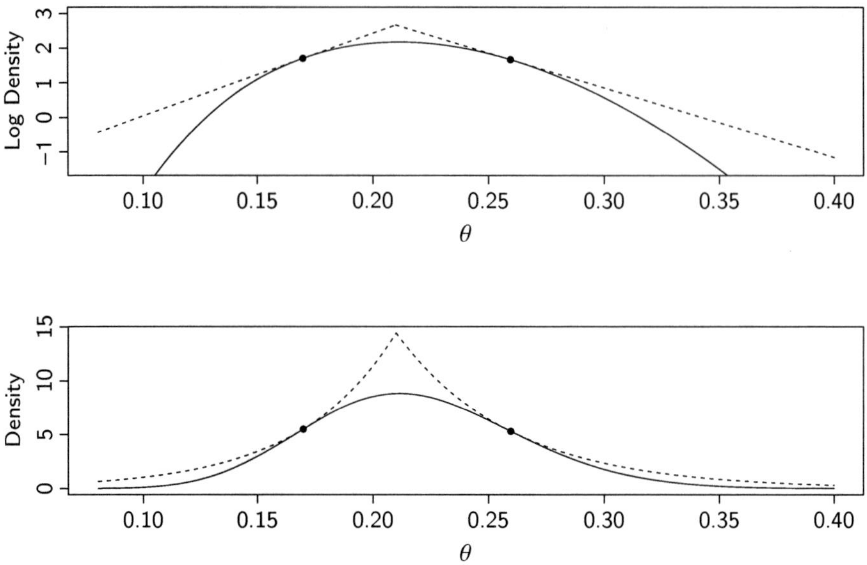

Fig. 2.2 Rejection sampling from the Beta(18.0, 64.4) distribution. The top panel shows the construction of a piecewise linear envelope function on the log scale; the bottom panel is on the raw (density) scale.

where $\Pr(\cdot)$ here is based on the (frequentist) Monte Carlo randomness inherent in $\bar{\theta}^*$. By the CLT, for large m in repeated sampling $\bar{\theta}^*$ is approximately normal with mean μ and variance σ^2/m; this leads to the inequality

$$m \geq \frac{\sigma^2 \left[\Phi^{-1}\left(1 - \frac{\epsilon}{2}\right) \right]^2}{T^2}, \tag{2.24}$$

where $\Phi^{-1}(\cdot)$ is the standard normal inverse CDF. To pin down three significant figures (sigfigs) in the posterior mean in this example with high Monte Carlo accuracy I might take $T = 0.0005$ and $\epsilon = 0.05$, which yields a recommended IID sample size of $(0.04505^2)(1.96)^2/0.0005^2 \doteq 31,200$. So I take another sample of 30,700 (which is virtually instantaneous at contemporary computing speeds) and merge it with the 500 draws I already have; this yields $\bar{\theta}^* = 0.21827$ and $\hat{\sigma} = 0.04528$, meaning that the MCSE of this estimate of μ is $0.04528/\sqrt{31200} \doteq 0.00026$. I might announce that I think $E(\theta \mid \boldsymbol{y})$ is about 0.2183, give or take about 0.0003, which accords well with the true value 0.2184.

Of course, other aspects of $p(\theta \mid \boldsymbol{y})$ are equally easy to monitor; for example, if I want a Monte Carlo estimate of $\Pr(\theta \leq q \mid \boldsymbol{y})$ for some q, as noted above I just work out the proportion of the sampled θ^* values that are no larger than q. Or, even better, I recall that $\Pr(\mathcal{A}) = E[I(\mathcal{A})]$ for any event or proposition $\mathcal{A}$, so to the MC data set consisting of 31,200 rows and one column (the θ_t^*) I add a column monitoring the values of the *derived variable* which is 1 whenever $\theta_t^* \le q$ and 0 otherwise; the mean of this derived variable is the Monte Carlo estimate of $\Pr(\theta \le q \mid \boldsymbol{y})$, and I can attach an MCSE to it in the same way I did with $\bar{\theta}^*$. By this approach, for instance, the Monte Carlo estimate of $\Pr(\theta \le 0.15 \mid \boldsymbol{y})$ based on the 31,200 draws examined above comes out $\hat{p} = 0.0556$ with an MCSE of 0.0013. Percentiles are typically harder to pin down with equal Monte Carlo accuracy (in terms of sigfigs) than means or SDs, because the 0/1 scale on which they're based is less information-rich than the θ^* scale itself; if I wanted an MCSE for $\hat{p}$ of 0.0001 I would need an IID sample of more than 5 million draws (which would still only take a few seconds at contemporary workstation speeds).

2.3.2 Metropolis-Hastings and Gibbs Sampling

As mentioned above, IID sampling of $p(\boldsymbol{\theta} \mid \boldsymbol{y})$ for $\boldsymbol{\theta}$ of length k is fine as far as it goes but can be difficult to implement when k is large. Metropolis et al. [54] accepted this unpleasant truth and proposed relaxing independence of the draws in favor of the next simplest random behavior—allowing the draws to form a (first-order) *Markov chain*—in combination with von Neumann's idea of rejection sampling, which had itself only been published a few years earlier in 1951.

Here's a quick review of all necessary facts about Markov chains to appreciate the basic Metropolis et al. idea. A *stochastic process* is just a collection of random variables $\{\boldsymbol{\theta}_t^*, t \in \mathcal{T}\}$ for some *index set* $\mathcal{T}$; when $\mathcal{T}$ stands for time the resulting process is a time series. In practice $\mathcal{T}$ can be either discrete, e.g., $\{0, 1, \ldots\}$, or continuous, e.g., $[0, \infty)$. *Markov chains* are a special kind of stochastic process that can either unfold in discrete or continuous time; discrete-time Markov chains are all that's needed for MCMC. The possible values that a stochastic process can take on are collectively called the *state space* $\mathcal{S}$ of the process—in the simplest case $\mathcal{S}$ is real-valued and can also either be discrete or continuous. Intuitively speaking, a Markov chain [e.g., 32, 35, 62] is a time series unfolding in such a way that the *past and future states of the process are independent given the present state*—in other words, to figure out where the chain is likely to go next you don't need to pay attention to where it's been, you just need to consider where it is now. More formally, a stochastic process $\{\boldsymbol{\theta}_t^*, t \in \mathcal{T}\}$, $\mathcal{T} = \{0, 1, \ldots\}$, with state space $\mathcal{S}$ is a Markov chain if, for any set $\mathcal{A}$ contained in $\mathcal{S}$,

$$\Pr(\boldsymbol{\theta}_{t+1}^* \in \mathcal{A} \mid \boldsymbol{\theta}_0^*, \ldots, \boldsymbol{\theta}_t^*) = \Pr(\boldsymbol{\theta}_{t+1}^* \in \mathcal{A} \mid \boldsymbol{\theta}_t^*).$$

The most nicely behaved Markov chains satisfy three properties:

- They're *irreducible*, which basically means that no matter where it starts the chain has to be able to reach any other state in a finite number of iterations with positive probability;
- They're *aperiodic*, meaning that for all states i the set of possible *sojourn times*, to get back to i having just left it, can have no divisor bigger than 1. This forces the chain to *mix freely* among its possible states rather than oscillating back and forth within a subset of $\mathcal{S}$; and
- They're *positive recurrent*, meaning that (a) for all states i, if the process starts at i it will return to i with probability 1, and (b) the expected length of waiting time until the first return to i is finite. Notice that this is a bit delicate: wherever the chain is now, we insist that it must certainly come back here, but we don't expect to have to wait forever for this to happen.

Imagine running a "nice" Markov chain (which satisfies the three properties above) for a long time, and look at the distribution of the states it visits—over time this distribution should settle down (converge) to a kind of limiting, steady-state behavior. Formally, a positive recurrent and aperiodic chain is called *ergodic*, and it turns out that chains of this type which are also irreducible possess a unique *stationary* (or *equilibrium*, or *invariant*) distribution π, characterized (in the case of discrete state spaces) by the relation

$$\pi(j) = \sum_i \pi(i)\, P_{ij}(t)$$

for all states j and times $t \geq 0$, where $P_{ij}(t) = \Pr(\theta_t^* = j \mid \theta_0^* = i)$ is the *transition matrix* of the chain. Informally, the stationary distribution summarizes the behavior that the chain will settle into after it's been run for a long time, regardless of its initial state.

I bring all of this up because Metropolis et al. were driven by the difficulty of creating IID samplers from complex probability distributions to seek a solution among the class of samplers with a Markov character (hence the name *Markov chain Monte Carlo* (MCMC)). Given a parameter vector $\boldsymbol{\theta}$ and a data vector $\boldsymbol{y}$, the Metropolis et al. idea is to simulate random draws from the posterior distribution $p(\boldsymbol{\theta} \mid \boldsymbol{y})$ by constructing a Markov chain with the following three properties:

- It should have the same state space as $\boldsymbol{\theta}$,
- It should be easy to simulate from, and
- Its stationary distribution should be $p(\boldsymbol{\theta} \mid \boldsymbol{y})$.

If you can do this, you can run the Markov chain for a long time, generating a huge sample from the posterior, and then (as noted at the beginning of Section 2.3) use simple descriptive summaries (means, SDs, correlations, histograms or kernel density estimates) to extract any features of the posterior you want. The Markov aspect of the sampler will induce a (typically positive) autocorrelation in the random draws, but (as noted above) this affects only the *efficiency* of the sampling scheme, not its *validity*: if $\{\theta_t^*, t = 1, 2, \ldots\}$ is a stationary time series then, e.g., the sample mean $\bar{\theta}^* = \frac{1}{m}\sum_{t=1}^{m}\theta_t^*$ is a consistent estimate of the mean μ of the stationary distribution, the sample SD

$$\hat{\sigma} = \sqrt{\frac{1}{m-1}\sum_{t=1}^{m}(\theta_t^* - \bar{\theta}^*)^2}$$

is consistent for the SD of the stationary distribution, and so on. The only change from IID sampling is that if the draws from the target distribution $p(\theta)$ are positively autocorrelated, you'll learn about p via MCMC at a slower rate than you would have if you could have figured out how to sample from p in an IID fashion (intuitively if the θ_t^* are positively autocorrelated then each time you get a new observation you're getting a bit of new information and some old information over again, so the effective sample size (in IID terms) of m positively correlated draws will be smaller than m).

Metropolis et al. were able to create what people would now call a successful MCMC algorithm by the following means (see the excellent book edited by Gilks et al. [40] for many more details about the MCMC approach). Consider the rejection sampling method given above in Algorithm 2.1 as a mechanism for generating realizations of a time series (where as above time indexes iteration number). At any time t in this process you make a draw $\boldsymbol{\theta}^*$ from the proposal distribution $g(\boldsymbol{\theta})$ (the normalized version of the envelope function G) and either accept a "move" to $\boldsymbol{\theta}^*$ or reject it, according to the acceptance probability $p(\boldsymbol{\theta}^*)/G(\boldsymbol{\theta}^*)$; if accepted the process moves to $\boldsymbol{\theta}^*$, if not you draw again until you do make a successful move. The stochastic process thus generated is an IID (white noise) series of draws from the target distribution $p(\boldsymbol{\theta})$. Metropolis et al. had the following beautifully simple idea for how this may be generalized to situations where IID sampling is difficult: *they allowed the proposal distribution at time t to depend on the current value* $\boldsymbol{\theta}_t$ *of the process*, and then—to make things work out right—if a proposed move is rejected, instead of discarding it *the process is forced to stay where it is for one iteration before trying again.* The resulting process is a Markov chain, because (a) the draws are now dependent but (b) all you need to know in determining where to go next is where you are now.

Letting $\boldsymbol{\theta}_t$ stand for where you are now and $\boldsymbol{\theta}^*$ for where you're thinking of going, in this approach there is enormous flexibility in the choice of the proposal distribution $g(\boldsymbol{\theta}^* \mid \boldsymbol{\theta}_t, \boldsymbol{y})$, even more so than in ordinary rejection sampling. The original Metropolis et al. idea was to work with *symmetric* proposal distributions, in the sense that $g(\boldsymbol{\theta}^* \mid \boldsymbol{\theta}_t, \boldsymbol{y}) = g(\boldsymbol{\theta}_t \mid \boldsymbol{\theta}^*, \boldsymbol{y})$, but Hastings [46] pointed out that this could easily be generalized; the resulting method is the *Metropolis-Hastings* (MH) algorithm. Building on the Metropolis et al. results, Hastings showed that you'll get the correct stationary distribution

$p(\boldsymbol{\theta} \mid \boldsymbol{y})$ for your Markov chain[29] by making the following choice for the acceptance probability:

$$\alpha_{\text{MH}}(\boldsymbol{\theta}^* \mid \boldsymbol{\theta}_t, \boldsymbol{y}) = \min\left\{1, \frac{p(\boldsymbol{\theta}^* \mid \boldsymbol{y})/g(\boldsymbol{\theta}^* \mid \boldsymbol{\theta}_t, \boldsymbol{y})}{p(\boldsymbol{\theta}_t \mid \boldsymbol{y})/g(\boldsymbol{\theta}_t \mid \boldsymbol{\theta}^*, \boldsymbol{y})}\right\}. \tag{2.25}$$

A summary of the method is as follows.

Algorithm 2.2 (Metropolis-Hastings sampling) To construct a Markov chain whose equilibrium distribution is $p(\boldsymbol{\theta} \mid \boldsymbol{y})$, choose a proposal distribution $g(\boldsymbol{\theta}^* \mid \boldsymbol{\theta}_t, \boldsymbol{y})$, define the acceptance probability $\alpha_{\text{MH}}(\boldsymbol{\theta}^* \mid \boldsymbol{\theta}_t, \boldsymbol{y})$ by (2.25), and

```
Initialize θ_0; t ← 0
Repeat {
    Sample θ* ~ g(θ | θ_t, y)
    Sample u ~ Uniform(0, 1)
    If u ≤ α_MH(θ* | θ_t, y) then θ_{t+1} ← θ*
       else θ_{t+1} ← θ_t
    t ← t + 1
}
```

It's instructive to compare Algorithms 2.1 and 2.2 to see how heavily the MH algorithm borrows from ordinary rejection sampling, with the key difference that the proposal distribution is allowed to change over time. Notice how (2.25) generalizes von Neumann's acceptance probability ratio $p(\boldsymbol{\theta}^* \mid \boldsymbol{y})/G(\boldsymbol{\theta}^* \mid \boldsymbol{y})$ for ordinary rejection sampling: the crucial part of the new MH acceptance probability becomes the ratio of two von-Neumann-like ratios, one for where you are now and one for where you're thinking of going (it's equivalent to work with g or G since the normalizing constant cancels in the ratio). When the proposal distribution is symmetric in the Metropolis et al. sense, the acceptance probability ratio reduces to $p(\boldsymbol{\theta}^* \mid \boldsymbol{y})/p(\boldsymbol{\theta}_t \mid \boldsymbol{y})$, which is easy to motivate intuitively: whatever the target density is at the current point $\boldsymbol{\theta}_t$, you want to visit points of higher density more often and points of lower density less often, and (2.25) does this for you in the natural and appropriate way.

A Metropolis-Hastings Example

As an example of the MH algorithm in action, consider one of the simplest possible Gaussian non-multilevel models: normal data with known mean μ

[29] The proposal distribution $g(\boldsymbol{\theta}^* \mid \boldsymbol{\theta}_t, \boldsymbol{y})$ can be virtually anything and you'll get the right equilibrium distribution using the acceptance probability (2.25); see, e.g., Roberts [62] and Tierney [72] for the mild regularity conditions necessary to support this statement.

and unknown variance σ^2. The likelihood function for σ^2, derived from the sampling model $(y_i \mid \sigma^2) \stackrel{iid}{\sim} \mathcal{N}(\mu, \sigma^2)$ for $i = 1, \dots, n$, is

$$\begin{aligned} l(\sigma^2 \mid \boldsymbol{y}) &= c \prod_{i=1}^{n} (\sigma^2)^{-1/2} \exp\left[-\frac{(y_i - \mu)^2}{2\sigma^2} \right] \\ &= c\,(\sigma^2)^{-n/2} \exp\left[-\frac{\sum_{i=1}^{n}(y_i - \mu)^2}{2\sigma^2} \right]. \end{aligned}$$

This is recognizable as a member of the *Scaled Inverse* χ^2 family $\chi^{-2}(\nu, s^2)$ [e.g., 37] of distributions, which is a rescaled version of the Inverse Gamma family[30] chosen so that s^2 is an estimate of σ^2 based upon ν "observations": if $\theta \sim \chi^{-2}(\nu, s^2)$ then θ has density

$$p(\theta) = c\,\theta^{-(\nu/2+1)} \exp\left(-\frac{\nu s^2}{2\theta} \right),$$

so that

$$l(\sigma^2 \mid \boldsymbol{y}) = \chi^{-2}\left[n - 2, \frac{\sum_{i=1}^{n}(y_i - \mu)^2}{n-2} \right]. \tag{2.26}$$

You can now convince yourself that if the prior for σ^2 in this model is taken to be $\chi^{-2}(\nu, s^2)$, then the posterior for σ^2 will also be Scaled Inverse χ^2: with this choice of prior

$$p(\sigma^2 \mid \boldsymbol{y}) = \chi^{-2}\left[\nu + n, \frac{\nu s^2 + \sum_{i=1}^{n}(y_i - \mu)^2}{\nu + n} \right]. \tag{2.27}$$

This makes good intuitive sense: the prior estimate s^2 of σ^2 receives ν votes and the sample estimate $\hat{\sigma}^2 = \frac{1}{n} \sum_{i=1}^{n} (y_i - \mu)^2$ receives n votes in the posterior weighted average estimate $(\nu s^2 + n\hat{\sigma}^2)/(\nu + n)$.

Equation (2.27) provides a satisfying closed-form solution to the Bayesian updating problem in this model (e.g., it's easy to compute posterior moments analytically, and you can use numerical integration or well-known approximations to the CDF of the Gamma distribution to compute percentiles). For illustration purposes suppose instead that you want to use MH sampling to summarize this posterior. Then your main choice as a user of the algorithm is the specification of the proposal distribution (PD) $g(\sigma^2 \mid \sigma_t^2, \boldsymbol{y})$. The goal in choosing the PD is getting a chain that *mixes well* (moves freely and fluidly among all of the possible values of $\theta = \sigma^2$), and nobody has (yet) come up with a sure-fire strategy for always succeeding at this task. Having said that, here are two basic ideas that often tend to promote good mixing:

[30] The simplest way to sample from the Scaled Inverse χ^2 distribution is to use any of a variety of methods for drawing from Gamma distributions, since if $\theta \sim \chi^{-2}(\nu, s^2)$ then $1/\theta \sim \Gamma(\frac{1}{2}\nu, \frac{1}{2}\nu s^2)$.

(1) Pick a PD that looks like a somewhat overdispersed version of the posterior you're trying to sample from [e.g., 72]. Some work is naturally required to overcome the circularity inherent in this choice: If I fully knew $p(\theta \mid \boldsymbol{y})$ and all of its properties, why would I be using this algorithm in the first place?
(2) Set up the PD so that the expected value of where you're going to move to (θ^*), given that you accept a move away from where you are now (θ_t), is to stay where you are now:[31] $E_g(\theta^* \mid \theta_t, \boldsymbol{y}) = \theta_t$. That way, when you do make a move, there will be an approximate left-right balance, so to speak, in the direction you move away from θ_t, which will encourage rapid exploration of the whole space.

Using idea (1), a decent choice for the PD in the Gaussian model with unknown variance might well be the Scaled Inverse χ^2 distribution: $g(\sigma^2 \mid \sigma_t^2, \boldsymbol{y}) = \chi^{-2}(\nu_*, \sigma_*^2)$. This distribution has mean $\sigma_*^2 \, \nu_* / (\nu_* - 2)$ for $\nu_* > 2$. To use idea (2), then, I can choose any ν_* greater than 2 that I want, and as long as I take $\sigma_*^2 = \sigma_t^2 \, (\nu_* - 2)/\nu_*$ that will center the PD at σ_t^2 as desired. So I'll use

$$g(\sigma^2 \mid \sigma_t^2, \boldsymbol{y}) = \chi^{-2}\left(\nu_*, \sigma_t^2 \, \frac{\nu_* - 2}{\nu_*} \right) .$$

This leaves ν_* as a kind of potential *tuning constant*—the hope is that I can vary ν_* to improve the mixing of the chain.

Figure 2.3, motivated by an analogous plot in Gilks et al. [40], presents some typical output of the MH sampler with $\nu_* = 2.5, 20, 500$. The acceptance probabilities with these values of ν_* are 0.07, 0.44, and 0.86, respectively. The SD of the $\chi^{-2}\left(\nu_*, \sigma_t^2 \, (\nu_* - 2)/\nu_*\right)$ distribution is proportional to $\nu_*^2 / \left[(\nu_*^2 - 2)^2 \sqrt{\nu_* - 4} \right]$, which decreases as ν_* increases, and this turns out to be crucial: when the proposal distribution SD is large (small ν_*, as in the top panel in Fig. 2.3), the algorithm tries to make big jumps around θ space (good), but almost all of them get rejected (bad), so there are long periods of no movement at all, whereas when the PD SD is small (large ν_*; see the bottom panel of the figure), the algorithm accepts most of its proposed moves (good), but they're so tiny that it takes a long time to fully explore the space (bad). Gelman et al. [37] have shown that in simple canonical problems with approximately normal target distributions the optimal acceptance rate for MH samplers like the one illustrated here is about 44% when the vector of unknowns is one-dimensional, and this can serve as a rough guide: you can modify the proposal distribution SD until the acceptance rate is around the Gelman et al. target figure. The central panel of Fig. 2.3 displays the best possible MH behavior in this problem in the family of PDs chosen. Even with this optimization you can see that the mixing is not wonderful, but contemporary computing speeds enable huge numbers of draws to be collected in a short period of time, compensating

[31] This makes the output of the MCMC sampler a *martingale*; see, e.g., Feller [32].

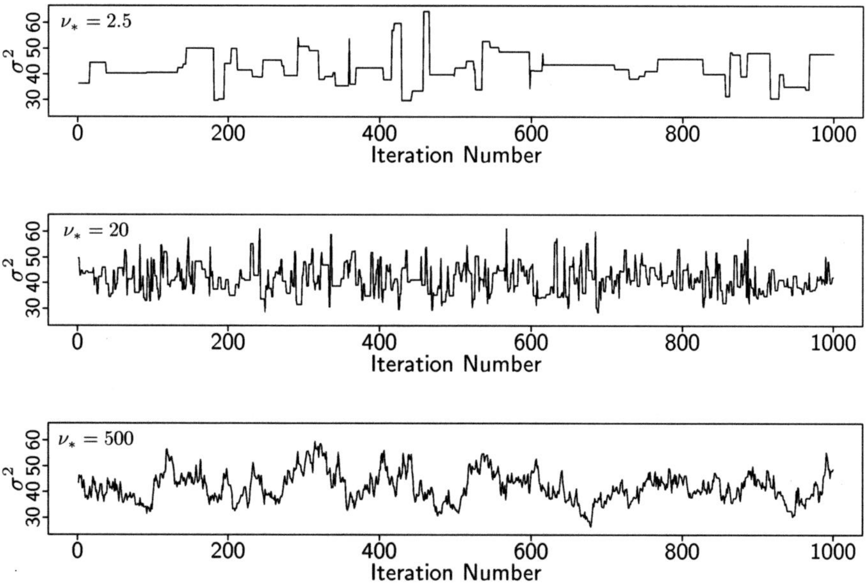

Fig. 2.3 Metropolis-Hastings sampling in the Gaussian model with known mean and unknown variance, and using a Scaled Inverse χ^2 proposal distribution with tuning constant ν_*. The top, middle, and bottom panels give typical output of the sampler for $\nu_* = 2.5, 20, 500$, respectively.

for the comparatively slow rate at which the MH algorithm learns about the posterior distribution of interest.

In this example the unknown quantity $\theta = \sigma^2$ was real-valued, but there's nothing in the MH method that requires this; in principle it works equally well when $\boldsymbol{\theta}$ is a vector of any finite dimension (look back at Algorithm 2.2 to verify this). Notice, crucially, that to implement this algorithm you only need to know how to calculate $p(\boldsymbol{\theta} \mid \boldsymbol{y})$ up to a constant multiple, since any such constant will cancel in computing the acceptance probability (2.25). Thus you're free to work with unnormalized versions of $p(\boldsymbol{\theta} \mid \boldsymbol{y})$, and this solves the final high-dimensional integration problem not already addressed above by the general Monte Carlo approach.

There's even more flexibility in this algorithm than might first appear: it's often possible to identify a set $\boldsymbol{A}$ of *auxiliary variables*—typically these are *latent* (unobserved) quantities—to be sampled along with the parameters, which have the property that they improve the mixing of the MCMC output (even though extra time is spent in sampling them). When the set $(\boldsymbol{\theta}, \boldsymbol{A})$ of quantities to be sampled is a vector of length k, there is additional flexibility: you can *block update* all of $(\boldsymbol{\theta}, \boldsymbol{A})$ at once, or with appropriate modifications of the acceptance probability you can divide $(\boldsymbol{\theta}, \boldsymbol{A})$ up into components, say

$(\boldsymbol{\theta}, \boldsymbol{A}) = (\boldsymbol{\lambda}_1, \ldots, \boldsymbol{\lambda}_l)$, and update the components one at a time (as Metropolis et al. originally proposed in 1953). As an example, consider data from the Junior School Project [e.g., 11], a longitudinal study of $N = 887$ students chosen randomly from $J = 48$ randomly sampled Inner London Education Authority (ILEA) primary schools in 1980 (see Mortimore et al. [56] for the original, and larger, data set). One focus of interest in this project was the relationship between mathematics test scores at year 3 and year 5 (x_{ij} and y_{ij}, respectively, for student i in school j). School-level scatterplots of these two variables indicated approximate bivariate normality, but with a fair amount of variation in the slopes and intercepts of the school-specific regression lines; moreover, the numbers n_j of pupils per school varied from 5 to 62 in this data set, with about a third of the schools having 12 pupils or less, so many of the school-level regressions were quite unstably estimated. It's natural to seek a balance between global regression fitting (which incorrectly ignores the cluster sampling) and noisy local linear estimation, by fitting a *random-slopes regression model* such as

$$y_{ij} = (\beta_0 + u_{0j}) + (\beta_1 + u_{1j})(x_{ij} - \bar{x}) + e_{ij} \,,$$
$$\boldsymbol{u}_j = \begin{pmatrix} u_{0j} \\ u_{1j} \end{pmatrix} \stackrel{iid}{\sim} \mathcal{N}_2(\boldsymbol{\emptyset}, \boldsymbol{V}_u), \quad \boldsymbol{V}_u = \begin{pmatrix} \sigma_{u0}^2 & \gamma_{01} \\ \gamma_{01} & \sigma_{u1}^2 \end{pmatrix}, \quad e_{ij} \stackrel{iid}{\sim} \mathcal{N}(0, \sigma_e^2), \tag{2.28}$$

where $j = 1, \ldots, J$, $i = 1, \ldots, n_j$, $\sum_{j=1}^J n_j = N$, and $\bar{x}$ is the mean of the math scores at year 3 over all N pupils. Centering the predictor in this way improves MCMC fitting by reducing the positive autocorrelation of the sampled draws. This model accounts properly for the clustering by regarding the schools as having been drawn randomly from the population of ILEA schools, each having its own slope and intercept, and the result of fitting (2.28) will be to *shrink* the local estimates of these parameters toward the global (population) regression. The parameter vector in this model is $\boldsymbol{\theta} = (\beta_0, \beta_1, \sigma_{u0}^2, \gamma_{01}, \sigma_{u1}^2, \sigma_e^2)$, but it will become clear below that, in models like (2.28) involving random effects (such as the u_{0j} and u_{1j}) at levels other than the subjects at the bottom of the nesting structure, it can greatly aid the MCMC sampling to treat the random effects as latent auxiliary variables to be sampled along with the parameters. An efficient division of the quantities $(\boldsymbol{\theta}, \boldsymbol{A}) = (\beta_0, \beta_1, \sigma_{u0}^2, \gamma_{01}, \sigma_{u1}^2, \sigma_e^2, \boldsymbol{u}_1, \ldots, \boldsymbol{u}_J)$ to be sampled in this model has been shown (see, e.g., Browne and Draper [11]) to involve $l = 4$, with $\boldsymbol{\lambda}_1 = (\beta_0, \beta_1)$; $\boldsymbol{\lambda}_2 = (\boldsymbol{u}_1, \ldots, \boldsymbol{u}_J)$; $\boldsymbol{\lambda}_3 = \boldsymbol{V}_u$ (as a matrix); and $\lambda_4 = \sigma_e^2$.

The idea in this component-by-component version of the algorithm, which Gilks et al. [40] call *single-component* MH sampling, is to have l different proposal distributions, one for each component of $\boldsymbol{\theta}$. Each iteration of the algorithm (indexed as usual by t) has l steps, indexed by i; at the beginning of iteration t you scan along, updating $\boldsymbol{\lambda}_1$ first, then $\boldsymbol{\lambda}_2$, and so on until you've updated $\boldsymbol{\lambda}_l$, which concludes iteration t. Let $\boldsymbol{\lambda}_{t,i}$ stand for the current state of component i at the end of iteration t, and let $\boldsymbol{\lambda}_{-i}$ stand for the $(\boldsymbol{\theta}, \boldsymbol{A})$ vector with component i omitted. (The notation gets awkward here; it can't be helped.) The proposal distribution $g_i(\boldsymbol{\lambda}_i^* \mid \boldsymbol{\lambda}_{t,i}, \boldsymbol{\lambda}_{t,-i}, \boldsymbol{y})$ for component i is allowed to depend on the most recent versions of all components of $(\boldsymbol{\theta}, \boldsymbol{A})$; here $\boldsymbol{\lambda}_{t,-i}$ is the current state of $\boldsymbol{\lambda}_{-i}$ after step $i-1$ of iteration $t+1$ is finished, so that components 1 through $i-1$ have been updated but not the rest. The acceptance probability for the proposed move to $\boldsymbol{\lambda}_i^*$ that creates the correct equilibrium distribution turns out to be

$$\alpha_{\text{MH}}(\boldsymbol{\lambda}_i^* \mid \boldsymbol{\lambda}_{t,-i}, \boldsymbol{\lambda}_{t,i}, \boldsymbol{y}) = \min\left[1, \frac{p(\boldsymbol{\lambda}_i^* \mid \boldsymbol{\lambda}_{t,-i}, \boldsymbol{y})\, g_i(\boldsymbol{\lambda}_{t,i} \mid \boldsymbol{\lambda}_i^*, \boldsymbol{\lambda}_{t,-i}, \boldsymbol{y})}{p(\boldsymbol{\lambda}_{t,i} \mid \boldsymbol{\lambda}_{t,-i}, \boldsymbol{y})\, g_i(\boldsymbol{\lambda}_i^* \mid \boldsymbol{\lambda}_{t,i}, \boldsymbol{\lambda}_{t,-i}, \boldsymbol{y})}\right]. \quad (2.29)$$

The distribution $p(\boldsymbol{\lambda}_i \mid \boldsymbol{\lambda}_{-i}, \boldsymbol{y})$ appearing in (2.29), which is called the *full conditional* distribution for $\boldsymbol{\lambda}_i$, has a natural interpretation: it represents the posterior distribution for the relevant portion of $(\boldsymbol{\theta}, \boldsymbol{A})$ *given* $\boldsymbol{y}$ *and the rest of* $(\boldsymbol{\theta}, \boldsymbol{A})$. The full conditional distributions act like building blocks in constructing the complete posterior distribution $p(\boldsymbol{\theta} \mid \boldsymbol{y})$, in the sense that any multivariate distribution is uniquely determined by its set of full conditionals [3].

Gibbs Sampling in Gaussian Multilevel Models

An important special case of single-component MH sampling arises when the proposal distribution $g_i(\boldsymbol{\lambda}_i^* \mid \boldsymbol{\lambda}_{t,i}, \boldsymbol{\lambda}_{t,-i}, \boldsymbol{y})$ for component i is chosen to be the full conditional $p(\boldsymbol{\lambda}_i^* \mid \boldsymbol{\lambda}_{t,-i}, \boldsymbol{y})$ for $\boldsymbol{\lambda}_i$: you can see from (2.29) that when this choice is made a glorious cancellation occurs and the acceptance probability is 1. This is *Gibbs sampling*, independently (re)discovered by Geman and Geman [39]: the Gibbs recipe is to sample from the full conditionals and accept all proposed moves. Even though it's just a version of MH, Gibbs sampling is important enough to merit a summary of its own. *Single-element* Gibbs sampling, in which each real-valued coordinate $\theta_1, \ldots, \theta_k$ gets updated in turn, is probably the most frequent way Gibbs sampling gets used, so that's what I'll summarize.[32]

Algorithm 2.3 (Single-element Gibbs sampling) To construct a Markov chain whose equilibrium distribution is $p(\boldsymbol{\theta} \mid \boldsymbol{y})$ with $\boldsymbol{\theta} = (\theta_1, \ldots, \theta_k)$,

`Initialize` $\theta_{0,1}^*, \ldots, \theta_{0,k}^*$; $t \leftarrow 0$

`Repeat {`

[32] Algorithm 2.3 details Gibbs sampling in the case with no auxiliary variables $\boldsymbol{A}$, but the algorithm works equally well when $\boldsymbol{\theta}$ is replaced by $(\boldsymbol{\theta}, \boldsymbol{A})$ in the summary.

Table 2.3 The MCMC data set generated by single-element Gibbs sampling applied to the variance-components model (2.22).

Phase of Sampling	Iteration t	Simulated Quantity β_0	a_1^H	$\cdots$	a_J^H	σ_H^2	σ_P^2
Initialization	0	$(\beta_0)_0^*$	$(a_1^H)_0^*$	$\cdots$	$(a_J^H)_0^*$	$(\sigma_H^2)_0^*$	$(\sigma_P^2)_0^*$
	1	$(\beta_0)_1^*$	$(a_1^H)_1^*$	$\cdots$	$(a_J^H)_1^*$	$(\sigma_H^2)_1^*$	$(\sigma_P^2)_1^*$
Burn-in	$\vdots$	$\vdots$	$\vdots$	$\ddots$	$\vdots$	$\vdots$	$\vdots$
	b	$(\beta_0)_b^*$	$(a_1^H)_b^*$	$\cdots$	$(a_J^H)_b^*$	$(\sigma_H^2)_b^*$	$(\sigma_P^2)_b^*$
	$b+1$	$(\beta_0)_{b+1}^*$	$(a_1^H)_{b+1}^*$	$\cdots$	$(a_J^H)_{b+1}^*$	$(\sigma_H^2)_{b+1}^*$	$(\sigma_P^2)_{b+1}^*$
Monitoring	$\vdots$	$\vdots$	$\vdots$	$\ddots$	$\vdots$	$\vdots$	$\vdots$
	$b+m$	$(\beta_0)_{b+m}^*$	$(a_1^H)_{b+m}^*$	$\cdots$	$(a_J^H)_{b+m}^*$	$(\sigma_H^2)_{b+m}^*$	$(\sigma_P^2)_{b+m}^*$

`Sample` $\theta_{t+1,1}^* \sim p(\theta_1 \mid \boldsymbol{y}, \theta_{t,2}^*, \theta_{t,3}^*, \theta_{t,4}^*, \ldots, \theta_{t,k}^*)$
`Sample` $\theta_{t+1,2}^* \sim p(\theta_2 \mid \boldsymbol{y}, \theta_{t+1,1}^*, \theta_{t,3}^*, \theta_{t,4}^*, \ldots, \theta_{t,k}^*)$
`Sample` $\theta_{t+1,3}^* \sim p(\theta_3 \mid \boldsymbol{y}, \theta_{t+1,1}^*, \theta_{t+1,2}^*, \theta_{t,4}^*, \ldots, \theta_{t,k}^*)$
$\vdots \quad \vdots \quad \vdots$
`Sample` $\theta_{t+1,k}^* \sim p(\theta_k \mid \boldsymbol{y}, \theta_{t+1,1}^*, \theta_{t+1,2}^*, \theta_{t+1,3}^*, \ldots, \theta_{t+1,k-1}^*)$
$t \leftarrow (t+1)$

}

To really see what's going on it's instructive to visualize the MC data set, which from now on I'll call the *MCMC data set*. Table 2.3 illustrates this data set when Gibbs sampling is applied to the variance-components model (2.22). The MH algorithm creates a Markov chain whose stationary distribution is $p(\boldsymbol{\theta} \mid \boldsymbol{y})$, but you have to start the chain off somewhere and there's no guarantee that the chain will already be in equilibrium at the beginning of the sampling. The usual way to run an MH sampler is to try to (i) start it off at a vector of *initial values* which is close to a measure of center for the target distribution, such as the posterior mean or mode; (ii) run the chain until it's shrugged off its dependence on the initial values and reached equilibrium (this is called the *burn-in* phase); and then (iii) monitor the quantities of interest for a long enough period of time to get whatever Monte Carlo accuracy you want in the descriptive summaries of the MCMC draws. Thus MCMC sampling can be divided into three phases, which are usually called (i) *initialization* (iteration 0), (ii) *burn-in* (iterations $1, \ldots, b$), and (iii) *monitoring* (iterations $b+1, \ldots, b+m$). The draws in phases (i) and (ii) (rows 0 through b in the MCMC data set) are typically discarded.

Returning to the variance-components model (2.22), as was the case with the random-slopes regression model (2.28), Gibbs sampling proceeds most smoothly by treating the hospital random effects a_j^H as latent auxiliary variables to be sampled along with the parameters $(\beta_0, \sigma_H^2, \sigma_P^2)$. For the algorithm to work correctly it doesn't matter in what order the elements of $(\boldsymbol{\theta}, \boldsymbol{A})$ are updated; the ordering from left to right in Table 2.3 is as good as any. Having chosen initial values $(\beta_0)_0^*$, $(a_1^H)_0^*$, $\ldots$, $(a_J^H)_0^*$, $(\sigma_H^2)_0^*$, $(\sigma_P^2)_0^*$ in some way (I'll address this in more detail in the section below on MCMC diagnostics), row $t = 1$ in the MCMC data set is filled in as follows:

- Sample $(\beta_0)_1^*$ from $p\big[\beta_0 \mid \boldsymbol{y}, (a_1^H)_0^*, \ldots, (a_J^H)_0^*, (\sigma_H^2)_0^*, (\sigma_P^2)_0^*\big]$,
- Sample $(a_1^H)_1^*$ from $p\big[a_1^H \mid \boldsymbol{y}, (\beta_0)_1^*, (a_2^H)_0^*, \ldots, (a_J^H)_0^*, (\sigma_H^2)_0^*, (\sigma_P^2)_0^*\big]$,
- Sample $(a_2^H)_1^*$ from $p\big[a_2^H \mid \boldsymbol{y}, (\beta_0)_1^*, (a_1^H)_1^*, (a_3^H)_0^*, \ldots, (a_J^H)_0^*, (\sigma_H^2)_0^*, (\sigma_P^2)_0^*\big]$,

  and so on down to

- Sample $(a_J^H)_1^*$ from $p\big[a_I^H \mid \boldsymbol{y}, (\beta_0)_1^*, (a_1^H)_1^*, \ldots, (a_{J-1}^H)_1^*, (\sigma_H^2)_0^*, (\sigma_P^2)_0^*\big]$,
- Sample $(\sigma_H^2)_1^*$ from $p\big[\sigma_H^2 \mid \boldsymbol{y}, (\beta_0)_1^*, (a_1^H)_1^*, \ldots, (a_J^H)_1^*, (\sigma_P^2)_0^*\big]$, and
- Sample $(\sigma_P^2)_1^*$ from $p\big[\sigma_H^2 \mid \boldsymbol{y}, (\beta_0)_1^*, (a_1^H)_1^*, \ldots, (a_J^H)_1^*, (\sigma_H^2)_1^*\big]$.

The key idea is always to use the most recent value of each component of $(\boldsymbol{\theta}, \boldsymbol{A})$, which will always be either in the current row in the MCMC data set or the one above it.

An important practical detail not yet addressed is how to calculate the full conditional distributions. In the VC model (2.22), for example, taking β_0 first and letting $\boldsymbol{a}^H = (a_1^H, \ldots, a_J^H)$, the definition of conditional probability gives

$$p(\beta_0 \mid \boldsymbol{y}, \boldsymbol{a}^H, \sigma_H^2, \sigma_P^2) = \frac{p(\beta_0, \boldsymbol{y}, \boldsymbol{a}^H, \sigma_H^2, \sigma_P^2)}{p(\boldsymbol{y}, \boldsymbol{a}^H, \sigma_H^2, \sigma_P^2)} . \tag{2.30}$$

Notice, however, from the acceptance probability (2.29) that the full conditionals only need to be computed up to a constant multiple, as was true with the complete posterior distribution in (2.25). This means that anything that doesn't involve β_0 in the right-hand side of (2.30), such as the denominator, can simply be absorbed into a generic constant:

$$p(\beta_0 \mid \boldsymbol{y}, \boldsymbol{a}^H, \sigma_H^2, \sigma_P^2) = c\, p(\beta_0, \boldsymbol{y}, \boldsymbol{a}^H, \sigma_H^2, \sigma_P^2).$$

Next, again using the definition of conditional probability, and thinking about the hierarchical nature of how the model (2.22) defines its knowns and unknowns,

$$p(\beta_0 \mid \boldsymbol{y}, \boldsymbol{a}^H, \sigma_H^2, \sigma_P^2) = c\, p(\beta_0, \sigma_H^2, \sigma_P^2)\, p(\boldsymbol{a}^H \mid \beta_0, \sigma_H^2, \sigma_P^2) \\ \times p(\boldsymbol{y} \mid \beta_0, \boldsymbol{a}^H, \sigma_H^2, \sigma_P^2).$$

Now (a) little of value is usually lost in multilevel modeling by taking the fixed effects and the random-effects variances to be independent[33] in the prior, so that $p(\beta_0, \sigma_H^2, \sigma_P^2) = p(\beta_0)\, p(\sigma_H^2)\, p(\sigma_P^2)$, which can be taken to be $c\, p(\beta_0)$ in this calculation; (b) the conditional distribution of $\boldsymbol{a}^H$ given $(\beta_0, \sigma_H^2, \sigma_P^2)$ depends only on σ_H^2 and again can be taken to be constant in this calculation; and (c) the conditional (sampling) distribution[34] of y_{ij} given $(\beta_0, \boldsymbol{a}^H, \sigma_H^2, \sigma_P^2)$ is $\mathcal{N}(\beta_0 + a_j^H, \sigma_P^2)$. Thus, after a bit more simplification,

$$\begin{aligned} p(\beta_0 \mid \boldsymbol{y}, \boldsymbol{a}^H, \sigma_H^2, \sigma_P^2) &= p(\beta_0 \mid \boldsymbol{y}, \boldsymbol{a}^H, \sigma_P^2) \\ &= c\, p(\beta_0) \exp\left[-\frac{1}{2\sigma_P^2} \sum_{j=1}^{J} \sum_{i=1}^{n_j} (y_{ij} - \beta_0 - a_j^H)^2 \right]. \end{aligned} \quad (2.31)$$

The full conditional likelihood for β_0—the exponential expression in (2.31), viewed as a distribution in β_0 for fixed $(\boldsymbol{y}, \boldsymbol{a}^H, \sigma_P^2)$—is

$$\begin{aligned} l(\beta_0 \mid \boldsymbol{y}, \boldsymbol{a}^H, \sigma_P^2) &= c \exp\left[-\frac{1}{2\sigma_P^2} \sum_{j=1}^{J} \sum_{i=1}^{n_j} (y_{ij} - \beta_0 - a_j^H)^2 \right] \\ &= \mathcal{N}\left[\frac{1}{N} \sum_{j=1}^{J} \sum_{i=1}^{n_j} (y_{ij} - a_j^H), \frac{\sigma_P^2}{N} \right]. \end{aligned} \quad (2.32)$$

This demonstrates that the conditional conjugate choice for the prior distribution for β_0 in this model, as far as Gibbs sampling is concerned, is normal: with this choice you can verify that the full conditional for β_0 will also be normal. Prior distributions in multilevel modeling will be discussed more fully below; for now it's enough to note that if you want to specify a diffuse prior for β_0 you can do so in a conditionally conjugate way by choosing a normal distribution with any mean you like and a huge variance $\sigma_{\beta_0}^2$. In the limit as $\sigma_{\beta_0}^2 \to \infty$ (the ultimate in diffuseness) the prior distribution β_0 would tend to a constant and the full conditional for β_0 would just be the Gaussian distribution in (2.32). Of course there's no such thing as a proper distribution which is constant on $(-\infty, \infty)$, because the area under such a curve would be infinite.

---

[33] Typically the data set will be sufficiently informative that the appropriate degree of correlation between these parameters in the posterior will be learned via the likelihood.

[34] You can begin to see why it's useful in multilevel MCMC to sample the random effects $\boldsymbol{a}^H$ along with the parameters: computing the sampling distribution of the y_{ij} without conditioning on the a_j^H would require integrating over the random effects, and while this can be done analytically in Gaussian random-effects models (because a mixture of Gaussians is still Gaussian) it gives a hint of how difficult things can become in non-Gaussian random-effects models if you don't sample the random effects.

$p(\beta_0) = c$ is another example of an improper prior (like the Beta(0, 0) prior in Section 2.1.2); this topic will be examined in more detail in Section 2.3.3.

A similar calculation reveals that the a_j^H are conditionally independent given $(\boldsymbol{y}, \beta_0, \sigma_H^2, \sigma_P^2)$ and that

$$p(a_j^H \mid \boldsymbol{y}, \beta_0, \sigma_H^2, \sigma_P^2) = \mathcal{N}\left[\frac{V_j}{\sigma_P^2} \sum_{i=1}^{n_j} (y_{ij} - \beta_0), V_j\right],$$

where $V_j = (n_j / \sigma_P^2 + 1/\sigma_H^2)^{-1}$. As for σ_H^2, logic similar to that underlying the full conditional for β_0 (and again assuming independence of β_0, σ_H^2, and σ_P^2 in the prior) yields

$$p(\sigma_H^2 \mid \boldsymbol{y}, \beta_0, \boldsymbol{a}^H, \sigma_P^2) = c\, p(\sigma_H^2, \boldsymbol{y}, \beta_0, \boldsymbol{a}^H, \sigma_P^2) = c\, p(\sigma_H^2)\, p(\boldsymbol{a}^H \mid \sigma_H^2)$$

(because $p(\boldsymbol{y} \mid \beta_0, \boldsymbol{a}^H, \sigma_H^2, \sigma_P^2) = \mathcal{N}(\beta_0 + a_j^H, \sigma_P^2)$ doesn't depend on σ_H^2), and this is

$$\begin{aligned} p(\sigma_H^2 \mid \boldsymbol{y}, \beta_0, \boldsymbol{a}^H, \sigma_P^2) &= p(\sigma_H^2 \mid \boldsymbol{a}^H) \\ &= c\, p(\sigma_H^2) \prod_{j=1}^{J} (\sigma_H^2)^{-1/2} \exp\left[-\frac{(a_j^H)^2}{2\sigma_H^2}\right] \\ &= c\, p(\sigma_H^2)\, (\sigma_H^2)^{-J/2} \exp\left[-\frac{\sum_{j=1}^{J} (a_j^H)^2}{2\sigma_H^2}\right]. \end{aligned} \tag{2.33}$$

Leaving aside the prior $p(\sigma_H^2)$ for the moment, the rest of (2.33)—the full conditional likelihood for σ_H^2 given $\boldsymbol{a}^H$—is recognizable as a member of the Scaled Inverse χ^2 family:

$$l(\sigma_H^2 \mid \boldsymbol{a}^H) = \chi^{-2}\left[J - 2, \frac{\sum_{j=1}^{J} (a_j^H)^2}{J - 2}\right].$$

As noted below (2.26), the Scaled Inverse χ^2 family is conditionally conjugate for Gibbs sampling in Gaussian models of this type; taking $p(\sigma_H^2) = \chi^{-2}(\nu_H, s_H^2)$, the full conditional for σ_H^2 becomes

$$p(\sigma_H^2 \mid \boldsymbol{a}^H) = \chi^{-2}\left[\nu_H + J, \frac{\nu_H s_H^2 + \sum_{j=1}^{J} (a_j^H)^2}{\nu_H + J}\right].$$

Finally, a similar calculation shows that

$$\begin{aligned} & p(\sigma_P^2 \mid \beta_0, \boldsymbol{y}, \boldsymbol{a}^H, \sigma_H^2) \\ & \quad = c\, p(\sigma_P^2)\, (\sigma_P^2)^{-N/2} \exp\left[-\frac{\sum_{j=1}^{J} \sum_{i=1}^{n_j} (y_{ij} - \beta_0 - a_j^H)^2}{2\sigma_P^2}\right] \\ & \quad = c\, p(\sigma_P^2)\, \chi^{-2}\left[N - 2, \frac{\sum_{j=1}^{J} \sum_{i=1}^{n_j} (y_{ij} - \beta_0 - a_j^H)^2}{N - 2}\right], \end{aligned}$$

so that the conditional conjugate prior for σ_P^2 is also Scaled Inverse χ^2, and with the choice $p(\sigma_P^2) = \chi^{-2}(\nu_P, s_P^2)$ the full conditional for σ_P^2 is[35]

$$p(\sigma_P^2 \mid \beta_0, \boldsymbol{y}, \boldsymbol{a}^H) = \chi^{-2}\left[\nu_P + N, \frac{\nu_P s_P^2 + \sum_{j=1}^{J} \sum_{i=1}^{n_j} (y_{ij} - \beta_0 - a_j^H)^2}{\nu_P + N}\right].$$

It would evidently not be pleasant to be forced routinely to make detailed calculations of full conditional distributions to perform Gibbs sampling in multilevel models. Fortunately, at least two rather general-purpose computer programs are available at this writing which make these calculations for you automatically: `WinBUGS` [70] and `MLwiN` [60].[36] `WinBUGS` can fit a broader class of Bayesian models than `MLwiN`, but the coding in `MLwiN` has been optimized in such a way that it often takes less CPU time than `WinBUGS` to achieve the same level of MCMC accuracy (some efficiency comparisons will be given below).

It's clear from Algorithm 2.3 and the discussion surrounding it that the single-component Metropolis-Hastings (MH) sampler offers immense flexibility in implementation: for example, you're free to use Gibbs updating for some components of the vector $(\boldsymbol{\theta}, \boldsymbol{A})$ of unknowns-plus-auxiliary-variables, and Metropolis or MH updating for other components. This is sometimes referred to as a *hybrid* Metropolis-Gibbs approach (even though it's all really MH sampling). `WinBUGS` generally attempts to use Gibbs sampling whenever possible, often employing *adaptive rejection sampling* (ARS), a method developed by Gilks and Wild [41], to sample from the full conditionals. If the distributions needed for Gibbs sampling are concave on the log scale, envelope functions can be created in a piecewise linear fashion, using tangents to the log full conditionals as in Fig. 2.2. ARS proceeds adaptively to create an increasingly tighter envelope by adding a new tangent line at each sampled point, so that the rejection probability goes down as the sampling unfolds. If some of the full conditionals are not log concave, `WinBUGS` uses a hybrid approach based on Gibbs sampling via ARS when possible and MH sampling otherwise. `MLwiN` typically uses a different hybrid strategy, which I'll now describe in the context of random-effects logistic regression (RELR) models.

---

[35] The conditionally conjugate choices for the prior distributions for β_0, σ_H^2, and σ_P^2 examined here have been motivated by computational convenience; other priors could of course be used, but (a) diffuse prior choices are easy to make in the conditionally conjugate families and (b) these families will often also be adequate approximations in a wide variety of situations when stronger prior information is available.

[36] Bill Browne (at the University of Bristol) and I were the co-developers of the Bayesian MCMC capabilities in version 1.0 of `MLwiN` in 1998. Bill has since gone on to greatly enhance the range of models that can be fit via MCMC in `MLwiN`; see Browne et al. [14] for details.

Metropolis and Hybrid Sampling Strategies in Multilevel Models with Dichotomous Responses

When the outcome variable in a multilevel investigation is binary and random-effects models are called for—an example is the RELR model (2.3)—the MCMC fitting process becomes more involved, because Gibbs sampling in RELR models is not straightforward. For instance, in the even simpler version of (2.3) in which no predictor variable x is available, and taking all prior distributions to be uniform for simplicity, the full conditional distribution for β_0 is

$$p(\beta_0 \mid \boldsymbol{y}, \boldsymbol{u}, \sigma_u^2) = c \prod_{ij} \left(1 + e^{-\beta_0 - u_j}\right)^{-y_{ij}} \left(1 + e^{\beta_0 + u_j}\right)^{y_{ij}-1} .$$

This distribution does not lend itself readily to direct sampling. Rejection sampling [75] is possible, and (as mentioned above) `WinBUGS` employs adaptive rejection sampling. `MLwiN` uses a hybrid Metropolis-Gibbs approach in RELR models which involves two steps: (a) a particular form of adaptive Metropolis sampling for the fixed effects, such as β_0 in (2.3), and the random effects u_j, treated as latent auxiliary variables as usual, and (b) Gibbs sampling for the random-effects variances.

Since the fixed and random effects live on the whole real line, the simplest choice for the proposal distributions (PDs) in the Metropolis sampling, if all of the fixed and random effects are to be updated one at a time, is a series of univariate Gaussian distributions, but it still remains to specify the location and scale of these PDs. Consider (as an example) the intercept β_0 in model (2.3), and imagine that the sampler is at some value $\beta_{0(t)}$ at time t. A simple way to specify a Gaussian PD for this parameter would be to use the normal distribution $\mathcal{N}(\beta_0^* \mid \beta_{0(t)}, \sigma_{\beta_0}^2)$ centered at where the sampler is now, $\beta_{0(t)}$, and with some PD variance $\sigma_{\beta_0}^2$. This PD has the property that it depends on β_0^* and $\beta_{0(t)}$ only through the distance $|\beta_0^* - \beta_{0(t)}|$ between them, which is the defining characteristic of a *random-walk* Metropolis sampler, and this idea fixes the location of the PDs.

`MLwiN` uses random-walk Metropolis on the fixed and random effects in RELR models, and chooses the PD variances adaptively to avoid the extremes illustrated by Fig. 2.3: in the top panel of this figure the PD variance is too big, leading to an acceptance probability that's too low, and in the bottom panel the scale of the PD is too small, resulting in an acceptance probability that's too high. I mentioned in Section 2.3.2 that the optimal acceptance rate for one-dimensional (random-walk) MH samplers with Gaussian PDs when the target distribution is approximately normal is about 44%; Browne and Draper [11, 12, 13] used this fact to equip `MLwiN` with the following simple adaptive method for choosing the PD variances. From starting values based on the estimated covariance matrices of the MLEs for the parameters in the given model, the method first employs a sampling period of random length (but with an upper bound) during which the proposal distribution variances are adaptively tuned and eventually fixed for the remainder of the run; this is followed by a burn-in period (see Section 2.4.1); and then the main monitoring run from which posterior summaries are calculated occurs. The tuning of the proposal distribution variances is based on achieving an acceptance rate for each parameter that lies within a specified tolerance interval $(r - \delta, r + \delta)$.

The algorithm examines empirical acceptance rates in batches of 100 iterations, comparing them for each parameter with the tolerance interval and modifying the proposal distribution appropriately before going on to the next batch of 100. With r^* as the acceptance rate in the most recent batch and σ_p as the proposal distribution SD for a given parameter, the modification performed at the end of each batch is as follows:

$$\texttt{If} \quad r^* \geq r \quad \texttt{then} \quad \sigma_p \to \sigma_p \left(2 - \frac{1 - r^*}{1 - r} \right) \quad \texttt{else} \quad \sigma_p \to \frac{\sigma_p}{2 - r^*/r} \, .$$

This modifies the proposal standard deviation by a greater amount the farther the empirical acceptance rate is from the target r. If r^* is too low, the proposed moves are too big, so σ_p is decreased; if r^* is too high, the parameter space is being explored with moves that are too small, and σ_p is increased. If the r^* values are within the tolerance interval during three successive batches of 100 iterations, the parameter is marked as satisfying its tolerance condition, and once all parameters have been marked the overall tolerance condition is satisfied and adapting stops. After a parameter has been marked it's still modified as before until all parameters are marked, but each parameter only needs to be marked once for the algorithm to end. To limit the time spent in the adapting procedure an upper bound is set (the `MLwiN` default is 5,000 iterations) and after this time the adapting period ends regardless of whether the tolerance conditions are met (in practice this occurs rarely). Values of $(r, \delta) = (0.5, 0.1)$ appear to give near-optimal univariate-update Metropolis performance for a wide variety of multilevel models [11, 12, 13].

To give some examples of MCMC efficiency comparisons, Browne and Draper [11, 12, 13] and Browne [10] have gathered information in a wide variety of multilevel models that can be fit both by `WinBUGS` using Gibbs sampling via ARS[37] and by `MLwiN` using adaptive hybrid Metropolis-Gibbs sampling. In addition to univariate Metropolis updating in multilevel modeling, as noted in the discussion surrounding (2.28), it's also possible to update parameters in L sets of blocks with multivariate proposal distributions, where L is the number of levels in the model. In models with Gaussian responses, `MLwiN` uses (a) Gibbs sampling on the random-effects variances and (b) multivariate normal

[37] `WinBUGS` version 1.4.1 allows the user to specify MH sampling instead of Gibbs sampling via ARS, but the latter is still the default.

PDs on everything else, with the fixed effects forming one block and the other $L-1$ groups of n_l blocks of size n_{rl} comprising all of the random effects at level $2, \ldots, L$, respectively, where n_l is the number of blocks at level l and n_{rl} is the number of random effects per block at level l.

- In the two-level RELR models examined in the work summarized here, Gibbs sampling via ARS was the most efficient method per MCMC iteration (in the sense of producing MCMC output with smaller levels of positive autocorrelation), but ARS was much slower per iteration than Metropolis; the winner in CPU time to achieve the same level of MCMC accuracy (as measured by the default *Raftery-Lewis* MCMC diagnostic; see Section 2.4.2) was multivariate Metropolis, by factors ranging from 1.7 to 9.0;
- In multilevel models involving *heteroscedasticity* (unequal random-effects variances, which may be modeled as a function of predictor variables) at one or more levels of the hierarchy, Metropolis sampling was 4.1–9.1 times faster than ARS to achieve default Raftery-Lewis accuracy in the examples studied;
- In multilevel models with *multivariate Gaussian responses*, `MLwiN`'s approach based on Gibbs sampling with block updating was 3.3 times faster than ARS; and
- In one particular example involving a multilevel *measurement error* model, `MLwiN`'s version of Gibbs sampling was 67 times faster than the `WinBUGS` Gibbs implementation (the clock time comparison was 1 hour 14 minutes versus 1.1 minutes to obtain 50,000 monitoring iterations on a 3 GHz PC).

These results are anecdotal but are typical of many examples studied. The principal reason for these efficiency findings appears to be that ARS's generality is bought at the price of considerable computational overhead in creating and adaptively improving the rejection-sampling envelope function. (Of course, none of these comparisons reflect the fact that the class of models that can currently be fit with `WinBUGS` is considerably larger than the range of models available at present via `MLwiN`; see `www.mrc-bsu.cam.ac.uk` and `www.cmm.bristol.ac.uk/MLwiN/index.shtml` for details.)

2.3.3 Prior Distributions for Multilevel Analysis

As with all Bayesian inference, broadly speaking two classes of prior distributions are available for multilevel models: (a) diffuse and (b) non-diffuse, corresponding to situations in which (a) little is known about the quantities of interest *a priori* or (b) substantial prior information is available, for instance from previous studies judged relevant to the current data set. In situation (a), on which I'll focus here, it seems natural to seek prior specifications that lead to well-calibrated inferences [e.g., 21], which I'll take to mean point estimates with little bias and interval estimates whose actual coverage is close to the nominal level (in both cases in repeated sampling). As mentioned in footnote 33, when the goal is a diffuse prior specification it's customary (and often does little harm) to take all parameters to be independent in the prior for convenience (on the ground that the likelihood will provide the appropriate correlation structure in the posterior), and—with one exception to be discussed below—I'll follow that practice here.

There is an extensive literature on the specification of diffuse priors [e.g., 2, 37, 70], leading in some models to more than one intuitively reasonable approach. It's sometimes stated in this literature that the performance of the resulting Bayesian estimates is broadly insensitive, with moderate to large sample sizes, to how the diffuse prior is specified. In preliminary studies in joint work with Bill Browne, we found this to be the case for fixed effects in a wide variety of multilevel models; as a result `MLwiN` uses (improper) priors that are uniform on the real line $\mathbb{R}$ for such parameters (these are functionally equivalent to proper Gaussian priors with huge variances). As others [e.g., 29] have elsewhere noted, however, we found noticeable differences in performance across plausible attempts to construct diffuse priors for random-effects variances in both model classes. Intuitively (as mentioned toward the end of Section 2.2) this is because the effective sample size for the level-2 variance in a two-level analysis with J level-2 units (e.g., hospitals) and N total level-1 units (e.g., patients; typically $J \ll N$) is often much closer to J than to N; in other words, in the language of the example near (2.22), even with data on hundreds of patients the likelihood information about the between-hospital variance can be fairly weak when the number of hospitals is modest, so that prior specification can make a real difference in such cases.

The off-the-shelf (improper) choice for a diffuse prior on a variance in many Bayesian analyses is $p(\sigma^2) = c/\sigma^2$, which is equivalent to assuming that $\log(\sigma^2)$ is uniform on $\mathbb{R}$. This is typically justified by noting that the posterior for σ^2 will be proper even for very small sample sizes; but [e.g., 30] this choice can lead to improper posteriors in random-effects models. `MLwiN` avoids this problem by using two alternative diffuse (but proper) priors, both of which produce proper posteriors:

- A locally uniform prior for σ^2 on $(0, 1/\epsilon)$ for small positive ϵ [17, 38], which is equivalent to a Pareto$(1, \epsilon)$ prior for the *precision* $\tau = 1/\sigma^2$ [70]; and
- A $\Gamma^{-1}(\epsilon, \epsilon)$ prior for σ^2 [70], for small positive ϵ.

Both of these priors are members of the $\chi^{-2}(\nu, s^2)$ family: the Uniform and Inverse Gamma priors just mentioned are formally specified by the choices $(\nu, s^2) = (-2, 0)$ and $(2\epsilon, 1)$, respectively (in the former case in the limit as $\epsilon \to 0$). We have found that results are generally insensitive to the specific choice of ϵ in the region of 0.001. (Earlier versions of the examples manual for `WinBUGS` [69] frequently employed $\Gamma(0.001, 0.001)$ marginal priors for quantities (such as precisions) which live on the positive part of the real line, and more recent versions still give results with this prior for comparison. Evidence is beginning to emerge that uniform priors on the standard deviation scale may have even better calibration properties than those described below; see [12] and the discussion therein.)

The exception I mentioned above to the idea of making all the parameters independent in the prior arises in models, such as the random-slopes regression formulation (2.28), in which random effects at more than one level are jointly modeled with a covariance matrix $\boldsymbol{V}$. In the same way that priors for a variance are typically either expressed on the scale of σ^2 or its reciprocal, priors for covariance matrices are usually either specified in terms of $\boldsymbol{V}$ or $\boldsymbol{V}^{-1}$. With Gaussian random effects the conditionally conjugate prior choice for the inverse of a covariance matrix is the *Wishart* family, a multivariate generalization of the Gamma distribution. In the parameterization used, for example, by Gelman et al. [37], the Wishart distribution $W_k(\nu, \boldsymbol{S})$ for a $k \times k$ matrix $\boldsymbol{W}$ has density

$$p(\boldsymbol{W}) = c\,|\boldsymbol{W}|^{(\nu-k-1)/2} \exp\left[-\frac{\mathrm{tr}(\boldsymbol{S}^{-1}\boldsymbol{W})}{2}\right];$$

in this expression $|\boldsymbol{A}|$ and $\mathrm{tr}(\boldsymbol{A})$ denote the determinant and trace of the matrix $\boldsymbol{A}$, respectively, and the density is only defined over positive definite matrices $\boldsymbol{W}$ and for $\nu \geq k$. This distribution has mean $E(\boldsymbol{W}) = \nu \boldsymbol{S}$, so (by analogy with the $\chi^{-2}(\nu, s^2)$ distribution) specifying $W_k(\nu, \boldsymbol{S})$ as a prior for $\boldsymbol{V}^{-1}$ is roughly equivalent to supplying $\boldsymbol{S}$ as a prior estimate of $\boldsymbol{V}$ based on ν prior "observations." Small values of ν thus lead to relatively diffuse specifications; for example, the default prior for a $k \times k$ covariance matrix $\boldsymbol{V}$ in `MLwiN` is $p(\boldsymbol{V}^{-1}) = W_k(k, \hat{\boldsymbol{V}})$, where $\hat{\boldsymbol{V}}$ is the MLE for $\boldsymbol{V}$. This is gently data-determined, but usually the effective sample size in the data for learning about $\boldsymbol{V}$ is so much larger than k that all reasonable choices of $\boldsymbol{S}$ in the $W_k(k, \boldsymbol{S})$ distribution as a prior for $\boldsymbol{V}$ yield essentially the same conclusions.

There's relatively little information in the literature about the calibration performance of diffuse priors in multilevel modeling. An exception is Browne and Draper [12], which presents results from large simulation studies in two multilevel settings: the variance-components model (2.22) and the three-level RELR model[38]

$$\begin{aligned} (y_{ijk} \mid p_{ijk}) &\sim \mathrm{Bernoulli}(p_{ijk}), \quad \text{with} \\ \mathrm{logit}(p_{ijk}) &= \beta_0 + \beta_1 x_{1k} + \beta_2 x_{2jk} + \beta_3 x_{3ijk} + v_k + u_{jk}\,, \end{aligned} \tag{2.34}$$

where y_{ijk} is a binary outcome variable and in which $v_k \sim \mathcal{N}(0, \sigma_v^2)$ and $u_{jk} \sim \mathcal{N}(0, \sigma_u^2)$. The RELR model in this work involved a design configuration based on a medical study (Rodríguez and Goldman [63]) of 2,449 births by

[38] Browne and Draper [11] offers similar results in random-slopes regression models.

1,558 women living in 161 communities in Guatemala; the VC simulation study, motivated by an educational example with pupils nested in schools, permitted the number of schools to range from 6 to 48 and the numbers of pupils per school to vary from 5 to 62 in such a way that the average number of pupils per school was 18 (the resulting total numbers of pupils varied from 108 to 864). The parameters in (2.34) in the RELR simulation were set to values similar to those in the original Rodríguez-Goldman study; the random-effects variances in the VC simulations were chosen to span a wide range of intraclass correlation values from 0.012 to 0.5.

A comparison was made between likelihood-based and Bayesian diffuse-prior methods, using bias of point estimates and nominal versus actual coverage of interval estimates in repeated sampling as evaluation criteria. As mentioned in Section 2.1, maximum likelihood estimates (and *restricted ML* (REML) estimates which attempt to achieve approximate unbiasedness) are readily found in VC models but are considerably more difficult to compute in RELR models, because numerical integration is required over the random effects to evaluate the likelihood function; as a result the likelihood-based methods in most frequent use at present are *quasi-likelihood* techniques based on linear approximations to the nonlinear RELR model. The results of the simulations were as follows.

- In two-level VC models (a) both likelihood-based and Bayesian approaches can be made to produce approximately unbiased estimates, although the automatic manner in which REML achieves this is an advantage, but (b) both approaches had difficulty achieving nominal coverage of interval estimates in small samples and with small values of the intraclass correlation.
- With the three-level RELR model examined, (c) quasi-likelihood methods for estimating random-effects variances perform badly with respect to bias and coverage in the example studied, and (d) Bayesian diffuse-prior methods lead to well-calibrated point and interval RELR estimates.

One important likelihood-Bayesian comparison I've not yet addressed is computational speed, where likelihood-based approaches have a distinct advantage (for example, quasi-likelihood fitting of model (2.34) to the original Rodríguez-Goldman data set takes 2.7 seconds on a 3 GHz PC versus 1.8 minutes using MCMC with 25,000 monitoring iterations). It's common practice in statistical modeling to examine a variety of models on the same data set before choosing a small number of models for reporting purposes (although this practice by itself encourages underpropagation of model uncertainty, e.g., Draper [25]). The results in Browne and Draper [12] suggest a hybrid modeling strategy, in which likelihood-based methods like those described here are used in the model exploration phase and Bayesian diffuse-prior methods are used for the reporting of final inferential results. Other analytic strategies based on less approximate likelihood methods are also possible but would benefit from further study of the type summarized above.

2.4 MCMC Diagnostics

While MCMC methods offer a promising path toward a solution to the numerical integration problem at the heart of Bayesian computations, it's evident from Section 2.3.2 that some new technical challenges arise in implementing such methods: What should you use for starting values? How long should the burn-in period be? (Equivalently, how do you know when the Markov chain has reached equilibrium?) How long do you need to monitor the chain to get results of sufficient Monte Carlo accuracy? A burgeoning literature on *MCMC diagnostics* to help answer these questions has developed over the last 15 years; see, e.g., Cowles and Carlin [19] and Brooks and Roberts [9] for good reviews. A number of the most promising diagnostics have been distributed by Best et al. [4] in a collection of programs called `CODA` (written in the `S-PLUS`® [47] and `R` languages). I'll confine my coverage of this topic here to a discussion of a few of the most useful diagnostic methods in multilevel modeling; these tend to be methods (a subset of the techniques in `CODA`) available in software such as `MLwiN` and `WinBUGS`.

2.4.1 Starting Values and the Length of the Burn-In Period

On the subjects of where to start the chain and how long the burn-in phase should be (which are of course related: the worse the starting values, the longer the burn-in needs to be), it helps to reach equilibrium quickly if you can initialize the chain somewhere near a measure of center of the relevant posterior distribution, such as its mean or mode. With diffuse priors this suggests using maximum-likelihood estimates for initial values, since the mode of the posterior and the maximum of the likelihood distribution will then be close. This strategy is particularly well suited to a package like `MLwiN` which permits both maximum-likelihood and MCMC fitting of multilevel models, and in fact the MLEs are the default starting values in `MLwiN`. Browne and Draper [11] have demonstrated anecdotally that a short burn-in period of only 500 iterations from MLE starting values is more than adequate to reach equilibrium in a remarkably wide variety of multilevel models; this is the default burn-in behavior in `MLwiN`. `WinBUGS` (and other software) users who do not have ready access to the MLEs often try generic starting values, such as 0 for any parameter whose range is the entire real line (fixed-effects regression coefficients, for example) and 1 for any parameter which lives only on $(0, \infty)$ (such as random-effects variances or precisions); this approach may fail in the sense that the MCMC software is unable to begin sampling from such a poor starting place, and even when this type of failure does not occur a longer burn-in period may be necessary.

The only real possibility, in Bayesian inference via MCMC, of obtaining a truly inaccurate summary of the correct posterior distribution arises when the posterior is multimodal and, for whatever reason, the sampler is run in such a way that not all of the modes are discovered. Suppose, for example, that the posterior has two modes which are far apart in parameter space, and your sampling strategy (Metropolis, say) is as follows: you start the chain off near one of the modes and (unknown to you) the proposal distribution standard deviation you're using is too small to permit discovery of the other mode quickly. Theoretically your sampler will still (with probability 1) find the other mode eventually, but this may not occur until millions of iterations have been performed, and nothing in the output of the sampler in the first 50,000 or 100,000 iterations will give you any clue that there's anything wrong. Gelman and Rubin [38] have developed a simple diagnostic (implemented in `CODA` and `WinBUGS`) for detecting multimodality based upon the idea of running multiple chains from widely dispersed starting values and performing an analysis of variance to see if the between-chain variability is large in relation to the average variation within the chains (if so this would indicate more than one mode). Fortunately multimodality is rare when fitting multilevel models in situations which typically arise in practice; this problem should not arise when the data provide substantial likelihood information about the parameters of interest and the prior information is relatively diffuse in relation to the likelihood.

2.4.2 The Required Length of the Monitoring Run

Once equilibrium has been reached it may still be true that an optimized version of an approach such as Metropolis sampling will produce output with sufficiently high autocorrelation that tens of thousands of iterations may be needed to achieve respectable Monte Carlo accuracy. I'll examine one possible approach to determining how long the monitoring run should be by looking at some output of the `MLwiN` package in a simple variance components model. Figure 2.4 presents results from MCMC fitting of the model

$$
\begin{aligned}
& y_{ij} = \beta_0 + \beta_1 x_{ij} + a_j^S + a_{ij}^P, \quad j = 1, \ldots, J, \quad i = 1, \ldots, n_j, \\
& \sum_{j=1}^{J} n_j = N, \quad a_j^S \stackrel{iid}{\sim} \mathcal{N}(0, \sigma_S^2), \quad a_{ij}^P \stackrel{iid}{\sim} \mathcal{N}(0, \sigma_P^2)
\end{aligned} \tag{2.35}
$$

to a 2-level data set from education obtained by Goldstein et al. [44]; these authors chose a random sample of $J = 65$ schools (factor S) from the Inner London Education Authority in the late 1980s and then sampled a total of 4,059 16-year-old pupils (factor P) at random from the chosen schools (a

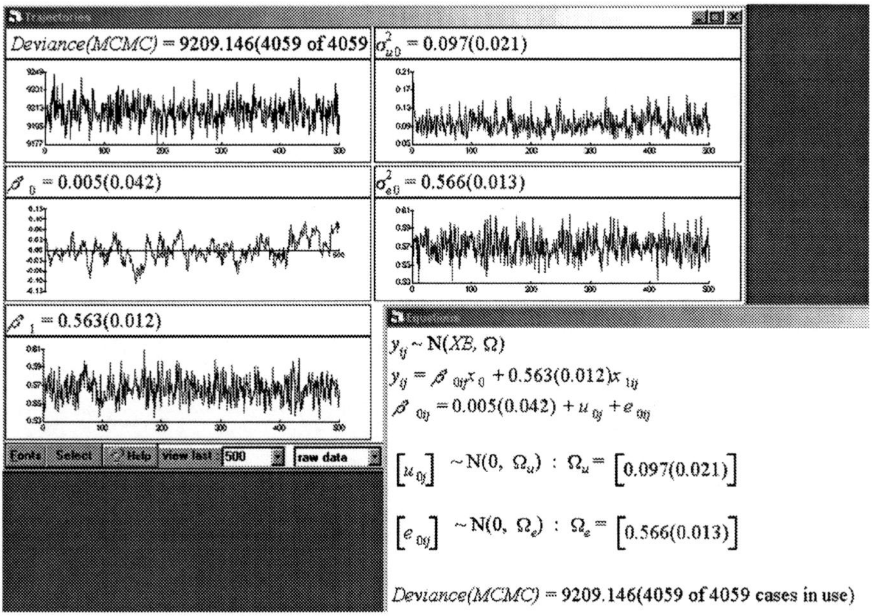

Fig. 2.4 Output of `MLwiN` initial MCMC fitting of model (2.35) to the Goldstein et al. [44] educational data.

single-stage cluster sample). Here y_{ij} is a normalized examination score at age 16 and x_{ij} is a score on a standardized reading test at age 11 (both variables were linearly transformed to have mean 0 and SD 1). Figure 2.4 summarizes where things stand after `MLwiN`'s default hybrid Metropolis-Gibbs sampler (described in Section 2.3.2) has performed a burn-in of 500 iterations from the maximum-likelihood estimates in (2.35) and an initial monitoring run of 5,000 iterations. The `Equations` window in the lower right corner gives the current posterior means and SDs of the model parameters (based on the 5,000 monitoring iterations): in the notation of (2.35), and regarding the posterior means as point estimates, the current values (with posterior SDs in parentheses) are $\hat{\beta}_0 = 0.005$ (0.042), $\hat{\beta}_1 = 0.563$ (0.012), $\hat{\sigma}_S^2 = 0.097$ (0.021), and $\hat{\sigma}_P^2 = 0.566$ (0.013). The `Trajectories` window in the upper left corner presents time series traces of the most recent 500 iterations for each of the four parameters in the model (plus the *deviance*—a value based on the log likelihood of the model evaluated at the current parameter estimates—which can be used to assess the fit of the model; see, e.g., Spiegelhalter et al. [68] for applications of the deviance to this task, some of which are controversial). It's evident that some parameters are mixing better than others; the rate of MCMC learning about β_0 is particularly slow, but that doesn't matter here

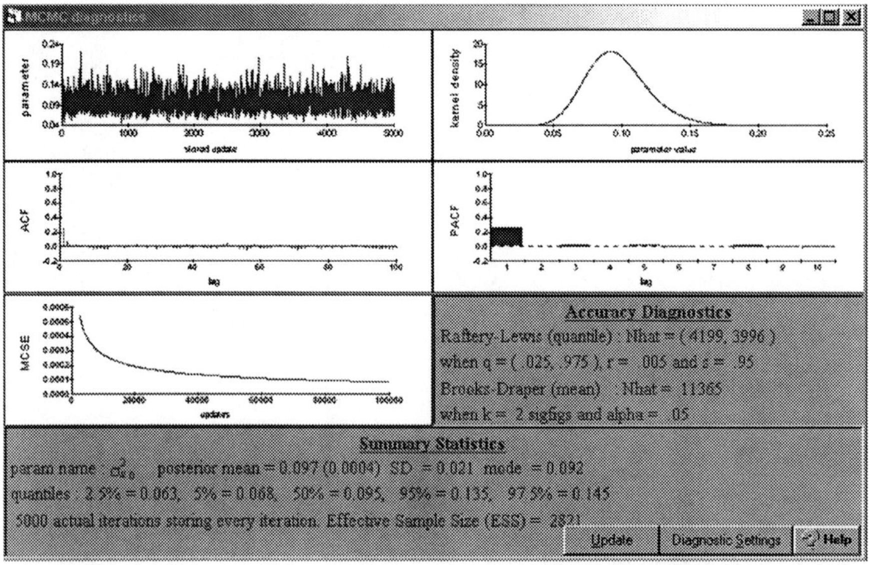

Fig. 2.5 `MLwiN` MCMC diagnostics for σ_S^2 in model (2.35), when fit to the educational data set examined in Fig. 2.4.

because the linear transformations of both x and y in this model mean that β_0 should be zero, rendering this parameter effectively ignorable.

Consider σ_S^2, which is referred to in `MLwiN` as σ_{u0}^2. Clicking on this parameter in the `Trajectories` window yields the default set of `MLwiN` MCMC diagnostics about σ_S^2, presented in Fig. 2.5. Five plots are given, together with a variety of numerical diagnostics and summaries. The upper left plot is a time series trace of all 5,000 monitoring iterations for σ_S^2, and the upper right plot is a *kernel density trace* (an estimate of the marginal posterior $p(\sigma_S^2 \mid \boldsymbol{y})$; see, e.g., Silverman [67] for details), which has a gentle degree of skewness that's about what you'd expect for a variance parameter (note that the posterior mode, median, and mean for σ_S^2, all of which are given in the `Summary Statistics` part of the display, are estimated as 0.092, 0.095, and 0.097, respectively, a pattern that's consistent with modest right skewness). The left-hand graph in the second row of the figure is a plot of the estimated autocorrelation function (ACF) for σ_S^2; as indicated in footnote 27, this measures the degree to which σ_S^2 at lag t is correlated with itself at lags $t-1$, $t-2$, and so on (a purely IID time series would have autocorrelation 1 at lag 0 and estimated autocorrelations near zero at all other lags).

The remaining two plots require a brief digression on the subject of *autoregressive* models for time series [e.g., 18]. Letting e_t denote an IID (white noise or purely random) process with mean 0 and variance σ_e^2, the time series θ_t^* is said to be an *autoregressive process of order p* (AR(p)) if

$$\theta_t^* = \alpha_1 \, \theta_{t-1}^* + \cdots + \alpha_p \, \theta_{t-p}^* + e_t. \tag{2.36}$$

Equation (2.36) is like a multiple regression model except that θ_t^* is being regressed on past values of itself instead of on other predictor variables; this gives rise to the term autoregressive. The right-hand graph in the second row of Fig. 2.5 is the estimated *partial autocorrelation function* (PACF) for σ_S^2. The PACF for a time series θ_t^* measures the excess correlation between θ_t^* and θ_{t+k}^* not accounted for by the autocorrelations at lags 1 through $k-1$, and is useful in diagnosing the order of an AR(p) process: if θ_t^* is AR(p) then the PACF at lags $1, \ldots, p$ will be significantly different from 0 and then close to 0 at lags larger than p. Evidently in this case the series for σ_S^2 behaves like an AR(1) process with first-order serial correlation of about $\hat{\rho}_1 \doteq +0.3$ (the height of the spike at lag 1 in both the ACF and PACF). The theoretical ACF for an AR(1) series exhibits geometric decay, with the autocorrelation at lag 2 related to that at lag 1 by $\rho_2 = \rho_1^2$, followed by $\rho_3 = \rho_1^3$, and so on, and you can see that the estimated ACF does have this behavior here.

The reason I bring this up is that when the output of an MCMC sampler for any given variable is at least approximately AR(1), as will often be the case, a simple generalization of (2.24)—the expression giving the required length of the monitoring run in *IID* Monte Carlo sampling, when the goal is to specify a target for the accuracy of the posterior mean—is available. It's a standard result from time series [e.g., 6] that if θ_t^* is a stationary process with variance σ_θ^2 and autocorrelation ρ_k at lag k, then in repeated sampling the standard error (the square root of the variance) of the sample mean $\bar{\theta}^* = \frac{1}{m}\sum_{t=1}^m \theta_t^*$ is

$$\mathrm{se}(\bar{\theta}^*) = \frac{\sigma_\theta}{\sqrt{m}} \sqrt{1 + 2 \sum_{k=1}^{m-1} \left(1 - \frac{k}{m}\right) \rho_k} \, , \tag{2.37}$$

and a good approximation to this for large m is given by

$$\mathrm{se}(\bar{\theta}^*) \doteq \frac{\sigma_\theta}{\sqrt{m}} \sqrt{\tau},$$

where

$$\tau = 1 + 2 \sum_{k=1}^{\infty} \rho_k \tag{2.38}$$

is called the *autocorrelation time* for the series. In the special case of an AR(1) process (2.37) reduces to

$$\mathrm{se}(\bar{\theta}^*) \doteq \frac{\sigma_\theta}{\sqrt{m}} \sqrt{\frac{1+\rho}{1-\rho}}, \tag{2.39}$$

where $\rho = \rho_1$ is the autocorrelation at lag 1. This formula wraps up the bad news arising from a poorly mixing chain in a neat package: as ρ approaches

+1 (i.e., as less and less new information is learned with each new Monte Carlo draw from the posterior), (2.39) goes to $+\infty$. The final plot in Fig. 2.5, the left-hand graph in the third row of the figure, gives an estimated version of (2.39)—using estimates of the posterior SD and first-order autocorrelation of σ_S^2 based on the 5,000 monitoring iterations so far—plotted against m, to indicate how long the chain needs to be run to achieve any particular target value of $\widehat{\text{se}}(\bar{\theta}^*)$.

MLwiN also estimates the autocorrelation time $\hat{\tau}$ (by summing the $\hat{\rho}_k$ in (2.38) from $k = 1$ forward until they're no longer statistically significantly different from 0) and uses this to create a quantity called the *effective sample size* $\widehat{ESS} = m/\hat{\tau}$, which measures the efficiency of the MCMC sampler in current use versus IID sampling; here the value $\widehat{ESS} = 2{,}821$ (printed in the Summary Statistics section of the display) means that MLwiN's default sampler for this model has achieved a level of accuracy with 5,000 monitoring iterations that's equivalent to what would have been achieved with an IID sample from the posterior of size 2,821. (You can use the relation $\hat{\tau} = (1 + \hat{\rho})/(1 - \hat{\rho})$ for AR(1) processes to solve backwards from the $\widehat{ESS}$ value, obtaining $\hat{\rho} = (m - \widehat{ESS})/(m + \widehat{ESS})$. Here this yields $\hat{\rho} \doteq 0.28$, which agrees well with the graphs in Fig. 2.5.)

The current MCSE for the posterior mean of σ_S^2 with $m = 5{,}000$ is 0.0004 (this can be read off the graph, and is also printed in the Summary Statistics section of the display). Given that the current posterior mean for σ_S^2 is 0.097, this turns out not to be accurate enough to pin down the posterior mean $\mu_{\sigma_S^2}$ to two sigfigs with high Monte Carlo probability; a calculation based on the CLT approximation to the repeated-sampling distribution of $\frac{1}{m}\sum_{i=1}^{m}(\sigma_S^2)_i^*$ reveals that a Monte Carlo confidence interval for $\mu_{\sigma_S^2}$ of the form $(0.0965, 0.0975)$ only has approximate confidence level 79%. A quantity referred to by MLwiN as the *Brooks-Draper* (BD) diagnostic [8, 11] estimates the required length $\hat{m}$ of a monitoring run to achieve at least k sigfigs, with Monte Carlo probability at least $1 - \alpha$, in the posterior mean estimate for a quantity θ with current sample mean $\bar{\theta}^*$ which can be written $a \cdot 10^b$ for $1 \le a < 10$; if the current estimate of the posterior SD of θ is $\hat{\sigma}_\theta$, and if the time series for θ behaves like an AR(1) process with estimated first-order autocorrelation $\hat{\rho}$, then $\hat{m}$ must satisfy

$$\hat{m} \ge 4\left[\Phi^{-1}\left(1 - \frac{\alpha}{2}\right)\right]^2 \left(\frac{\hat{\sigma}_\theta}{10^{b-k+1}}\right)^2 \left(\frac{1+\hat{\rho}}{1-\hat{\rho}}\right). \tag{2.40}$$

Here to achieve $k = 2$ sigfigs with at least 95% Monte Carlo probability MLwiN evaluates (2.40) and obtains $\hat{m} \ge 11{,}365$ (this is referred to as Nhat in the Accuracy Diagnostics section of the display, next to the phrase Brooks-Draper (mean)).

The final accuracy diagnostic routinely printed by MLwiN was developed by Raftery and Lewis [59] to address a different aspect of the posterior than that covered by the BD diagnostic: Raftery and Lewis were interested in the accuracy of the quantiles defining (say) the 95% central interval estimate for a quantity θ, obtained by quoting the 2.5% and 97.5% points in the empirical distribution for θ based on the m monitoring iterations so far (`MLwiN` denotes this choice of relevant quantiles by $q = (q_1, q_2) = (0.025, 0.975)$). The Raftery-Lewis (RL) diagnostic is expressed not on the scale of the data but on the probability scale; in the case of the q_1 point, for example, the diagnostic—when used with its default settings—indicates how long the monitoring run needs to be so that the actual amount of probability to the left of the quoted q_1 point in the true posterior distribution is within $r = 0.005$ of the nominal value q_1 with at least $s = 0.95$ Monte Carlo probability. With these choices of q and r the goal is for the actual probability content of the nominal 95% central interval estimate for θ to be somewhere between 94% and 96% with at least 95% Monte Carlo probability. By default `MLwiN` reports RL values for both of the q_1 and q_2 points; a natural way to use this output is to take the larger of the two values as the recommended length of monitoring run from the RL point of view. Here the default settings produce $\hat{m} = (4199, 3996)$ (`MLwiN` also calls these values `Nhat`), so the 95% central interval $(0.063, 0.145)$ for σ_S^2 reported in the `Summary Statistics` portion of the display exceeds the default RL accuracy standards with the monitoring run of 5,000 iterations already performed.

A natural strategy in `MLwiN` for choosing a final length of monitoring run to produce publishable findings when the posterior distribution is multivariate is to decide on the desired BD and RL accuracy standards for each parameter (the defaults are easy to change) and to run the sampler for m^* iterations, where m^* is at least as large as the maximum across the resulting BD and RL recommendations for all parameters. In model (2.35) with the Goldstein et al. data, ignoring the irrelevant parameter β_0 and using the settings chosen above, this yields $m^* = 11{,}365$ (it turns out that the mixing for σ_S^2 is the worst for the three main parameters in the model); rounding up to 12,000 and running the sampler for an additional 7,000 iterations (which takes just a few seconds at 3 GHz) yields good MCMC diagnostics and leads to final reportable values (posterior means, with posterior SDs in parentheses) of $\hat{\beta}_0 = 0.005$ (0.041), $\hat{\beta}_1 = 0.563$ (0.013), $\hat{\sigma}_S^2 = 0.097$ (0.020), and $\hat{\sigma}_P^2 = 0.566$ (0.013). These estimates differ little from the earlier values; in this model with this data set, the initial monitoring run of 5,000 draws was already highly informative.

2.5 The Case Study Revisited

I'll conclude this chapter by applying the ideas discussed in the previous four sections to the Berkeley traffic case study in Section 2.1.1. Figure 2.6 illustrates the MCMC fitting of the random-effects logistic regression (RELR)

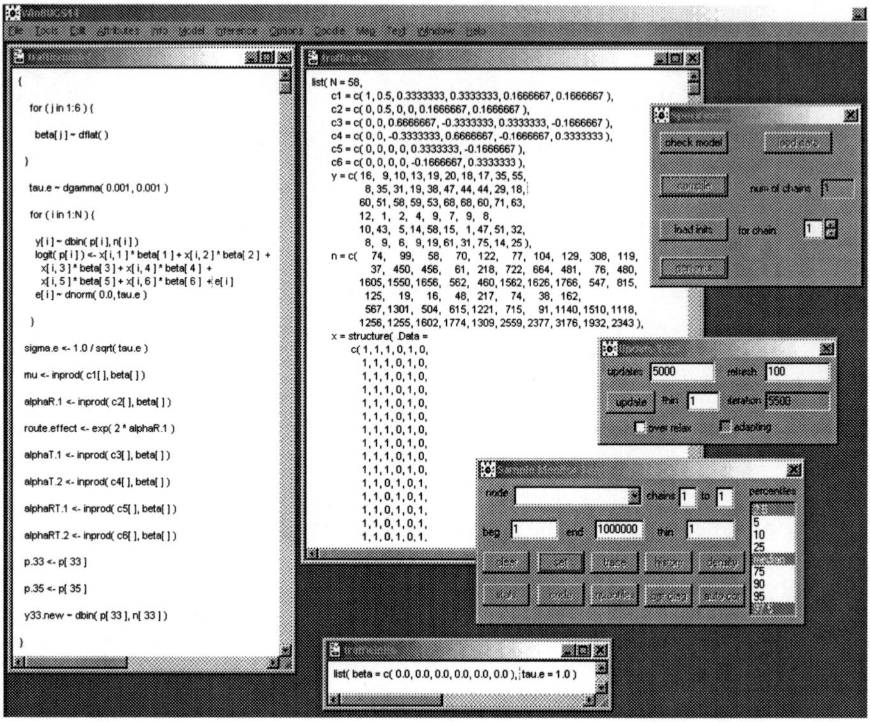

Fig. 2.6 `WinBUGS` fitting of a regression reformulation of model (2.5) to the Berkeley traffic data in Section 2.1.1.

model (2.5) to the data summarized in Tables 2.1 and 2.2, using `WinBUGS` release 1.4.1. The fixed effects in (2.5) are written using an analysis-of-variance over-parameterization which employs slightly awkward side conditions (such as $\sum_{l=1}^{L} \alpha_l^T = 0$) to make the model identifiable; it's arguably more natural to fit a model like this by appealing to the duality between ANOVA and regression, which is what I've done. There are $K = 6$ fixed-effects degrees of freedom to fit in this model (in ANOVA language, one for the grand mean, one for the bike-route effect, two for street type, and two for the interaction between the two fixed factors), so I defined 6 dummy or indicator variables: x_1 is 1 for all $N = 82$ rows in the data set (the intercept); x_2 is 1 for streets with a bike route (and 0 otherwise); x_3 and x_4 are 1 if street type is residential and fairly busy, respectively; and $x_5 = x_2\, x_3$ and $x_6 = x_2\, x_4$ carry the interaction information. With these definitions, in regression notation (2.5) becomes (for $i = 1, \ldots, N$)

$$
\begin{aligned}
(y_i \mid p_i) &\stackrel{\text{indep}}{\sim} \text{Binomial}(n_i, p_i), \quad \text{where} \\
\text{logit}(p_i) &= \sum_{k=1}^{K} \beta_k \, x_{ik} + e_i, \quad e_i \stackrel{iid}{\sim} \mathcal{N}(0, \sigma_e^2),
\end{aligned} \tag{2.41}
$$

with the "errors" or residuals e_i corresponding to the block-level random effects a_{jkl}^B. The left-most window in Fig. 2.6 specifies this regression model in `WinBUGS` syntax, and the larger middle window in the figure gives the data. I used diffuse priors for the β_k (in `WinBUGS` the specification `dflat( )` corresponds effectively to a normal distribution with mean 0 and huge variance, or, equivalently, tiny precision), and I used a $\Gamma(0.001, 0.001)$ diffuse prior for the residual precision $\tau_e = 1/\sigma_e^2$ (similar results are obtained here with other diffuse priors). I also initialized all the parameters in (2.41) in a default manner (see the small window at the bottom of Fig. 2.6), by starting the Markov chain off at 0 for all the β_k and 1 for τ_e; this was lazy (and `WinBUGS` had to generate initial values for all 82 of the residuals from this relatively inaccurate starting point), but exploration revealed that the chain reached equilibrium well within a quick burn-in of 500 iterations, so my laziness was not punished in this case.

The dummy coding described above makes it easy to fit the model, but some work is required to translate between the β (regression) and (μ, α) (ANOVA) parameterizations. The top part of Table 2.4 gives the representation of each of the cells in the 2×3 table in this example in terms of the x_k (for example, membership in the (no bike route, busy) cell corresponds to $(x_1, \ldots, x_6) = (1, 0, 0, 0, 0, 0)$), and the representations of the row means, column means, and the grand mean (on the logit scale) are then available by simple averaging of the indicator codings in the cells (this ignores the slight imbalance in the design in Table 2.1, with the two missing blocks in the (no bike route, residential) cell, but a more precise analysis that corrects for this yields almost exactly the same results). The bottom part of Table 2.4 records the expected $\text{logit}(p_{jkl})$ values in model (2.5), in the ANOVA parameterization and bearing in mind the usual side conditions. Given that the expected $\text{logit}(p_i)$ values in (2.41) are of the form $\sum_{k=1}^{K} \beta_k \, x_{ik}$ with the x_k values presented in the top part of Table 2.4, it's now possible to work out what functions of the β_k in the regression parameterization need to be monitored to obtain estimates of the parameter values in the ANOVA representation. For example, the grand mean μ in (2.5), which corresponds to the entry $c_1' = (1, \frac{1}{2}, \frac{1}{3}, \frac{1}{3}, \frac{1}{6}, \frac{1}{6})$ in the top part of the table, can be estimated by monitoring the linear combination $c_1' \beta = \sum_{k=1}^{K} c_{1k} \beta_k$ of the β_k; similarly $\alpha_1^R = (\mu + \alpha_1^R) - \mu$ corresponds to $c_2' = (1, 1, \frac{1}{3}, \frac{1}{3}, \frac{1}{3}, \frac{1}{3}) - (1, \frac{1}{2}, \frac{1}{3}, \frac{1}{3}, \frac{1}{6}, \frac{1}{6}) = (0, \frac{1}{2}, 0, 0, \frac{1}{6}, \frac{1}{6})$; and so on. The data file in Fig. 2.6 displays the weight vectors c_3 through c_6 which permit the monitoring of the other fixed-effect parameters (α_1^T, α_2^T, α_1^{RT}, and α_2^{RT}) in

Table 2.4 The top table gives the dummy coding of the cells in the 2 × 3 table for the Berkeley traffic study of Section 2.1.1; the bottom table records the expected logit(p_{jkl}) values in model (2.5), in the ANOVA parameterization and bearing in mind the usual side conditions.

Bike Route?	Street Type Residential	Fairly Busy	Busy	Mean
Yes	$(1,1,1,0,1,0)$	$(1,1,0,1,0,1)$	$(1,1,0,0,0,0)$	$(1,1,\frac{1}{3},\frac{1}{3},\frac{1}{3},\frac{1}{3})$
No	$(1,0,1,0,0,0)$	$(1,0,0,1,0,0)$	$(1,0,0,0,0,0)$	$(1,0,\frac{1}{3},\frac{1}{3},0,0)$
Mean	$(1,\frac{1}{2},1,0,\frac{1}{2},0)$	$(1,\frac{1}{2},0,1,0,\frac{1}{2})$	$(1,\frac{1}{2},0,0,0,0)$	$(1,\frac{1}{2},\frac{1}{3},\frac{1}{3},\frac{1}{6},\frac{1}{6})$

Bike Route?	Street Type Residential	Fairly Busy	Busy	Mean
Yes	$\mu+\alpha_1^R+\alpha_1^T+\alpha_{11}^{RT}$	$\mu+\alpha_1^R+\alpha_2^T+\alpha_{12}^{RT}$	$\mu+\alpha_1^R+\alpha_3^T+\alpha_{13}^{RT}$	$\mu+\alpha_1^R$
No	$\mu+\alpha_2^R+\alpha_1^T+\alpha_{21}^{RT}$	$\mu+\alpha_2^R+\alpha_2^T+\alpha_{22}^{RT}$	$\mu+\alpha_2^R+\alpha_3^T+\alpha_{23}^{RT}$	$\mu+\alpha_2^R$
Mean	$\mu+\alpha_1^T$	$\mu+\alpha_2^T$	$\mu+\alpha_3^T$	μ

the ANOVA version of the model (the `WinBUGS` function `inprod` can be used to create the desired linear combinations).

The online MCMC diagnostics in `WinBUGS` are not as extensive as those in `MLwiN`, although it's easy in `WinBUGS` to store the MCMC data set for offline analysis with `CODA`; here I'll illustrate a simple approach to determining how long the Markov chain should be monitored using only the online diagnostics. As an example of the method I'm describing, Fig. 2.7 presents a variety of plots and numerical summaries for the parameter α_1^R in (2.5), which measures (on the logit scale) the amount—on streets with a bicycle route—by which bicycle traffic is more likely than average; the figure is based on an initial monitoring run of 5,000 iterations after the burn-in of length 500 mentioned above. The dynamic trace in the lower left corner of the figure (which tracks the last 500 iterations in the current monitoring run) shows how slowly this parameter is mixing with the default `WinBUGS` sampling strategy,[39] and the time series trace of all 5,000 monitored iterations (in the center of Fig. 2.7) also looks like that of an AR(1) series with a high first-order autocorrelation, an impression that's confirmed by the autocorrelation plot in the lower right corner. The running quantiles for α_1^R in the right center of the figure show that the estimates of the 2.5% and 97.5% points of the marginal posterior distribution for this parameter at about iteration 700 (200 iterations into

[39] Other MCMC strategies are available in `WinBUGS`; I did not explore them.

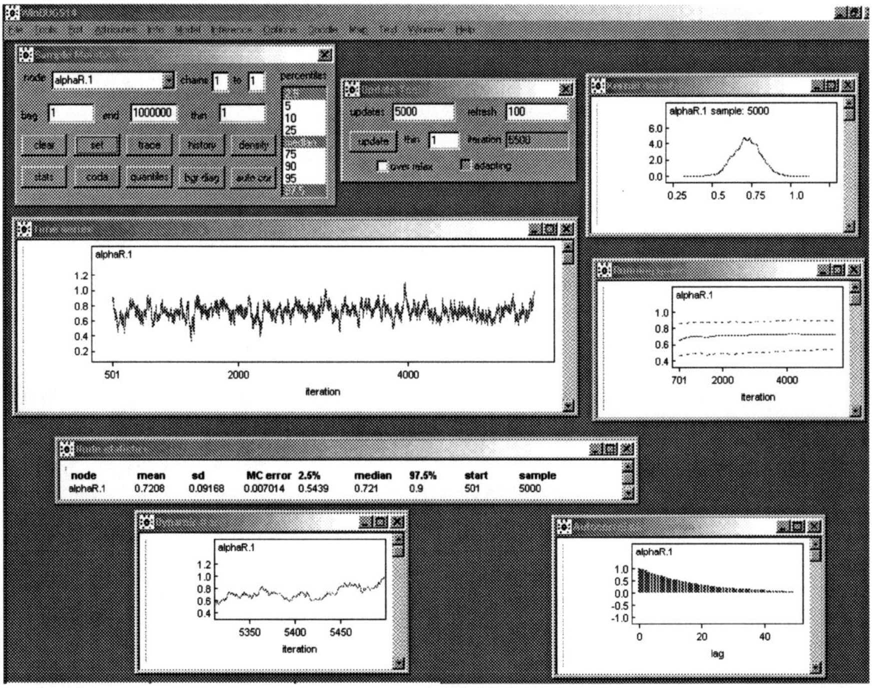

Fig. 2.7 `WinBUGS` online summaries and MCMC diagnostics for the parameter α_1^R in model (2.5), as fit to the Berkeley traffic data.

the current monitoring run) are quite different than their values at iteration 5,500, and the kernel density trace in the upper right corner is quite jagged with the `WinBUGS` default choice for smoothing of the density estimate; all of this reinforces the sense that a longer monitoring run is needed.

Columns 2–6 in Table 2.5 summarize where things stand after the initial 5,000 monitoring iterations for a variety of unknown quantities (not all of which I've yet discussed) in models (2.5) and (2.41), including α_1^R; column 5 estimates ρ_1 for each of these quantities ($\hat{\rho}_1$ values are available in `WinBUGS`), and column 6 records the corresponding estimated BD $\hat{m}$ values, with target numbers of sigfigs in parentheses. The first-order autocorrelations for the first eight quantities in the table are all above +0.9, a typical result in RELR models; across all the unknowns in the table this leads to $\hat{m}$ recommendations ranging from about 1,100 to more than 115,000, depending on the autocorrelation and the number of sigfigs desired. A parameter like α_2^{RT}, which has an estimated posterior mean near 0 but which nevertheless has substantial values for both the posterior SD and ρ_1, will need a fairly long monitoring run just to be able to quote a single sigfig with decent Monte Carlo accuracy.

Table 2.5 Numerical summaries for a variety of unknown quantities (11 parameters and a future observable) in the Berkeley traffic case study. `r.e` is short for `route.effect`.

	After 5,000 Iterations						After 120,000 Iterations		
Unknown	Mean	(MCSE)	SD	$\hat{\rho}_1$	BD $\hat{m}$	(sigfig)	Mean	(MCSE)	SD
α_1^R	0.72	(.007)	0.092	0.934	37800	(2)	0.72	(.002)	0.093
α_1^T	0.88	(.009)	0.13	0.907	56400	(2)	0.87	(.002)	0.14
α_2^T	0.02	(.009)	0.13	0.921	61100	(1)	−0.01	(.002)	0.13
α_1^{RT}	−0.24	(.01)	0.13	0.940	85800	(2)	−0.26	(.003)	0.14
α_2^{RT}	0.04	(.01)	0.13	0.953	115900	(1)	0.08	(.003)	0.13
β_5	−0.88	(.04)	0.41	0.947	9610	(1)	−0.87	(.01)	0.46
μ	−2.85	(.006)	0.092	0.912	28200	(3)	−2.84	(.001)	0.092
`r.e`	4.3	(.06)	0.80	0.934	28500	(2)	4.3	(.02)	0.80
p_{33}	0.10	(.001)	0.047	0.537	1140	(2)	0.10	(.0003)	0.048
p_{35}	0.048	(.0002)	0.013	0.361	5640	(2)	0.048	(.00005)	0.013
σ_e	0.63	(.002)	0.074	0.673	4290	(2)	0.63	(.0005)	0.074
y_{33}^{new}	1.6	(.02)	1.42	0.108	3850	(2)	1.6	(.006)	1.42

After looking at the results in Table 2.5 I decided to aim for a total monitoring run of 120,000 iterations (by merging 115,000 new simulated draws from the posterior with the previous 5,000); this took about 1 minute at 3 GHz. The last three columns in Table 2.5 summarize the posterior means (with MCSEs) and SDs of the monitored quantities after 120,000 iterations. All of the Monte Carlo estimates were quite stable in passing from the shorter to the longer monitoring run except α_2^{RT}, whose posterior mean doubled in size (while still, of course, remaining close to 0). The following substantive and statistical conclusions, some of which echo and reinforce the preliminary impressions from Table 2.2 mentioned in Section 2.1.1, may be drawn.

- Averaging over street type, a randomly chosen vehicle on a street with a bike route is far likelier to be a bicycle than if the street had no bike route: α_1^R, the main effect of the bike route variable, has a posterior mean of +0.72 on the logit scale, with a posterior SD of 0.093, meaning that the mean difference (in logits) between the bike-route-yes and bike-route-no blocks was $2\alpha_1^R = 1.43$. Given that the average block without a bike route had only about 4% bicycle traffic (from Table 2.2), the "effect" of adding a bike route would be to roughly quadruple[40] the

[40] Starting with a no-bike-route PBT of $p = 0.041$, for which $l = \text{logit}(p) \doteq -3.15$, and (naively) adding 1.43 yields $l^* = l + 1.43 = -1.72$, from which $p^* = [1 + \exp(-l^*)]^{-1} \doteq 0.15$.

PBT (proportion of bicycle traffic; "effect" is in quotes because this is an observational study, from which it would be bold to draw such a strong causal conclusion). As a rough confirmation of this, I monitored the quantity[41] `r.e` = `route.effect` = $\exp(2\alpha_1^R)$ along with everything else in Table 2.5; its posterior mean (which does not adjust for bias arising from the nonlinear transformation) was about 4.3, with a posterior SD of 0.80.

- Averaging over presence or absence of a bike route, street type also has a strong effect on PBT; for example, α_1^T, which contrasts residential streets (on the logit scale) with average behavior, had a posterior mean and SD of +0.87 and 0.14, respectively. Based on calculations similar to those given above, residential streets were about four times as likely to have bicycles on them as busy streets, and about twice as likely as fairly busy streets.
- However, fairly large interactions between the two fixed effects complicate the picture; the "effect" of bike route on PBT is different according to street type. This may be seen by looking at posterior summaries of the interaction parameters (for example, on the logit scale α_1^{RT} has posterior mean (SD) -0.26 (0.14)), but the interaction comes into focus even more clearly by contrasting what happens when you go from no-bike-route to bike-route for each of the residential and busy street types. With (i, j) denoting the cell in row i and column j of the basic 2×3 table, this involves making a cell-means comparison of the form $[(1,1)-(2,1)]-[(1,3)-(2,3)]$. The relevant linear combination of the β_k from Table 2.4 turns out to be $(0,0,0,0,1,0)$; in other words, it suffices to monitor β_5 to address this question. From Table 2.5 you can see that its posterior mean and SD came out -0.87 and 0.46, respectively. Since $e^{0.87} \doteq 2.4$, the interpretation would be that the "effect" of bike route on PBT is more than twice as large for busy streets as it is for residential ones.
- There is substantial unexplained heterogeneity between city blocks within the cells of the 2×3 layout: the SD of the random effects at the block level (σ_e in model (2.41) and σ_B in (2.5)) has a posterior mean on the logit scale of 0.63 (with a posterior SD of 0.074). In an average cell with a typical PBT of about 9%, this means (using calculations similar to that in footnote 40) that it would not be surprising to see block-level PBT values ranging from about 3% to 26%. Clearly, while model (2.5)/(2.41) has made progress in explaining why some city blocks in Berkeley have a lot of bicycle traffic and others do not, there is still some way to go in achieving full causal understanding.
- In addition to parameters in both the ANOVA and regression formulations of the RELR model at issue here, I also monitored the underlying p_i values in (2.41) for two city blocks in the (residential, no bike route) cell of the

[41] See footnote 2 for the reasoning behind this choice.

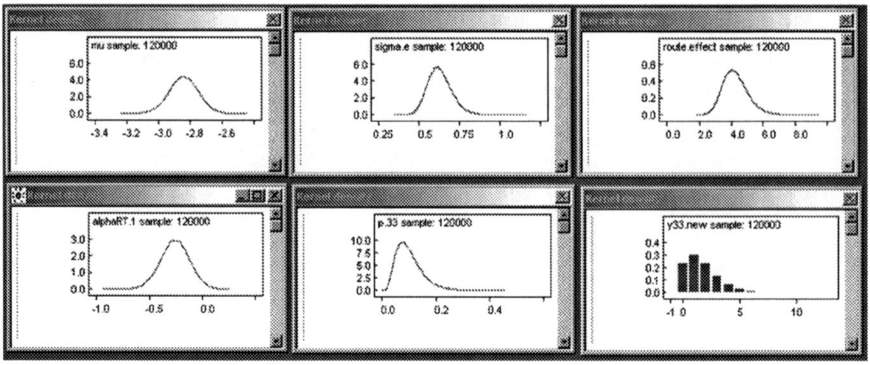

Fig. 2.8 Marginal posterior kernel density traces for six of the unknown quantities in Table 2.5.

2×3 table. These blocks, numbered 33 and 35 in the data set, had observed PBT and n_j values (written in the PBT/n_j format of Table 2.1) of 0.125/16 and 0.041/217, respectively. Since the two blocks were both in the same cell you might have thought that the posterior means of p_{33} and p_{35} would be similar, but they're not: from Table 2.5, in posterior (mean $\pm$ SD) notation, $p_{33} \doteq 0.10 \pm 0.048$ and $p_{35} \doteq 0.048 \pm 0.013$. Recalling from Table 2.2 that the overall PBT rate in this cell of the table was 0.097, some reflection clarifies what's going on here: under the random-effects formulation in the model, the posterior means of both p_{33} and p_{35} will *shrink* toward the cell mean 0.097 (this is the same phenomenon noted in Section 2.3.2), but the amount of the shrinkage will depend strongly on how much data is available for each city block. Block 33, which only had 16 vehicles, experiences substantial shrinkage, from a data value of 0.125 all the way to 0.10 (nearly the entire distance to 0.097), whereas block 35, with 217 vehicles, hardly shrinks at all (the posterior mean moves only to 0.048 from a data value of 0.041).

- Finally, Fig. 2.8 captures marginal posterior density estimates for a variety of the unknowns of interest in Table 2.5, including a predictive distribution for a future observable. Using the method described in Section 2.3, by adding the line `y33.new ~ dbin( p[ 33 ], n[ 33 ] )` to the `WinBUGS` model (as in Fig. 2.6), the MCMC approach correctly calculates the predictive distribution for the number of bicycles in a future sample of $n_{33} = 16$ vehicles from a location like city block 33 as a mixture of binomial distributions, with the posterior distribution for p_{33} providing the mixing weights. Some of the marginal posteriors in Fig. 2.8 are approximately Gaussian (the plots for μ and α_1^{RT} on the left) and some are skewed (the density traces for σ_e, p_{33}, and `route.effect`); some are discrete (the predictive for y_{33}^{new}) and some are continuous; and all are correct up to a small (and controllable) amount of Monte Carlo noise. Asymptotic posterior calculations, of the type on which Bayesian multilevel analyses were necessarily based until the mid-1990s, would provide poor approximations to the right answers here. In the last several years MCMC has clearly opened a new door, and the many Bayesian multilevel modeling applications that are now within easy reach are just the beginning.

Acknowledgement I'm grateful to Bill Browne for specific helpful comments on this chapter and for a long, productive, and enjoyable collaboration, without which a good bit of the work described here would not have been possible.

References

1. M. Aitkin. A general maximum likelihood analysis of variance components in generalized linear models. *Biometrics*, 55:117–128, 1999.
2. J. M. Bernardo and A. F. M. Smith. *Bayesian Theory*. Wiley, New York, 1994.
3. J. Besag. Spatial interaction and the statistical analysis of lattice systems. *Journal of the Royal Statistical Society, Series B*, 36:192–236, 1974. (with discussion)
4. N. G. Best, M. K. Cowles, and S. K. Vines. CODA: Convergence diagnosis and output analysis software for Gibbs sampling output, version 0.40, 1997. (Available at `www.mrc-bsu.cam.ac.uk`)
5. P. J. Bickel and K. A. Doksum. *Mathematical Statistics: Basic Ideas and Selected Topics*, 2nd edition, volume 1. Prentice Hall, Upper Saddle River, NJ, 2001.
6. G. E. P. Box, G. M. Jenkins, and G. C. Reinsel. *Time Series Analysis: Forecasting and Control*, 3rd edition. Prentice-Hall, Englewood Cliffs, NJ, 1994.
7. N. E. Breslow and D. G. Clayton. Approximate inference in generalized linear mixed models. *Journal of the American Statistical Association*, 88:9–25, 1993.
8. S. P. Brooks and D. Draper. Comparing the efficiency of MCMC samplers. Manuscript, 2007.
9. S. P. Brooks and G. O. Roberts. Convergence assessment techniques for Markov chain Monte Carlo. *Statistics and Computing*, 8:319–335, 1998.
10. W. J. Browne. Personal communication, 2000.
11. W. J. Browne and D. Draper. Implementation and performance issues in the Bayesian and likelihood fitting of multilevel models. *Computational Statistics*, 15:391–420, 2000.
12. W. J. Browne and D. Draper. A comparison of Bayesian and likelihood methods for fitting multilevel models. *Bayesian Analysis*, 1:473–550, 2006. (with discussion)
13. W. J. Browne, D. Draper, H. Goldstein, and J. Rasbash. Bayesian and likelihood methods for fitting multilevel models with complex level-1 variation. *Computational Statistics & Data Analysis*, 39:203–225, 2002.
14. W. J. Browne, J. Rasbash, and E. W. S. Ng. *MCMC Estimation in MLwiN (version 2.0)*. Centre for Multilevel Modelling, University of Bristol, Bristol, UK, 2005.

15. A. S. Bryk and S. W. Raudenbush. *Hierarchical Linear Models: Applications and Data Analysis Methods.* Sage, Newbury Park, CA, 1992.
16. E. Butkov. *Mathematical Physics.* Addison-Wesley, Reading, MA, 1968.
17. B. Carlin. Discussion of "Hierarchical models for combining information and for meta-analysis", by C. N. Morris and S. L. Normand. In J. M. Bernardo, J. O. Berger, A. P. Dawid, and A. F. M. Smith, editors, *Bayesian Statistics 4*, pages 336–338. Clarendon Press, Oxford, UK, 1992.
18. C. Chatfield. *The Analysis of Time Series*, 6th edition. Chapman & Hall, London, 2003.
19. M. K. Cowles and B. P. Carlin. Markov chain Monte Carlo convergence diagnostics: A comparative review. *Journal of the American Statistical Association*, 91:883–904, 1996.
20. H. E. Daniels. Saddlepoint approximations in statistics. *Annals of Mathematical Statistics*, 25:631–650, 1954.
21. A. P. Dawid. Calibration-based empirical probability. *Annals of Statistics*, 13: 1251–1274, 1985.
22. B. de Finetti. Funzione caratteristica di un fenomeno aleatorio. *Memorie della Accademia Nazionale dei Lincei*, 4:86–133, 1930. (Reprinted in B. de Finetti, *Scritti (1926–1930)*, Padova, Cedam, 1981.)
23. B. de Finetti. *Theory of Probability*, volumes 1–2. Wiley, New York, 1974/1975.
24. P. Diaconis and D. Ylvisaker. Quantifying prior opinion. In J. M. Bernardo, M. H. DeGroot, D. V. Lindley, and A. F. M. Smith, editors, *Bayesian Statistics 2*, pages 133–156. North-Holland, Amsterdam, 1985. (with discussion)
25. D. Draper. Assessment and propagation of model uncertainty. *Journal of the Royal Statistical Society, Series B*, 57:45–97, 1995. (with discussion)
26. D. Draper. Inference and hierarchical modeling in the social sciences. *Journal of Educational and Behavioral Statistics*, 20:115–147, 233–239, 1995. (with discussion)
27. D. Draper. *Bayesian Modeling, Inference and Prediction.* Springer, New York, forthcoming.
28. D. Draper, J. Hodges, C. Mallows, and D. Pregibon. Exchangeability and data analysis. *Journal of the Royal Statistical Society, Series A*, 156:9–37, 1993. (with discussion)
29. W. DuMouchel. Bayesian meta-analysis. In D. Berry, editor, *Statistical Methodology in the Pharmaceutical Sciences*, pages 509–529. Marcel Dekker, New York, 1990.
30. W. DuMouchel and C. Waternaux. Discussion of "Hierarchical models for combining information and for meta-analysis", by C. N. Morris and S. L. Normand. In J. M. Bernardo, J. O. Berger, A. P. Dawid, and A. F. M. Smith, editors, *Bayesian Statistics 4*, pages 338–341. Clarendon Press, Oxford, UK, 1992.
31. W. L. Edwards, H. Lindman, and L. J. Savage. Bayesian statistical inference for psychological research. *Psychological Review*, 70:193–242, 1963.
32. W. Feller. *An Introduction to Probability Theory and its Applications*, volumes 1–2. Wiley, New York, 1968/1971.

33. R. A. Fisher. On the mathematical foundations of theoretical statistics. *Philosophical Transactions of the Royal Society of London, Series A*, 222:309–368, 1921.
34. R. A. Fisher. Studies in crop variation: II. The manurial response of different potato varieties. *Journal of Agricultural Science*, 13:311–320, 1923.
35. D. Gamerman. *Markov Chain Monte Carlo: Stochastic Simulation for Bayesian Inference.* Chapman & Hall, London, 1997.
36. C. F. Gauss. *Theoria Motus Corporum Celestium.* Perthes et Besser, Hamburg, 1809.
37. A. Gelman, J. B. Carlin, H. S. Stern, and D. B. Rubin. *Bayesian Data Analysis*, 2nd edition. Chapman & Hall/CRC, Boca Raton, FL, 2004.
38. A. Gelman and D. B. Rubin. Inference from iterative simulation using multiple sequences. *Statistical Science*, 7:457–511, 1992. (with discussion)
39. S. Geman and D. Geman. Stochastic relaxation, Gibbs distributions, and the Bayesian restoration of images. *IEEE Transactions on Pattern Analysis and Machine Intelligence*, 6:721–741, 1984.
40. W. R. Gilks, S. Richardson, and D. J. Spiegelhalter, editors. *Markov Chain Monte Carlo in Practice.* Chapman & Hall, London, 1996.
41. W. R. Gilks and P. Wild. Adaptive rejection sampling for Gibbs sampling. *Applied Statistics*, 41:337–348, 1992.
42. H. Goldstein. Multilevel mixed linear model analysis using iterative generalized least squares. *Biometrika*, 73:43–56, 1986.
43. H. Goldstein. *Multilevel Statistical Models*, 3rd edition. Edward Arnold, London, 2003.
44. H. Goldstein, J. Rasbash, M. Yang, G. Woodhouse, H. Pan, D. Nutall, and S. Thomas. A multilevel analysis of school examination results. *Oxford Review of Education*, 19:425–433, 1993.
45. H. Goldstein and D. J. Spiegelhalter. League tables and their limitations: Statistical issues in comparisons of institutional performance. *Journal of the Royal Statistical Society, Series A*, 159:385–444, 1996. (with discussion)
46. W. K. Hastings. Monte Carlo sampling methods using Markov chains and their applications. *Biometrika*, 57:97–109, 1970.
47. Insightful Corporation. `S-PLUS` data analysis software, 2006. (Available at `www.insightful.com`)
48. P.-S. Laplace. Mémoire sur la probabilité des causes par les évenements. *Mémoires de l'Academie Royale de Science de Paris*, 6:621–656, 1774.
49. P.-S. Laplace. Mémoire sur les approximations des formules qui sont fonctions de très grands nombres (suite). *Mémoires de l'Academie Royale de Science de Paris*, 18:423–467, 1786.
50. Y. Lee and J. A. Nelder. Hierarchical generalised linear models: A synthesis of generalised linear models, random-effect models and structured dispersions. *Biometrika*, 88:987–1006, 2001.
51. A. M. Legendre. *Nouvelles méthodes pour la détermination des orbites des comètes.* Courcier, Paris, 1805.

52. N. T. Longford. A fast scoring algorithm for maximum likelihood estimation in unbalanced mixed models with nested random effects. *Biometrika*, 74:817–827, 1987.
53. N. T. Longford. On estimating standard errors in multilevel analysis. *The Statistician*, 49:389–398, 2000.
54. N. Metropolis, A. W. Rosenbluth, M. N. Rosenbluth, A. H. Teller, and E. Teller. Equation of state calculations by fast computing machines. *Journal of Chemical Physics*, 21:1087–1092, 1953.
55. N. Metropolis and S. Ulam. The Monte Carlo method. *Journal of the American Statistical Association*, 44:335–341, 1949.
56. P. Mortimore, P. Sammons, L. Stoll, D. Lewis, and R. Ecob. *School Matters, the Junior Years*. Open Books, Wells, UK, 1988.
57. J. C. Pinheiro and D. M. Bates. Approximations to the log-likelihood function in the nonlinear mixed-effects model. *Journal of Computational and Graphical Statistics*, 4:12–35, 1995.
58. R Development Core Team. *R: A Language and Environment for Statistical Computing*. R Foundation for Statistical Computing, Vienna, Austria, 2006. URL `http://www.r-project.org`
59. A. E. Raftery and S. M. Lewis. How many iterations in the Gibbs sampler? In J. M. Bernardo, J. O. Berger, A. P. Dawid, and A. F. M. Smith, editors, *Bayesian Statistics 4*, pages 763–773. Oxford University Press, Oxford, UK, 1992.
60. J. Rasbash, F. Steele, W. J. Browne, and B. Prosser. *A User's Guide to MLwiN. Version 2.0.* Centre for Multilevel Modelling, University of Bristol, Bristol, UK, 2005.
61. S. W. Raudenbush, M.-L. Yang, and M. Yosef. Maximum likelihood for generalized linear models with nested random effects via high-order, multivariate Laplace approximation. *Journal of Computational and Graphical Statistics*, 9: 141–157, 2000.
62. G. O. Roberts. Markov chain concepts related to sampling algorithms. In W. R. Gilks, S. Richardson, and D. J. Spiegelhalter, editors, *Markov Chain Monte Carlo in Practice*, pages 45–58. Chapman & Hall, London, 1996.
63. G. Rodríguez and N. Goldman. An assessment of estimation procedures for multilevel models with binary response. *Journal of the Royal Statistical Society, Series A*, 158:73–89, 1995.
64. SAS Institute. Statistical Analysis System, 2006. (Software available at `www.sas.com`)
65. H. Scheffé. *The Analysis of Variance*. Wiley, New York, 1959.
66. Scientific Software International. HLM6: Hierarchical linear and nonlinear modeling, 2006. (Software available from `www.ssicentral.com/hlm/index.htm`)
67. B. W. Silverman. *Density Estimation for Statistics and Data Analysis*. Chapman & Hall, London, 1986.
68. D. J. Spiegelhalter, N. G. Best, B. P. Carlin, and A. van der Linde. Bayesian measures of model complexity and fit. *Journal of the Royal Statistical Society, Series B*, 64:583–639, 2002. (with discussion)

69. D. J. Spiegelhalter, A. Thomas, N. G. Best, and W. R. Gilks. *BUGS: Bayesian Inference Using Gibbs Sampling, Version 0.60.* Medical Research Council Biostatistics Unit, Cambridge, UK, 1997.
70. D. J. Spiegelhalter, A. Thomas, N. G. Best, and W. R. Gilks. *BUGS: Bayesian Inference Using Gibbs Sampling, Version 1.4.1.* Medical Research Council Biostatistics Unit, Cambridge, UK, 2004.
71. Stata Corporation. `Stata` statistical software, 2006. (Software available at `www.stata.com`)
72. L. Tierney. Introduction to general state-space Markov chain theory. In W. R. Gilks, S. Richardson, and D. J. Spiegelhalter, editors, *Markov Chain Monte Carlo in Practice*, pages 59–74. Chapman & Hall, London, 1996.
73. J. von Neumann. Various techniques used in connection with random digits. In A. S. Householder, G. E. Forsythe, and H. H. Germond, editors, *Monte Carlo Method*, pages 36–38. National Bureau of Standards, Washington, DC, 1951.
74. Waterloo Maple. `Maple` computer algebra system, 2006. (Software available at `www.maplesoft.com`)
75. S. L. Zeger and R. M. Karim. Generalized linear models with random effects: a Gibbs sampling approach. *Journal of the American Statistical Association*, 86: 79–86, 1991.

3

Diagnostic Checks for Multilevel Models

Tom A. B. Snijders[1,2] and Johannes Berkhof[3]

[1] University of Oxford
[2] University of Groningen
[3] VU University Medical Center, Amsterdam

3.1 Specification of the Two-Level Model

This chapter focuses on diagnostics for the two-level Hierarchical Linear Model (HLM). This model, as defined in Chapter 1, is given by

$$\underline{\boldsymbol{y}}_j = \boldsymbol{X}_j\boldsymbol{\beta} + \boldsymbol{Z}_j\underline{\boldsymbol{\delta}}_j + \underline{\boldsymbol{\epsilon}}_j\,, \quad j = 1, \ldots, m, \tag{3.1a}$$

with

$$\begin{pmatrix} \underline{\boldsymbol{\epsilon}}_j \\ \underline{\boldsymbol{\delta}}_j \end{pmatrix} \sim \mathcal{N}\left(\begin{pmatrix} \boldsymbol{\emptyset} \\ \boldsymbol{\emptyset} \end{pmatrix}, \begin{pmatrix} \boldsymbol{\Sigma}_j(\boldsymbol{\theta}) & \boldsymbol{\emptyset} \\ \boldsymbol{\emptyset} & \boldsymbol{\Omega}(\boldsymbol{\xi}) \end{pmatrix} \right) \tag{3.1b}$$

and

$$(\underline{\boldsymbol{\epsilon}}_j, \underline{\boldsymbol{\delta}}_j) \perp (\underline{\boldsymbol{\epsilon}}_\ell, \underline{\boldsymbol{\delta}}_\ell) \tag{3.1c}$$

for all $j \neq \ell$. The lengths of the vectors $\boldsymbol{y}_j$, $\boldsymbol{\beta}$, and $\boldsymbol{\delta}_j$, respectively, are n_j, r, and s. Like in all regression-type models, the explanatory variables $\boldsymbol{X}$ and $\boldsymbol{Z}$ are regarded as fixed variables, which can also be expressed by saying that the distributions of the random variables $\underline{\boldsymbol{\epsilon}}$ and $\underline{\boldsymbol{\delta}}$ are conditional on $\boldsymbol{X}$ and $\boldsymbol{Z}$. The random variables $\underline{\boldsymbol{\epsilon}}$ and $\underline{\boldsymbol{\delta}}$ are also called the vectors of residuals at levels 1 and 2, respectively. The variables $\underline{\boldsymbol{\delta}}$ are also called random slopes. Level-2 units are also called clusters.

The standard and most frequently used specification of the covariance matrices is that level-1 residuals are i.i.d., i.e.,

$$\boldsymbol{\Sigma}_j(\boldsymbol{\theta}) = \sigma^2 \boldsymbol{I}_{n_j}\,, \tag{3.1d}$$

where $\boldsymbol{I}_{n_j}$ is the n_j-dimensional identity matrix; and that either all elements of the level-2 covariance matrix $\boldsymbol{\Omega}$ are free parameters (so one could identify $\boldsymbol{\Omega}$ with $\boldsymbol{\xi}$), or some of them are constrained to 0 and the others are free parameters.

J. de Leeuw, E. Meijer (eds.), *Handbook of Multilevel Analysis*,
© Springer 2008

Questioning this model specification can be aimed at various aspects: the choice of variables included in $\boldsymbol{X}$, the choice of variables for $\boldsymbol{Z}$, the residuals having expected value 0, the homogeneity of the covariance matrices across clusters, the specification of the covariance matrices, and the multivariate normal distributions. Note that in our treatment, the explanatory variables $\boldsymbol{X}$ and $\boldsymbol{Z}$ are regarded as being deterministic; the assumption that the expected values of the residuals (for fixed explanatory variables!) are zero is analogous to the assumption, in a model with random explanatory variables, that the residuals are uncorrelated with the explanatory variables.

The various different aspects of the model specification are entwined, however: Problems with one may be solved by tinkering with one of the other aspects, and model misspecification in one respect may lead to consequences in other respects. For example, unrecognized level-1 heteroscedasticity may lead to fitting a model with a significant random slope variance, which then disappears if the heteroscedasticity is taken into account; non-linear effects of some variables in $\boldsymbol{X}$, when unrecognized, may show up as heteroscedasticity at level 1 or as a random slope; and non-zero expected residuals sometimes can be dealt with by transformations of variables in $\boldsymbol{X}$.

This presentation of diagnostic techniques starts with techniques that can be represented as model checks remaining within the framework of the HLM. This is followed by a section on model checking based on various types of residuals. An important type of misspecification can reside in non-linearity of the effects of explanatory variables. The last part of the chapter presents methods to identify such misspecifications and estimate the non-linear relationships that may obtain.

3.2 Model Checks Within the Framework of the Hierarchical Linear Model

The HLM is itself already a quite general model, a generalization of the General Linear Model, the latter often being used as a point of departure in modeling or conceptualizing effects of explanatory on dependent variables. Accordingly, checking and improving the specification of a multilevel model in many cases can be carried out while staying within the framework of the multilevel model. This holds to a much smaller extent for the General Linear Model. This section treats some examples of model specification checks that do not have direct parallels in the General Linear Model.

3.2.1 Heteroscedasticity

The comprehensive nature of most algorithms for estimating the HLM makes it relatively straightforward to include some possibilities for modeling heteroscedasticity, i.e., non-constant variances of the random effects. (This is sometimes indicated by the term "complex variation", which however does not imply any thought of the imaginary number $i = \sqrt{-1}$.)

As an example, the iterated generalized least squares (IGLS) algorithm implemented in MLwiN [18, 19] accommodates variances depending as linear or quadratic functions of variables. For level-1 heteroscedasticity, this is carried out formally by writing

$$\underline{\epsilon}_{ij} = \boldsymbol{v}_{ij} \underline{\boldsymbol{\epsilon}}_{ij}^{0} ,$$

where $\boldsymbol{v}_{ij}$ is a $1 \times t$ variable and $\underline{\boldsymbol{\epsilon}}_{ij}^{0}$ is a $t \times 1$ random vector with

$$\underline{\boldsymbol{\epsilon}}_{ij}^{0} \sim \mathcal{N}\big(\boldsymbol{\emptyset}, \boldsymbol{\Sigma}^{0}(\boldsymbol{\theta})\big).$$

This implies

$$\mathrm{Var}(\underline{\epsilon}_{ij}) = \boldsymbol{v}_{ij} \boldsymbol{\Sigma}^{0}(\boldsymbol{\theta}) \boldsymbol{v}_{ij}' . \tag{3.2}$$

The standard homoscedastic specification is obtained by letting $t = 1$ and $\boldsymbol{v}_{ij} \equiv 1$.

The IGLS algorithm works only with the expected values and covariance matrices of $\underline{\boldsymbol{y}}_j$ implied by the model specification; see Goldstein [18, pp. 49–51]. A sufficient condition for model (3.1a)–(3.1c) to be a meaningful representation is that (3.2) is nonnegative for all i, j—clearly less restrictive than $\boldsymbol{\Sigma}^0$ being positive definite. Therefore, it is not required that $\boldsymbol{\Sigma}^0$ be positive definite, but it is sufficient that (3.2) is positive for all observed $\boldsymbol{v}_{ij}$. For example, a level-1 variance function depending linearly on $\boldsymbol{v}$ is obtained by defining

$$\boldsymbol{\Sigma}^{0}(\boldsymbol{\theta}) = \big(\sigma_{hk}(\boldsymbol{\theta})\big)_{1 \le h,k \le t},$$

with

$$\begin{aligned} \sigma_{h1}(\boldsymbol{\theta}) &= \sigma_{1h}(\boldsymbol{\theta}) = \theta_h, & h &= 1, \ldots, t, \\ \sigma_{hk}(\boldsymbol{\theta}) &= 0, & \min\{h,k\} &\ge 2, \end{aligned}$$

where $\boldsymbol{\theta}$ is a $t \times 1$ vector. Quadratic variance functions can be represented by letting $\boldsymbol{\Sigma}^0$ be a symmetric matrix, subject only to a positivity restriction for (3.2).

In exactly the same way, variance functions for the level-2 random effects depending linearly or quadratically on level-2 variables are obtained by including these level-2 variables in the matrix $\boldsymbol{Z}$. The usual interpretation of a "random slope" then is lost, although this term continues to be used in this type of model specification.

Given that among multilevel modelers random slopes tend to be more popular than heteroscedasticity, unrecognized heteroscedasticity may show up in the form of a fitted model with a random slope of the same or a correlated variable, which then may disappear if the heteroscedasticity is modeled.

Therefore, when a researcher is interested in a random slope of some variable Z_k and thinks to have found a significant slope variance, it is advisable to test for the following two kinds of heteroscedasticity: The level-1 residual variance may depend (e.g., linearly or quadratically) on the variable Z_k, or the level-2 intercept variance may depend on the cluster mean of Z_k, i.e., on the variable defined by

$$\bar{z}_{.jk} = \frac{1}{n_j} \sum_{i=1}^{n_j} z_{ijk} \ .$$

Given that one uses software that can implement models with these types of heteroscedasticity, this is an easy (and sometimes disconcerting) model check. Some examples of checking for heteroscedasticity can be found in Goldstein [18, Chapter 3] and Snijders and Bosker [51, Chapter 8].

3.2.2 Random or Fixed Coefficients

A basic question in applying the HLM is whether a random coefficient model is appropriate at all for representing the differences between the level-2 units. In other words, is it appropriate indeed to treat the variables $\underline{\boldsymbol{\delta}}_j$ in (3.1) as random variables, or should they rather be treated as fixed parameters $\boldsymbol{\delta}_j$?

On a conceptual level, this depends on the purpose of the statistical inference. If the level-2 units j may be regarded as a sample from some population (which in some cases will be hypothetical or hard to circumscribe, but nevertheless conceptually meaningful) and the statistical inference is directed at this population, then a random coefficient model is in principle appropriate; cf. Hsiao [30]. This is the case, e.g., when one wishes to test the effect of an explanatory variable that is defined at level 2, i.e., it is a function of the level-2 units only. Then testing this variable has to be based on some way of comparing the variation accounted for by this variable to the total residual variation between level-2 units, and it is hard to see how this could be done meaningfully without assuming that the level-2 units are a sample from a population.

If, on the other hand, the statistical inference aims only at the particular set of units j included in the data set at hand, then a fixed effects model is appropriate. Note that in the fixed effects model, the only random effects are the level-1 residuals $\underline{\epsilon}_j$; under the usual assumption (3.1d) of homoscedasticity, this model can be analysed by ordinary least squares (OLS) regression, so that the analysis is very straightforward except perhaps for the large number of dummy variables. When the cluster sizes are very large, there is hardly a difference between the fixed effects and the random effects specification for the estimation of parameters that they have in common.

If the differences between the level-2 units are a nuisance factor rather than a point of independent interest, so that there is interest only in the within-cluster effects, the analysis could in principle be done either way. Then the fixed effects estimates of the within-cluster regression coefficients, obtainable by OLS regression, achieve a better control for unexplained differences between the level-2 units, because they do not need the assumption that the explanatory variables $\boldsymbol{X}$ are uncorrelated with the level-2 random effects $\underline{\boldsymbol{\delta}}$. More generally, the fixed effects estimates have the attractive robustness property that they are not influenced at all by the specification of the level-2 model. This can, of course, be generalized to models with more than two levels. This robustness property is elaborated with a lot of detailed matrix calculus in Kim and Frees [32].

On a practical level, the choice between random and fixed effects depends strongly on the tenability of the model assumptions made for the random coefficients and the properties of the statistical procedures available under the two approaches. Such practical considerations will be especially important if the differences between level-2 units are a nuisance factor only. The assumptions in model (3.1) for the random effects $\underline{\boldsymbol{\delta}}_j$ are their zero expectations, homogeneous variances, and normal distributions. The normality of the distributions can be checked to some extent by plots of residuals (see below). If normality seems untenable, one could use models with other distributions for the random effects such as t-distributions (e.g., Seltzer, Wong, and Bryk [48]) or mixtures of normal distributions (the heterogeneity model of Verbeke and Lesaffre [54]; also see Verbeke and Molenberghs [55]). Homogeneity of the variances is very close to the assumption that the level-2 units are indeed a random sample from a population; in the preceding section, it was discussed how to model variances depending on level-2 variables, which can occur, e.g., if the level-2 units are a sample from a stratified population and the variances depend on the stratum-defining variables.

To understand the requirement that the expected values of the level-2 residuals are zero, we first focus on the simplest case of a random intercept model, where $\boldsymbol{Z}_j$ contains only the constant vector with all its n_j entries equal to 1, expressed as $\boldsymbol{Z}_j = \mathbf{1}_{n_j}$. Subsequently, we will give a more formal treatment of a more general case.

The level-2 random effects $\underline{\boldsymbol{\delta}}_j$ consist of only one variable, the random intercept $\underline{\delta}_j$. Suppose that the expected value of $\underline{\delta}_j$ is given by

$$E\,\underline{\delta}_j = \boldsymbol{z}_{2j}\boldsymbol{\gamma},$$

for $1 \times u$ vectors $\boldsymbol{z}_{2j}$ and a regression coefficient $\boldsymbol{\gamma}$. Accordingly, $\underline{\delta}_j$ is written as $\underline{\delta}_j = \boldsymbol{z}_{2j}\boldsymbol{\gamma} + \underline{\tilde{\delta}}_j$. Note that a term in $\mathbf{1}_{n_j}\,E\,\underline{\delta}_j$ that is a linear combination of $\boldsymbol{X}_j$ will be absorbed into the model term $\boldsymbol{X}_j\boldsymbol{\beta}$, so this misspecification is nontrivial only if $\mathbf{1}_{n_j}\boldsymbol{z}_{2j}$ cannot be written as a linear combination $\boldsymbol{X}_j\boldsymbol{A}$ for some weight matrix $\boldsymbol{A}$ independent of j.

The question now is in the first place, how the parameter estimates are affected by the incorrectness of the assumption that $\underline{\delta}_j$ has a zero expected value, corresponding to the omission of the term $\boldsymbol{z}_{2j}\boldsymbol{\gamma}$ from the model equation.

It is useful to split the variable $\boldsymbol{X}_j$ into its cluster mean $\bar{\boldsymbol{X}}_j$ and the within-cluster deviation variable $\tilde{\boldsymbol{X}}_j = \boldsymbol{X}_j - \bar{\boldsymbol{X}}_j$:

$$\boldsymbol{X}_j = \bar{\boldsymbol{X}}_j + \tilde{\boldsymbol{X}}_j \,,$$

where

$$\bar{\boldsymbol{X}}_j = \mathbf{1}_{n_j}(\mathbf{1}'_{n_j}\mathbf{1}_{n_j})^{-1}\mathbf{1}'_{n_j}\boldsymbol{X}_j \,.$$

Then the data-generating model can be written as

$$\underline{\boldsymbol{y}}_j = \bar{\boldsymbol{X}}_j\boldsymbol{\beta} + \tilde{\boldsymbol{X}}_j\boldsymbol{\beta} + \mathbf{1}_{n_j}\boldsymbol{z}_{2j}\boldsymbol{\gamma} + \mathbf{1}_{n_j}\underline{\tilde{\delta}}_j + \underline{\boldsymbol{\epsilon}}_j \ ,$$

for random effects $\underline{\tilde{\delta}}_j$ that do satisfy the condition that they have zero expected values.

A bias in the estimation of $\boldsymbol{\beta}$ will be caused by lack of orthogonality of the matrices $\boldsymbol{X}_j = \bar{\boldsymbol{X}}_j + \tilde{\boldsymbol{X}}_j$ and $\mathbf{1}_{n_j}\boldsymbol{z}_{2j}$. Since the definition of $\tilde{\boldsymbol{X}}_j$ implies that $\tilde{\boldsymbol{X}}_j$ is orthogonal to $\mathbf{1}_{n_j}\boldsymbol{z}_{2j}$, it is clear that $\bar{\boldsymbol{X}}_j$ is the villain of the piece: Analogous to the situation of a misspecified General Linear Model, there will be a bias if the cluster mean of $\boldsymbol{X}$ is non-zero, $\bar{\boldsymbol{X}}'_j\mathbf{1}_{n_j} \neq \boldsymbol{\emptyset}$. If it is non-zero, there is an obvious solution: extend the fixed part by giving separate fixed parameters $\boldsymbol{\beta}_1$ to the cluster means $\bar{\boldsymbol{X}}$ and $\boldsymbol{\beta}_2$ to the deviation variables $\tilde{\boldsymbol{X}}$, so that the working model reads

$$\underline{\boldsymbol{y}}_j = \bar{\boldsymbol{X}}_j\boldsymbol{\beta}_1 + \tilde{\boldsymbol{X}}_j\boldsymbol{\beta}_2 + \mathbf{1}_{n_j}\underline{\delta}_j + \underline{\boldsymbol{\epsilon}}_j,$$

(taking out the zero columns from $\bar{\boldsymbol{X}}_j$ and $\tilde{\boldsymbol{X}}_j$, which are generated by columns in $\boldsymbol{X}_j$ which themselves are within-cluster deviation variables or level-2 variables, respectively). An equivalent working model is obtained by adding to (3.1) the fixed effects of the non-constant cluster means $\bar{\boldsymbol{X}}_j$. In this way, the bias in the fixed effect estimates due to ignoring the term $\boldsymbol{z}_{2j}\boldsymbol{\gamma}$ is absorbed completely by the parameter estimate for $\boldsymbol{\beta}_1$, and this misspecification does not affect the unbiasedness of the estimate for $\boldsymbol{\beta}_2$. The estimate for the level-2 variance $\mathrm{Var}(\underline{\delta}_j)$ will be affected, which is inescapable if there is no knowledge about $\boldsymbol{z}_{2j}$, but the estimate for the level-1 variance σ^2 will be consistent.

In the practice of multilevel analysis, it is known that the cluster means often have a substantively meaningful interpretation, different from the level-1 variables from which they are calculated (cf. the discussion in Sections 3.6 and 4.5 of Snijders and Bosker [51] about within- and between-group regressions). This often leads to a substance-matter-related rationale for including the cluster means among the variables with fixed effects.

It can be concluded that in a two-level random intercept model, the sensitive part of the assumption that the level-2 random effects have a zero expected value is the orthogonality of these expected values to the cluster means of the variables $\boldsymbol{X}$ with fixed effects. This orthogonality can be tested simply by testing the effects of these cluster means included as additional variables in the fixed part of the model. This can be interpreted as testing the equality between the within-cluster regression coefficient and the between-cluster coefficient. This test—or at least a test with the same purpose—is often referred to as the Hausman test. (Hausman [26] proposed a general procedure for tests of model specification, of which the test for equality of the within-cluster and between-cluster coefficients is an important special case. Also see Baltagi [3], who shows on p. 69 that this case of the Hausman test is equivalent to testing the effect of the cluster means $\bar{\boldsymbol{X}}$.)

In econometrics, the Hausman test for the difference between the within-cluster and between-cluster regression coefficients is often seen as a test for deciding whether to use a random or fixed coefficient model for the level-2 residuals $\boldsymbol{\delta}_j$. The preceding discussion shows that this is slightly beside the point. If there is a difference between the within-cluster and between-cluster regression coefficients, which is what this Hausman test intends to detect, then unbiased estimates for the fixed within-cluster effects can be obtained also with random coefficient models, provided that the cluster means of the explanatory variables are included among the fixed effect variables $\boldsymbol{X}$. Including the cluster means will lead to an increase of the number of fixed effects by at most r, which normally is much less than the $m-1$ fixed effects required for including fixed main effects of the clusters. Whether or not to use a random coefficient model depends on other considerations, as discussed earlier in this section. Fielding [16] gives an extensive discussion of this issue and warns against the oversimplification of using this Hausman test without further thought to decide between random effects and fixed effects models.

Now consider the general case that $\boldsymbol{Z}$ has some arbitrary positive dimension s. Let the expected value of the level-2 random effects $\underline{\boldsymbol{\delta}}_j$ in the data-generating model be given by

$$E\,\underline{\boldsymbol{\delta}}_j = \boldsymbol{Z}_{2j}\gamma,$$

instead of the assumed value of $\boldsymbol{\emptyset}$. It may be assumed that $\boldsymbol{Z}_j\boldsymbol{Z}_{2j}$ cannot be expressed as a linear combination $\boldsymbol{X}_j\boldsymbol{A}$ for some matrix $\boldsymbol{A}$ independent of j, because otherwise the contribution caused by $E\,\underline{\boldsymbol{\delta}}_j$ could be absorbed into $\boldsymbol{X}_j\boldsymbol{\beta}$.

Both $\boldsymbol{X}_j$ and $\boldsymbol{y}_j$ are split in two terms, the within-cluster projections $\vec{\boldsymbol{X}}_j$ and $\vec{\boldsymbol{y}}_j$ on the linear space spanned by the variables $\boldsymbol{Z}_j$,

$$\vec{\boldsymbol{X}}_j = \boldsymbol{Z}_j(\boldsymbol{Z}_j'\boldsymbol{Z}_j)^{-1}\boldsymbol{Z}_j'\boldsymbol{X}_j \quad \text{and} \quad \vec{\boldsymbol{y}}_j = \boldsymbol{Z}_j(\boldsymbol{Z}_j'\boldsymbol{Z}_j)^{-1}\boldsymbol{Z}_j'\boldsymbol{y}_j\,,$$

and the difference variables

$$\tilde{\boldsymbol{X}}_j = \boldsymbol{X}_j - \vec{\boldsymbol{X}}_j \quad \text{and} \quad \tilde{\boldsymbol{y}}_j = \boldsymbol{y}_j - \vec{\boldsymbol{y}}_j \,.$$

The projection $\vec{\boldsymbol{X}}_j$ can be regarded as the prediction of $\boldsymbol{X}_j$, produced by the ordinary least squares (OLS) regression of $\boldsymbol{X}_j$ on $\boldsymbol{Z}_j$ for cluster j separately, and the same for $\vec{\boldsymbol{y}}_j$. The data-generating model now is written as

$$\underline{\boldsymbol{y}}_j = \boldsymbol{X}_j\boldsymbol{\beta} + \boldsymbol{Z}_j\boldsymbol{Z}_{2j}\boldsymbol{\gamma} + \boldsymbol{Z}_j\underline{\tilde{\boldsymbol{\delta}}}_j + \underline{\boldsymbol{\epsilon}}_j \;,$$

where, again, the $\underline{\tilde{\boldsymbol{\delta}}}_j$ do have zero expected values.

The distribution of $\underline{\boldsymbol{y}}_j$ is the multivariate normal

$$\underline{\boldsymbol{y}}_j \sim \mathcal{N}(\boldsymbol{X}_j\boldsymbol{\beta} + \boldsymbol{Z}_j\boldsymbol{Z}_{2j}\boldsymbol{\gamma}, \boldsymbol{V}_j),$$

where

$$\boldsymbol{V}_j = \sigma^2\boldsymbol{I}_{n_j} + \boldsymbol{Z}_j\boldsymbol{\Omega}(\boldsymbol{\xi})\boldsymbol{Z}_j' \,. \tag{3.3}$$

Hence, the log-likelihood function of the data-generating model is given by

$$-\tfrac{1}{2}\sum_j\Big(\log\det(\boldsymbol{V}_j) + (\boldsymbol{y}_j - \boldsymbol{X}_j\boldsymbol{\beta} - \boldsymbol{Z}_j\boldsymbol{Z}_{2j}\boldsymbol{\gamma})'\boldsymbol{V}_j^{-1}(\boldsymbol{y}_j - \boldsymbol{X}_j\boldsymbol{\beta} - \boldsymbol{Z}_j\boldsymbol{Z}_{2j}\boldsymbol{\gamma})\Big).$$

The inverse of $\boldsymbol{V}_j$ can be written as [41, 44]

$$\boldsymbol{V}_j^{-1} = \sigma^{-2}\boldsymbol{I}_{n_j} - \boldsymbol{Z}_j\boldsymbol{A}_j\boldsymbol{Z}_j' \,, \tag{3.4}$$

for a matrix

$$\boldsymbol{A}_j = \sigma^{-2}(\boldsymbol{Z}_j'\boldsymbol{Z}_j)^{-1} - (\boldsymbol{Z}_j'\boldsymbol{Z}_j)^{-1}\big(\sigma^2(\boldsymbol{Z}_j'\boldsymbol{Z}_j)^{-1} + \boldsymbol{\Omega}(\boldsymbol{\xi})\big)^{-1}(\boldsymbol{Z}_j'\boldsymbol{Z}_j)^{-1}.$$

This implies that

$$\begin{aligned}(\boldsymbol{y}_j - \boldsymbol{X}_j\boldsymbol{\beta} - \boldsymbol{Z}_j\boldsymbol{Z}_{2j}\boldsymbol{\gamma})'\boldsymbol{V}_j^{-1}(\boldsymbol{y}_j - \boldsymbol{X}_j\boldsymbol{\beta} - \boldsymbol{Z}_j\boldsymbol{Z}_{2j}\boldsymbol{\gamma})\\ = (\vec{\boldsymbol{y}}_j - \vec{\boldsymbol{X}}_j\boldsymbol{\beta} - \boldsymbol{Z}_j\boldsymbol{Z}_{2j}\boldsymbol{\gamma})'\boldsymbol{V}_j^{-1}(\vec{\boldsymbol{y}}_j - \vec{\boldsymbol{X}}_j\boldsymbol{\beta} - \boldsymbol{Z}_j\boldsymbol{Z}_{2j}\boldsymbol{\gamma})\\ + \sigma^{-2}\|\tilde{\boldsymbol{y}}_j - \tilde{\boldsymbol{X}}_j\boldsymbol{\beta}\|^2,\end{aligned}$$

where $\|\cdot\|$ denotes the usual Euclidean norm. The log-likelihood is

$$\begin{aligned}-\tfrac{1}{2}\sum_j\Big(&\log\det(\boldsymbol{V}_j)\\ &+ (\vec{\boldsymbol{y}}_j - \vec{\boldsymbol{X}}_j\boldsymbol{\beta} - \boldsymbol{Z}_j\boldsymbol{Z}_{2j}\boldsymbol{\gamma})'\boldsymbol{V}_j^{-1}(\vec{\boldsymbol{y}}_j - \vec{\boldsymbol{X}}_j\boldsymbol{\beta} - \boldsymbol{Z}_j\boldsymbol{Z}_{2j}\boldsymbol{\gamma})\\ &+ \sigma^{-2}\|\tilde{\boldsymbol{y}}_j - \tilde{\boldsymbol{X}}_j\boldsymbol{\beta}\|^2\Big).\end{aligned} \tag{3.5}$$

This shows that the omission from the model of $\boldsymbol{Z}_j\boldsymbol{Z}_{2j}\boldsymbol{\gamma}$ will affect the estimates only through the term $\vec{\boldsymbol{X}}_j\boldsymbol{\beta}$. If now separate fixed parameters are given to $\vec{\boldsymbol{X}}$ and $\tilde{\boldsymbol{X}}$ so that the working model is

$$\underline{y}_j = \vec{X}_j\beta_1 + \tilde{X}_j\beta_2 + Z_j\underline{\delta}_j + \underline{\epsilon}_j\,,$$

the bias due to neglecting the term $Z_{2j}\gamma$ in the expected value of $\underline{\delta}_j$ will be absorbed into the estimate of β_1, and β_2 will be an unbiased estimate for the fixed effect of X. The log-likelihood (3.5) shows that the ML and REML estimates of β_2 are equal to the OLS estimate based on the deviation variables $\vec{y}_j$, and also equal to the OLS estimate in the model obtained by replacing the random effects $\underline{\delta}_j$ by fixed effects.

This discussion shows that in the general case, if one is uncertain about the validity of the condition that the level-2 random effects have zero expected values, and one wishes to retain a random effects model rather than work with a model with a large number (viz. ms) of fixed effects, it is advisable to add to the model the fixed effects of the variables

$$\vec{X}_j = Z_j(Z_j'Z_j)^{-1}Z_j'X_j\,, \tag{3.6}$$

i.e., the predictions of the variables in X by within-cluster OLS regression of X_j on Z_j. The model term $Z_j\,E\,\underline{\delta}_j$ will be entirely absorbed into the fixed effects of $\vec{X}_j$, and the estimates of β_2 will be unbiased for the corresponding elements of β in (3.1). Depending on the substantive context, there may well be a meaningful interpretation of the constructed level-2 variables (3.6).

3.3 Residuals

Like in other regression-type models, residuals (this term is now also used to refer to estimates of the residuals $\underline{\epsilon}$ and $\underline{\delta}$ in (3.1)) play an important exploratory role for model checking in multilevel models. For each level, there is a set of residuals and a residual analysis can be executed. One of the practical questions is whether residual checking should be carried out upward—starting with level 1, then continuing with level 2, etc.—or downward—starting from the highest level and continuing with each subsequent lower level. The literature contains different kinds of advice. For example, Raudenbush and Bryk [45] suggest an upward approach for model construction, whereas Langford and Lewis [35] propose a downward approach for the purpose of outlier inspection. In our view, the argument given by Hilden-Minton [27] is convincing: Level-1 residuals can be studied unconfounded by the higher-level residuals, but the reverse is impossible. Therefore, the upward approach is preferable for the careful checking of model assumptions. However, if one wishes to carry out a quick check for outliers, a downward approach may be very efficient.

This section first treats the "internal" standardization of the residuals. Externally standardized residuals, also called deletion residuals, are treated in Section 3.3.5.

3.3.1 Level-1 Residuals

In this section we assume that level-1 residuals are i.i.d. Residuals at level 1 that are unconfounded by the higher-level residuals can be obtained, as remarked by Hilden-Minton [27], as the OLS residuals calculated separately within each level-2 cluster. These are just the same as the estimated residuals in the OLS analysis of the fixed effects model, where all level-2 (or higher-level, if there are any higher levels) residuals are treated as fixed rather than random. These will be called here the OLS within-cluster residuals. Consider again model (3.1) with the further specification (3.1d). When $\check{\boldsymbol{X}}_j$ is the matrix containing all non-redundant columns in $(\boldsymbol{X}_j\ \boldsymbol{Z}_j)$ and $\boldsymbol{P}_j$ is the corresponding projection matrix (the "hat matrix")

$$\boldsymbol{P}_j = \check{\boldsymbol{X}}_j(\check{\boldsymbol{X}}_j'\check{\boldsymbol{X}}_j)^{-1}\check{\boldsymbol{X}}_j' ,$$

the OLS within-cluster residuals are given by

$$\hat{\underline{\boldsymbol{\epsilon}}}_j = \left(\boldsymbol{I}_{n_j} - \boldsymbol{P}_j\right)\underline{\boldsymbol{y}}_j .$$

The model definition implies that

$$\hat{\underline{\boldsymbol{\epsilon}}}_j = \left(\boldsymbol{I}_{n_j} - \boldsymbol{P}_j\right)\underline{\boldsymbol{\epsilon}}_j , \tag{3.7}$$

which shows that indeed these residuals depend only on the level-1 residuals $\underline{\boldsymbol{\epsilon}}_j$ without confounding by the level-2 residuals $\underline{\boldsymbol{\delta}}_j$.

These level-1 residuals can be used for two main purposes. In the first place, for investigating the specification of the within-cluster model, i.e., the choice of the explanatory variables contained in $\boldsymbol{X}$ and $\boldsymbol{Z}$. Linearity of the dependence on these variables can be checked by plotting the residuals $\hat{\underline{\boldsymbol{\epsilon}}}_j$ against the variables in $\boldsymbol{X}$ and $\boldsymbol{Z}$. The presence of outliers and potential effects of omitted but available variables can be studied analogously.

In the second place, the homoscedasticity assumption (3.1d) can be checked. Equation (3.7) implies that, if the model assumptions are correct,

$$\hat{\underline{\epsilon}}_{ij} \sim \mathcal{N}\big(0, \sigma^2(1 - h_{ij})\big), \tag{3.8}$$

where h_{ij} is the i-th diagonal element of the hat matrix $\boldsymbol{P}_j$. This implies that the "semi-standardized residuals"

$$\check{\underline{\epsilon}}_{ij} = \frac{\hat{\underline{\epsilon}}_{ij}}{\sqrt{1 - h_{ij}}},$$

have a normal distribution with mean 0 and variance σ^2. For checking homoscedasticity, the squared semi-standardized residuals can be plotted against explanatory variables or in a meaningful order. This is informative only under the assumption that the expected value of the residuals is indeed 0. Therefore, these heteroscedasticity checks should be performed only after having ascertained the linear dependence of the fixed part on the explanatory variables.

To check linearity and homoscedasticity as a function of explanatory variables, if the plot of the residuals just shows a seemingly chaotic mass of scatter, it often is helpful to smooth the plots of residuals against explanatory variables, e.g., by moving averages or by spline smoothers. We find it particularly helpful to use smoothing splines [cf. 21], choosing the smoothing parameter so as to minimize the cross-validatory estimated prediction error.

If there is evidence of inhomogeneity of level-1 variances, the level-1 model is in doubt and attempts to improve it are in order. The analysis of level-1 residuals might suggest non-linear transformations of the explanatory variables, as discussed in the second half of this chapter, or a heteroscedastic level-1 model. Another possibility is to apply a non-linear transformation to the dependent variable. Atkinson [2] has an illuminating discussion of non-linear transformations of the dependent variable in single-level regression models. Hodges [28, p. 506] discusses Box-Cox transformations for multilevel models.

As an example, consider the data set provided with the MLwiN software [19] in the worksheet `tutorial.ws`. This includes data for 4059 students in 65 schools; we use the normalized exam score (`normexam`) (mean 0, variance 1) as the dependent variable and only the standardized reading test (`standlrt`) as an explanatory variable. The two mentioned uses of the OLS level-1 residuals will be illustrated.

When the OLS within-cluster residuals are plotted against the explanatory variable `standlrt`, an unilluminating cloud of points is produced. Therefore, only the smoothed residuals are plotted in Fig. 3.1.

This figure shows a smooth curve suggestive of a cubic polynomial. The shape of the curve suggests including the square and cube of `standlrt` as extra explanatory variables. The resulting model estimates are presented as Model 2 in Table 3.1. Indeed, the model improvement is significant ($\chi^2 = 11.0$, d.f. $= 2$, $p < .005$).

As a check of the level-1 homoscedasticity, the semi-standardized residuals (3.8) are calculated for Model 2. The smoothed squared semi-standardized residuals are plotted against `standlrt` in Fig. 3.2.

This figure suggests that the level-1 variance decreases linearly with the explanatory variable. A model with this specification (cf. Section 3.2.1),

$$\mathrm{Var}(\underline{\epsilon}_{ij}) = \sigma^2 + \theta_2\, \texttt{standlrt}_{ij} \ ,$$

is presented as Model 3 in Table 3.1. The heteroscedasticity is a significant model improvement ($\chi^2 = 4.8$, d.f. $= 1$, $p < .05$).

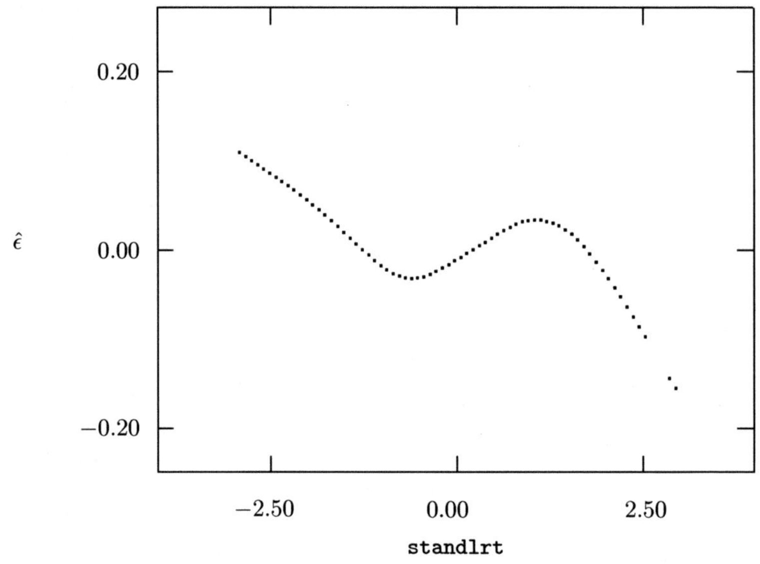

Fig. 3.1 Smoothing spline approximation for OLS within-cluster residuals ($\hat{\epsilon}$) under Model 1 against standardized reading test (`standlrt`).

3.3.2 Homogeneity of Variance Across Clusters

The OLS within-cluster residuals can also be used in a test of the assumption that the level-1 variance is the same in all level-2 units against the specific alternative hypothesis that the level-1 variance varies across the level-2 units. Formally, this means that the null hypothesis (3.1d) is tested against the alternative

$$\boldsymbol{\Sigma}_j(\boldsymbol{\theta}) = \sigma_j^2 \boldsymbol{I}_{n_j},$$

where the σ_j^2 are unspecified and not identical.

Table 3.1 Parameter estimates for models fitted to normalized exam scores.

	Model 1		Model 2		Model 3	
Fixed part						
Constant term	0.002	(0.040)	−0.017	(0.041)	−0.017	(0.041)
`standlrt`	0.563	(0.012)	0.604	(0.021)	0.605	(0.021)
`standlrt`^{2}			0.017	(0.009)	0.017	(0.008)
`standlrt`^{3}			−0.013	(0.005)	−0.013	(0.005)
Random part						
Level 2: ω_{11}	0.092	(0.018)	0.093	(0.018)	0.095	(0.019)
Level 1: σ^2	0.566	(0.013)	0.564	(0.013)	0.564	(0.013)
Level 1: θ_2					−0.007	(0.003)
Deviance	9357.2		9346.2		9341.4	

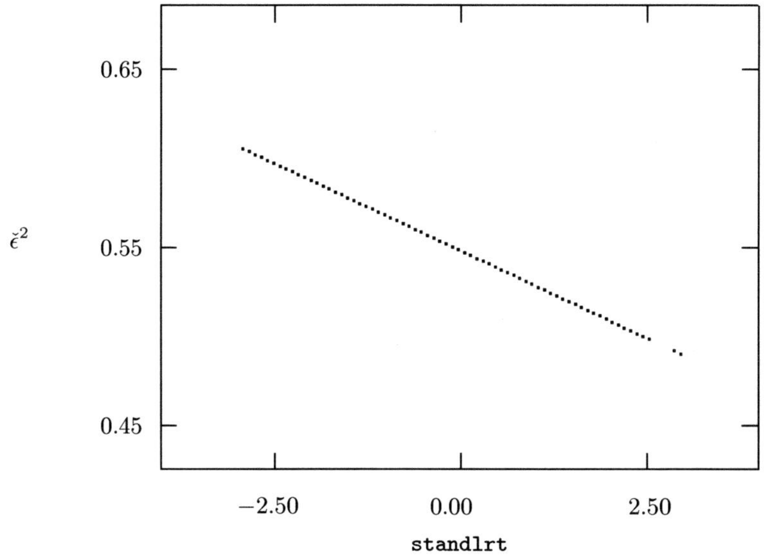

Fig. 3.2 Smoothing spline approximation for the squared semi-standardized OLS within-cluster residuals ($\check{\epsilon}^2$) under Model 2 against the standardized reading test (`standlrt`).

Indicating the rank of $\check{\boldsymbol{X}}_j$ defined in Section 3.3.1 by r_j, the within-cluster residual variance is

$$\underline{s}_j^2 = \frac{1}{n_j - r_j}\, \hat{\underline{\boldsymbol{\epsilon}}}_j' \hat{\underline{\boldsymbol{\epsilon}}}_j .$$

If model (3.1d) is correct, $(n_j - r_j)\underline{s}_j^2/\sigma^2$ has a chi-squared distribution with $(n_j - r_j)$ degrees of freedom. The homogeneity test of Bartlett and Kendall [4] can be applied here (it is also proposed in Raudenbush and Bryk [45, p. 264] and Snijders and Bosker [51, p. 127]). Denoting $\sum n_j = n_+$, $\sum r_j = r_+$ and

$$\underline{ls}_{\text{pooled}} = \frac{1}{n_+ - r_+} \sum_j (n_j - r_j) \log(\underline{s}_j^2), \tag{3.9}$$

the test statistic is given by

$$\underline{H} = \sum_j \frac{n_j - r_j}{2} \left(\log(\underline{s}_j^2) - \underline{ls}_{\text{pooled}}\right)^2 . \tag{3.10}$$

Under the null hypothesis, this statistic has approximately a chi-squared distribution with $\tilde{m} - 1$ degrees of freedom, where $\tilde{m}$ is the number of clusters included in the summation (this could be less than m because some small clusters might be skipped).

This chi-squared approximation is valid if the degrees of freedom $n_j - r_j$ are large enough. If this approximation is in doubt, a Monte Carlo test can be used. This test is based on the property that, under the null hypothesis, $(n_j - r_j)\underline{s}_j^2/\sigma^2$ has an exact chi-squared distribution, and the unknown parameter σ^2 does not affect the distribution of $\underline{H}$ because its contribution in (3.10) cancels out. This implies that under the null hypothesis the distribution of $\underline{H}$ does not depend on any unknown parameters, and a random sample from its distribution can be generated by randomly drawing random variables $\underline{c}_j^2$ from chi-squared distributions with $(n_j - r_j)$ d.f. and applying formulae (3.9) and (3.10) to $\underline{s}_j^2 = \underline{c}_j^2/(n_j - r_j)$. By simulating a sufficiently large sample from the null distribution of $\underline{H}$, the p-value of an observed value can be approximated to any desired precision.

3.3.3 Level-2 Residuals

There are two main ways for predicting[1] the level-2 residuals $\underline{\boldsymbol{\delta}}_j$: the OLS method (based on treating them as fixed effects $\boldsymbol{\delta}_j$) and the empirical Bayes (EB) method. The empirical Bayes "estimate" of $\underline{\boldsymbol{\delta}}_j$ can be defined as its conditional expected value given the observations $\underline{\boldsymbol{y}}_1, \ldots, \underline{\boldsymbol{y}}_m$, plugging in the parameter estimates for $\boldsymbol{\beta}$, $\boldsymbol{\theta}$, and $\boldsymbol{\xi}$. (In the name, "Bayes" refers to the conditional expectation and "empirical" to plugging in the estimates.)

The advantage of the EB method is that it is more precise, but the disadvantage is its stronger dependence on the model assumptions. The two approaches were compared by Waternaux et al. [59] and Hilden-Minton [27]. Their conclusion was that, provided the level-1 model (i.e., the assumptions about the level-1 predictors included in $\boldsymbol{X}$ and about the level-1 residuals $\underline{\boldsymbol{\epsilon}}_j$) is adequate, it is advisable to use the EB estimates.

Basic properties of the multivariate normal distribution imply that the EB level-2 residuals are given by

$$\begin{aligned}
\hat{\underline{\boldsymbol{\delta}}}_j &= E\{\underline{\boldsymbol{\delta}}_j \mid \underline{\boldsymbol{y}}_1, \ldots, \underline{\boldsymbol{y}}_m\} \qquad \text{(using parameter estimates } \underline{\hat{\boldsymbol{\beta}}}, \underline{\hat{\boldsymbol{\theta}}}, \underline{\hat{\boldsymbol{\xi}}}\text{)} \\
&= \underline{\hat{\boldsymbol{\Omega}}} \boldsymbol{Z}_j' \underline{\hat{\boldsymbol{V}}}_j^{-1} (\underline{\boldsymbol{y}}_j - \boldsymbol{X}_j \underline{\hat{\boldsymbol{\beta}}}) \\
&= \underline{\hat{\boldsymbol{\Omega}}} \boldsymbol{Z}_j' \underline{\hat{\boldsymbol{V}}}_j^{-1} \big(\boldsymbol{Z}_j \underline{\boldsymbol{\delta}}_j + \underline{\boldsymbol{\epsilon}}_j - \boldsymbol{X}_j (\underline{\hat{\boldsymbol{\beta}}} - \boldsymbol{\beta})\big),
\end{aligned}$$

where

$$\boldsymbol{V}_j = \mathrm{Cov}(\underline{\boldsymbol{y}}_j) = \boldsymbol{Z}_j \boldsymbol{\Omega} \boldsymbol{Z}_j' + \boldsymbol{\Sigma}_j\,, \tag{3.11a}$$

$$\underline{\hat{\boldsymbol{V}}}_j = \boldsymbol{Z}_j \underline{\hat{\boldsymbol{\Omega}}} \boldsymbol{Z}_j' + \underline{\hat{\boldsymbol{\Sigma}}}_j\,, \tag{3.11b}$$

---

[1] Traditional statistical terminology is to reserve the word "estimation" for empirical ways to obtain reasonable values for parameters, and use "prediction" for ways to empirically approximate unobserved outcomes of random variables. We will not consistently respect this terminology, since almost everybody writes about *estimation* of residuals.

with $\underline{\hat{\boldsymbol{\Omega}}} = \boldsymbol{\Omega}(\underline{\hat{\boldsymbol{\xi}}})$ and $\underline{\hat{\boldsymbol{\Sigma}}}_j = \boldsymbol{\Sigma}_j(\underline{\hat{\boldsymbol{\theta}}})$.

Some more insight into the properties of these estimated residuals may be obtained by defining the estimated reliability matrix

$$\underline{\hat{\boldsymbol{R}}}_j = \underline{\hat{\boldsymbol{\Omega}}}\boldsymbol{Z}_j'\underline{\hat{\boldsymbol{V}}}_j^{-1}\boldsymbol{Z}_j\,.$$

This matrix is the multivariate generalization of the reliability of estimation of $\underline{\boldsymbol{\delta}}_{jq}$, the ratio of the true variance of $\underline{\boldsymbol{\delta}}_{jq}$ to the variance of its OLS estimator based on cluster j (not taking into account the component of variability due to the estimation of $\boldsymbol{\beta}$), as defined by Raudenbush and Bryk [45, p. 49].

The EB residuals can be expressed as

$$\underline{\hat{\boldsymbol{\delta}}}_j = \underline{\hat{\boldsymbol{R}}}_j\underline{\boldsymbol{\delta}}_j + \underline{\hat{\boldsymbol{\Omega}}}\boldsymbol{Z}_j'\underline{\hat{\boldsymbol{V}}}_j^{-1}\underline{\boldsymbol{\epsilon}}_j - \underline{\hat{\boldsymbol{\Omega}}}\boldsymbol{Z}_j'\underline{\hat{\boldsymbol{V}}}_j^{-1}\boldsymbol{X}_j(\underline{\hat{\boldsymbol{\beta}}} - \boldsymbol{\beta}). \tag{3.12}$$

The first term can be regarded as a shrinkage transform of $\underline{\boldsymbol{\delta}}_j$, the second term is the confounding due to the level-1 residuals $\underline{\boldsymbol{\epsilon}}_j$, and the third term is the contribution due to the estimation of the fixed parameters $\boldsymbol{\beta}$.

Ignoring the contribution to the variances and covariances due to the estimation of $\boldsymbol{\xi}$ and $\boldsymbol{\theta}$, the covariance matrix of the EB residuals is

$$\mathrm{Cov}\big(\underline{\hat{\boldsymbol{\delta}}}_j\big) = \boldsymbol{\Omega}\boldsymbol{Z}_j'\boldsymbol{V}_j^{-1}\left(\boldsymbol{V}_j - \boldsymbol{X}_j\Big(\sum_{\ell=1}^{m}\boldsymbol{X}_\ell'\boldsymbol{V}_\ell^{-1}\boldsymbol{X}_\ell\Big)^{-1}\boldsymbol{X}_j'\right)\boldsymbol{V}_j^{-1}\boldsymbol{Z}_j\boldsymbol{\Omega}. \tag{3.13}$$

The second term in the large parentheses is due to the third term in (3.12) and will be negligible if the number m of clusters is large. The resulting simpler expression is

$$\mathrm{Cov}\big(\underline{\hat{\boldsymbol{\delta}}}_j\big) \approx \boldsymbol{\Omega}\boldsymbol{Z}_j'\boldsymbol{V}_j^{-1}\boldsymbol{Z}_j\boldsymbol{\Omega}. \tag{3.14}$$

Another relevant covariance matrix contains the variances and covariances of the prediction errors. The same approximation leading to (3.14) yields

$$\mathrm{Cov}\big(\underline{\hat{\boldsymbol{\delta}}}_j - \underline{\boldsymbol{\delta}}_j\big) \approx \boldsymbol{\Omega} - \boldsymbol{\Omega}\boldsymbol{Z}_j'\boldsymbol{V}_j^{-1}\boldsymbol{Z}_j\boldsymbol{\Omega}. \tag{3.15}$$

If all n_j become very large, (3.15) tends to $\boldsymbol{\emptyset}$. Expression (3.14) is the asymptotic covariance matrix for fixed n_j, which tends to $\boldsymbol{\Omega}$ if n_j tends to infinity. The variances in (3.13) and (3.14) are relevant for diagnosing properties of the residuals $\underline{\boldsymbol{\delta}}_j$ and are called *diagnostic variances* by Goldstein [18]. The variances in (3.15) are relevant for comparing residuals $\underline{\boldsymbol{\delta}}_j$ and are called *comparative* (or *conditional*) *variances*.

It may be noted that the predictions $\underline{\hat{\boldsymbol{\delta}}}_j$ are necessarily uncorrelated with the errors $(\underline{\hat{\boldsymbol{\delta}}}_j - \underline{\boldsymbol{\delta}}_j)$, because otherwise a better prediction could be made. This implies

$$\mathrm{Cov}\big(\underline{\boldsymbol{\delta}}_j\big) = \mathrm{Cov}\big(\underline{\boldsymbol{\delta}}_j - \underline{\hat{\boldsymbol{\delta}}}_j\big) + \mathrm{Cov}\big(\underline{\hat{\boldsymbol{\delta}}}_j\big),$$

which indeed is evident from the formulae.

For each of the s level-2 random effects separately, various diagnostic plots can be made. The explanation of the level-2 random effects by level-2 variables, as reflected by the fixed main effects of level-2 variables and their cross-level interaction effects with the variables contained in $\boldsymbol{Z}$, can be diagnosed for linearity by plots of the raw residuals $\hat{\underline{\boldsymbol{\delta}}}_j$ against the level-2 explanatory variables. The normality and homoscedasticity assumptions for $\underline{\boldsymbol{\delta}}_j$ can be checked by normal probability plots for the s residuals separately, standardized by dividing them by the diagnostic standard deviations obtained as the square roots of the diagonal elements of (3.13) or (3.14), and by plotting the squares of these standardized residuals against the level-2 variables. Such plots were proposed and discussed by Lange and Ryan [34]. Examples of these plots are given in Goldstein [18], Snijders and Bosker [51], and Lewis and Langford [37].

Eberly and Thackeray [13] showed that it is very well possible that when such a plot shows deviations from normality, the cause is a misspecification of the fixed effects model rather than of the distribution of the random effects. This is in accordance with the general caveat that different aspects of the specification of statistical models are entwined, and the particular importance of this issue for assessing fit of multilevel models. It also supports the principle to first try achieve a good specification of the level-1 model, and assess the level-2 specification only after this has been done.

A diagnostic for the entire vector of level-2 residuals for cluster j can be based on the standardized value

$$\hat{\underline{\boldsymbol{\delta}}}_j' \left\{ \widehat{\mathrm{Cov}}(\hat{\underline{\boldsymbol{\delta}}}_j) \right\}^{-1} \hat{\underline{\boldsymbol{\delta}}}_j \,. \tag{3.16}$$

If one neglects the fact that the estimated rather than the true covariance matrix is used, this statistic has a chi-squared distribution with s degrees of freedom.

With some calculations, using formula (3.4) and the approximate covariance matrix (3.14), the standardized value (3.16) is seen to be given by

$$\hat{\underline{\boldsymbol{\delta}}}_j' \left\{ \widehat{\mathrm{Cov}}(\hat{\underline{\boldsymbol{\delta}}}_j) \right\}^{-1} \hat{\underline{\boldsymbol{\delta}}}_j \approx \hat{\underline{\boldsymbol{\delta}}}_j^{(\mathrm{OLS})\prime} \left(\hat{\underline{\sigma}}^2 (\boldsymbol{Z}_j' \boldsymbol{Z}_j)^{-1} + \hat{\underline{\boldsymbol{\Omega}}} \right)^{-1} \hat{\underline{\boldsymbol{\delta}}}_j^{(\mathrm{OLS})} \,, \tag{3.17}$$

where

$$\hat{\underline{\boldsymbol{\delta}}}_j^{(\mathrm{OLS})} = (\boldsymbol{Z}_j' \boldsymbol{Z}_j)^{-1} \boldsymbol{Z}_j' (\underline{\boldsymbol{y}}_j - \boldsymbol{X}_j \hat{\underline{\boldsymbol{\beta}}}_j)$$

is the OLS estimate of $\boldsymbol{\delta}_j$, estimated from the OLS within-cluster residuals $\underline{\boldsymbol{y}}_j - \boldsymbol{X}_j \hat{\underline{\boldsymbol{\beta}}}_j$. This illustrates that the standardized value can be based on the OLS residuals as well as the EB residuals, if one uses for standardization the covariance matrix $\hat{\sigma}^2 (\boldsymbol{Z}_j' \boldsymbol{Z}_j)^{-1} + \hat{\boldsymbol{\Omega}}$, of which the first part is the sampling variance (level-1 variance) and the second part the true variance (level-2 variance) of the OLS residuals. The name *standardized level-2 residual* is therefore more appropriate for (3.17) than the name standardized EB or OLS residual, since the latter terminology suggests a non-existing distinction.

The ordered standardized level-2 residuals can be plotted against the corresponding quantiles of the chi-squared distribution with s d.f., as a check for outliers and for the multivariate normality of the level-2 random effects.

3.3.4 Multivariate Residuals

The fit of the model for level-2 cluster j is expressed by the multivariate residual

$$\underline{\boldsymbol{y}}_j - \boldsymbol{X}_j \hat{\underline{\boldsymbol{\beta}}} . \tag{3.18}$$

The covariance matrix of this residual, if we neglect the use of the estimated parameter $\hat{\underline{\boldsymbol{\beta}}}$ instead of the unknown true $\boldsymbol{\beta}$, is given by $\boldsymbol{V}_j$ in (3.11a). Accordingly, the *standardized multivariate residual* is defined by

$$\underline{M}_j^2 = (\underline{\boldsymbol{y}}_j - \boldsymbol{X}_j \hat{\underline{\boldsymbol{\beta}}})' \hat{\underline{\boldsymbol{V}}}_j^{-1} (\underline{\boldsymbol{y}}_j - \boldsymbol{X}_j \hat{\underline{\boldsymbol{\beta}}}).$$

This residual has, when the model is correct, approximately a chi-squared distribution with n_j degrees of freedom.

If all variables with fixed effects also have random effects, then $\boldsymbol{X}_j = \boldsymbol{Z}_j = \check{\boldsymbol{X}}_j$ as defined in Section 3.3.1, and $r_j = r = s$. Using (3.4), it can be proved that in this case

$$\underline{M}_j^2 = (n_j - r) \frac{\underline{s}_j^2}{\hat{\underline{\sigma}}^2} + \hat{\underline{\boldsymbol{\delta}}}_j' \left\{ \widehat{\mathrm{Cov}}(\hat{\underline{\boldsymbol{\delta}}}_j) \right\}^{-1} \hat{\underline{\boldsymbol{\delta}}}_j . \tag{3.19}$$

In words, the standardized multivariate residual (with n_j d.f.) is the sum of the scaled within-cluster residual sum of squares (with $n_j - r$ d.f.) and the standardized level-2 residual (with $r = s$ d.f.). If some of the variables with fixed effects do not have a random effect, then the difference between the left-hand side and the right-hand side of (3.19) is a test statistic for the null hypothesis that the variables in $\boldsymbol{X}_j$ indeed have the effect expressed by the overall parameter estimate $\hat{\underline{\boldsymbol{\beta}}}$, i.e., the hypothesis that the variables in $\boldsymbol{X}$ and not in $\boldsymbol{Z}$ have only fixed (and not random) effects. This then approximately is a chi-squared variate with $r_j - s$ d.f.

This split implies that if the standardized multivariate residual for some cluster j is unexpectedly large, it will be informative to consider its two (or three) components and investigate whether the high value can be traced to one of these components separately.

3.3.5 Deletion Residuals

To assess the fit of the model and the possibility of outliers, it is better to calculate and standardize residuals for cluster j using parameter estimates of $\boldsymbol{\beta}$ and $\boldsymbol{V}_j$ calculated on the basis of the data set from which cluster j has been omitted. Such measures are called externally studentized residuals [11] or deletion residuals [2]. This means using the fixed parameter estimate $\underline{\hat{\boldsymbol{\beta}}}_{(\text{-}j)}$ obtained by estimating $\boldsymbol{\beta}$ from the data set from which cluster j has been omitted and estimating (3.11a) by

$$\underline{\hat{\boldsymbol{V}}}_{(\text{-}j)} = \boldsymbol{Z}_j \underline{\hat{\boldsymbol{\Omega}}}_{(\text{-}j)} \boldsymbol{Z}_j' + \underline{\hat{\boldsymbol{\Sigma}}}_{(\text{-}j)}, \tag{3.20}$$

where $\underline{\hat{\boldsymbol{\Omega}}}_{(\text{-}j)} = \boldsymbol{\Omega}(\underline{\hat{\boldsymbol{\xi}}}_{(\text{-}j)})$ and $\underline{\hat{\boldsymbol{\Sigma}}}_{(\text{-}j)} = \boldsymbol{\Sigma}_j(\underline{\hat{\boldsymbol{\theta}}}_{(\text{-}j)})$, while $\underline{\hat{\boldsymbol{\xi}}}_{(\text{-}j)}$ and $\underline{\hat{\boldsymbol{\theta}}}_{(\text{-}j)}$ are the estimates of $\boldsymbol{\xi}$ and $\boldsymbol{\theta}$ based on the data set from which cluster j has been omitted.

Using these ingredients, the *deletion standardized multivariate residual* is defined by

$$\underline{M}^2_{(\text{-}j)} = \left(\underline{\boldsymbol{y}}_j - \boldsymbol{X}_j \underline{\hat{\boldsymbol{\beta}}}_{(\text{-}j)}\right)' \underline{\hat{\boldsymbol{V}}}^{-1}_{(\text{-}j)} \left(\underline{\boldsymbol{y}}_j - \boldsymbol{X}_j \underline{\hat{\boldsymbol{\beta}}}_{(\text{-}j)}\right). \tag{3.21}$$

The *deletion standardized level-2 residual* (for a model where $\boldsymbol{\Sigma}_j(\boldsymbol{\theta}) = \sigma^2 \boldsymbol{I}_{n_j}$) is defined by

$$\underline{\hat{\boldsymbol{\delta}}}^{(\text{OLS})\prime}_{(\text{-}j)} \left(\underline{\hat{\sigma}}^2_{(\text{-}j)} (\boldsymbol{Z}_j' \boldsymbol{Z}_j)^{-1} + \underline{\hat{\boldsymbol{\Omega}}}_{(\text{-}j)}\right)^{-1} \underline{\hat{\boldsymbol{\delta}}}^{(\text{OLS})}_{(\text{-}j)}, \tag{3.22}$$

where

$$\underline{\hat{\boldsymbol{\delta}}}^{(\text{OLS})}_{(\text{-}j)} = (\boldsymbol{Z}_j' \boldsymbol{Z}_j)^{-1} \boldsymbol{Z}_j' \left(\underline{\boldsymbol{y}}_j - \boldsymbol{X}_j \underline{\hat{\boldsymbol{\beta}}}_{(\text{-}j)}\right)$$

and $\underline{\hat{\sigma}}^2_{(\text{-}j)}$ is the estimate for σ^2 calculated from the data set from which cluster j was omitted.

The general idea of model diagnostics is that they should be easy, or at least quick, to compute. Elegant computational formulae have been derived for deletion residuals in the General Linear Model (see Atkinson [2]), and recently by Zewotir and Galpin [63] and Haslett and Dillane [23] also for random coefficient models with uncorrelated random coefficients. This yields the possibility of quick calculations of level-2 deletion residuals. In the HLM, the assumption of uncorrelated higher-level residuals is trivially satisfied for the random intercept model where $\underline{\boldsymbol{\delta}}_j$ is a column vector, but not if there are random slopes. Therefore, these formulae are not generally applicable for random slope models.

Re-estimation of a multilevel model for a lot of different data sets, as implied by the definition of deletion residuals, is not very attractive from the point of view of quick computations. Two alternatives to full computation have been proposed in the literature: Lesaffre and Verbeke [36] proposed influence statistics using an analytic approximation based on second-order Taylor expansions, and Snijders and Bosker [51] proposed a computational approximation based on a one-step estimator. The latter approximation will be followed here because of its simple generalizability to other situations. This approximation is defined as follows.

An iterative estimation algorithm is used, viz. Fisher scoring or (R)IGLS. The initial value for the estimation algorithm is the estimate obtained from the full data set. The one-step estimate is the result of a single step of the algorithm, using the data set reduced by omitting all data for cluster j. It is known from general statistical theory that such one-step estimates are asymptotically efficient. They can be quickly estimated by software that implements Fisher scoring or (R)IGLS. Therefore, all estimates denoted here with the suffix $(-j)$ can be implemented as such one-step estimates obtained with the full-data estimate as the initial value.

3.4 Influence Diagnostics of Higher-Level Units

Next to the direct study of residuals as proposed in the previous section, another approach to model checking is to investigate the influence of individual data points, or sets of data points, on the parameter estimates. In OLS regression, the most widely known technique in this approach is Cook's distance, explained, e.g., in Cook and Weisberg [11], Atkinson [2], and Weisberg [60]. A natural way of performing such checks in multilevel models is to investigate the separate influence of each higher-level unit. This means that the estimates obtained from the total data set are compared to the estimates obtained from the data set from which a particular higher-level unit is omitted.

An influence measure of level-2 unit j on the estimation of the parameters should reflect the importance of the influence of the data for this unit on the parameter estimates. First, consider the regression coefficients $\boldsymbol{\beta}$. Recall that $\underline{\hat{\boldsymbol{\beta}}}$ is the estimate obtained from the full data set, and $\underline{\hat{\boldsymbol{\beta}}}_{(-j)}$ the estimate obtained from the data set from which unit j has been omitted, or an approximation to this estimate. The difference between these two estimates should be standardized on the basis of the inherent imprecision expressed by the covariance matrix of these estimates. In Lesaffre and Verbeke [36] and Snijders and Bosker [51] it was proposed to use the estimated covariance matrix of the estimators obtained from the full data set. Since the diagnostic measure has the aim to detect unduly influential units, it should be taken into account, however, that the unit under scrutiny also might have an undue influence on this estimated covariance matrix. Therefore, it is more appropriate to use the estimated covariance matrix of the estimator obtained from the reduced data set. It may be noted that the computation of this matrix is straightforward in the computational approach of Snijders and Bosker [51], but does not fit well in the analytic approach of Lesaffre and Verbeke [36].

Denote by $\underline{\hat{\boldsymbol{S}}}_{F(-j)}$ the estimated covariance matrix of $\underline{\hat{\boldsymbol{\beta}}}_{(-j)}$ as calculated from the data set from which level-2 unit j has been omitted. Then a standardized measure of the influence of this unit on the fixed parameter estimates is

$$\underline{C}_j^F = \frac{1}{r}\left(\underline{\hat{\boldsymbol{\beta}}} - \underline{\hat{\boldsymbol{\beta}}}_{(\text{-}j)}\right)' \underline{\hat{\boldsymbol{S}}}_{F(\text{-}j)}^{-1}\left(\underline{\hat{\boldsymbol{\beta}}} - \underline{\hat{\boldsymbol{\beta}}}_{(\text{-}j)}\right). \tag{3.23}$$

This formula is analogous to Cook's distance for the General Linear Model.

For the parameters $\boldsymbol{\theta}$ and $\boldsymbol{\xi}$ of the random part of the model, the same procedure can be followed. Indicating these parameters jointly by $\boldsymbol{\eta} = (\boldsymbol{\theta}, \boldsymbol{\xi})$, this leads to the influence measure

$$\underline{C}_j^R = \frac{1}{p}\left(\underline{\hat{\boldsymbol{\eta}}} - \underline{\hat{\boldsymbol{\eta}}}_{(\text{-}j)}\right)' \underline{\hat{\boldsymbol{S}}}_{R(\text{-}j)}^{-1}\left(\underline{\hat{\boldsymbol{\eta}}} - \underline{\hat{\boldsymbol{\eta}}}_{(\text{-}j)}\right), \tag{3.24}$$

where the analogous definitions are used for $\underline{\hat{\boldsymbol{\eta}}}_{(\text{-}j)}$ and $\underline{\hat{\boldsymbol{S}}}_{R(\text{-}j)}$, and p is the total number of parameters in $\boldsymbol{\eta}$. Since the parameters of the fixed and random parts are asymptotically uncorrelated [40], these two influence measures can be combined in the overall influence measure

$$\underline{C}_j = \frac{1}{r+p}\left(r\underline{C}_j^F + p\underline{C}_j^R\right). \tag{3.25}$$

Comparisons with alternative definitions for diagnostics of the type of Cook's distance are given in Verbeke and Molenberghs [55] and Skrondal and Rabe-Hesketh [49].

The influence of a part of the data set on the parameter estimates depends on the *fit* of the model to this part of the data together with the *leverage* of this part, i.e., its potential to influence the parameters as determined from the amount of data and the distribution of the explanatory variables $\boldsymbol{X}$ and $\boldsymbol{Z}$. For a level-2 unit, its size n_j and the distribution of $\boldsymbol{X}_j$ and $\boldsymbol{Z}_j$ determine the leverage. The fit can be measured by the deletion standardized multivariate residual (3.21). A poorly fitting cluster with small leverage will not do much damage to the results of the data analysis. If the model fits well, while there are no systematic differences between the clusters in the distribution of $\boldsymbol{X}_j$ and $\boldsymbol{Z}_j$, and the n_j are small compared to $\sum_j n_j$, the diagnostics (3.23)–(3.25) will have expected values that are roughly proportional to the cluster sizes n_j. A plot of these diagnostics against n_j may draw the attention toward clusters that have an undue influence on the parameter estimates. This information can be combined with the p-values for the deletion standardized multivariate residuals (3.21) obtained from the chi-squared distribution with n_j degrees of freedom, which give information on the fit of the clusters independently of their leverage.

3.5 Simulation-Based Assessment of Model Specification

It was shown above that the specification of the level-1 model can be investigated by considering within-cluster relations between variables or, equivalently, by fixed effect models. These are analyses that effectively reduce the HLM to the General Linear Model, for which distributional properties of many statistics have been derived. These properties can be found in the ample literature of model diagnostics in such models. Properties of higher-level diagnostics cannot be derived by going back to the General Linear Model, and tend to be approximate or unknown. Longford [42] elaborates how simulations can be used to assess p-values of arbitrary statistics based, e.g., on residuals or influence measures. This is done by repeatedly simulating the data under the tested model assumptions and considering the resulting distribution of the statistic under consideration; such a procedure is also called the parametric bootstrap, cf. Van der Leeden et al. [53].

Among such simulation-based procedures, the Monte Carlo test proposed at the end of Section 3.3.2 illustrates the relative simplicity of checking the level-1 model by the fact that the distribution of the statistic considered is independent of any unknown parameters (it is said to be *pivotal*), contrasting to the general case for higher-level diagnostics.

3.6 Non-linear Transformations in the Fixed Part

One of the purposes for which one can use the residuals discussed in the preceding sections is to give guidance of an informal kind when investigating possible non-linear effects of explanatory variables. The remainder of this chapter presents methods to examine non-linear fixed effects of explanatory variables by incorporating them formally into the model.

We consider multilevel models for analyzing the effect of a predictor x on a response variable $\underline{y}$ under the assumption that this effect is a non-linear function $f(x)$ with an unknown functional form. The latter situation is common, e.g., when x refers to time in longitudinal studies, since the effect of time on the response is usually complex and not well understood. Then it seems sensible to approximate $f(x)$ by a flexible function that requires only minimal prior knowledge about $f(x)$ and still provides insight into the dependence between $\underline{y}$ and x.

In the following sections, we will consecutively discuss multilevel models in which the non-linear function $f(x)$ is approximated by a polynomial function, a regression spline, and a smoothing spline. As a guiding model in the discussion, we will use a two-level model for normal responses. Since longitudinal data offer the main (but not only) applications of this approach, clusters will be regarded as individual subjects and level-1 units as repeated measurements of the subjects. We assume that the responses of subject j are generated by

$$\underline{\boldsymbol{y}}_j = f(\boldsymbol{x}_j) + \boldsymbol{X}_{2j}\boldsymbol{\beta} + \boldsymbol{Z}_j\underline{\boldsymbol{\delta}}_j + \underline{\boldsymbol{\epsilon}}_j \,. \tag{3.26}$$

The difference with respect to model (3.1) is that the fixed part is split into, first, a real-valued variable $\boldsymbol{x}$ with a non-linear effect and, second, variables $\boldsymbol{X}_2$ with linear effects.

3.7 Polynomial Model

The polynomial model for multilevel data was put forward by many authors including Goldstein [17], Bryk and Raudenbush [8], and Snijders [50]. The use of a polynomial approximation seems quite natural since it can be regarded as a Taylor expansion of the true unknown function. The Q-th degree polynomial equals

$$f_{\text{pol}}(x) = \alpha_0 + \alpha_1 x + \cdots + \alpha_Q x^Q.$$

The smoothness of $f_{\text{pol}}(x)$ is controlled by the degree Q. The function $f_{\text{pol}}(x)$ is a linear combination of polynomial terms x, x^2, ..., x^Q and therefore this model remains within the confines of the Hierarchical Linear Model, and can be estimated straighforwardly like any other such model. The number of parameters only depends on the degree Q so that the polynomial model is easy to estimate also when $\boldsymbol{x}_j$ differs among subjects. However, estimation problems may arise when $\boldsymbol{x}_j$ is badly scaled. In that case, a simple solution that works for many data sets is to subtract the subject mean from $\boldsymbol{x}_j$. A slightly more elaborate solution is to orthogonalize the polynomial terms using the Gram-Schmidt method.

An attractive feature of the polynomial model is that the regression coefficients can be interpreted as growth parameters, which often are of substantive interest. The effect α_1, for instance, can be interpreted as the rate of change in the response at $x = 0$, which may be a useful parameter of a growth process.

The function $f(x)$ is not always well approximated by a low-degree polynomial, however. In human growth studies, for example, polynomials may fail to produce a smooth and accurate fit because of strong growth during the first year of age and early adolescence [5]. The underlying problem is that a polynomial exhibits non-local behavior, which means that a change in one of the regression coefficients α_q leads to a change in the estimated $f_{\text{pol}}(x)$ for (nearly) all values of x. A consequence of non-local behavior is that when the fit at a certain value of x is improved by increasing Q, the fit may become poorer at other values of x. In general, a polynomial with a high value of Q tends to fit accurately in intervals of x with many observations, but this may be achieved at the cost of a poor fit at other values of x.

3.8 Regression Spline Model

A regression spline [61] consists of piecewise polynomials that are joined at locations on the x-axis named knots. At each knot, two Q-th degree polynomials are connected such that the $(Q-1)$-st derivative of the resulting function exists and is itself a continuous function of x. A popular regression spline in practical data analysis is the cubic or third-degree regression spline, the second derivative of which is continuous at the knots. Regression splines are more flexible than polynomials and often provide a better fit in the presence of strong local non-linearity. However, regression splines are more difficult to specify than polynomials because the number of knots and the positions of the knots need to be determined. For selection of the number of knots, an ad hoc approach can be adopted in which the number of knots is increased until an accurate fit is obtained. This approach may lead to overfitting because there is no penalty for model complexity. To limit the number of knots, a possible approach is to optimize a model summary such as Akaike's Information Criterion (AIC) or the cross-validated log-likelihood [47]. Regarding the positions of the knots on the x-axis, common choices are equally spaced points or quantile points of the empirical distribution of x.

A Q-th degree regression spline with L knots at $a_1, \ldots, a_L$ can be constructed by extending a Q-th degree polynomial with L truncated polynomial terms $(x-a_l)_+^Q$ $(l=1,\ldots,L)$, where the truncated term $(x-a_l)_+^Q$ is equal to $(x-a_l)^Q$ if $x > a_l$ and zero otherwise. The resulting function $f_{\text{reg}}(x)$ can be written as

$$f_{\text{reg}}(x) = \sum_{q=0}^{Q} \alpha_q x^q + \sum_{l=1}^{L} \alpha_{Q+l}(x-a_l)_+^Q \,. \tag{3.27}$$

This representation is easy to understand and the α_q's have a clear interpretation. It shows that the regression spline is a linear function of polynomial terms and therefore easy to handle, as it remains within a finite-dimensional linear function space. For numerical reasons, however, the use of truncated polynomials is not recommendable especially not when the knots are chosen close together. It often is better to work with a different set of basis functions. If $f_{\text{reg}}(x)$ is a cubic regression spline, it is recommendable to write $f_{\text{reg}}(x)$ as a linear combination of so-called B-splines, which are a specific set of piecewise cubic splines. Computation is stable because B-splines take nonzero values over an interval with at most five knots [12]. If $f_{\text{reg}}(x)$ contains one knot at position a only (i.e., $L = 1$ in (3.27)), a simple method to improve scaling of the design matrix is to replace the term x^q in the truncated polynomial formulation of $f_{\text{reg}}(x)$ by the term $(x-a)_-^q$ which equals $(x-a)^q$ if $x < a$ and 0 otherwise [51, p. 189]. Because the data columns of values of $(x-a)_-^q$ and $(x-a)_+^q$ are orthogonal, estimation is stable.

The regression spline is more flexible than the polynomial and tends to exhibit less non-local behavior. The knots are determined outside the model and good placement on the x-axis may require some trial and error. Furthermore, if only a small number of knots is used, the regression spline will not be free from non-local behavior, whereas using too many knots is undesirable since it induces non-smooth behavior. To prevent the spline from being either non-smooth or insufficiently flexible, a possible strategy is to include a large number of knots and at the same time penalize the regression coefficients so that a smooth fit is obtained [14]. A limiting case is a function in which a knot is placed at each distinct value of x in the data set. Splines of the latter type are discussed in the next section.

3.9 Smoothing Spline Model

Suppose that the data set contains T ordered distinct values $x_1, \ldots, x_T$. The cubic smoothing spline, denoted by $f_{\text{css}}(x)$, then is a cubic regression spline with knots at $x_1, \ldots, x_T$ and it is a linear function outside the interval $[x_1, x_T]$. The degree of smoothness is regulated by extending the log-likelihood function with a roughness penalty that penalizes functions for having strong curvature, that is, a large absolute second derivative $|f''_{\text{css}}(x)|$. The definition of the roughness penalty is

$$-\tfrac{1}{2}\lambda \int_{x_0}^{x_{T+1}} \{f''(x)\}^2 \, \mathrm{d}x, \tag{3.28}$$

where λ is a nonnegative smoothing parameter determining the degree of smoothing, and $x_0 < x_1$ and $x_{T+1} > x_T$.

The following basic properties of smoothing splines can be found in the literature on this topic, such as Green and Silverman [21]. The fitted cubic smoothing spline is obtained by maximizing the penalized log-likelihood, that is, the sum of the log-likelihood and the roughness penalty. An additional constraint to ensure that $f_{\text{css}}(x)$ is a cubic smoothing spline does not have to be included because among all functions $f_{\text{css}}(x)$ with continuous second derivatives, the unique minimizer of the penalized log-likelihood is the cubic smoothing spline. If we substitute the cubic smoothing spline $f_{\text{css}}(x)$ with knots at $x_1, \ldots, x_T$ in (3.28), we can evaluate the roughness penalty as

$$-\tfrac{1}{2}\lambda \boldsymbol{f}'_{\text{css}} \boldsymbol{K} \boldsymbol{f}_{\text{css}} \,,$$

where $\boldsymbol{f}_{\text{css}}$ is the vector of values of $f_{\text{css}}(x)$ at $x_1, \ldots, x_T$. The $T \times T$ matrix $\boldsymbol{K}$ equals

$$\boldsymbol{K} = \boldsymbol{Q}\boldsymbol{R}^{-1}\boldsymbol{Q}',$$

where $\boldsymbol{Q}$ is a $T \times (T-2)$ matrix having entries $q_{i,i} = 1/(x_{i+1} - x_i)$, $q_{i+2,i} = 1/(x_{i+2} - x_{i+1})$, $q_{i+1,i} = -(q_{i,i} + q_{i+2,i})$ for $i = 1, \ldots, T-2$, and zero otherwise. The $(T-2) \times (T-2)$ matrix $\boldsymbol{R}$ is symmetric tridiagonal with diagonal entries $r_{i,i} = \frac{1}{3}(x_{i+2} - x_i)$ for $i = 1, \ldots, T-2$. The non-zero off-diagonal entries are $r_{i,i+1} = r_{i+1,i} = \frac{1}{6}(x_{i+2} - x_{i+1})$ for $i = 1, \ldots, T-3$.

3.9.1 Estimation

The model for the responses of subject j is obtained by substituting $\boldsymbol{N}_j \boldsymbol{f}_{\mathrm{css}}$ for $f(\boldsymbol{x}_j)$ in (3.26), where $\boldsymbol{N}_j$ is an $n_j \times T$ matrix of zeros and ones. Each row of $\boldsymbol{N}_j$ contains a single one at the entry t for which $x_{ij} = x_t$. The resulting equation is

$$\underline{\boldsymbol{y}}_j = \boldsymbol{N}_j \boldsymbol{f}_{\mathrm{css}} + \boldsymbol{X}_{2j}\boldsymbol{\beta} + \boldsymbol{Z}_j \underline{\boldsymbol{\delta}}_j + \underline{\boldsymbol{\epsilon}}_j\,, \quad j = 1, \ldots, m. \tag{3.29}$$

The model parameters to be estimated are the vector of spline values $\boldsymbol{f}_{\mathrm{css}}$, the fixed regression coefficients $\boldsymbol{\beta}$, the level-1 variance σ^2, and the level-2 variance parameters $\boldsymbol{\xi}$. Given σ^2 and $\boldsymbol{\xi}$, the penalized log-likelihood is maximized by

$$\underline{\hat{\boldsymbol{f}}}_{\mathrm{css}} = \left(\sum_{j=1}^{m} \boldsymbol{N}_j' \boldsymbol{U}_{X_2,j} \boldsymbol{N}_j + \lambda \boldsymbol{K} \right)^{-1} \sum_{j=1}^{m} \boldsymbol{N}_j' \boldsymbol{U}_{X_2,j}\, \underline{\boldsymbol{y}}_j \tag{3.30}$$

and

$$\underline{\hat{\boldsymbol{\beta}}} = \left(\sum_{j=1}^{m} \boldsymbol{X}_{2j}' \boldsymbol{U}_{N,j} \boldsymbol{X}_{2,j} \right)^{-1} \sum_{j=1}^{m} \boldsymbol{X}_{2j}' \boldsymbol{U}_{N,j} \underline{\boldsymbol{y}}_j\,, \tag{3.31}$$

where

$$\boldsymbol{U}_{N,j} = \boldsymbol{V}_j^{-1} - \boldsymbol{V}_j^{-1} \boldsymbol{N}_j \Big(\sum_j \boldsymbol{N}_j' \boldsymbol{V}_j^{-1} \boldsymbol{N}_j + \lambda \boldsymbol{K} \Big)^{-1} \boldsymbol{N}_j' \boldsymbol{V}_j^{-1}$$

and

$$\boldsymbol{U}_{X_2,j} = \boldsymbol{V}_j^{-1} - \boldsymbol{V}_j^{-1} \boldsymbol{X}_{2j} \Big(\sum_j \boldsymbol{X}_{2j}' \boldsymbol{V}_j^{-1} \boldsymbol{X}_{2j} \Big)^{-1} \boldsymbol{X}_{2j}' \boldsymbol{V}_j^{-1}\,,$$

with $\boldsymbol{V}_j$ given in (3.3).

The parameters $\boldsymbol{f}$ and $\boldsymbol{\beta}$ can also be estimated by the Expectation Maximization (EM) algorithm. The EM algorithm is an iterative procedure for locating the mode of the likelihood, or in Bayesian modeling for determining the posterior mode, see Appendix 1.D. In our case, we need to maximize the penalized likelihood rather than the likelihood itself. From a Bayesian viewpoint, this does not substantially alter the problem but is merely a choice of the prior. Note that the modes of the log posterior and the penalized log-likelihood coincide if a flat prior is taken for $\boldsymbol{\beta}$, and the log-prior of $\boldsymbol{f}_{\mathrm{css}}$ is, except for a constant, equal to $-\frac{1}{2}\lambda \boldsymbol{f}_{\mathrm{css}}' \boldsymbol{K} \boldsymbol{f}_{\mathrm{css}}$.

The EM algorithm consists of an E-step and an M-step. To carry out the E-step, we define the complete-data log-likelihood of $\boldsymbol{y}$ and the random coefficients $\underline{\boldsymbol{\delta}}$ (treated in this algorithm as missing data) given the model parameters, i.e., $\log p(\underline{\boldsymbol{y}}, \underline{\boldsymbol{\delta}} \mid \boldsymbol{f}_{\mathrm{css}}, \boldsymbol{\beta}, \sigma^2, \boldsymbol{\xi})$. We penalize the complete-data log-likelihood with roughness penalty (3.28) and we further define the conditional distribution of missing data $\underline{\boldsymbol{\delta}}$ given $\underline{\boldsymbol{y}}$ and the model parameters, i.e.,

$p(\underline{\boldsymbol{\delta}} \mid \underline{\boldsymbol{y}}, \tilde{\underline{\boldsymbol{f}}}_{\text{css}}, \tilde{\underline{\boldsymbol{\beta}}}, \sigma^2, \boldsymbol{\xi})$. Here, $\boldsymbol{f}_{\text{css}}$ and $\boldsymbol{\beta}$ have been replaced by their current estimates $\tilde{\underline{\boldsymbol{f}}}_{\text{css}}$ and $\tilde{\underline{\boldsymbol{\beta}}}$. The variance components σ^2 and $\boldsymbol{\xi}$ are assumed to be known. The E-step consists of taking the expectation of the penalized complete-data log-likelihood with respect to the conditional distribution of the missing data. This involves computing the conditional expectations of $\underline{\boldsymbol{\delta}}$ and $\underline{\boldsymbol{\delta}}\,\underline{\boldsymbol{\delta}}'$, where the former expectation is the empirical Bayes estimator of the random effects.

In the M-step, we maximize the expected penalized complete-data log-likelihood (retrieved from the E-step) with respect to the model parameters $\boldsymbol{f}_{\text{css}}$ and $\boldsymbol{\beta}$. The M-step is computationally expensive if the number of distinct time points T is large because it involves inverting a $T \times T$ matrix. In that case, it is better to update the estimates of $\boldsymbol{f}_{\text{css}}$ and $\boldsymbol{\beta}$ sequentially. First, we maximize with respect to $\boldsymbol{f}_{\text{css}}$ and obtain the updated estimate

$$\tilde{\underline{\boldsymbol{f}}}_{\text{css}} = \Big(\sum_j (\boldsymbol{N}_j'\boldsymbol{N}_j) + \sigma^2\lambda\boldsymbol{K}\Big)^{-1} \sum_j (\underline{\boldsymbol{y}}_j - \boldsymbol{X}_{2j}\tilde{\underline{\boldsymbol{\beta}}} - \boldsymbol{Z}_j\tilde{\underline{\boldsymbol{\delta}}}_j),$$

where $\tilde{\underline{\boldsymbol{\delta}}}_j$ is the empirical Bayes estimate of $\underline{\boldsymbol{\delta}}_j$ at the current estimates of $\boldsymbol{f}_{\text{css}}$ and $\boldsymbol{\beta}$. Second, we maximize with respect to $\boldsymbol{\beta}$ only and obtain the update

$$\tilde{\underline{\boldsymbol{\beta}}} = \Big(\sum_j \boldsymbol{X}_{2j}'\boldsymbol{X}_{2j}\Big)^{-1} \sum_j (\underline{\boldsymbol{y}}_j - \boldsymbol{N}_j\tilde{\underline{\boldsymbol{f}}}_{\text{css}} - \boldsymbol{Z}_j\tilde{\underline{\boldsymbol{\delta}}}_j).$$

These two steps are computationally cheap: The number of numerical operations to update the estimates of $\boldsymbol{f}_{\text{css}}$ and $\boldsymbol{\beta}$ is of order T. Although the expression for $\tilde{\underline{\boldsymbol{f}}}_{\text{css}}$ contains the inverse of a $T \times T$ matrix, efficient computation is possible using the Cholesky factorization method as described, for example, in Green and Silverman [21]. This algorithm where the M-step is replaced by two sequential steps is known as the EC(onditional)M algorithm [43]. The two sequential steps can also be viewed as steps of the backfitting algorithm as described by Hastie and Tibshirani [24, p. 91].

An EM algorithm can also be constructed after having reparametrized the model according to Green [20]. Using features of cubic splines, we can write $\boldsymbol{f}_{\text{css}}$ in (3.29) via a one-to-one transformation as

$$\boldsymbol{f}_{\text{css}} = \gamma_0 \mathbf{1}_T + \gamma_1 \boldsymbol{x}^* + \boldsymbol{Q}(\boldsymbol{Q}'\boldsymbol{Q})^{-1}\boldsymbol{L}\boldsymbol{\eta}, \tag{3.32}$$

where γ_0 and γ_1 are scalars, $\boldsymbol{x}^* = (x_1, \ldots, x_T)'$, $\boldsymbol{\eta}$ is a $(T-2) \times 1$ parameter vector, and $\boldsymbol{L}$ satisfies $\boldsymbol{L}\boldsymbol{L}' = \boldsymbol{R}$. For the definition of $\boldsymbol{Q}$ and $\boldsymbol{R}$, see Section 3.9. Because the columns of $\boldsymbol{Q}$ are orthogonal to $\mathbf{1}_T$ and $\boldsymbol{x}^*$, it follows that $\boldsymbol{\eta}'\boldsymbol{\eta}$ is equal to $\boldsymbol{f}_{\text{css}}'\boldsymbol{K}\boldsymbol{f}_{\text{css}}$. Hence, the penalized log-likelihood of model (3.29) with $\boldsymbol{f}_{\text{css}}$ replaced by (3.32) is equal to the sum of the log-likelihood and the term $-\frac{1}{2}\lambda\boldsymbol{\eta}'\boldsymbol{\eta}$. The E-step and M-step can be derived as before. When T is large, the computational burden can again be lowered by replacing the M-step by sequential steps.

So far, we have regarded the variance components σ^2 and $\boldsymbol{\xi}$ as known. Simple estimators of σ^2 and $\boldsymbol{\xi}$ are obtained by fitting an overelaborated model with in the fixed part T dummy predictors, one for each distinct time point [55, p. 123]. If the model with dummy effects is estimated by restricted IGLS, unbiased estimates are obtained for σ^2 and $\boldsymbol{\xi}$ also when $\underline{y}$ in the true model is associated to x by a smooth function $f(x)$. For reasons of efficiency, it may be preferable to use estimators that depend on the external smoothing parameter λ. Several authors have suggested considering $\boldsymbol{\eta}$ as a vector of random effects $\underline{\boldsymbol{\eta}}$ and fitting a crossed random effects model with model parameters γ_0, γ_1, $\boldsymbol{\beta}$, σ^2, and $\boldsymbol{\xi}$ and random effects $\underline{\boldsymbol{\delta}}$ and $\underline{\boldsymbol{\eta}}$ [52, 58, 64]. The formulation of the crossed random effects model is attractive because it allows us to estimate $\boldsymbol{f}_{\mathrm{css}}$ using existing software. Estimates can be obtained with the restricted IGLS algorithm implemented in MLwiN [19] and SAS [39]. Here, the variance of $\underline{\boldsymbol{\eta}}$ is set equal to the inverse of the roughness penalty λ. The restricted IGLS estimator of σ^2 performs well in simulation studies [64]. Besides, in a single-level situation (e.g., longitudinal data of one subject), this estimator is equal to the classical estimate of σ^2 described, for example, by Green and Silverman [21, p. 39]. The estimation of the crossed random effects model via restricted IGLS is computationally demanding if the number of distinct values $x_1, \ldots, x_T$ is large because the number of crossed random effects is equal to $T - 2$.

3.9.2 Inferences

A common approach to drawing inferences about $\boldsymbol{f}_{\mathrm{css}}$ is to construct pointwise correct confidence intervals at $x_1, \ldots, x_T$. This requires an estimate of the variance of $\underline{\hat{\boldsymbol{f}}}_{\mathrm{css}}$. Two common estimates will be discussed. The first estimate is obtained by assuming that $\underline{\hat{\boldsymbol{f}}}_{\mathrm{css}}$ is an estimate of the fixed, unknown $\boldsymbol{f}_{\mathrm{css}}$. From (3.30), where $\underline{\hat{\boldsymbol{f}}}_{\mathrm{css}}$ is written as a linear function of $\underline{\boldsymbol{y}}$, it follows that the covariance matrix is given by

$$\mathrm{Cov}_{\mathrm{F}}(\underline{\hat{\boldsymbol{f}}}_{\mathrm{css}}) = \boldsymbol{W}^{-1} \left(\sum_{j=1}^{m} \boldsymbol{N}_j' \boldsymbol{U}_{X_2,j} \boldsymbol{N}_j \right) \boldsymbol{W}^{-1}, \tag{3.33}$$

where

$$\boldsymbol{W} = \sum_{j=1}^{m} \boldsymbol{N}_j' \boldsymbol{U}_{X_2,j} \boldsymbol{N}_j + \lambda \boldsymbol{K}.$$

The second estimate of the variance is the posterior variance obtained from a Bayesian model where the logarithm of the prior of $\boldsymbol{f}_{\mathrm{css}}$ is equal to $-\frac{1}{2}\lambda \boldsymbol{f}_{\mathrm{css}}' \boldsymbol{K} \boldsymbol{f}_{\mathrm{css}}$ except for a constant. The posterior covariance matrix has a simple form

$$\mathrm{Cov}_{\mathrm{B}}(\underline{\hat{\boldsymbol{f}}}_{\mathrm{css}}) = \boldsymbol{W}^{-1}. \tag{3.34}$$

Zhang et al. [64] and Lin and Zhang [38] compare the frequentist and Bayesian estimator in a simulation study in which a fixed nonparametric function $f(x)$ is postulated. The main conclusion in these studies is that both estimators are accurate but that the Bayesian estimator sometimes performs slightly better because it accounts for the bias in $\hat{\underline{\boldsymbol{f}}}_{\text{css}}$. The Bayesian variances can also be obtained from the model with crossed random effects. Software packages such as MLwiN yield estimates of the variances of $\hat{\underline{\gamma}}_0$ and $\hat{\underline{\gamma}}_1$ and the comparative variance of the empirical Bayes estimator $\hat{\underline{\boldsymbol{\eta}}}$. The covariance between $\hat{\underline{\boldsymbol{\eta}}}$ and $(\hat{\underline{\gamma}}_0, \hat{\underline{\gamma}}_1)$ is not always produced. However, the design matrices of (γ_0, γ_1) and $\underline{\boldsymbol{\eta}}$ are orthogonal if the points at which the measurements are taken are common to all subjects. Therefore, the precision of the estimator of $\text{Cov}_{\text{B}}(\hat{\underline{\boldsymbol{f}}}_{\text{css}})$ is in general not substantially affected by the omission of the covariance between $(\hat{\underline{\gamma}}_0, \hat{\underline{\gamma}}_1)$ and $\hat{\underline{\boldsymbol{\eta}}}$.

Besides the Bayesian model with a finite-dimensional prior for $\boldsymbol{f}_{\text{css}}$, a model with an infinite-dimensional prior for the continuous spline $f_{\text{css}}(x)$ exists as well [58, 64]. This model was put forward by Wahba [56] and is appealing because a smoothing spline $f_{\text{css}}(x)$ is defined for all x and not only for the observed values. The finite- and infinite-dimensional formulation lead to the same posterior variance of $\hat{\underline{\boldsymbol{f}}}_{\text{css}}$.

3.9.3 Smoothing Parameter Selection

Several methods exist for selecting the smoothing parameter λ. In this section, three are discussed. The first method is to maximize the cross-validated log-likelihood as a function of λ. The cross-validated log-likelihood is an approximation to the expectation of the predictive log-likelihood, which is the expected log-likelihood of a new vector of observations $\underline{\boldsymbol{y}}^*$ at the penalized likelihood estimators of the model parameters $\boldsymbol{f}_{\text{css}}$, $\boldsymbol{\beta}$, σ^2, and $\boldsymbol{\xi}$. The prediction process is imitated by leaving out one subject at a time and predicting the omitted subject on the basis of the other subjects' data [46].

A drawback of cross-validation is that it is computationally expensive. An alternative strategy is to estimate the expected predictive log-likelihood by the sum of the log-likelihood and the trace of the matrix $\boldsymbol{A}$ that maps $\underline{\boldsymbol{y}}$ on the estimator $\hat{\underline{\boldsymbol{f}}}_{\text{css}} = \boldsymbol{A}\underline{\boldsymbol{y}}$ [21, p. 37; 24, p. 52]. This estimator, named Mallows' C_p, is cheap and unbiased if the (co)variance parameters σ^2 and $\boldsymbol{\xi}$ are known. For uncorrelated data, the unbiasedness proof is provided by Hastie and Tibshirani [24, p. 48]. The proof in the case of multilevel data is analogous. In practice, σ^2 and $\boldsymbol{\xi}$ are unknown and can be estimated by the restricted IGLS estimators in the overelaborated model with dummy time predictors (see Section 3.9.1).

A limitation of applying criteria like the cross-validated log-likelihood or Mallows' C_p is that λ is not treated as a model parameter but as an external variable. The smoothing parameter becomes a model parameter if we adopt the crossed random effects model and estimate the variance of $\underline{\eta}$ freely instead of constraining the variance to be equal to the inverse of λ. It can be shown that if the crossed random effects model is estimated by restricted IGLS implemented in MLwiN [19], then the estimate of λ is the generalized maximum likelihood (GML) estimate [57, 64], which has good performance in simulation studies [33]. It may also be sensible to examine whether a model with smoothing spline $f_{\text{css}}(x)$ fits better than a model with a linear effect for x. Hastie and Tibshirani [25, p. 65] provide some approximate F-tests based on residual sums of squares and Cantoni and Hastie [9] and Guo [22] present likelihood ratio tests for $H_0 : \lambda^{-1} = 0$, which is equivalent to $H_0 : \underline{\eta} = 0$ (3.32). Instead of the likelihood ratio test, the score test may also be considered. The score test is computationally cheap because estimates of the model with crossed random effects are not required. The test is based on the one-step estimator that is obtained when we start from the estimate of the null model. The ratio of the one-step estimator to its standard error has an asymptotic standard normal null distribution. The score test also has good power properties in a small sample setting [6]. For testing against an unspecified but monotonic effect of x, this test against a linear effect may be expected to have good power against most non-linear effects.

3.10 Example: Effect of IQ on a Language Test

We fitted the three different functions that were discussed so far, i.e., the polynomial function, the regression spline function, and the cubic smoothing spline function, to a real data set. The estimations were done using MLwiN 1.1 [19] and Gauss 3.2 [1]. The data set is described in Snijders and Bosker [51]. It contains language test scores of 2287 pupils within 131 elementary schools. We modeled the test score (Y) as a function of the grand-mean centered IQ of the pupil (`IQ`), the gender of the pupil (`SEX`), the school average of IQ ($\overline{\texttt{IQ}}$), and the socio-economic status (`SES`) of the pupil. We assumed a non-linear effect for `IQ` and linear effects for the other predictors. Note that in most applications of models with functional non-linear effects, time is the ordering principle, but an ordering according to any other unidimensional variable is possible as well. Between-school differences were modeled by including a random intercept and a random slope of `IQ` at level 2. Finally, we assumed that the level-1 measurement errors are homoscedastic and uncorrelated. The model can be written as

$$\underline{y}_{ij} = f(\texttt{IQ}_{ij}) + \beta_1 \, \texttt{SES}_{ij} + \beta_2 \, \texttt{SEX}_{ij} + \beta_3 \, \overline{\texttt{IQ}}_j + \underline{\delta}_{0j} + \underline{\delta}_{1j}\texttt{IQ}_{ij} + \underline{\epsilon}_{ij} \, .$$

The estimated polynomial function, regression spline function, and cubic smoothing spline function are presented in Fig. 3.3. The chosen polynomial function is of order three. We also considered a fourth-degree polynomial, but this did not yield a further improvement in fit. The chosen regression spline is a quadratic spline with a knot at zero. This function was considered by Snijders and Bosker [51, p. 113] as a flexible and parsimonious alternative for the polynomial function. We determined the smoothness of the cubic smoothing spline by maximizing GML. We also considered optimization of the cross-validated log-likelihood and Mallows' C_p, but the three methods rendered similar values for the smoothing parameter: $\lambda_{\text{GML}} = 1.6$, $\lambda_{C_p} = 1.6$, $\lambda_{\text{CV}} = 2.0$.

The three fitted functions lead to similar predictions: The effect of `IQ` on Y is larger in the middle than in the tails of the distribution of `IQ`. The smoothing spline performs slightly better than the other two functions since it is monotonically increasing, whereas the polynomial function and the regression spline have a negative slope at low and high values of `IQ`.

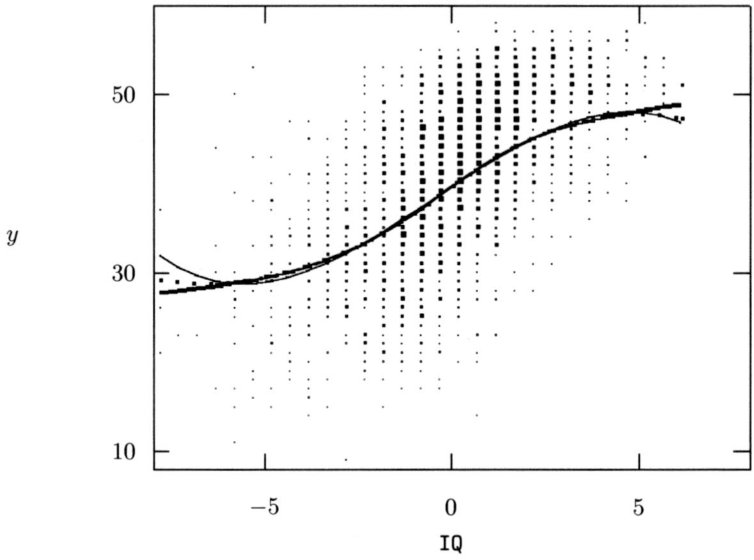

Fig. 3.3 Language test score (y) against centered `IQ` score: raw data, cubic polynomial estimate (thin), quadratic regression spline estimate (dashed), and smoothing spline estimate (bold).

We also estimated the pointwise standard errors of the fitted functions. These are presented in Fig. 3.4. We see that the standard errors of the fitted functions are very similar. Data are sparse at the left and right ends of the window (Fig. 3.4) and the standard errors are large there compared to the middle part. We further see that the Bayesian standard error of the cubic smoothing spline estimate is slightly larger than its frequentist counterpart, as it should be according to (3.33) and (3.34) [cf. 64].

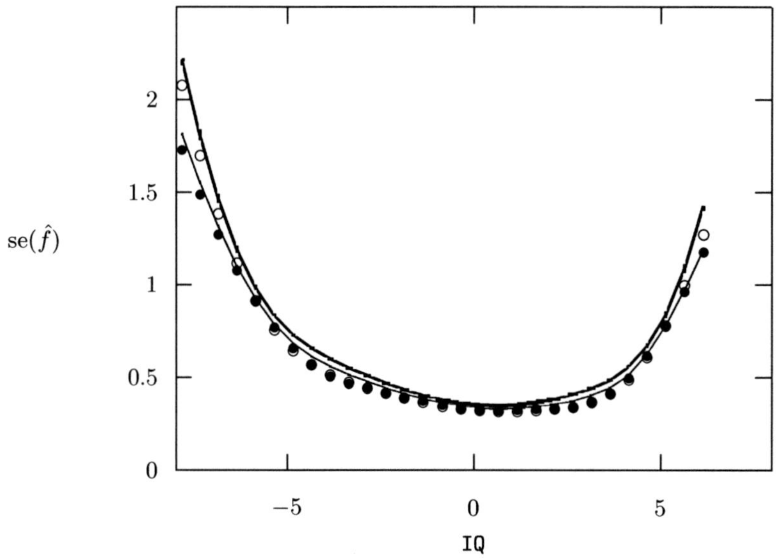

Fig. 3.4 Standard errors of the cubic polynomial estimate (open circle) and the quadratic regression spline estimate (closed circle), and Bayesian (bold line) and frequentist (thin line) standard errors of the smoothing spline estimate.

3.11 Extensions

The model can be extended to a model with more than two levels or a model with non-normal responses in the same way as multilevel models without a functional effect can be extended. Another direction is to specify a model with two functional effects, $f(x)$ and $g(v)$. This model is called an additive model and is put forward by Hastie and Tibshirani [24]. Algorithms for estimating additive multilevel models with cubic smoothing splines are provided by Lin and Zhang [38]. A related model is a model in which the effect of predictor w on y is described by function $h(x) \times w$. This model is known as the varying coefficient model and has been used to describe time-varying effects of predictors in longitudinal studies [25]. A multilevel extension of the model is presented by Hoover et al. [29]. The additive and varying coefficient models can be formulated as random effects models with a separate random effect for each functional effect. The estimation can be done in MLwiN but becomes demanding if we have many functional effects. For varying coefficient models, less demanding estimators are available [10, 15].

We have discussed functional effects to describe the mean pattern. Functional effects for the random part of the model have been proposed as well. In multilevel modeling, a common, simple choice is to include polynomial functions in the random part of the model [cf. 8, 17, 50]. When adding spline functions instead of polynomial functions to the random part, a possible approach is to define a separate smoothing spline for each level-2 unit and to use the mixed effects formulation to define a nested sample of curves [7, 22]. The mixed effects approach is appealing, but it is computationally demanding when the number of distinct points is large. A somewhat different approach is to explore the covariance structure by a principal components analysis yielding functions that describe the main sources of variation among the individual curves. These methods are particularly attractive when studying variability between individual curves. Rice and Silverman [46] propose a principal components model where the differences among individuals are described by cubic smoothing splines. The model is applicable only when the points at which measurements are taken are common to all level-2 units. Rice and Wu [47] and James et al. [31] use B-spline functions to allow for irregular spacing of the data. Yao et al. [62] present a model for irregular data with functions retrieved from a smooth estimate of the (continuous) covariance surface. For the underlying functions, they also provide asymptotic confidence bounds.

Acknowledgement The research by Johannes Berkhof was financially supported by grant ESR 510-78-501 of the Dutch Organization for Scientific Research (NWO).

References

1. Aptech Systems. *Gauss.* Aptech Systems, Maple Valley, WA, 1994.
2. A. C. Atkinson. *Plots, Transformations, and Regression.* Clarendon Press, Oxford, UK, 1985.
3. B. H. Baltagi. *Econometric Analysis of Panel Data.* Wiley, New York, 1995.
4. M. S. Bartlett and D. G. Kendall. The statistical analysis of variance-heterogeneity and the logarithmic transformation. *Supplement to the Journal of the Royal Statistical Society*, 8:128–138, 1946.
5. C. S. Berkey and R. L. Kent Jr. Longitudinal principal components and non-linear regression models of early childhood growth. *Annals of Human Biology*, 10:523–536, 1983.
6. J. Berkhof and T. A. B. Snijders. Variance component testing in multilevel models. *Journal of Educational and Behavioral Statistics*, 26:133–152, 2002.
7. B. A. Brumback and J. A. Rice. Smoothing spline models for the analysis of nested and crossed samples of curves. *Journal of the American Statistical Association*, 93:961–994, 1998. (with discussion)

8. A. S. Bryk and S. W. Raudenbush. Application of hierarchical linear models to assessing change. *Psychological Bulletin*, 101:147–158, 1987.
9. E. Cantoni and T. Hastie. Degrees-of-freedom tests for smoothing splines. *Biometrika*, 89:251–263, 2002.
10. C.-T. Chiang, J. A. Rice, and C. O. Wu. Smoothing spline estimation for varying coefficient models with repeatedly measured dependent variables. *Journal of the American Statistical Association*, 96:605–619, 2001.
11. R. D. Cook and S. Weisberg. *Residuals and Influence in Regression*. Chapman & Hall, London, 1982.
12. C. de Boor. *A Practical Guide to Splines*. Springer, New York, 1978.
13. L. E. Eberly and L. M. Thackeray. On Lange and Ryan's plotting technique for diagnosing non-normality of random effects. *Statistics & Probability Letters*, 75: 77–85, 2005.
14. P. H. C. Eilers and B. D. Marx. Flexible smoothing with splines and penalties. *Statistical Science*, 11:89–121, 1996. (with discussion)
15. J. Fan and J. T. Zhang. Two-step estimation of functional linear models with applications to longitudinal data. *Journal of the Royal Statistical Society, Series B*, 62:303–322, 2000.
16. A. Fielding. Role of the Hausman test and whether higher level effects should be treated as random or fixed. *Multilevel Modelling Newsletter*, 16(2):3–9, 2004.
17. H. Goldstein. Efficient statistical modelling of longitudinal data. *Annals of Human Biology*, 13:129–141, 1986.
18. H. Goldstein. *Multilevel Statistical Models*, 3rd edition. Edward Arnold, London, 2003.
19. H. Goldstein, J. Rasbash, I. Plewis, D. Draper, W. Browne, M. Yang, G. Woodhouse, and M. Healy. *A User's Guide to MLwiN*. Multilevel Models Project, Institute of Education, University of London, London, 1998.
20. P. J. Green. Penalized likelihood for general semi-parametric regression models. *International Statistical Review*, 55:245–259, 1987.
21. P. J. Green and B. W. Silverman. *Nonparametric Regression and Generalized Linear Models*. Chapman & Hall, London, 1994.
22. W. Guo. Functional mixed effects model. *Biometrics*, 58:121–128, 2002.
23. J. Haslett and D. Dillane. Application of 'delete = replace' to deletion diagnostics for variance component estimation in the linear mixed model. *Journal of the Royal Statistical Society, Series B*, 66:131–143, 2004.
24. T. Hastie and R. Tibshirani. *Generalized Additive Models*. Chapman & Hall, London, 1990.
25. T. Hastie and R. Tibshirani. Varying-coefficient models. *Journal of the Royal Statistical Society, Series B*, 55:757–796, 1993. (with discussion)
26. J. A. Hausman. Specification tests in econometrics. *Econometrica*, 46:1251–1271, 1978.
27. J. A. Hilden-Minton. *Multilevel Diagnostics for Mixed and Hierarchical Linear Models*. PhD thesis, Department of Mathematics, University of California, Los Angeles, 1995.

28. J. S. Hodges. Some algebra and geometry for hierarchical linear models, applied to diagnostics. *Journal of the Royal Statistical Society, Series B*, 60:497–536, 1998.
29. D. R. Hoover, J. A. Rice, C. O. Wu, and L.-P. Yang. Nonparametric smoothing estimates of time-varying coefficient models with longitudinal data. *Biometrika*, 85:809–822, 1998.
30. C. Hsiao. Random coefficient models. In L. Mátyás and P. Sevestre, editors, *The Econometrics of Panel Data*, 2nd edition, pages 77–99. Kluwer, Dordrecht, The Netherlands, 1996.
31. G. James, T. Hastie, and C. A. Sugar. Principal component models for sparse functional data. *Biometrika*, 87:587–602, 2000.
32. J.-S. Kim and E. W. Frees. Omitted variables in multilevel models. *Psychometrika*, 71:659–690, 2006.
33. R. Kohn, C. F. Ansley, and D. Tharm. The performance of cross-validation and maximum likelihood estimators of spline smoothing parameters. *Journal of the American Statistical Association*, 86:1042–1050, 1991.
34. N. Lange and L. Ryan. Assessing normality in random effects models. *Annals of Statistics*, 17:624–642, 1989.
35. I. H. Langford and T. Lewis. Outliers in multilevel data. *Journal of the Royal Statistical Society, Series A*, 161:121–160, 1998.
36. E. Lesaffre and G. Verbeke. Local influence in linear mixed models. *Biometrics*, 54:570–582, 1998.
37. T. Lewis and I. H. Langford. Outliers, robustness and the detection of discrepant data. In A. H. Leyland and H. Goldstein, editors, *Multilevel Modelling of Health Statistics*, pages 75–91. Wiley, New York, 2001.
38. X. Lin and D. Zhang. Inference in generalized additive mixed models by using smoothing splines. *Journal of the Royal Statistical Society, Series B*, 61:381–400, 1999.
39. R. C. Littell, G. A. Milliken, W. W. Stroup, and R. D. Wolfinger. *SAS System for Mixed Models.* SAS Institute, Cary, NC, 1996.
40. N. T. Longford. A fast scoring algorithm for maximum likelihood estimation in unbalanced mixed models with nested random effects. *Biometrika*, 74:817–827, 1987.
41. N. T. Longford. *Random Coefficient Models.* Oxford University Press, Oxford, UK, 1993.
42. N. T. Longford. Simulation-based diagnostics in random-coefficient models. *Journal of the Royal Statistical Society, Series A*, 164:259–273, 2001.
43. X. L. Meng and D. B. Rubin. Maximum likelihood estimation via the ECM algorithm: A general framework. *Biometrika*, 80:267–278, 1993.
44. C. R. Rao. *Linear Statistical Inference and its Applications*, 2nd edition. Wiley, New York, 1973.
45. S. W. Raudenbush and A. S. Bryk. *Hierarchical Linear Models: Applications and Data Analysis Methods*, 2nd edition. Sage, Thousand Oaks, CA, 2002.
46. J. A. Rice and B. W. Silverman. Estimating the mean and covariance structure nonparametrically when the data are curves. *Journal of the Royal Statistical Society, Series B*, 53:233–243, 1991.

47. J. A. Rice and C. O. Wu. Nonparametric mixed effects models for unequally sampled noisy curves. *Biometrics*, 57:253–259, 2001.
48. M. H. Seltzer, W. H. Wong, and A. S. Bryk. Bayesian analysis in applications of hierarchical models: Issues and methods. *Journal of Educational and Behavioral Statistics*, 21:131–167, 1996.
49. A. Skrondal and S. Rabe-Hesketh. *Generalized Latent Variable Modeling: Multilevel, Longitudinal, and Structural Equation Models.* Chapman & Hall/CRC, Boca Raton, FL, 2004.
50. T. A. B. Snijders. Analysis of longitudinal data using the hierarchical linear model. *Quality & Quantity*, 30:405–426, 1996.
51. T. A. B. Snijders and R. J. Bosker. *Multilevel Analysis: An Introduction to Basic and Advanced Multilevel Modeling.* Sage, London, 1999.
52. T. P. Speed. Comment on "That BLUP is a good thing: the estimation of random effects" (by G. K. Robinson). *Statistical Science*, 6:44, 1991.
53. R. Van der Leeden, E. Meijer, and F. M. T. A. Busing. Resampling multilevel models. In J. de Leeuw and E. Meijer, editors, *Handbook of Multilevel Analysis*, Chapter 11. Springer, New York, 2008. (this volume)
54. G. Verbeke and E. Lesaffre. A linear mixed-effects model with heterogeneity in the random-effects population. *Journal of the American Statistical Association*, 91:217–221, 1996.
55. G. Verbeke and G. Molenberghs. *Linear Mixed Models for Longitudinal Data.* Springer, New York, 2000.
56. G. Wahba. Bayesian "confidence" intervals for the cross-validated smoothing spline. *Journal of the Royal Statistical Society, Series B*, 45:133–150, 1983.
57. G. Wahba. A comparison of GCV and GML for choosing the smoothing parameter in the generalized spline smoothing problem. *Annals of Statistics*, 4: 1378–1402, 1985.
58. Y. Wang. Mixed effects smoothing spline analysis of variance. *Journal of the Royal Statistical Society, Series B*, 60:159–174, 1998.
59. C. Waternaux, N. M. Laird, and J. H. Ware. Methods for analysis of longitudinal data: Blood lead concentrations and cognitive development. *Journal of the American Statistical Association*, 84:33–41, 1989.
60. S. Weisberg. *Applied Linear Regression*, 3rd edition. Wiley, New York, 2005.
61. S. Wold. Spline functions in data analysis. *Technometrics*, 16:1–11, 1974.
62. F. Yao, H.-G. Müller, and J.-L. Wang. Functional data analysis for sparse longitudinal data. *Journal of the American Statistical Association*, 100:577–590, 2005.
63. T. Zewotir and J. S. Galpin. Influence diagnostics for linear mixed models. *Journal of Data Science*, 3:153–177, 2005.
64. D. Zhang, X. Lin, J. Raz, and M. Sowers. Semiparametric stochastic mixed models for longitudinal data. *Journal of the American Statistical Association*, 93:710–719, 1998.

4

Optimal Designs for Multilevel Studies

Mirjam Moerbeek[1], Gerard J. P. Van Breukelen[2], and Martijn P. F. Berger[2]

[1] Utrecht University, Department of Methodology and Statistics
[2] Maastricht University, Department of Methodology and Statistics

4.1 Introduction

The analysis of multilevel data with individuals nested within clusters is complicated by the correlation between outcomes of individuals within the same cluster. Ignoring this correlation and the use of traditional analysis methods, like ordinary least squares regression, may sometimes lead to biased parameter estimates and will generally lead to incorrect standard errors and, consequently, to incorrect tests and conclusions on effect sizes. The presence of an intraclass correlation also complicates the design of multilevel studies. Optimal designs calculated from standard formulae for non-nested data [5] may be far from optimal for multilevel data. Moreover, these formulae only specify the total number of individuals needed to gain a certain power on statistical tests, and cannot specify the number of clusters and the number of individuals per cluster.

Experiments and observational studies in the social and medical sciences often involve large amounts of time, money, and labor. These efforts could be somewhat wasted if the study was not designed optimally. Therefore, guidelines for the optimal design of multilevel studies are asked for. During the last two decades, a number of papers on the design of multilevel studies has been published. Most have focussed on the optimal sample sizes for cluster randomized trials [9, 10, 12, 18, 23, 25, 29, 34, 35, 36, 39, 42, 49, 55], and multisite randomized trials where randomization to treatment conditions is done at the patient level and treatment by site interaction may be present [50]. A comparison of cluster randomized trials and multisite trials with person randomization shows that the latter are more efficient [30, 34, 35, 36]. However, control group contamination may destroy this advantage of person randomization and call for cluster randomization [32]. Snijders and Bosker

J. de Leeuw, E. Meijer (eds.), *Handbook of Multilevel Analysis*,
© Springer 2008

[57] derive sample size formulae for two-level designs with any number of explanatory variables at each level. Cohen [7] derives optimal sample size formulae for surveys based on several optimality criteria for the fixed and random part. Afshartous [1] and Mok [41] compare designs with different sample sizes at both levels by means of simulation studies.

For multilevel experiments, four design issues may arise. The first three that are listed may also arise for surveys with nested data. The first design issue concerns the optimal allocation of units, or, in other words, the optimal sample sizes at each level of the multilevel data structure. The optimal sample sizes are restricted by the actual sample sizes in the study population, since the number of clusters that are enrolled in the study cannot be larger than the number of clusters that are available for the study. Likewise, the number of individuals per cluster in the study cannot be larger than the actual cluster size. Sampling individuals within an already selected cluster may be less expensive than sampling in a new cluster. This can be expressed by a cost function that is used as a precondition in the derivation of the optimal sample sizes.

The second design issue concerns the required budget to obtain a specified power on the test of a certain parameter given the true value of that parameter and a type I error rate. As we will see in the next section, the power of the test of a certain parameter is inversely related to the variance of that parameter, which depends on the sample sizes at each level of the multilevel data structure. Thus, the second design issue is closely related to the first one.

The third design issue concerns the robustness of optimal designs. A prior specification of the values of the model parameters, in particular the intraclass correlation coefficient, must be given to calculate optimal sample sizes, and one may wonder if the optimal design is robust against incorrect prior specifications.

A fourth design issue that may be considered is the efficiency of cluster randomization versus randomization at the individual level. Although individual-level randomization gives a higher power on statistical tests of a treatment effect, randomization is often done in practice at the cluster level and one may wonder what the loss in efficiency for this level of randomization is. Reasons to favor a cluster randomized trial are often of an ethical, practical, logistical, or administrative nature. Examples are the need to reduce costs and the need to avoid control group contamination, which occurs when information leaks from the intervention to the control group.

In this chapter we will give some guidelines for designing multilevel experiments and surveys (observational studies). The contents of this chapter are as follows. The next section focuses on optimality criteria and power calculation. Section 4.3 deals with the optimal design of multilevel experiments. Thereafter we focus on optimal experimental designs for models with covariates (Section 4.4), and for multilevel logistic models (Section 4.5). Section 4.6 gives results for optimal experimental designs with longitudinal data. Sections 4.7 and 4.8 deal with optimal designs for surveys and variance parameters, respectively. In Section 4.9 the robustness of optimal designs against an incorrect prior specification of the values of the model parameters is dealt with. This chapter concludes with some remarks on the use of the optimal designs in practice. Optimal designs will be derived for two levels of nesting; optimal designs for three levels of nesting can be found elsewhere [30, 34]. For the sake of concreteness, units at level 1 and 2 are called pupils and schools in this chapter, but, of course, any other terminology may be substituted. We will focus on optimal designs that minimize one optimality criterion at a time; multiple-objective optimal designs are presented elsewhere [40].

4.2 Optimality and Power

4.2.1 Optimality Criteria

Choosing an optimal design means to choose the design ξ^* among all designs ξ in the design space χ that provides maximum information on the parameters $\boldsymbol{\theta}$ in the model. This information may be captured by the Fisher information matrix $\boldsymbol{M}(\boldsymbol{X},\boldsymbol{\theta})$, where $\boldsymbol{X}$ is the design matrix that contains the measures on the predictor variables, and depends on the chosen regression model η and the design ξ: $\boldsymbol{X} = \boldsymbol{X}(\eta,\xi)$. The Fisher information matrix is defined as minus the expectation of the second-order derivatives of the logarithm of the likelihood function $\underline{L}(\boldsymbol{X}(\eta,\xi),\boldsymbol{\theta})$ [e.g., 56]:

$$\boldsymbol{M}(\boldsymbol{X}(\eta,\xi),\boldsymbol{\theta}) = -\,E\left(\frac{\partial^2 \log \underline{L}(\boldsymbol{X}(\eta,\xi),\boldsymbol{\theta})}{\partial\boldsymbol{\theta}\,\partial\boldsymbol{\theta}'}\right),$$

and contains information on each parameter $\boldsymbol{\theta}$ and each combination of parameters $\boldsymbol{\theta}$ and $\boldsymbol{\theta}'$. The limit of its inverse is equal to the asymptotic covariance matrix of the parameter estimators, and maximizing the Fisher information matrix is equal to minimizing the covariance matrix of the parameter estimators. Since matrices cannot be ordered in a unique way, different functions Φ of the matrix $\boldsymbol{M}$, which at least have to be convex and differentiable, have been proposed as optimality criteria. Examples are A-, D-, and c-optimal designs [e.g., 2, 56].

In this chapter we will use the variance of one single parameter θ as optimality criterion, since minimal variance leads to maximal statistical power of the test of $H_0 : \theta = 0$. This variance will be minimized subject to the precondition that the number of schools, n_2, and the number of pupils per school, n_1, are at least equal to 2, to be able to estimate the variance components at both levels. Furthermore, the budget C for sampling schools and pupils may not be exceeded by the costs for sampling, which are assumed to be equal to the total number of schools times the costs c_2 for sampling a school, plus the total number of pupils times the costs c_1 for sampling an pupil, i.e.,

$$C \geq c_1 n_1 n_2 + c_2 n_2 \qquad (c_l > 0,\ n_l \geq 2 \text{ for } l = 1, 2). \tag{4.1}$$

An optimal design that does not depend on the model parameters $\boldsymbol{\theta}$ is called a globally optimal design, whereas a locally optimal design is a design that does depend on one or more of the model parameters. For the latter, a prior specification of the values of these model parameters needs to be given to calculate the optimal sample sizes. The robustness of optimal designs against incorrect prior specifications is discussed in Section 4.9. The optimal design ξ^* may not always be feasible in practice and an alternative design ξ may be chosen instead. The efficiency of the alternative design relative to the optimal design is given by

$$\text{relative efficiency} = \frac{\Phi(\boldsymbol{M}^{-1}(\boldsymbol{X}(\eta, \xi^*), \boldsymbol{\theta}))}{\Phi(\boldsymbol{M}^{-1}(\boldsymbol{X}(\eta, \xi), \boldsymbol{\theta}))}, \tag{4.2}$$

and this ratio is between 0 and 1. When the variance of one single parameter is used as optimality criterion, the inverse of the relative efficiency gives the number of times the suboptimal design ξ needs to be replicated to be as efficient as the optimal design ξ^*.

4.2.2 Power Calculation

Suppose that we want to test the null hypothesis $H_0 : \theta = 0$, where θ is a model parameter, against an alternative H_1 that its value differs from zero. This hypothesis may be tested with the test statistic $z = \underline{\hat{\theta}}/\sqrt{\text{Var}(\underline{\hat{\theta}})}$. If θ is a regression coefficient and the error variance is assumed known, this test statistic is asymptotically standard normally distributed under H_0. For one-sided alternatives $H_1 : \theta > 0$ and $H_1 : \theta < 0$, the power $1 - \gamma$, type I error rate α, $\text{Var}(\underline{\hat{\theta}})$, and the true value of θ are related by

$$1 - \gamma = \Phi\left(\frac{\theta}{\sqrt{\text{Var}(\underline{\hat{\theta}})}} - z_{1-\alpha}\right), \tag{4.3}$$

where Φ is the standard normal cumulative distribution function and $z_{1-\alpha}$ is the $100(1-\alpha)$ standard normal percentile. For two-sided alternatives $H_1 : \theta \neq 0$, α is replaced by $\alpha/2$. For a two-sided alternative hypothesis, the power is derived from only one of the critical regions. The probability of a rejection at the wrong side is always less than $\alpha/2$, and is negligibly small in relation to a rejection at the correct side. The only exception is an effect size (i.e., difference between H_0 and H_1) that is so small that the power is as large as α. In trial designs, we aim at power levels equal to 80% or higher, and the error of the approximation is negligible.

When the error variance is unknown, the test statistic has approximately a t-distribution under the null hypothesis, and the standard normal approximation works well for large degrees of freedom. As follows from (4.3), the power increases with the true value of θ, since a large θ is easier to detect than a small one. Furthermore, the power also increases with the type I error rate, and decreases with the variance $\text{Var}(\underline{\hat{\theta}})$, which in its turn is a function of the sample size. Thus minimizing $\text{Var}(\underline{\hat{\theta}})$ implies maximizing the power. Formula (4.3) contains four unknowns. Once three of these are specified, the fourth can be calculated. In practice, a researcher often wishes to calculate the number of individuals needed to obtain a certain power, which means that $\text{Var}(\underline{\hat{\theta}})$ has to be calculated from (4.3). For non-nested data, the relationship between sample size and variance is well known and can be found in, for example, Cochran [5, Section 4.1]. For nested data, this relation depends on the sample sizes at both the school and pupil level and will be presented in the next sections.

4.3 Optimal Designs for Experiments

In this section we focus on the comparison of two treatment conditions, for example, an intervention and a control. Randomization to these treatment conditions may be done at the pupil or the school level. The latter is often referred to as cluster randomization. We will assume a balanced design: The number of pupils per school is constant across schools and denoted by n_1, whereas the number of schools is denoted by n_2. If randomization is done at the school level, $\frac{1}{2}n_2$ schools are randomized to the intervention group and the others are randomized to the control group, assuming that n_2 is even. Likewise, $\frac{1}{2}n_1$ pupils per school are randomized to each treatment condition for pupil-level randomization, assuming that n_1 is even. The model that relates the outcome $\underline{y}_{ij}$ for pupil i in school j to treatment condition x_{ij} is given by

$$\underline{y}_{ij} = \beta_0 + \beta_1 x_{ij} + \underline{\delta}_{0j} + \underline{\delta}_{1j} x_{ij} + \underline{\epsilon}_{ij}, \tag{4.4}$$

where the treatment condition has values -1 for the control group and $+1$ for the intervention group since this will simplify the formulae on optimal sample sizes if covariates are added to the model. The random error terms $\underline{\delta}_{0j} \sim \mathcal{N}(0, \tau_0^2)$, $\underline{\delta}_{1j} \sim \mathcal{N}(0, \tau_1^2)$, and $\underline{\epsilon}_{ij} \sim \mathcal{N}(0, \sigma^2)$ are assumed to be independent of the treatment condition, and the covariance between $\underline{\delta}_{0j}$ and $\underline{\delta}_{1j}$ is denoted by τ_{01}. Note that x_{ij} may be replaced with x_j for school-level randomization since all pupils within a school will then receive the same treatment condition. In that case, τ_0^2 and τ_1^2 cannot be estimated separately. Instead, their sum $\tau^2 = \tau_0^2 + \tau_1^2$ is estimated. The covariance τ_{01} can be consistently estimated for school-level randomization, which is remarkable since τ_0^2 and τ_1^2 are not identified. This is because τ_{01} is identified by the variances of the outcome in both conditions. Since x_{ij} is coded by -1 and $+1$, β_1 is estimated unbiasedly by $\frac{1}{2}(\bar{y}_t - \bar{y}_c)$, where $\bar{y}_t$ and $\bar{y}_c$ are the mean outcomes in the intervention and control group, respectively, and thus β_1 is equal to *half* the treatment effect.

For both levels of randomization, the variance $\text{Var}(\hat{\underline{\beta}}_1)$ is given in the second column of Table 4.1. For school-level randomization, this variance is larger than would have been obtained when ignoring the nested data structure:

$$\text{Var}(\hat{\underline{\beta}}_1) = \frac{n_1\tau^2 + \sigma^2}{n_1 n_2} = \frac{\tau^2 + \sigma^2}{n_1 n_2}[1 + (n_1 - 1)\rho],$$

where $\rho = \tau^2/(\tau^2 + \sigma^2)$ is the intra-school correlation coefficient, which measures the amount of variation at the school level. The factor $[1 + (n_1 - 1)\rho]$ is called the *design effect* and increases with n_1 and ρ. Even for small ρ, this factor may already be considerable. For example, if $\rho = 0.05$ and $n_1 = 30$, the design effect is equal to 2.45. On the other hand, when randomization is done at the pupil level and there is no treatment by school interaction, the $\text{Var}(\hat{\underline{\beta}}_1)$ obtained when ignoring the multilevel data structure is larger than that obtained with the multilevel model. For randomization at the pupil level and models with a random slope, it may be smaller or larger than that obtained with the multilevel model, depending on the number of pupils per school and the values of the variance components [37].

Optimal designs are calculated under the precondition that the prespecified budget for sampling is not exceeded by the total costs for sampling; see Section 4.2. When n_1 is fixed to a constant, the optimal n_2 can directly be calculated from (4.1) and the $\text{Var}(\hat{\underline{\beta}}_1)$ follows from the second column of Table 4.1. The same method may be applied when n_2 is fixed to a constant. When both n_1 and n_2 are unrestricted, the optimal sample sizes can be obtained by expressing n_2 in terms of n_1 and the costs and budget using (4.1), substituting into the formula for $\text{Var}(\hat{\underline{\beta}}_1)$ and solving for n_1. The optimal sample sizes n_1 and n_2 thus obtained are given in the third and fourth columns of Table 4.1, and the $\text{Var}(\hat{\underline{\beta}}_1)$ obtained with these optimal sample sizes is given in the last column of this table. Note that the optimal number of schools n_2 should be larger than or equal to 2 in order to maintain the multilevel data structure. In some studies the number of schools or pupils per school may be limited. If the limited number of schools or pupils per school is smaller than the optimal number, then this limited number should be used. Note that the optimal sample sizes and the $\text{Var}(\hat{\underline{\beta}}_1)$ for pupil-level randomization and $\tau_1^2 > 0$ do not reduce to those for pupil-level randomization and $\tau_1^2 = 0$. This is a consequence of the fact that the optimal sample sizes for the latter case were calculated such that both n_1 and n_2 are at least 2. Otherwise, $\tau_1^2 \to 0$ would lead to $n_2 \to 0$.

From Table 4.1 it follows that a higher budget C results in sampling more schools, except when randomization is done at the pupil level and there is

Table 4.1 Var($\hat{\underline{\beta}}_1$), optimal sample sizes and Var($\hat{\underline{\beta}}_1$) given the optimal sample sizes for two levels of nesting and a random slope. $\tau^2 = \tau_0^2 + \tau_1^2$.

Level of randomization	Var($\hat{\underline{\beta}}_1$)	n_1	n_2	Optimal Var($\hat{\underline{\beta}}_1$)
Pupil ($\tau_1^2 > 0$)	$\frac{n_1\tau_1^2+\sigma^2}{n_1 n_2}$	$\sqrt{\frac{\sigma^2 c_2}{\tau_1^2 c_1}}$	$\frac{C}{\sqrt{\frac{\sigma^2 c_1 c_2}{\tau_1^2}}+c_2}$	$\frac{\left(\sqrt{\sigma^2 c_1}+\sqrt{\tau_1^2 c_2}\right)^2}{C}$
Pupil ($\tau_1^2 = 0$)	$\frac{\sigma^2}{n_1 n_2}$	$\frac{C-2c_2}{2c_1}$	2	$\frac{\sigma^2 c_1}{C-2c_2}$
School	$\frac{n_1\tau^2+\sigma^2}{n_1 n_2}$	$\sqrt{\frac{\sigma^2 c_2}{\tau^2 c_1}}$	$\frac{C}{\sqrt{\frac{\sigma^2 c_1 c_2}{\tau^2}}+c_2}$	$\frac{\left(\sqrt{\sigma^2 c_1}+\sqrt{\tau^2 c_2}\right)^2}{C}$

no treatment by school interaction, since then the optimal number of pupils per school increases with the budget. Furthermore, this table shows that the number of pupils to be sampled per school reaches its maximum in case of pupil-level randomization and $\tau_1^2 = 0$. This is obvious because when school-by-treatment interaction is assumed to be absent, there is no point in adding more schools. In fact, the optimal design is reached when just one school is sampled, but in that case, the variance component τ_0^2 cannot be estimated and therefore the number of schools is restricted to be at least 2. Of course, τ_0^2 cannot be estimated very well when just two schools are sampled, but on the other hand, the Var($\hat{\underline{\beta}}_1$) does not depend on this variance component in case of pupil randomization with $\tau_0^2 = 0$. For school-level randomization and for pupil-level randomization with $\tau_1^2 > 0$, the optimal number of schools will generally be larger than 2. For these cases, the number of pupils per school increases with the pupil-level variance component σ^2, which is obvious since more pupils are needed in the experiment when there is much variation in the outcome at the pupil level. Also, the optimal n_1 increases with the costs of sampling an extra school relative to the costs of sampling a pupil because generally less schools will be sampled in favor of sampling more pupils per school when it is relatively expensive to sample a school.

Table 4.1 shows that the pupil level is the optimal level of randomization. The relative efficiency of school-level versus pupil-level randomization is given by the ratio of the reciprocal of their optimal variances as given in the last column of Table 4.1, which for models with a fixed slope (i.e., $\tau_1^2 = 0$) is approximated by

$$RE \approx \frac{\sigma^2 c_1}{\left(\sqrt{\sigma^2 c_1} + \sqrt{\tau_0^2 c_2}\,\right)^2} = \frac{1-\rho}{\left(\sqrt{1-\rho} + \sqrt{\rho c_2/c_1}\,\right)^2}, \tag{4.5}$$

and this approximation works well when $C > 40c_2$. Equation (4.5) shows that the relative efficiency decreases when ρ and/or the cost ratio c_2/c_1 increase. The inverse of the relative efficiency gives the number of times the optimal design for randomization at the school level needs to be replicated to be as efficient as the optimal design for randomization at the pupil level assuming $\tau_0^2 = 0$. Figure 4.1 shows the relative efficiency as a function of the intra-school correlation coefficient and for $c_2/c_1 = 10, 20, 40$. As follows from this figure, the decrease in the relative efficiency is already considerable for small ρ. When $\rho = 0.05$, it is equal to 0.34, 0.24, and 0.17 for $c_2/c_1 = 10$, 20, and 40, respectively. When ρ approaches unity, the relative efficiency goes to zero. The relative efficiency is larger when treatment by school interaction is present (i.e., $\tau_1^2 > 0$).

Figure 4.2 gives an impression of the difference in power of two-sided tests with significance level $\alpha = 0.05$ obtained with randomization at the school and pupil level, as a function of the effect size, which is calculated as $ES = 2\beta_1/\sqrt{\sigma^2 + \tau_0^2 + \tau_1^2}$, where $2\beta_1$ is the true value of the treatment effect and

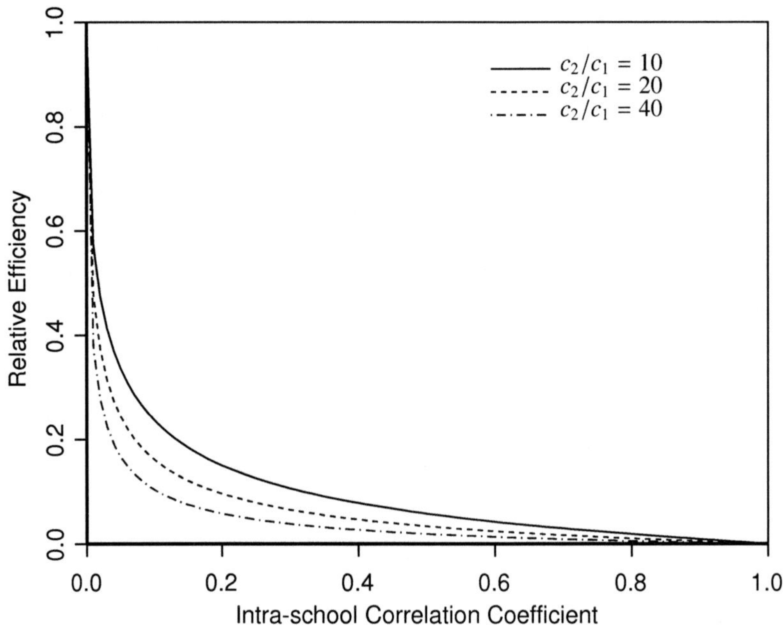

Fig. 4.1 Approximate relative efficiency of school- versus pupil-level randomization as a function of the intra-school correlation coefficient and the cost ratio c_2/c_1. For both levels of randomization, the optimal $\text{Var}(\underline{\hat{\beta}}_1)$ is used.

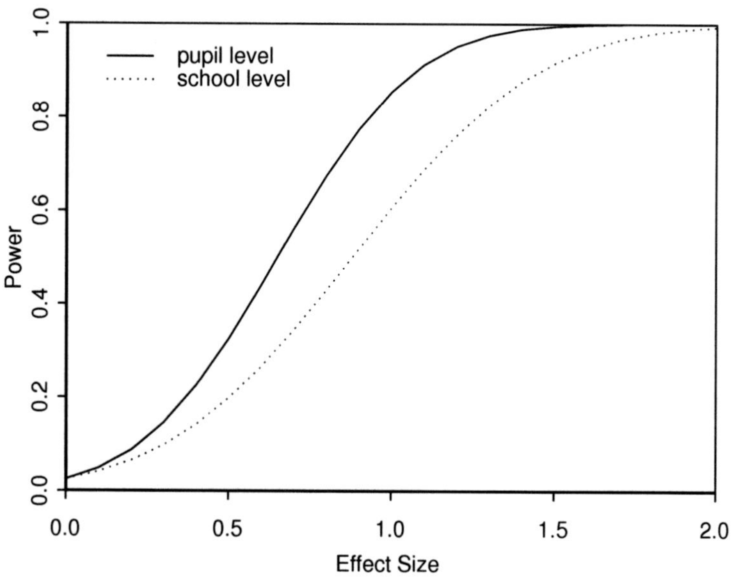

Fig. 4.2 Power of two-sided tests as a function of the effect size.

the denominator gives the standard deviation of the outcome y_{ij}. Values 0.2, 0.5, and 0.8 correspond to small, medium, and large effects, respectively [6]. From Fig. 4.2 it follows that the difference in power is especially large for effect sizes near 1. In order to draw this figure, we used the following values for the costs and budget: $c_1 = 10$, $c_2 = 200$, and $C = 8000$, which reflect the fact that the costs for sampling schools are often larger than the costs for sampling a pupil in an already sampled school. Furthermore, there is often more variation in the outcome at the pupil level than at the school level, which is reflected by the parameter values $\sigma^2 = 24$, $\tau_0^2 = 2$, and $\tau_1^2 = 1$. For these parameter values, the optimal sample sizes for school-level randomization are $n_1 = 12.6$, $n_2 = 24.5$, and $\mathrm{Var}(\underline{\hat{\beta}}_1) = 0.200$. Rounding to even n_2 such that the budget C is not exceeded gives $n_1 = 13$, $n_2 = 24$, and $\mathrm{Var}(\underline{\hat{\beta}}_1) = 0.202$. For randomization at the pupil level, $n_1 = 21.9$, $n_2 = 19.1$, and $\mathrm{Var}(\underline{\hat{\beta}}_1) = 0.110$. Rounding to even n_1 such that the budget is not exceeded results in optimal sample sizes $n_1 = 22$, $n_2 = 19$, and $\mathrm{Var}(\underline{\hat{\beta}}_1) = 0.110$, which is about half of the variance that is obtained with school-level randomization.

It should be noted that the comparison of randomization at the cluster level and randomization at the person level as presented in this example is based on the assumption that control group contamination is absent. This assumption is not always true in practice. It is easily violated in intervention studies where the clustering is such that persons within the same cluster meet regularly, such as families, classes within schools, and work sites. When the degree of the contamination is known, the two designs can still be compared on the basis of their relative efficiencies [32]. For large degrees of control group contamination, a cluster randomized trial may be favored over a trial that randomizes persons within clusters.

For both levels of randomization, the power levels of proposed designs can be evaluated and compared using specialized software, such as the OPTDES program by Raudenbush et al. [52]. This program allows one to plot the power levels as a function of the sample size per school (n_1), number of sampled schools (n_2), intra-school correlation coefficient, and effect size. It also calculates optimal sample sizes for equal and unequal costs across the treatment conditions.

4.4 Optimal Experimental Designs for Models with Covariates

4.4.1 Effect of Including Covariates on the Optimality Criterion

So far, we have considered optimal experimental designs for models without covariates. In practice, however, covariates are often included into the multilevel model to decrease variances of treatment effect estimators and thus to increase statistical power, and in observational studies also to correct for confounding. For uncorrelated outcomes, the formula for the variance of the treatment effect estimator is equal to

$$\mathrm{Var}(\underline{\hat{\beta}}_1) = \frac{\sigma_r^2}{N(1 - r_{xc}^2)},$$

if x is denoted -1 and 1 for the control and intervention group, respectively, and both treatment groups are of equal size [e.g., 24, 47]. $\sigma_r^2 = \sigma^2 + \tau_0^2 + \tau_1^2$ is the total residual variance in $\underline{y}_{ij}$ and r_{xc}^2 is the squared multiple correlation coefficient between the treatment condition x and all covariates c. The term $1/(1 - r_{xc}^2)$ is often called the *variance inflation factor*, abbreviated VIF. Of course, σ_r^2 will decrease when a covariate is added to the model, leading to a smaller variance of the treatment effect estimator, at least if treatment condition and the covariates are uncorrelated.

Similar formulae have been derived for multilevel data [36]. Following Neuhaus and Kalbfleisch [47], a grand-mean centered covariate c_{ij} can be split into a component $\bar{c}_{.j}$ that varies at the school level and a component $(c_{ij} - \bar{c}_{.j})$ that varies at the pupil level. The fixed-slope multilevel model for pupil i within school j is then given by

$$\underline{y}_{ij} = \beta_0^* + \beta_1^* x_{ij} + \beta_2^* \bar{c}_{.j} + \beta_3^* (c_{ij} - \bar{c}_{.j}) + \underline{\delta}_{0j}^* + \underline{\epsilon}_{ij}^*, \tag{4.6}$$

where $\underline{\delta}_{0j}^* \sim \mathcal{N}(0, \tau_0^{*2})$ and $\underline{\epsilon}_{ij}^* \sim \mathcal{N}(0, \sigma^{*2})$. Note that the regression coefficients and random terms are superscribed with asterisks to stress that their values may differ from those of the parameters in the model (4.4) without the covariate. In the analysis stage, we condition on the values of the treatment effect and the covariate, and these variables are treated as fixed. When the covariate only varies at the school level, the term $\beta_3^*(c_{ij} - \bar{c}_{.j})$ is equal to zero and may be removed from the model. Likewise, the term $\beta_2^* \bar{c}_{.j}$ may be removed when the covariate only varies at the pupil level. For the fixed-slope model in (4.6), $\mathrm{Var}(\underline{\hat{\beta}}_1)$ as given in the second column of Table 4.1 needs to be multiplied by the factor $1/(1 - r_{xc}^2)$ [36], and thus $\mathrm{Var}(\underline{\hat{\beta}}_1)$ is minimal for $r_{xc}^2 = 0$. For school-level randomization, r_{xc}^2 is equal to the correlation between x_{ij} and $\bar{c}_{.j}$ since x_{ij} and $(c_{ij} - \bar{c}_{.j})$ are automatically orthogonal, and for pupil-level randomization, r_{xc} is equal to the sample correlation between x_{ij} and $(c_{ij} - \bar{c}_{.j})$. For a binary distributed variable $\underline{x}_{ij}$ and a normally or binary distributed variable $\underline{c}_{ij}$, $\underline{r}_{xc}$ is approximately normally distributed with zero mean and variance $1/n$ and thus $\underline{r}_{xc}^2 \in [0, 4/n]$ with 95% probability [19], where n is equal to $n_1 n_2$ or n_2 for pupil- or school-level randomization, respectively. Thus, r_{xc}^2 is small for large sample sizes, especially for pupil-level randomization. A zero sample correlation between treatment condition and covariate can also be achieved by pre-stratification. For school-level randomization, pre-stratification needs to be done on $\bar{c}_{.j}$ and thus for each value of $\bar{c}_{.j}$, half of the schools must be randomized to the treatment condition and the others to the control group. Similarly, pre-stratification needs to be done on $(c_{ij} - \bar{c}_{.j})$ for pupil-level randomization. In the remainder, we will assume that $r_{xc}^2 = 0$, due to pre-stratification or large sample sizes. Then, the optimal sample size formulae and the $\mathrm{Var}(\underline{\hat{\beta}}_1)$ as given in Table 4.1 hold when τ_0^2 and σ^2 are replaced with τ_0^{*2} and σ^{*2}, respectively, and it can be shown that this is also true for models with a random slope with τ_1^2 replaced with τ_1^{*2} [30].

4.4.2 Effect of Including Covariates on the Values of the Variance Components

The inclusion of a covariate will also lead to a change in the values of the estimated variance components, given the total variance of the outcome. Suppose that both components of the covariate are added to model (4.4) with $\tau_1^2 = 0$ so that we obtain model (4.6). The changes in the estimated variance components for the fixed-slope model can be established as follows [58]. Note that we turn from the data and the estimators to the population. The total variance of an outcome $\underline{y}_{ij}$ and the covariance of two outcomes $\underline{y}_{ij}$ and $\underline{y}_{i'j}$ within the same school are equal to

Table 4.2 Changes in variance components due to the inclusion of a covariate to the two-level model with a random intercept and a fixed slope.

Changes due to the inclusion of $\bar{c}_{.j}$	Changes due to the inclusion of $c_{ij} - \bar{c}_{.j}$
$\underline{\hat{\tau}}_0^2 - \underline{\hat{\tau}}_0^{*2} = \underline{\hat{\beta}}_2^{*2}\,\mathrm{Var}(\bar{c}_{.j}) > 0$	$\underline{\hat{\tau}}_0^2 - \underline{\hat{\tau}}_0^{*2} \approx 0$
$\underline{\hat{\sigma}}^2 - \underline{\hat{\sigma}}^{*2} = 0$	$\underline{\hat{\sigma}}^2 - \underline{\hat{\sigma}}^{*2} \approx \underline{\hat{\beta}}_3^{*2}\,\mathrm{Var}(c_{ij} - \bar{c}_{.j}) > 0$

Note. It is assumed that n_1 is not too small and that $r_{xc}^2 = 0$.

$$\begin{aligned} \mathrm{Var}(\underline{y}_{ij}) &= \mathrm{Var}(\beta_1 \underline{x}_{ij} + \underline{\delta}_{0j} + \underline{\epsilon}_{ij}), \\ \mathrm{Cov}(\underline{y}_{ij}, \underline{y}_{i'j}) &= \mathrm{Cov}(\beta_1 \underline{x}_{ij} + \underline{\delta}_{0j}, \beta_1 \underline{x}_{i'j} + \underline{\delta}_{0j}) \end{aligned} \tag{4.7}$$

for model (4.4) with $\tau_1^2 = 0$, and

$$\begin{aligned} \mathrm{Var}(\underline{y}_{ij}) &= \mathrm{Var}(\beta_1^* \underline{x}_{ij} + \beta_2^* \underline{\bar{c}}_{.j} + \beta_3^* (\underline{c}_{ij} - \underline{\bar{c}}_{.j}) + \underline{\delta}_{0j}^* + \underline{\epsilon}_{ij}^*), \\ \mathrm{Cov}(\underline{y}_{ij}, \underline{y}_{i'j}) &= \mathrm{Cov}\Big[\beta_1^* \underline{x}_{ij} + \beta_2^* \underline{\bar{c}}_{.j} + \beta_3^* (\underline{c}_{ij} - \underline{\bar{c}}_{.j}) + \underline{\delta}_{0j}^*, \\ &\qquad \beta_1^* \underline{x}_{i'j} + \beta_2^* \underline{\bar{c}}_{.j} + \beta_3^* (\underline{c}_{i'j} - \underline{\bar{c}}_{.j}) + \underline{\delta}_{0j}^*\Big] \end{aligned} \tag{4.8}$$

for model (4.6). Since $\mathrm{Var}(\underline{y}_{ij})$ and $\mathrm{Cov}(\underline{y}_{ij}, \underline{y}_{i'j})$ are given by the data and are therefore independent of the chosen model, the $\mathrm{Var}(\underline{y}_{ij})$ and $\mathrm{Cov}(\underline{y}_{ij}, \underline{y}_{i'j})$ given by (4.7) can be set equal to the $\mathrm{Var}(\underline{y}_{ij})$ and $\mathrm{Cov}(\underline{y}_{ij}, \underline{y}_{i'j})$ given by (4.8). From these two equations, the changes in the estimated variance components can be derived, and for both levels of randomization, these are given in Table 4.2. We assume that $r_{xc}^2 = 0$, so that $\underline{\hat{\beta}}_1 = \underline{\hat{\beta}}_1^*$, and that n_1 is large. The total change due to the inclusion of both components of the covariate is equal to the sum of the change due to the inclusion of the separate components. From Table 4.2 it follows that only the estimated variance component at the level at which the covariate varies decreases when a covariate is added to the model. Likewise, it can be shown that for models with a random slope of x_{ij} only $\underline{\hat{\tau}}_0^2$ or $\underline{\hat{\sigma}}^2$ change when the school- or pupil-level component of the covariate are added to the model, respectively, under the assumption that $r_{xc}^2 = 0$ within each school [30, Chapter 4].

4.5 Optimal Experimental Designs for Multilevel Logistic Models

When the responses y_{ij} are measured on a binary scale, the multilevel logistic model applies, see Chapter 6. Assuming treatment by school interaction (i.e., a random slope), it is equal to

$$\underline{y}_{ij} = \underline{\pi}_{ij} + \underline{\epsilon}_{ij} = \frac{1}{1 + \exp[-(\beta_0 + \beta_1 x_{ij} + \underline{\delta}_{0j} + \underline{\delta}_{1j} x_{ij})]} + \underline{\epsilon}_{ij},$$

where $\underline{\pi}_{ij}$ is the probability of a response $\underline{y}_{ij} = 1$, and $\underline{\epsilon}_{ij}$ has zero mean and variance $\underline{\pi}_{ij}(1 - \underline{\pi}_{ij})$. The independent variable x_{ij} is coded -1 for the control group and $+1$ for the intervention group. Again, we use the $\mathrm{Var}(\hat{\underline{\beta}}_1)$ as optimality criterion. An analytical expression for it can only be obtained when the so-called first-order Marginal Quasi-Likelihood [MQL, 14] estimation method is used. It can then be shown [35] that the $\mathrm{Var}(\hat{\underline{\beta}}_1)$ is equal to the formulae given in Table 4.1 when σ^2 is replaced with

$$\tfrac{1}{2}(4 + e^{\beta_0+\beta_1} + e^{\beta_0-\beta_1} + e^{-\beta_0+\beta_1} + e^{-\beta_0-\beta_1}). \tag{4.9}$$

To calculate optimal sample sizes, the variance components must be known or a reasonable prior specification must be given.

First-order MQL, however, produces biased estimates [15, 53], while Penalized Quasi-Likelihood [PQL, 15] and estimation by means of numerical integration [16] perform better [54, and Chapter 9]. Only for second-order PQL, however, the test statistic to test the significance of β_1 was shown to follow the standard normal distribution [38]. Therefore, a simulation study was done [35] to investigate how the variance of the treatment effect estimator, $\mathrm{Var}(\hat{\underline{\beta}}_1)$, is affected when second-order PQL with unknown variance components is used instead of first-order MQL with known variance components.

For models with a fixed slope (i.e., $\tau_1^2 = 0$), data sets were generated for the following parameter values: $\beta_0 = 0$; $\beta_1 = 1.5$, 1, 0.5, or 0; and $\tau_0^2 = 1$, 0.5, 0.25, or 0. Three different allocations of units were used: $(n_1, n_2) = (10, 40)$, $(n_1, n_2) = (20, 20)$, and $(n_1, n_2) = (40, 10)$. Both levels of randomization were considered. Thus, there were 96 combinations of level of randomization, allocation of units, and parameters values, which will be called simulation combinations, and for each of these, 200 data sets were simulated. Second-order PQL as implemented in the computer program MLwiN [48] was used for parameter estimation. For each of the 96 simulation combinations, the sampling variance of $\hat{\underline{\beta}}_1$ was estimated by

$$\text{Sampling Variance}(\hat{\underline{\beta}}_1) = \frac{\sum_{r=1}^{200}(\hat{\beta}_{1r} - \sum_{s=1}^{200}\hat{\beta}_{1s}/200)^2}{199},$$

where $\hat{\beta}_{1r}$ is the estimate of β_1 from the r-th simulated data set. Furthermore, a correction factor was calculated, which is equal to the Sampling Variance($\hat{\underline{\beta}}_1$) divided by the $\mathrm{Var}(\hat{\underline{\beta}}_1)$ as obtained from first-order MQL with known variance components. This factor may be used as a multiplication factor to the analytical $\mathrm{Var}(\hat{\underline{\beta}}_1)$ as given in Table 4.1 with σ^2 replaced with (4.9) when second-order PQL is used instead of first-order MQL.

The results of the study show that the Sampling Variance($\hat{\underline{\beta}}_1$) increases with τ_0^2, β_1 , and the number of pupils per school (given the total sample size

n_1n_2) when randomization is done at the school level. For pupil-level randomization, it only increases with β_1. Furthermore, it is larger for randomization at the school level, especially when τ_0^2 and/or n_1 are large. These conclusions also hold for first-order MQL with known variance components, see Table 4.1 with σ^2 replaced with (4.9). The results of the simulation study suggest that, on average, the Var($\underline{\hat{\beta}}_1$) as obtained with second-order MQL needs to be multiplied by 1.2 to get the Var($\underline{\hat{\beta}}_1$) for first-order PQL. The correction factor is fairly constant across allocations of units, so that the optimal allocations of units obtained with the formulae for first-order MQL with known variance components are also optimal for second-order PQL.

A simulation study was also done for models with a random slope and randomization at the pupil level. The following parameter values and allocations of units were used: $\beta_0 = 0$; $\beta_1 = 1.5$, 1, 0.5, or 0; $\tau_0^2 = 1, 0.5, 0.25$, or 0; $\tau_1^2 = 0$, or 0.25; and $(n_1, n_2) = (10, 40)$, $(n_1, n_2) = (20, 20)$, or $(n_1, n_2) = (40, 10)$, so there were 96 simulation combinations. As for first-order MQL with known variance components, the Sampling Variance($\underline{\hat{\beta}}_1$) increases with τ_1^2 and the number of pupils per school, again given the total sample size n_1n_2. The results of the study suggest that the correction factor is about 1.2 for second-order PQL.

4.6 Optimal Experimental Designs for Longitudinal Data

4.6.1 Sample Sizes, Duration, and Power

In longitudinal intervention studies, persons are randomly assigned to the control or experimental condition, and their responses are measured at successive points in time. The multilevel model is an appropriate tool for the analysis of data obtained from experiments with longitudinal data. The model that relates the response y_{ij} of person j to time point i is given by

$$\underline{y}_{ij} = \underline{\beta}_{0j} + \underline{\beta}_{1j} t_i + \underline{\epsilon}_{ij}, \tag{4.10}$$

assuming a linear trend for the sake of simplicity. The intercept $\underline{\beta}_{0j}$ and slope $\underline{\beta}_{1j}$ vary across persons and are predicted from the treatment condition x_j:

$$\underline{\beta}_{0j} = \beta_{00} + \beta_{01} x_j + \underline{\delta}_{0j}, \tag{4.11a}$$

$$\underline{\beta}_{1j} = \beta_{10} + \beta_{11} x_j + \underline{\delta}_{1j}, \tag{4.11b}$$

where the treatment condition has values -1 and $+1$ for the control and intervention group, respectively. The random error terms $\underline{\delta}_{0j} \sim \mathcal{N}(0, \tau_0^2)$, $\underline{\delta}_{1j} \sim \mathcal{N}(0, \tau_1^2)$, and $\underline{\epsilon}_{ij} \sim \mathcal{N}(0, \sigma^2)$ are assumed to be independent of the treatment condition, and the covariance between $\underline{\delta}_{0j}$ and $\underline{\delta}_{1j}$ is denoted by τ_{01}. Substitution of (4.11) into (4.10) results in the single-equation model

$$\underline{y}_{ij} = \beta_{00} + \beta_{01}x_j + \beta_{10}t_i + \beta_{11}x_jt_i + \underline{\delta}_{0j} + \underline{\delta}_{1j}t_i + \underline{\epsilon}_{ij}. \tag{4.12}$$

The aim of a longitudinal intervention study is to detect whether the linear time effect varies across the two treatment conditions; that is, we want to test the cross-level interaction effect β_{11}. The variance of $\hat{\underline{\beta}}_{11}$ depends on the total number of persons n, the number of measurements per person m, and the duration of the study d. For equally spaced measurements between $t_1 = 0$ and $t_m = d$, it is equal to

$$\mathrm{Var}(\hat{\underline{\beta}}_{11}) = \frac{\sigma^2}{nms^2} + \frac{\tau_1^2}{n}, \qquad \text{with } s^2 = \frac{1}{m}\sum_{i=1}^{m}(t_i - \bar{t}\,)^2, \tag{4.13}$$

see Galbraith and Marschner [13]. The variance s^2 of the time points is an increasing function of the study duration d. From (4.13) it follows that the $\mathrm{Var}(\hat{\underline{\beta}}_{11})$ decreases with increasing m, n, and d. However, increasing n will have a larger effect on $\mathrm{Var}(\hat{\underline{\beta}}_{11})$ than increasing m and d, since m and s^2 only appear in the denominator of the first term of $\mathrm{Var}(\hat{\underline{\beta}}_{11})$.

The test statistic $\hat{\underline{z}} = \hat{\underline{\beta}}_{11}/\sqrt{\mathrm{Var}(\hat{\underline{\beta}}_{11})}$ is approximately normally distributed when the null hypothesis $H_0 : \beta_{11} = 0$ is true. The relation between study duration d, sample sizes m and n and power $1 - \gamma$ is given by

$$\frac{\sigma^2 + ms^2\tau_1^2}{nms^2} = \frac{\beta_{11}^2}{(z_{1-\alpha/2} + z_{1-\gamma})^2}. \tag{4.14}$$

The power for the test depends on the true effect β_{11}, of which a realistic value may be difficult to specify. The *standardized* effect size for a linear trend is defined as the group difference in the linear trend divided by the standard deviation of the linear trend: $ES = \beta_{11}/\tau_1$ [51]. Substitution of the standardized effect size into (4.14) results in

$$\frac{\sigma^2/\tau_1^2 + ms^2}{nms^2} = \frac{ES^2}{(z_{1-\alpha/2} + z_{1-\gamma})^2}, \tag{4.15}$$

which shows that only the ratio σ^2/τ_1^2 of the variances σ^2 and τ_1^2 needs to be known to calculate the power level of a proposed design. It should be noted that the comments on power calculations for tests with two-sided alternative hypothesis in Section 4.5 are also applicable to (4.15).

As an example, consider a study for which it is expected that $\sigma^2/\tau_1^2 = 5$ and for which the power to detect a size $ES = 0.5$ in a two-sided test with $\alpha = 0.05$ should be at least 0.8. Figure 4.3 shows the power levels as a function of n, for two different values of the study duration d ($d = 2$, 4) and for two different values of m ($m = 5$, 9). It follows that increasing m only has a small effect on power relative to increasing d and n. For $d = 4$, the number of persons to reach a power of 0.8 is about 42 ($m = 9$) and about 48 ($m = 5$). For $d = 2$, a much larger number of persons is needed.

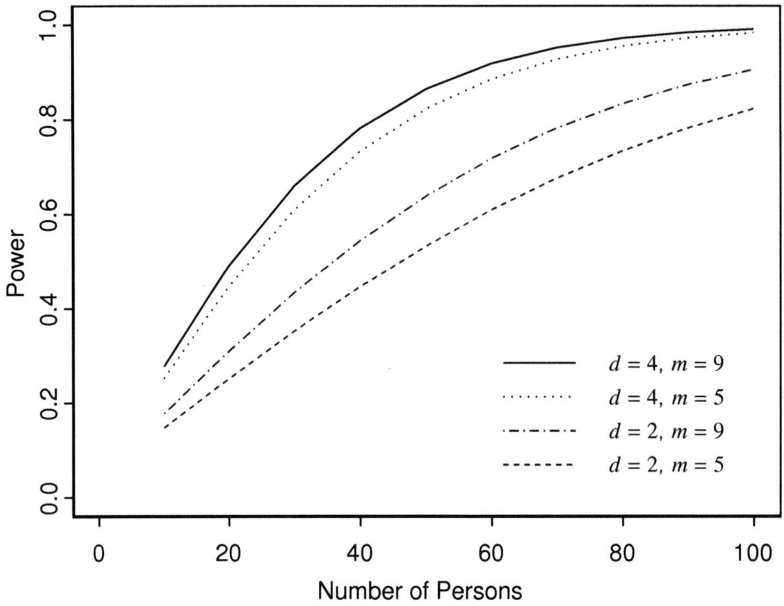

Fig. 4.3 Power as a function of m, n, and d.

It should be noted that these results hold for linear growth. Results for higher-order polynomial growth can be found elsewhere [51]. As shown by Laird and Wang [21], efficiency is gained by dropping in (4.12) the term $\beta_{01}x$, which reflects the group difference at time point 0 and is therefore equal to zero in the case of randomized trials. Model (4.12) implies a certain nonstationary covariance structure for the repeated measures. For optimal designs under different covariance structures, see, e.g., Winkens et al. [64]. Furthermore, the extensions to studies with drop-out and missing data are presented by others [13, 17, 33, 44, 45]. The computer programs M*plus* [46] and OPTDES [52] can be used to evaluate and compare alternative designs for studies with longitudinal data.

4.6.2 Some Other Results on Optimal Experimental Designs for Longitudinal Data

The multilevel models with covariates as given in Section 4.4 may also be used for longitudinal designs in which two treatment conditions are compared and the dependent variable is measured once at pre-test and once at post-test. The outcome variable in our model (4.6) is the post-test measurement, whereas the pre-test measurement is included as a covariate. One may also choose to use a model in which the dependent variable is equal to the difference between the pre- and post-test (i.e., the change score) and in which the pre-test measurement of the dependent variable is not included as a covariate. Due to randomization, both approaches yield the same expected treatment effect. The first approach, however, is preferred, since due to the inclusion of the pre-treatment measurement as a covariate, a lower residual variance in the outcome, and consequently more statistical power, is achieved. For a repeated measures formulation of both methods, ANCOVA and ANOVA of change, see Laird and Wang [21] and Van Breukelen [62].

The papers by Feldman and McKinlay [11] and McKinlay [29] also focus on longitudinal designs. These papers are restricted to cluster randomization and the change score is used as dependent variable. Two types of designs are considered in these papers: cohort designs and cross-sectional designs, whereas in the previous sections of this chapter we only focused on cohort designs. In both cohort and cross-sectional designs, a set of clusters is sampled. In a cohort design, the same individuals are measured at at least two time points. In a cross-sectional design, a new sample of individuals is drawn within each cluster at every time point. Cohort designs are favored above cross-sectional designs if the clusters contains relatively few subjects, if the population is stable throughout the intervention period, if the intervention period is short enough to prevent substantial dropout, and if the act of measurement does not influence the subjects' subsequent behavior.

The variance of the treatment effect estimator was used as optimality criterion in the papers mentioned above. The relative efficiency of cohort designs versus cross-sectional designs was studied [11, p. 68] as a function of the subject autocorrelation (i.e., the correlation over time between individual level means) denoted ρ_s. They show that a cohort design is more efficient than a cross-sectional design for any $\rho_s > 0$, but the ρ_s has to be unrealistically close to unity to provide noticeable gains in efficiency. Thus, for weak subject autocorrelation, the cross-sectional design may be preferred, since for this design memory effect and drop-out do not occur.

The optimal number of clusters per treatment condition and the optimal number of individuals per cluster were calculated by McKinlay [29] for both cohort and cross-sectional designs. The cost function that was used by McKinlay takes drop-outs and recovery of drop-outs in a cohort design into account, and allows the costs at the cluster level to vary across treatment conditions. In an example, McKinlay shows that cohort designs are more cost efficient for short trials and high autocorrelations at both the cluster and individual levels.

4.7 Optimal Designs for Surveys

In multilevel surveys, generally more than one parameter is of main interest. These parameters may be regression coefficients corresponding to level-1 or level-2 predictors or cross-level interactions, as well as variance components or the intra-school correlation coefficient. Designing multilevel surveys may be very complicated, since the values of the predictor variables are not under experimental control, whereas their means, variances, and covariances as expressed in the design matrix, as well as the covariance matrix of the random effects, need to be known in advance to design the survey optimally. We will first derive optimal sample size formulae when there is just one explanatory variable at either the pupil or school level, and thereafter focus on the case with more than one explanatory variable. Optimal designs for variance parameters are the subject of the next section.

Let us first assume that the multilevel model only contains a school-level explanatory variable x_j:

$$\underline{y}_{ij} = \beta_0 + \beta_1 x_j + \underline{\delta}_{0j} + \underline{\epsilon}_{ij}.$$

For this model, it can be shown [30, Chapter 4] that the $\mathrm{Var}(\underline{\hat{\beta}}_1)$ is equal to

$$\mathrm{Var}(\underline{\hat{\beta}}_1) = \frac{\sigma^2 + n_1\tau_0^2}{n_1 n_2 s_x^2},$$

where s_x^2 is the variance of x_j defined as $\sum_j (x_j - \bar{x}_.)^2/n_2$. This variance reduces to 1 if x_j is a treatment coded -1 and $+1$ with both values occurring with 50% probability, as in Table 4.1. The optimal sample sizes for estimating β_1 as efficiently as possible are equal to those for optimal experimental designs with school-level randomization as given in Table 4.1.

Now suppose that the explanatory variable is a pupil-level variable x_{ij} with school mean zero and that treatment by school interaction is absent. Then

$$\mathrm{Var}(\underline{\hat{\beta}}_1) = \frac{\sigma^2}{n_1 n_2 s_x^2}, \qquad \text{with } s_x^2 = \sum_j \sum_i (x_{ij} - \bar{x}_{.j})^2/n_1 n_2, \tag{4.16}$$

see Moerbeek [30, Chapter 4]. Again, the optimal sample sizes can be found in Table 4.1 and are equal to those for randomization at the pupil level and no treatment by school interaction.

Now suppose that the model also contains a random slope, i.e.,

$$\underline{y}_{ij} = \beta_0 + \beta_1 x_{ij} + \underline{\delta}_{0j} + \underline{\delta}_{1j} x_{ij} + \underline{\epsilon}_{ij}.$$

For this model, $\mathrm{Var}(\underline{\hat{\beta}}_1)$ is equal to

$$\mathrm{Var}(\underline{\hat{\beta}}_1) = \frac{\sigma^2 + n_1\tau_1^2 s_x^2}{n_1 n_2 s_x^2} \tag{4.17}$$

if the explanatory variable x_{ij} has school mean zero and its variance s_x^2 is the same within each cluster. The optimal sample sizes are given in Table 4.1 and are equal to those for randomization at the pupil level and a random slope if σ^2 is replaced with σ^2/s_x^2. Note that the variance s_x^2 in (4.16) and (4.17) is equal to 1 if x_{ij} is treatment coded -1 and $+1$ and both values occur with 50% probability within each cluster, as in Table 4.1.

Sample size formulae for the model with explanatory variables at the pupil and/or school level, with cross-level interaction terms and with fixed or random slopes for the pupil-level variables are presented by Snijders and Bosker [57]. Their computer program PinT (*P*ower *in* *T*wo-level designs) [4] calculates approximate standard errors of regression coefficients for different combinations of n_1 and n_2 using the cost constraint (4.1).

To illustrate the use of the program PinT, we work out the following example. Suppose we want to assess the relationship between a test score on the one side and the pupil's socio-economic status (SES) and school size on the other side. The data structure has two levels: Pupils are nested within schools. The budget C that is available for his study is equal to $500c_1$, whereas the costs c_2 for sampling a school are equal to $5c_1$, with c_1 the costs for sampling a pupil in an already sampled school. Figure 4.4 shows the total number of pupils and the number of schools as a function of the number of pupils sampled per school as calculated by PinT. As follows from this figure, the number of schools decreases as the number of pupils per school increases. This is obvious, since with large n_1, less money is available for sampling schools. For the same reason, the total number of pupils increases with the number of pupils per school, see Fig. 4.4.

In order to select the optimal sample sizes, a multilevel regression model must be specified and an optimality criterion must be chosen. Let us assume the effect of SES on the test score is constant across schools. The multilevel model then becomes

$$\underline{y}_{ij} = \beta_0 + \beta_1 SES_{ij} + \beta_2 SCHOOL\_SIZE_j + \underline{\delta}_{0j} + \underline{\epsilon}_{ij}, \tag{4.18}$$

where $\underline{y}_{ij}$ is the score of pupil i within school j. Standard errors of both estimated regression coefficients will be used as optimality criteria. To calculate approximate standard errors for these regression coefficients, a reasonable guess of the within- and between-school covariance matrices Σ^W and Σ^B of the predictor variables and of the variances of the random effects is needed. In the PinT manual, guidelines for obtaining such guesses are given. For convenience, it is assumed that all predictor variables have zero mean and variance 1. We assume that 80% of the variance in *SES* is located at the pupil level; thus, $\Sigma^W = (0.8)$. The remaining 20% is between-group variance. The covariance of *SES* and *SCHOOL_SIZE* is assumed to be equal to 0.2; thus,

$$\Sigma^B = \begin{pmatrix} 1 & 0.2 \\ 0.2 & 0.2 \end{pmatrix}.$$

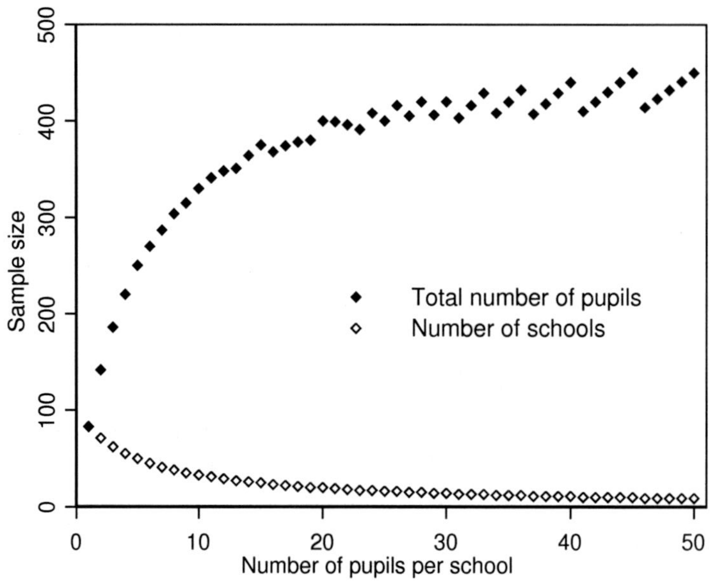

Fig. 4.4 Number of schools and total number of pupils as a function of the number of pupils per school.

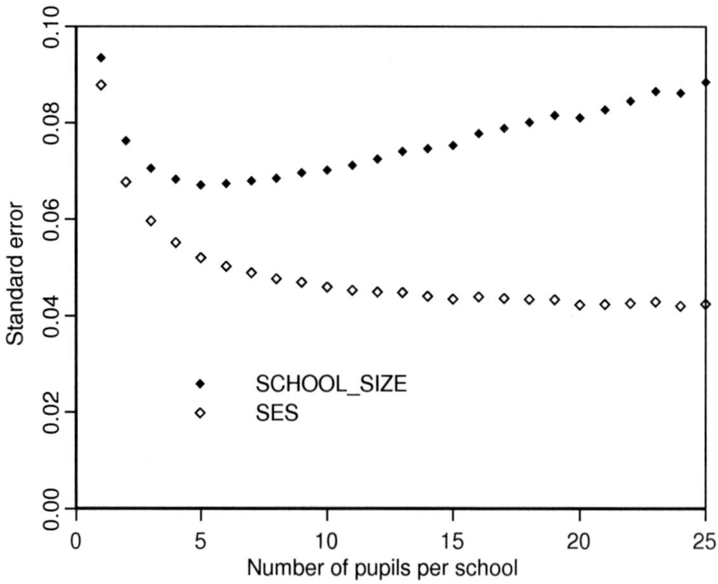

Fig. 4.5 Standard error of the effects of *SES* and *SCHOOL_SIZE* as a function of the number of pupils per school.

Furthermore, let us assume that $\sigma^2 = 0.6$ and that $\tau_0^2 = 0.1$. The standard errors of $\hat{\underline{\beta}}_1$ and $\hat{\underline{\beta}}_2$ for these values are plotted in Fig. 4.5. Note that se($\hat{\underline{\beta}}_1$) < se($\hat{\underline{\beta}}_2$) and that the se($\hat{\underline{\beta}}_2$) is a convex function of n_1, whereas se($\hat{\underline{\beta}}_1$) decreases with increasing n_1. Thus, the optimal design for estimating the effect of SES on the test score as efficiently as possible is achieved by taking n_1 as large as possible. The optimal design for estimating β_2 as efficiently as possible, however, is obtained for $n_1 = 5$. But since the function of se($\hat{\underline{\beta}}_2$) is quite flat near this formal minimum, values of n_1 close to this formal minimum may be chosen as an alternative.

4.8 Optimal Designs for Variance Parameters

Although the regression coefficients are generally of main interest, in some multilevel studies one may also be interested in estimating the variance parameters as efficiently as possible. For two levels of nesting asymptotic variances of the variance components in a linear multilevel model with a random intercept and a fixed slope are given by Longford [27] and shown in the first column of Table 4.3. The optimal sample sizes for estimating the variance components as efficiently as possible given the cost restriction (4.1) were derived by Cohen [7]. To obtain sample size formulae of practical use, Cohen used an approximation to the Var($\hat{\underline{\tau}}_0^2$) and showed that the optimal sample sizes for this approximated Var($\hat{\underline{\tau}}_0^2$) are usually the same as those for the true Var($\hat{\underline{\tau}}_0^2$). The optimal n_1 are also presented in Table 4.3; the optimal n_2 follow from the cost restriction.

The variance of $\hat{\underline{\rho}}$ was given by Donner [8] and is also shown in Table 4.3. For this parameter, the analytical formulae for the optimal sample sizes are too complex. Instead, one may substitute $n_2 = C/(c_1 n_1 + c_2)$ into the Var($\hat{\underline{\rho}}$), which is then a function of n_1 and ρ, c_1, c_2, and C. Once a reasonable prior

Table 4.3 Variance of variance components and intraclass correlation coefficient and optimal n_1.

Optimality criterion	Optimal n_1
$\mathrm{Var}(\hat{\underline{\sigma}}^2) = \dfrac{2\sigma^4}{n_2(n_1-1)}$	$\dfrac{C - 2c_2}{2c_1}$
$\mathrm{Var}(\hat{\underline{\tau}}_0^2) = \dfrac{2\sigma^4}{n_2 n_1}\left[\dfrac{1}{n_1-1} + 2\dfrac{\rho}{1-\rho} + n_1\left(\dfrac{\rho}{1-\rho}\right)^2\right]$	$\dfrac{\sqrt{c_1\left(c_1 + 8c_2\dfrac{\rho}{1-\rho}\right)} + c_1}{2c_1\dfrac{\rho}{1-\rho}}$
$\mathrm{Var}(\hat{\underline{\rho}}) = \dfrac{2[(1-\rho)(1+(n_1-1)\rho)]^2}{n_1(n_1-1)(n_2-1)}$	Analytical formula complex

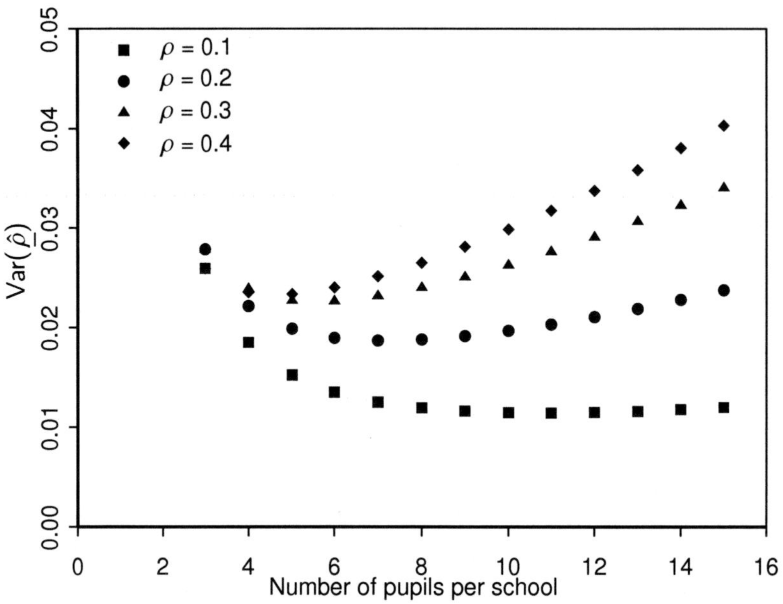

Fig. 4.6 Variance of $\hat{\underline{\rho}}$ as a function of the number of pupils per school and ρ.

specification of the value of ρ has been made, the Var($\hat{\underline{\rho}}$) may be plotted as a function of n_1 and the optimal n_1 may be established. Such a plot is given in Fig. 4.6, in which the Var($\hat{\underline{\rho}}$) is plotted as a function of the true ρ and the number of pupils per school for $c_1 = 100$, $c_2 = 200$, $C = 8000$. As follows from this figure, the Var($\hat{\underline{\rho}}$) increases with the true ρ. The value n_1 at which the Var($\hat{\underline{\rho}}$) is minimized decreases when the true value ρ increases.

Although we use the variance of the estimators $\hat{\underline{\sigma}}^2$, $\hat{\underline{\tau}}_0^2$, and $\hat{\underline{\rho}}$ as optimality criterion, it has to be noted that the estimators are skewed and thus statistical tests of and confidence intervals for these parameters do not only depend on the variance.

4.9 Robustness of Optimal Designs

The values of the parameters of the multilevel random effects regression model that is used to analyze the data have to be known in advance to plan multilevel studies as efficiently as possible. On the other hand, the study is implemented to get some knowledge of the values of these unknown parameters. To solve this problem, one may use a reasonable prior specification of these parameter values. Such values may be obtained from the results of comparable studies (see the references in Table 1 of Murray et al. [43]), from a pilot study or from theoretical opinions about the minimally relevant treatment effect. One may, however, wonder to what extent the optimal design is robust against misspecification of the model parameters. For each model parameter θ, the robustness can be expressed in terms of the relative efficiency as given by (4.2). If, for example, $\Phi(\boldsymbol{M}^{-1}(\boldsymbol{X}(\eta,\xi),\boldsymbol{\theta})) = \text{Var}(\hat{\underline{\beta}}_1)$ and randomization is done at the school level, the efficiency of the design $\xi = (n_1, n_2)$ obtained with an incorrect prior specification of the value of ρ relative to the efficiency of the design $\xi^* = (n_1^*, n_2^*)$ obtained with the true value of ρ is given by

$$\left(\frac{n_1^*\rho + (1-\rho)}{n_1\rho + (1-\rho)}\right)\left(\frac{n_1 n_2}{n_1^* n_2^*}\right),$$

where ρ is the true value of the intra-school correlation coefficient.

As an example, let us consider the model (4.4) with randomization at the school level and no treatment by school interaction (i.e., $\tau_1^2 = 0$). The optimal sample sizes for estimating β_1 are given in Table 4.1, whereas those for the variance components σ^2 and τ_0^2 can be found in Table 4.3. The intra-school correlation coefficient ρ needs to be known in order to calculate the optimal sample sizes for β_1 and τ_0^2. Let us assume that $C = 10000$, $c_1 = 2$, and $c_2 = 30$, and that the true $\rho = 0.07$. Then the robustness of the optimal designs for β_1 and τ_0^2 is plotted in Fig. 4.7 in terms of the relative efficiency as a function of ρ. As follows from this figure, the optimal design for β_1 is a bit more robust against incorrect prior guesses of ρ than the optimal design for τ_0^2. For $\rho = 0.07$, the relative efficiency especially decreases very rapidly when a too low prior guess for ρ is supplied. When the incorrect prior guess of ρ lies within the interval $[0.04, 0.15]$ the relative efficiency for both parameters is high (i.e., > 0.9). This is, however, not necessarily the case for each combination of C, c_1, and c_2.

Different approaches have been proposed to derive robust optimal designs for multilevel model. One such approach is the use of sample size re-estimation. The optimal sample sizes are calculated based on prior estimates of the model parameters as obtained from subject-matter knowledge or an educated guess. Then, a predefined proportion of the number of clusters or of the number of persons within clusters is sampled, the data are collected, and the model parameters are estimated on basis of the collected data. Then, the optimal sample sizes are re-estimated and the remainder of the data are collected. All data are used in the final analysis; hence, the pilot is referred to as an internal pilot. This approach has been evaluated for cluster randomized trials [22] and surveys with nested data [31] by means of simulation studies. The results showed that sample size re-estimation has large control over power and the costs of the study. Another approach is the use of Bayesian optimal designs [59, 61]. This approach allows taking uncertainty about the model parameters into account by specifying prior distributions of these parameters. Then, a large number of times the model parameters are sampled from their prior distributions, and the power levels of the test statistic of the model parameter

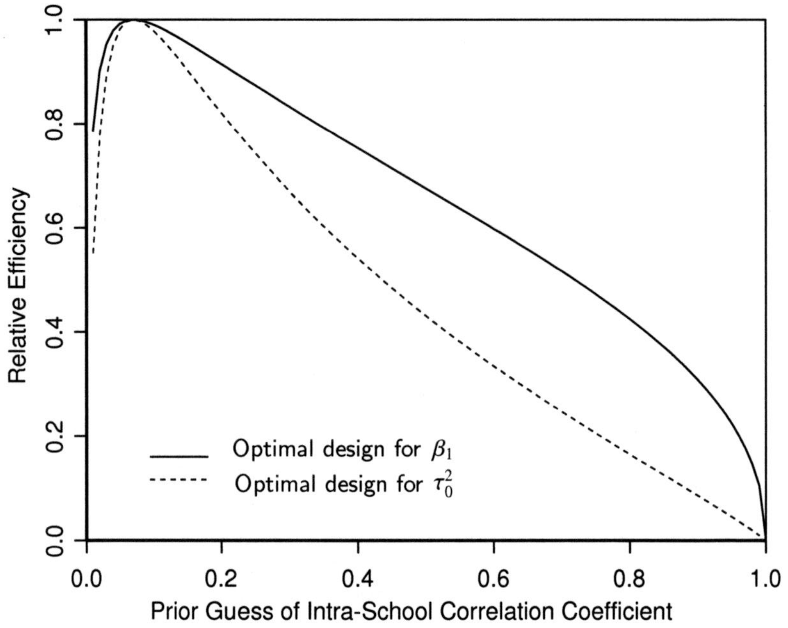

Fig. 4.7 Robustness of optimal design for β_1 and τ^2 against an incorrect prior guess of the intra-school correlation coefficient ρ_c. The true $\rho_c = 0.07$.

of interest are calculated. The power distribution that is thus obtained reflects the uncertainty in the model parameters. The computer program WINBUGS can be used to calculate Bayesian optimal designs [60]. Another approach to calculate robust optimal designs is the use of maximin optimal designs. A maximin optimal design is the design among all possible designs in the design space χ that maximizes the minimum relative efficiency over the parameter space. For an application of maximin optimal designs, we refer to Berger and Tan [3].

4.10 Concluding Remarks

In this chapter, four design issues for the design of multilevel experiments and surveys were considered: the required budget to obtain a specified power on the test of a certain parameter, the optimal sample sizes at each level of the multilevel data structure, the robustness of optimal designs, and the optimal level of randomization to treatment conditions. As optimality criterion, the variance of model parameters was used, since minimum variance leads to maximum power of statistical tests, at least assuming unbiased parameter estimation and an approximately normal distribution of the estimator. When designing multilevel experiments, the treatment effect is generally of main interest and so its variance is used as optimality criterion. For multilevel surveys, more than one parameter may be of interest and it may be worthwhile to derive a multiple-objective optimal design [40].

The statistical optimality criteria may or may not conflict with other criteria. In some circumstances, ethical criteria may be applied. For example, in some experiments it may be unethical to treat certain individuals within a certain cluster while others are not treated. In this case, randomization at the individual level will become impossible. Practical criteria consist of the need to reduce costs and administrative efforts, political and logistical reasons, and the need to avoid control group contamination, which occurs when information leaks from the intervention group to the control group. Sometimes there is no alternative to cluster randomization. This may occur in, for example, community-based interventions where the intervention will necessarily affect all members of a cluster.

The optimal sample sizes as given in this chapter may be considered as guidelines that have to be pursued as much as possible in designing multilevel studies. They were calculated under the assumption that cluster sizes do not vary and that the costs do not vary across treatment conditions, which is not always plausible in practice. For instance, school sizes in private schools may be smaller than those in public schools. And even if school sizes were equal, there will always be some non-response due to drop-out or for other reasons. The effects of unbalanced cluster sizes on the variance of the treatment effect estimator in cluster randomized trials are studied by Manatunga et al. [28], Kerry and Bland [20], and Van Breukelen et al. [63]. The assumption of equal costs across treatment conditions may also be unrealistic, and optimal sample size formulae for varying costs per treatment condition have been published by Liu [26].

References

1. D. Afshartous. Determination of sample size for multilevel model design. Unpublished manuscript, 1995.
2. A. C. Atkinson and A. N. Donev. *Optimum Experimental Designs.* Clarendon Press, Oxford, UK, 1996.
3. M. P. F. Berger and F. E. S. Tan. Robust designs for linear mixed effects models. *Applied Statistics*, 53:569–581, 2004.
4. R. J. Bosker, T. A. B. Snijders, and H. Guldemond. *PINT: Estimating Standard Errors of Regression Coefficients in Hierarchical Linear Models for Power Calculations. User's Manual Version 1.6.* University of Twente, Enschede, The Netherlands, 1999.
5. W. G. Cochran. *Planning and Analysis of Observational Studies.* Wiley, New York, 1983.

6. J. Cohen. A power primer. *Psychological Bulletin*, 112:155–159, 1992.
7. M. P. Cohen. Determining sample sizes for surveys with data analyzed by hierarchical linear models. *Journal of Official Statistics*, 14:267–275, 1998.
8. A. Donner. A review of inference procedures for the intraclass correlation coefficientin the one-way random effects model. *International Statistical Review*, 54:67–82, 1986.
9. A. Donner. Some aspects of the design and analysis of cluster randomization trials. *Applied Statistics*, 47:95–113, 1998.
10. A. Donner and N. Klar. *Design and Analysis of Cluster Randomization Trials in Health Research.* Edward Arnold, London, 2000.
11. H. A. Feldman and S. M. McKinlay. Cohort versus cross-sectional design in larage field trials: Precision, sample size, and a unifying model. *Statistics in Medicine*, 13:61–78, 1994.
12. Z. Feng and J. E. Grizzle. Correlated binomial variates: Properties of estimator of intraclass correlation and its effect on sample size calculation. *Statistics in Medicine*, 11:1607–1614, 1992.
13. S. Galbraith and I. C. Marschner. Guidelines for the design of clinical trials with longitudinal outcomes. *Controlled Clinical Trials*, 23:257–273, 2002.
14. H. Goldstein. Nonlinear multilevel models, with an application to discrete response data. *Biometrika*, 78:45–51, 1991.
15. H. Goldstein and J. Rasbash. Improved approximations for multilevel models with binary responses. *Journal of the Royal Statistical Society, Series A*, 159: 505–513, 1996.
16. D. Hedeker and R. D. Gibbons. A random-effects ordinal regression model for multilevel analysis. *Biometrics*, 50:933–944, 1994.
17. D. Hedeker, R. D. Gibbons, and C. Waternaux. Sample size estimation for longitudinal designs with attrition: Comparing time-related contrasts between two groups. *Journal of Educational and Behavioral Statistics*, 24:70–93, 1999.
18. F. Y. Hsieh. Sample size formulae for intervention studies with the cluster as unit of randomization. *Statistics in Medicine*, 8:1195–1201, 1988.
19. M. Kendall and A. Stuart. *The Advanced Theory of Statistics*, 4th edition, volume 2. Griffin, London, 1979.
20. S. M. Kerry and J. M. Bland. Unequal cluster sizes for trials in English and Welsh general practice: Implications for sample size calculations. *Statistics in Medicine*, 20:377–390, 2001.
21. N. M. Laird and F. Wang. Estimating rates of change in randomized clinical trials. *Controlled Clinical Trials*, 11:405–419, 1990.
22. S. Lake, E. Kammann, K. Klar, and R. A. Betensky. Sample size re-estimation in cluster randomization trials. *Statistics in Medicine*, 21:1337–1350, 2002.
23. E. W. Lee and N. Dubin. Estimation and sample size consideration for clustered binary responses. *Statistics in Medicine*, 13:1241–1252, 1994.
24. S. R. Lipsitz and M. Parzen. Sample size calculations for non-randomized studies. *The Statistician*, 44:81–90, 1995.
25. G. Liu and K.-Y. Liang. Sample size calculation for studies with correlated observations. *Biometrics*, 53:937–947, 1997.

26. X. Liu. Statistical power and optimum sample allocation ratio for treatment and control having unequal costs per unit of randomization. *Journal of Educational and Behavioral Statistics*, 28:231–248, 2003.
27. N. T. Longford. *Random Coefficient Models.* Oxford University Press, Oxford, UK, 1993.
28. A. K. Manatunga, M. G. Hudgens, and S. Chen. Sample size estimation in cluster randomized studies with varying cluster size. *Biometrical Journal*, 43: 75–86, 2001.
29. S. M. McKinlay. Cost-efficient designs of cluster unit trials. *Preventive Medicine*, 23:606–611, 1994.
30. M. Moerbeek. *Design and Analysis of Multilevel Intervention Studies.* PhD thesis, Maastricht University, Maastricht, 2000.
31. M. Moerbeek. The use of internal pilot studies to derive powerful and cost-efficient designs for studies with nested data. In *2004 Proceedings of the American Statistical Association.* American Statistical Association, Alexandria, VA, 2004.
32. M. Moerbeek. Randomization of clusters versus randomization of persons within clusters: Which is preferable? *The American Statistician*, 59:72–78, 2005.
33. M. Moerbeek. Powerful and cost-efficient designs for longitudinal intervention studies with two treatment groups. *Journal of Educational and Behavioral Statistics*, forthcoming.
34. M. Moerbeek, G. J. P. Van Breukelen, and M. P. F. Berger. Design issues for experiments in multilevel populations. *Journal of Educational and Behavioral Statistics*, 25:271–284, 2000.
35. M. Moerbeek, G. J. P. Van Breukelen, and M. P. F. Berger. Optimal experimental designs for multilevel logistic models. *The Statistician*, 50:1–14, 2001.
36. M. Moerbeek, G. J. P. Van Breukelen, and M. P. F. Berger. Optimal experimental designs for multilevel models with covariates. *Communications in Statistics, Theory and Methods*, 30:2683–2697, 2001.
37. M. Moerbeek, G. J. P. Van Breukelen, and M. P. F. Berger. A comparison between traditional methods and multilevel regression for the analysis of multicenter intervention studies. *Journal of Clinical Epidemiology*, 56:341–350, 2003.
38. M. Moerbeek, G. J. P. Van Breukelen, and M. P. F. Berger. A comparison of estimation methods for multilevel logistic models. *Computational Statistics*, 18: 19–37, 2003.
39. M. Moerbeek, G. J. P. Van Breukelen, and M. P. F. Berger. Optimal sample sizes in experimental designs with individuals nested within clusters. *Understanding Statistics*, 2:151–175, 2003.
40. M. Moerbeek and W. K. Wong. Multiple-objective optimal designs for the hierarchical linear model. *Journal of Official Statistics*, 18:291–303, 2002.
41. M. Mok. Sample size requirements for 2-level designs in educational research. *Multilevel Modelling Newsletter*, 7:11–16, 1996.
42. D. M. Murray. *Design and Analysis of Group-Randomized Trials.* Oxford University Press, New York, 1998.

43. D. M. Murray, S. P. Varnell, and J. L. Blitstein. Design and analysis of group-randomized trials: A review of recent methodological developments. *Public Health Matters*, 94:423–432, 2004.
44. B. O. Muthén and P. J. Curran. General longitudinal modeling of individual differences in experimental designs: A latent variable framework for analysis and power estimation. *Psychological Methods*, 2:371–402, 1997.
45. L. K. Muthén and B. O. Muthén. How to use a Monte Carlo study to decide on sample size and determine power. *Structural Equation Modeling*, 9:599–620, 2002.
46. L. K. Muthén and B. O. Muthén. *Mplus User's Guide.* Muthén and Muthén, Los Angeles, 2004.
47. J. M. Neuhaus and J. D. Kalbfleisch. Between- and within-cluster covariate effects in the analysis of clustered data. *Biometrics*, 54:638–645, 1998.
48. J. Rasbash, F. Steele, W. J. Browne, and B. Prosser. *A User's Guide to MLwiN. Version 2.0.* Centre for Multilevel Modelling, University of Bristol, Bristol, UK, 2005.
49. S. W. Raudenbush. Statistical analysis and optimal design for cluster randomized trials. *Psychological Methods*, 2:173–185, 1997.
50. S. W. Raudenbush and X. Liu. Statistical power and optimal design for multisite randomized trials. *Psychological Methods*, 5:199–213, 2000.
51. S. W. Raudenbush and X. Liu. Effects of study duration, frequency of observation, and sample size on power in studies of group differences in polynomial change. *Psychological Methods*, 6:387–401, 2001.
52. S. W. Raudenbush, J. Spybrook, X. Liu, and R. Congdon. *Optimal Design for Longitudinal and Multilevel Research: Documentation for the Optimal Design Software.* University of Michigan, Ann Arbor, 2004.
53. G. Rodríguez and N. Goldman. An assessment of estimation procedures for multilevel models with binary responses. *Journal of the Royal Statistical Society, Series A*, 158:73–89, 1995.
54. G. Rodríguez and N. Goldman. Improved estimation procedures for multilevel models with binary response: A case-study. *Journal of the Royal Statistical Society, Series A*, 164:339–355, 2001.
55. W. J. Shih. Sample size and power calculations for periodontal and other studies with clustered samples using the method of generalized estimation equations. *Biometrical Journal*, 39:899–908, 1997.
56. S. D. Silvey. *Optimal Design.* Chapman & Hall, London, 1980.
57. T. A. B. Snijders and R. J. Bosker. Standard errors and sample sizes for two-level research. *Journal of Educational Statistics*, 18:237–259, 1993.
58. T. A. B. Snijders and R. J. Bosker. Modeled variance in two-level models. *Sociological Methods & Research*, 22:342–363, 1994.
59. D. J. Spiegelhalter. Bayesian methods for cluster randomized trials with continuous responses. *Statistics in Medicine*, 20:435–452, 2001.
60. D. J. Spiegelhalter, A. Thomas, N. G. Best, and D. Lunn. *WinBUGS User Manual, Version 1.4.* MRC Biostatistics Unit, University of Cambridge, Cambridge, UK, 2003.

61. R. M. Turner, A. T. Prevost, and S. G. Thompson. Allowing for imprecision of the intracluster correlation coefficient in the design of cluster randomized trials. *Statistics in Medicine*, 23:1195–1214, 2004.
62. G. J. P. Van Breukelen. ANCOVA versus change from baseline: more power in randomized studies, more bias in nonrandomized studies. *Journal of Clinical Epidemiology*, 59:920–925, 2006.
63. G. J. P. Van Breukelen, M. J. J. M. Candel, and M. P. F. Berger. Relative efficiency of unequal *versus* equal cluster sizes in cluster randomized and multicentre trials. *Statistics in Medicine*, 26:2589–2603, 2007.
64. B. Winkens, H. J. A. Schouten, G. J. P. Van Breukelen, and M. P. F. Berger. Optimal time-points in clinical trials with linearly divergent treatment effects. *Statistics in Medicine*, 24:3743–3756, 2005.

5

Many Small Groups

Stephen W. Raudenbush

University of Chicago, Department of Sociology

5.1 Introduction

Hierarchical data from many small clusters arise by necessity and by design. They arise by necessity when the aim is to study married couples [1], identical twins [25], siblings [12], paired comparison tasks [2], cooperative learning groups [36], multiple informants of child social behavior [20], and studies of animal reproduction [35]. They arise by design in cross-sectional studies: cluster randomized trials [11, 18], multisite randomized trials [3, 6], and surveys that sample a small number of persons in each of many neighborhoods [14] or a small number of teachers in each of many schools [17]. In repeated measures studies, it is common to encounter small numbers of observations for each of many persons in short time-series designs, such as studies of student learning based on annual assessments [37], the extreme case being a pre-post design.

In my experience teaching methods for multilevel data, students and other workshop participants have often expressed dismay that their data involve many clusters but few cluster members. However, there are often good reasons for such design choices. If the primary aim of a study is to estimate fixed regression coefficients (as opposed to variance components or realizations of random effects), a design that minimizes cluster size, n, and maximizes the number of clusters, J, may be optimal (cf. Chapter 4 in this volume; also [7, 26, 30, 38]). Optimal n per cluster depends on the cost of sampling at each level, the magnitude of variation at each level, and research question at hand. Choosing a small n is wise when little variability exists within clusters or when it is comparatively expensive to assess each individual within a cluster (relative to the cost of sampling clusters).

Yet, under certain conditions, the "small n, large J" scenario can pose challenges to valid statistical inference and can create demanding computational

J. de Leeuw, E. Meijer (eds.), *Handbook of Multilevel Analysis*,
© Springer 2008

tasks as well as problems of statistical precision. The problems are likely to be less challenging in the case of linear models with normal random effects at each level and more challenging when non-linear link functions and non-normal data are involved. These problems are likely to be less challenging when the aim is to estimate fixed regression coefficients, and more challenging when the aim is to draw inferences about random regression coefficients (e.g., cluster-specific intercepts and slopes) or to estimate variance and covariance components at the second level of the hierarchy. I provide a brief overview of each scenario before considering each in more detail.

5.1.1 Linear Models, Normal Random Effects

Consider a two-level setting with J clusters and n members per cluster. One aim might be to study fixed regression coefficients. A second aim might be to study random coefficients, that is, cluster-specific coefficients defined as randomly varying over clusters. A third aim might entail inference about variation and covariation in such random coefficients defined as a universe of clusters. How the "small n, large J design" fares will depend on which of these three aims is of central interest in a given study.

For linear models with normal random effects at each level, having "small n and large J" generally creates no problems in statistical inference in estimating fixed regression coefficients. Such a design will be inefficient if it is far more expensive to sample clusters than to sample cluster members, especially if variation within clusters is large relative to variation between clusters. In other cases, such a design may be optimal. Either way, inferences about regression coefficients proceeds smoothly, and it is a simple matter to compute consistent and robust standard errors as a check on the sensitivity of inferences to model assumptions.

When the aim is to estimate cluster-specific intercepts or slopes, "the small n, large J" strategy tends to be more problematic unless the fit of the model at level 1 is very good. Holding constant the fit of the model, the optimal sample size per cluster for estimating random coefficients and second-level variance components will tend to be larger than when the aim is to estimate fixed regression coefficients. In part, the difficulty is simply one of obtaining adequate precision with available resources. However, a more subtle problem is that the likelihood for the second-level variance will sometimes tend to be skewed even if J is quite large. In this scenario, the maximum likelihood (ML) estimator may poorly represent the plausible values of the variance, and inferences based on large-sample normal theory for the ML estimator can be misleading. Moreover, empirical Bayes estimates [cf. 23] of cluster-specific intercepts (or slopes), which condition on the ML estimator of the second-level variances, may be accompanied by negatively biased standard errors.

In general, the researcher must keep the complexity of the model of the covariance structure at level-2 in check when n per cluster is small. In essence, the lack of data at level 1 requires the imposition of more assumptions (e.g., that certain slopes don't vary). The availability of robust standard errors minimizes the impact of these assumptions on inferences about the fixed regression coefficients, but this insurance does not apply to inferences about the random coefficients or the variance-covariance components.

5.1.2 Non-Linear Links and Non-Normal Random Effects

As mentioned, the problems that can afflict variance estimation in the "small n, large J" scenario do not seriously affect inference about the fixed regression coefficients in the case of linear models and normal random effects. This happy result, which derives from the asymptotic orthogonality of the variance estimates and the mean structure estimates under normality and linearity, does not extend to the case of non-linear link functions and non-normal random effects. For these models, beliefs about level-2 variability have potentially strong implications for beliefs about the mean as characterized by regression coefficients. The sensitivity of the fixed regression coefficients to inferences about variances is more pronounced under unit-specific than population-average models [15].

Moreover, concerns about inferences for random coefficients and variance components, mentioned above in the case of linear models and normal random effects, are, if anything, more pronounced in the case of non-linear models and non-normal random effects. Discrete data generally carry less information per cluster, holding constant n, than do continuous data. This tendency is especially pronounced when the outcome data are highly skewed; examples involve binary outcomes with small probabilities of occurrence and counts based on low event rates.

Finally, a computational problem arises in the non-linear case that is not present in the linear-normal case. In general, likelihood-based inference for hierarchical models requires integration of the random effects from the joint distribution of the random effects and the observed data. The required integral is available in closed form in the linear-normal case. In the non-linear and non-normal case, the integral is not available in closed form and must be approximated using numerical or Monte Carlo methods (see Chaps. 6 and 9 in this volume). If n per cluster is sufficiently large, approximating the integral is comparatively easy because the integrand tends toward normality. When n is small, the integration problem is more challenging, though this problem is clearly soluble given current knowledge and technology.

5.1.3 Focus and Organization of This Chapter

Attention is confined in this chapter to two-level models where the level-1 outcome, conditional on the random effects, is distributed according to an exponential family with canonical link function (e.g., continuous outcomes with identity link, binary outcomes with a logit link; Poisson-distributed count data with a log link). Some or all of the coefficients at level 1 vary over level-2 units according to a multivariate normal distribution. Although model checking and robust variance estimation are essential, assuming normality at level 2 is useful for planning research and for considering the issues that arise in the "small n, large J" scenario, the focus I have been assigned in this chapter.

I will generally be concerned with likelihood-based inference. This includes inference based on Bayesian methods, which converge to likelihood-based inference in the case of large J. I will comment briefly on the added value of the Bayesian perspective in certain contexts, but I refer the reader to Chapter 2 for a thorough discussion of that perspective.

After this introduction, the second section considers the model. The third section considers how the model might be tailored to specific applications when n per cluster is small by necessity or by design. The fourth section considers statistical issues that arise in linear models with normal random effects. The fifth section considers the additional issues that arise in the non-linear and non-normal case.

5.2 The Model

The linear models I will discuss have the form

$$\underline{\boldsymbol{y}}_j = \boldsymbol{X}_j\boldsymbol{\beta} + \boldsymbol{Z}_j\,\underline{\boldsymbol{\delta}}_j + \underline{\epsilon}_j\,, \tag{5.1}$$

where

- $\underline{\boldsymbol{y}}_j[n_j, 1]$ is the vector of outcomes with elements $\underline{y}_{ij}$, the outcome for the i-th level-1 unit within the j-th level-2 unit;
- $\boldsymbol{X}_j[n_j, f]$ is a known matrix of predictors associated with the fixed effects vector $\boldsymbol{\beta}[f, 1]$;
- $\boldsymbol{Z}_j[n_j, r]$ is a known matrix of predictors associated with the random effects vector $\underline{\boldsymbol{\delta}}_j[r, 1]$; and
- $\underline{\epsilon}_j[n_j, 1]$ is a vector of level-1 random effects having elements $\underline{\epsilon}_{ij}$.

The indices thus identify level-1 units $i = 1, \ldots, n_j$ nested within level-2 units $j = 1, \ldots, J$. In many applications, we will have n_j people nested within cluster j, but in some cases, the level-1 units will be repeated measurements nested within people.

For the expository purposes of this chapter, we will assume the level-1 random effects $\underline{\epsilon}_{ij}$ to be independently and identically distributed as $\mathcal{N}(0, \sigma^2)$ unless otherwise specified. These are also independent of the level-2 random effects vectors $\underline{\boldsymbol{\delta}}_j, j = 1, \ldots, J$, which are independently and identically distributed as r-variate $\mathcal{N}(\boldsymbol{\emptyset}, \boldsymbol{\Omega})$.

Equation (5.1) is the "mixed model" formulation. In specifying this model, it is often conceptually appealing to build it level by level. Thus, we will write the level-1 model as

$$\underline{\boldsymbol{y}}_j = \boldsymbol{U}_j \underline{\boldsymbol{\beta}}_j + \underline{\boldsymbol{\epsilon}}_j, \tag{5.2}$$

where $\boldsymbol{U}_j[n_j, p]$ is the matrix of level-1 predictors having rows $\boldsymbol{U}'_{ij}[1, p]$, and $\underline{\boldsymbol{\beta}}_j[p, 1]$ is a vector of level-1 coefficients. At level-2, the level-1 coefficients become outcomes:

$$\underline{\boldsymbol{\beta}}_j = \boldsymbol{H}_j \gamma + \underline{\boldsymbol{\delta}}_j, \tag{5.3}$$

where $\boldsymbol{H}_j[p, f]$ is the matrix of level-2 predictors and γ is the $f \times 1$ vector of level-2 coefficients. Substituting (5.3) into (5.2) produces a combined model

$$\underline{\boldsymbol{y}}_j = \boldsymbol{U}_j \boldsymbol{H}_j \gamma + \boldsymbol{U}_j \underline{\boldsymbol{\delta}}_j + \underline{\boldsymbol{\epsilon}}_j, \tag{5.4}$$

which is clearly a special case of the mixed model (5.1) with

$$\boldsymbol{X}_j = \boldsymbol{U}_j \boldsymbol{H}_j; \qquad \boldsymbol{Z}_j = \boldsymbol{U}_j.$$

The mixed model formulation is more general than the combined model (5.4) because (5.4) requires every level-1 coefficient to have a random effect at level 2 ($p = r$). However, the structure of the combined model (5.4) is quite useful for expository purposes. Particularly if we can assume $\boldsymbol{U}_j$ to be of full column rank p, we can gain insight by writing the combined model using the ordinary least squares estimator as the outcome:

$$(\boldsymbol{U}'_j \boldsymbol{U}_j)^{-1} \boldsymbol{U}'_j \underline{\boldsymbol{y}}_j = \hat{\underline{\boldsymbol{\beta}}}_j = \boldsymbol{H}_j \gamma + \underline{\boldsymbol{\delta}}_j + \underline{\bar{\boldsymbol{\epsilon}}}_j\,, \tag{5.5}$$

where $\underline{\bar{\boldsymbol{\epsilon}}}_j \sim \mathcal{N}[\boldsymbol{\emptyset}, \sigma^2 (\boldsymbol{U}'_j \boldsymbol{U}_j)^{-1}]$. Note that $\boldsymbol{U}'_j \boldsymbol{U}_j = \sum_{i=1}^{n_j} \boldsymbol{U}_{ij} \boldsymbol{U}'_{ij} \equiv n_j \boldsymbol{\Sigma}_u$. Note further that, given $\underline{\boldsymbol{\beta}}_j = \boldsymbol{\beta}_j$, the variance covariance matrix of $\hat{\underline{\boldsymbol{\beta}}}_j$ is

$$\text{Var}(\hat{\underline{\boldsymbol{\beta}}}_j \mid \underline{\boldsymbol{\beta}}_j = \boldsymbol{\beta}_j) = \frac{\sigma^2}{n_j} \boldsymbol{\Sigma}_u^{-1}.$$

This conditional variance will be small if (a) n_j is large; (b) σ^2, which measures the misfit of the level-1 model, is small; or (c) $\boldsymbol{\Sigma}_u$, the dispersion of the level-1 predictors, $\boldsymbol{U}_{ij}$, is large.

In clarifying how the level-1 design affects precision of estimation of model parameters, a useful concept is the multivariate reliability matrix

$$\boldsymbol{\Lambda}_j = \text{Cov}(\underline{\boldsymbol{\beta}}_j, \hat{\underline{\boldsymbol{\beta}}}_j)\left[\text{Var}(\hat{\underline{\boldsymbol{\beta}}}_j)\right]^{-1} = \boldsymbol{\Omega}\boldsymbol{\Delta}_j^{-1} = \boldsymbol{\Omega}\left(\boldsymbol{\Omega} + \frac{\sigma^2}{n_j}\boldsymbol{\Sigma}_u^{-1}\right)^{-1}. \quad (5.6)$$

Equation (5.6) defines a matrix of regression coefficients that emerge when the true random coefficients $\underline{\boldsymbol{\beta}}_j$ are regressed on the least squares estimates $\hat{\underline{\boldsymbol{\beta}}}_j$. Note that, holding constant $\boldsymbol{\Omega}$, $\boldsymbol{\Lambda}_j$ converges to the identity matrix $\boldsymbol{I}_r$ when (a) n_j becomes large, (b) σ^2 becomes small, or (c) $\boldsymbol{\Sigma}_u$ becomes large, meaning that $\boldsymbol{\Sigma}_u^{-1}$ converges to the null matrix.

The combined model (5.5) creates a useful framework within which we can study the properties of estimators of the three quantities of interest: the fixed regression coefficients, γ; the random coefficients, $\underline{\boldsymbol{\beta}}_j$, $j = 1, \ldots, J$; and the variance-covariance components, $\boldsymbol{\Omega}$.

5.2.1 Fixed Regression Coefficients

The variance-covariance matrix of $\hat{\underline{\boldsymbol{\beta}}}_j$, the outcome of (5.5), is

$$\text{Var}(\hat{\underline{\boldsymbol{\beta}}}_j) = \text{Var}(\underline{\boldsymbol{\delta}}_j + \underline{\bar{\boldsymbol{\epsilon}}}_j) = \boldsymbol{\Omega} + \frac{\sigma^2}{n_j}\boldsymbol{\Sigma}_u = \boldsymbol{\Delta}_j.$$

This leads immediately to the generalized least squares estimator

$$\hat{\underline{\gamma}} = \left(\sum_{j=1}^{J} \boldsymbol{H}_j' \boldsymbol{\Delta}_j^{-1} \boldsymbol{H}_j\right)^{-1} \sum_{j=1}^{J} \boldsymbol{H}_j' \boldsymbol{\Delta}_j^{-1} \hat{\underline{\boldsymbol{\beta}}}_j, \quad (5.7)$$

which has as its variance matrix

$$\text{Var}(\hat{\underline{\gamma}}) = \left(\sum_{j=1}^{J} \boldsymbol{H}_j' \boldsymbol{\Delta}_j^{-1} \boldsymbol{H}_j\right)^{-1}. \quad (5.8)$$

Equation (5.7) assumes $\boldsymbol{\Delta}_j$ to be known. In practice, it will equated to its ML or restricted ML estimator [see 5, Chapter 10].

A useful re-expression for (5.8) is

$$\text{Var}(\hat{\underline{\gamma}}) = \left(\sum_{j=1}^{J} \boldsymbol{H}_j' \boldsymbol{\Omega}^{-1} \boldsymbol{\Lambda}_j \boldsymbol{H}_j\right)^{-1}. \quad (5.9)$$

As (5.9) shows, the weight accorded each unit j in the estimation of γ is proportional to its reliability matrix $\boldsymbol{\Lambda}_j$.

Example 1. Suppose $\boldsymbol{H}_j$ is the identity matrix. Thus, γ is the population mean of $\underline{\boldsymbol{\beta}}_j$. Then the information in the data about γ is the precision of $\hat{\underline{\gamma}}$, that is, the inverse of its variance

$$[\mathrm{Var}(\underline{\hat{\gamma}})]^{-1} = \boldsymbol{\Omega}^{-1}\sum_{j=1}^{J}\boldsymbol{\Lambda}_j. \tag{5.10}$$

Increasing n_j will increase the information about $\boldsymbol{\gamma}$ only by pushing $\boldsymbol{\Lambda}_j$ toward $\boldsymbol{I}_r$. If $\boldsymbol{\Lambda}_j$ is already near $\boldsymbol{I}_r$ because σ^2 is small or $\boldsymbol{\Sigma}_u$ is large, increasing n_j will add little to the information about $\boldsymbol{\gamma}$. However, if σ^2 is appreciable and $\boldsymbol{\Sigma}_u$ is modest, and especially if it is comparatively inexpensive to increase n_j, doing so may add significantly to the information about $\boldsymbol{\gamma}$ at small cost.

Example 2. Suppose further that $\underline{\beta}_j$ is univariate and is, in fact, the mean of $\underline{y}$ in cluster j. Thus, $\underline{\hat{\beta}}_j$ is the sample mean $\underline{\overline{y}}_j$. Then (5.10) becomes

$$[\mathrm{Var}(\underline{\hat{\gamma}})]^{-1} = \sum_{j=1}^{J}\frac{\lambda_j}{\omega},$$

where $\omega = \mathrm{Var}(\underline{\beta}_j)$ and

$$\lambda_j = \frac{\omega}{\omega + \sigma^2/n_j}.$$

Here, λ_j is the ratio of the variance ω_j of the "true mean" $\underline{\beta}_j$ to the variance of its estimator, the sample mean $\underline{\overline{y}}_j$. We will use this expression several times in later discussions.

5.2.2 Random Regression Coefficients

The conditional mean of the random effect $\underline{\boldsymbol{\delta}}_j$ given the data $\underline{\boldsymbol{y}} = \boldsymbol{y}$ (and thus $\underline{\hat{\boldsymbol{\beta}}}_j = \hat{\boldsymbol{\beta}}_j$) and the parameters $(\boldsymbol{\gamma}, \sigma^2, \boldsymbol{\Omega})$ is

$$E(\underline{\boldsymbol{\delta}}_j \mid \boldsymbol{y}, \boldsymbol{\gamma}, \boldsymbol{\Omega}, \sigma^2) = \boldsymbol{\delta}_j^* = \boldsymbol{\Lambda}_j(\hat{\boldsymbol{\beta}}_j - \boldsymbol{H}_j\boldsymbol{\gamma}). \tag{5.11}$$

When ML estimates are substituted for the unknown parameters in (5.11), $\boldsymbol{\delta}_j^*$ is the empirical Bayes posterior mean commonly used as a point estimate of the unknown random effect $\underline{\boldsymbol{\delta}}_j$. Note that this posterior mean is simply the least squares residual $\hat{\boldsymbol{\beta}}_j - \boldsymbol{H}_j\boldsymbol{\gamma}$ "shrunk" toward a mean vector of zero. The amount of shrinkage is large when $\boldsymbol{\Lambda}_j$ is small, that is, when the least squares estimator $\underline{\hat{\boldsymbol{\beta}}}_j$ is unreliable, as will be the case when n_j is small unless the level-1 model fits the data well so that σ^2 is small or unless $\boldsymbol{\Sigma}_u$ is large. If one wishes to estimate the coefficient $\boldsymbol{\beta}_j$ rather than the random effect $\underline{\boldsymbol{\delta}}_j$, the corresponding expression is

$$E(\underline{\boldsymbol{\beta}}_j \mid \boldsymbol{y}, \boldsymbol{\gamma}, \boldsymbol{\Omega}, \sigma^2) = \boldsymbol{\beta}_j^* = \boldsymbol{\Lambda}_j\hat{\boldsymbol{\beta}}_j + (\boldsymbol{I}_r - \boldsymbol{\Lambda}_j)\boldsymbol{H}_j\boldsymbol{\gamma}. \tag{5.12}$$

This is the well-known weighted average of the data-based estimate $\hat{\boldsymbol{\beta}}_j$ and the prior mean $\boldsymbol{H}_j\boldsymbol{\gamma}$. Large weight is accorded the data-based estimater when

$\boldsymbol{\Lambda}_j$ is near $\boldsymbol{I}_r$, that is, the least squares estimator is highly reliable. Large weight is accorded the prior mean otherwise.

Given the parameters, the posterior variance of the random effect and the random coefficient are equal, that is,

$$\mathrm{Var}(\underline{\boldsymbol{\delta}}_j \mid \boldsymbol{y}, \sigma^2, \boldsymbol{\Omega}, \gamma) = \boldsymbol{\Omega}(\boldsymbol{I}_r - \boldsymbol{\Lambda}_j).$$

Thus, the posterior uncertainty about the random effect is the product of the prior uncertainty $\boldsymbol{\Omega}$ and $\boldsymbol{I}_r - \boldsymbol{\Lambda}_j$. If we consider $\boldsymbol{y}_j$ to constitute the observed data while $\boldsymbol{\delta}_j$ constitutes the missing data, we can define $\boldsymbol{I}_r - \boldsymbol{\Lambda}_j$ as the fraction of missing information in cluster j.

5.2.3 Variance-Covariance Components

For simplicity, let us assume that σ^2 is known. Given large J, the estimate of σ^2 will be precise in any case. At each iteration, the Fisher scoring estimate of $\boldsymbol{\Omega}$ will then be equal to the iterative generalized least squares estimator [see 29, Chapter 14]

$$\mathrm{vech}(\underline{\hat{\boldsymbol{\Omega}}}) = \left(\sum_{j=1}^{J} (\boldsymbol{X}^*)'(\boldsymbol{V}_j^*)^{-1}\boldsymbol{X}^* \right)^{-1} \sum_{j=1}^{J} (\boldsymbol{X}^*)'(\boldsymbol{V}_j^*)^{-1}\underline{\boldsymbol{Y}}_j^*, \tag{5.13}$$

where vech(·) denotes the vector of unique elements of a matrix, and

$$\begin{aligned} \boldsymbol{X}^* &= \frac{\partial\, \mathrm{vec}(\boldsymbol{\Omega})}{\partial(\mathrm{vech}(\boldsymbol{\Omega}))'}, \\ \boldsymbol{V}_j^* &= 2(\boldsymbol{\Delta}_j \otimes \boldsymbol{\Delta}_j), \\ \underline{\boldsymbol{Y}}_j^* &= \mathrm{vec}\left[(\underline{\hat{\boldsymbol{\beta}}}_j - \boldsymbol{H}_j\gamma)(\underline{\hat{\boldsymbol{\beta}}}_j - \boldsymbol{H}_j\gamma)' - \sigma^2(\boldsymbol{U}_j'\boldsymbol{U}_j)^{-1}\right]. \end{aligned}$$

Each term on the right side of (5.13) is evaluated at the parameter estimates from the previous iteration. The asymptotic variance matrix at convergence is the inverse of the expected information

$$\begin{aligned} \mathrm{Var}\left[\mathrm{vech}(\underline{\hat{\boldsymbol{\Omega}}})\right] &\approx \left(\sum_{j=1}^{J} (\boldsymbol{X}^*)'(\boldsymbol{V}_j^*)^{-1}\boldsymbol{X}^* \right)^{-1} \\ &= 2\left\{ (\boldsymbol{X}^*)' \left[\sum_{j=1}^{J} (\boldsymbol{\Delta}_j^{-1} \otimes \boldsymbol{\Delta}_j^{-1}) \right] \boldsymbol{X}^* \right\}^{-1} \\ &= 2\Big((\boldsymbol{X}^*)' (\boldsymbol{\Omega}^{-1} \otimes \boldsymbol{\Omega}^{-1}) \sum_{j=1}^{J} (\boldsymbol{\Lambda}_j \otimes \boldsymbol{\Lambda}_j)\, \boldsymbol{X}^* \Big)^{-1}. \end{aligned} \tag{5.14}$$

Once again, the multivariate reliability $\boldsymbol{\Lambda}_j$ of $\underline{\hat{\boldsymbol{\beta}}}_j$ plays a central role in understanding how n_j affects precision. This becomes clear in a simple example.

Example 3. Once again, let us consider the case in which $\underline{\beta}_j$ is a scalar, the mean of cluster j, so that $\boldsymbol{\Omega} = \omega$ is also a scalar. Then the Fisher information for ω is the inverse of (5.14), which becomes

$$[\text{Var}(\underline{\hat{\omega}})]^{-1} = \tfrac{1}{2} \sum_{j=1}^{J} \lambda_j^2/\omega^2,$$

where λ_j and ω are evaluated at the MLE. Thus, the information contained in each level-1 unit about the level-2 variance is the sum of squared reliability coefficients. It is because this sum of squares is likely to be small when the typical n is small that small n can sharply undermine precision of the estimation of the level-2 variance, even when J is fairly large.

5.3 Some Applications

The "small n, large J" setting arises by necessity and by design. It arises by necessity when the object is to study twins, siblings, or married couples, and when repeated measures studies by necessity involve few time points. It arises by design when "small n, large J" is desirable for statistical efficiency or cost considerations.

5.3.1 Small n, Large J of Necessity

Matched Pair Designs

Two types of matched pair designs may be distinguished: those in which pair members are exchangeable and those in which pair members are always distinguished by an observed characteristic. A paradigm case of exchangeability involves studies of twins. Although twin members may differ by gender, they often do not. A second case involves randomly selected pairs of observers chosen to assess a person or some other entity such as a classroom. In contrast, many other matched pair designs involve non-exchangeable pair members. Examples include pre-post designs, studies of heterosexual couples, or experiments in which one pair member is assigned randomly to an experimental group and a second is assigned to a control. The analytic model will be generally different in these two cases.

Exchangeable Pair Members

We might begin with a simple unconditional model

$$\underline{y}_{ij} = \beta_0 + \underline{\delta}_{0j} + \underline{\epsilon}_{ij}, \tag{5.15a}$$

$$\underline{\delta}_{0j} \overset{iid}{\sim} \mathcal{N}(0, \omega_{00}), \tag{5.15b}$$

$$\underline{\epsilon}_{ij} \overset{iid}{\sim} \mathcal{N}(0, \sigma^2), \tag{5.15c}$$

for $i = 1, 2$ and $j = 1, \ldots, J$. Here the i subscript is exchangeable. Given balanced data, the ML estimates of the population mean β_0, the within-pair variance σ^2, and between-pair variance ω_{00}, have an interesting structure:

$$\hat{\underline{\beta}}_0 = \frac{1}{2J} \sum_{j=1}^{J} \sum_{i=1}^{2} \underline{y}_{ij},$$

$$\hat{\underline{\sigma}}^2 = \frac{1}{2J} \sum_{j=1}^{J} (\underline{y}_{1j} - \underline{y}_{2j})^2,$$

$$\hat{\underline{\omega}}_{00} = \max \left\{ \frac{1}{J} \sum_{j=1}^{J} (\underline{y}_{1j} - \hat{\underline{\beta}}_0)(\underline{y}_{2j} - \hat{\underline{\beta}}_0), \quad 0 \right\}.$$

Thus, the between-pair variance estimate is the sample covariance between pair members while the within-pair variance estimate is the average squared difference between pair members.

The variance of the mean estimate depends on the reliability $\lambda = \omega_{00}/(\omega_{00} + \sigma^2/2)$ by

$$\mathrm{Var}(\hat{\underline{\beta}}_0) = \frac{\omega_{00}}{J\lambda}.$$

Clearly, increasing n_j is not an option here. If within-pair differences are small, λ may still be near 1.0, restricting how large J must be to obtain adequate precision.

In a similar vein, the variance of the between-pair variance estimate also depends strongly on λ. When σ^2 is unknown, we have (for $\omega_{00} > 0$)

$$\mathrm{Var}(\hat{\underline{\omega}}_{00}) = \frac{2\omega_{00}^2}{(J-1)\lambda^2} \left[1 + (1 - \lambda^2) \frac{J-1}{J(n-1)} \right].$$

In studies that compare monozygotic to dizygotic twins, one might compute the correlation $\hat{\rho} = \hat{\omega}_{00}/(\hat{\omega}_{00} + \hat{\sigma}^2)$ for each group. One might also compute a single model that constrains the means to be equal but allows the variance components to differ for the two types of twins. Other twin types of interest might be same-gender, both-male, or both-female pairs.

A likely goal in twin designs or other sibling designs is to compare pair members who have experienced some different treatment or environment. Such a design eliminates unobserved heterogeneity between twin pairs in the evaluation of causal effects. The model can easily be elaborated to include within-pair and between-pair covariates:

$$\underline{y}_{ij} = \beta_0 + \sum_{p=1}^{P} U_{pij}\beta_p + \sum_{q=1}^{Q} H_{qj}\beta_{P+q} + \underline{\delta}_{0j} + \underline{\epsilon}_{ij},$$
$$\underline{\delta}_{0j} \overset{iid}{\sim} \mathcal{N}(0, \omega_{00}),$$
$$\underline{\epsilon}_{ij} \overset{iid}{\sim} \mathcal{N}(0, \sigma^2),$$

where U_{pij} are within-pair covariates, H_{qj} are between-pair covariates, and the within- and between-variances (σ^2, ω_{00}) are now residual variances.

Non-Exchangeable Pair Members

Consider now a study of heterosexual couples. A level-1 variable—gender—thus discriminates between pair members within every pair. We might now modify the matched pairs model of (5.15) by adding a level-1 variable. However, such a model has only two variance components, which enforces the assumption that men and women have equal variances. A simple fix is to allow distinct level-1 variances, one for each gender:

$$\underline{y}_{ij} = \beta_{00} + \beta_1(\mathit{Female})_{ij} + \underline{\delta}_{0j} + \underline{\epsilon}_{ij},$$

where $(\mathit{Female})_{ij}$ is an indicator for females and $\underline{\epsilon}_{ij}$ has variance σ_F^2 for females and σ_M^2 for males. The marginal distribution of the pair of outcomes is thus bivariate normal:

$$\begin{pmatrix} \underline{y}_{Mj} \\ \underline{y}_{Fj} \end{pmatrix} \sim \mathcal{N}\left(\begin{pmatrix} \beta_0 \\ \beta_0 + \beta_1 \end{pmatrix}, \begin{pmatrix} \omega_{00} + \sigma_M^2 & \omega_{00} \\ \omega_{00} & \omega_{00} + \sigma_F^2 \end{pmatrix} \right). \tag{5.16}$$

This reveals that this two-level hierarchical model is equivalent to the multivariate model

$$\underline{y}_{ij} = (\mathit{Female})_{ij}(\beta_F + \underline{\delta}_F) + (\mathit{Male})_{ij}(\beta_M + \underline{\delta}_M), \tag{5.17}$$

where $(\mathit{Male})_{ij} = 1 - (\mathit{Female})_{ij}$ is an indicator for males, and the bivariate normal distribution is given by

$$\begin{pmatrix} \underline{y}_{Mj} \\ \underline{y}_{Fj} \end{pmatrix} \sim \mathcal{N}\left(\begin{pmatrix} \beta_M \\ \beta_F \end{pmatrix}, \begin{pmatrix} \omega_{MM} & \omega_{MF} \\ \omega_{FM} & \omega_{FF} \end{pmatrix} \right). \tag{5.18}$$

Equalities between (5.17) and (5.18) are clear. Level-1 and level-2 covariates can again be added as needed given the research problem at hand.

A limitation of model (5.18) and, equivalently, of (5.16) is that variances and covariances and, hence, correlations, are not adjusted for measurement error. By exploiting information about measurement errors, one can solve this problem, extending (5.17) to create

$$\underline{y}_{ij} = (\mathit{Female})_{ij}(\beta_F + \underline{\delta}_F + \underline{e}_{Fj}) + (\mathit{Male})_{ij}(\beta_M + \underline{\delta}_M + \underline{e}_{Mj}),$$

where $\underline{e}_{Fj}$ and $\underline{e}_{Mj}$ are measurement errors with variances assumed known, leading to the bivariate distribution

$$\begin{pmatrix} \underline{y}_{Mj} \\ \underline{y}_{Fj} \end{pmatrix} \sim \mathcal{N}\left(\begin{pmatrix} \beta_M \\ \beta_F \end{pmatrix}, \begin{pmatrix} \omega_{MM} + \sigma_e^2 & \omega_{MF} \\ \omega_{FM} & \omega_{FF} + \sigma_e^2 \end{pmatrix} \right).$$

Here the measurement error variances are assumed equal for males and females. This assumption can readily be abandoned if there is evidence that measurement error variances depend on gender. Applications of this model appear in Barnett et al. [1]. Raudenbush et al. [27] extend the model to include repeated measures.

Short Time Series

The "small n, large J" scenario arises in many studies of individual change. For example, researchers may use a school's annual testing program to construct child-specific records of cognitive growth during the elementary years [37]. Here n is the number of time points per child and generally will not exceed five or six. Modeling issues that arise in this scenario are discussed elsewhere in this volume (see Chapter 7). Small within-person errors lead to small σ^2. Moreover, individuals are often quite heterogeneous on growth parameters, which are the random coefficients $\underline{\beta}_j$. In this setting, multivariate reliabilities (5.6) are quite high. For example, in Bryk and Raudenbush's study of academic learning during pre-school [5, Chapter 6], least squares estimates of person-specific intercepts and growth rates displayed reliabilities of about .80. While one tends to recommend that the number of random effects per level-2 unit should be small when n is small, I have often found that time series as short as five points per person will often support quadratic or even cubic growth models with ease, producing 3 or 4 random coefficients per person. This result contrasts with data collected on persons nested within schools or neighborhoods where the level-1 fit is often poor and the clusters are not highly heterogeneous with respect to random coefficients of interest. In these cases, the number of random coefficients per cluster must be sharply curtailed when n is small.

In longitudinal studies, however, study duration will often be more important than n_j in influencing $\boldsymbol{\Lambda}_j$ and, therefore, the precision of estimation of model parameters. Let D denote the duration of the study in some meaningful metric (e.g., years) and let n denote the number of time-series observations. Then, assuming equally spaced observations starting at time 0, the frequency of observation will be $(n-1)/D$ observations per year. Consider, for example, a simple straight-line growth model for level 1 (time series $i = 1, \ldots, n$ within participant j):

$$\underline{y}_{ij} = \underline{\beta}_{0j} + \underline{\beta}_{1j} D_i + \underline{\epsilon}_{ij}.$$

Here D_i is the duration of the study at the time of observation i. Then we have a special case of (5.6) with

$$\mathrm{Var}\left(\underline{\hat{\beta}}_{1j} \mid \underline{\beta}_j = \beta_j\right) = \frac{\sigma^2}{n} \bigg/ \frac{D^2(n+1)}{12(n-1)},$$

so that the reliability of the least squares estimator becomes

$$\lambda_j = \frac{\omega_{11}}{\omega_{11} + \dfrac{\sigma^2}{n} \bigg/ \dfrac{D^2(n+1)}{12(n-1)}}.$$

Suppose, for example, we choose $n = 5$ time points with a frequency of one observation per year, so that the duration of the study is $D = 4$. Then the reliability will be $\omega_{11}/(\omega_{11}+\sigma^2/10)$. On the other hand, with the same number of time points ($n = 5$) but with frequency twice per year, the duration would be $D = 2$. Now the reliability is $\omega_{11}/(\omega_{11}+\sigma^2/2.50)$. Despite holding constant the number of time points, the second study produces a reliability that is likely substantially diminished because the duration of the study has been reduced, reducing the leverage in estimating the growth rate.

5.3.2 Small n, Large J by Design

In general, the researcher must keep the complexity of the model of the covariance structure at level-2 in check when n per cluster is small. With this caveat in mind, we will see that "small n, large J" can produce excellent statistical power for some but not all research questions. In particular, we will examine cases for which $2 < n < 12$ and $J = 100$.

Most two-level cross-sectional designs can be viewed as closely related to two classical experimental designs: the cluster-randomized trial and the multi-site randomized trial. In the cluster-randomized trial, the key contrast of interest is at level 2—the cluster level. For the multi-site randomized trial, the key contrast is at level 1. The basic design features extend to quasi-experimental designs, though small adjustments are needed to accommodate covariates.

Cluster-Randomized Trials

In this design, clusters rather than persons are randomly assigned to treatments. Random assignment of persons is not typically feasible or desirable given the nature of the treatment, which is crafted to operate on the entire cluster, or because of concerns about diffusion of the treatment within clusters. The design serves as an ideal type for many observational studies, including comparisons of public and private schools [4, 8].

The model may be written simply as

$$\underline{y}_{ij} = \beta_0 + \beta_1(\mathit{Treatment})_j + \underline{\delta}_{0j} + \underline{\epsilon}_{ij},$$

where we assume a balanced design with n participants per cluster and with $J/2$ clusters in each treatment. Here $(\mathit{Treatment})_j$ is coded as 0.5 for experimentals and -0.5 for controls; thus β_1 is the mean difference between treatments. To evaluate power for "small n, large J", we define

$$\mathrm{Var}(\underline{\delta}_{0j}) = \rho; \qquad \mathrm{Var}(\underline{\epsilon}_{ij}) = 1 - \rho,$$

where ρ is the intra-cluster correlation coefficient, and the treatment effect β_1 is now a standardized effect size. Assuming $\rho > 0$, the F-statistic for $H_0 : \beta_1 = 0$ is distributed under the alternative hypothesis as a non-central F with non-centrality parameter $J\beta_1^2\lambda/(4\rho)$ with

$$\lambda = \frac{\rho}{\rho + (1-\rho)/n}.$$

Thus, power will increase as the number of clusters J and the effect size β_1 increase and as ρ decreases. Here λ $(0 < \lambda < 1)$ may again be thought of as the penalty for small n; holding constant ρ, λ converges to 1.0 as n increases. Note therefore that as n increases, the non-centrality parameter converges to a limit of $J\beta_1^2/(4\rho)$ $(\rho > 0)$, while the non-centrality parameter increases without bound as J increases. In this sense, increasing J is more effective than increasing n in driving up the power as long as $\rho > 0$.

Figure 5.1 displays power for several values of the effect size and ρ as n varies from 2 to 10, holding J constant at 100. For a standardized effect size of 0.50, power is uniformly high. However, such an effect size is typically viewed as quite large in many social and educational interventions. For a much more modest effect size of $\beta_1 = .30$, only $n = 4$ participants per cluster are needed to achieve a power of .80 given small ρ (at .01) and only $n = 7$ participants are needed for $\rho = .15$, typically regarded as a fairly large ρ. Only in the worst-case scenario of a small effect size ($\beta_1 = .20$) and $\rho = .15$ is the power of the "small n, $J = 100$" design inadequate. And in that case, increasing J rather than increasing n is essential to achieving adequate power.

In sum, if the intra-cluster correlation is not too large and the effect size is not too small, it is possible to achieve substantial power in the "small n, $J = 100$" scenario. Skillful choice of covariates will typically reduce ρ to modest values in many studies in education and human development with only a small loss in degrees of freedom as the penalty. Thus, "small n, large J" designs should certainly not be dismissed out of hand for this design, cost considerations aside.

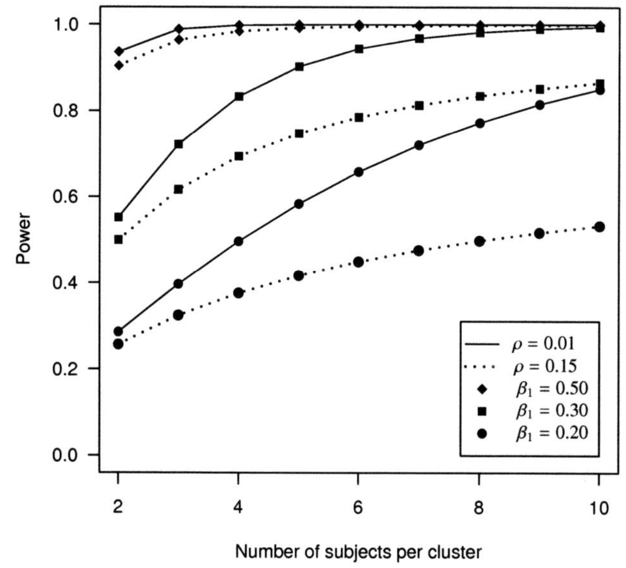

Fig. 5.1 Power to detect the treatment effect in a cluster randomized trial with $J = 100$ clusters (50 per treatment) as a function of n, the number of participants per cluster. Power is calculated at standardized effect sizes of 0.20, 0.30, and 0.50 and for intra-cluster correlations of 0.01 and 0.15. Significance level is always 0.05.

Multi-Site Randomized Trials

The second paradigm case of a two-level experimental design involves an experiment that is replicated in each of many clusters (often termed "sites"). Such experiments are common in medicine, and the Tennessee class size experiment provides an example in education [13]. We will consider the case in which $n/2$ participants are randomly assigned to an experimental or control site within each of J clusters. Thus, there are Jn participants overall.

The model may be written as

$$\underline{y}_{ij} = \beta_0 + (\beta_1 + \underline{\delta}_{1j})(\mathit{Treatment})_{ij} + \underline{\delta}_{0j} + \underline{\epsilon}_{ij}.$$

Note that the treatment contrast is now a level-1 variable taking on a value of $(\mathit{Treatment})_{ij} = 0.5$ for experimentals and -0.5 for controls. Therefore, the treatment effect $\beta_1 + \underline{\delta}_{1j}$ is potentially site-specific. We assume

$$\begin{pmatrix} \underline{\delta}_{0j} \\ \underline{\delta}_{1j} \end{pmatrix} \sim \mathcal{N}\left(\begin{pmatrix} 0 \\ 0 \end{pmatrix}, \begin{pmatrix} \omega_{00} & \omega_{01} \\ \omega_{10} & \omega_{11} \end{pmatrix} \right)$$

and

$$\underline{\epsilon}_{ij} \sim \mathcal{N}(0, 1).$$

Once again, β_1, the average treatment effect, is standardized, that is, expressed as a ratio of the average mean difference between treatments divided by the within-site standard deviation.

The aim now is not only to estimate the average treatment effect, β_1, but also to estimate the treatment-by-site variance component ω_{11} in the case of small n with J again held constant at $J = 100$. Following Raudenbush and Liu [30], we view ω_{11} as a standardized treatment-by-site variance and consider values of .01, .05, and .15 to be small, medium, and moderately large. For β_1, the non-centrality parameter may be expressed as $nJ\beta_1^2/(n\omega_{11} + 4)$.

Figure 5.2 provides an idea of the power afforded by small n designs when $J = 100$. For large effect size, even $n = 4$ provides high power at every value of ω_{11}. For a small effect size of 0.20, $n = 8$, $n = 10$, and $n = 12$ are required to achieve power of .80 at small, medium, and large values of ω_{11}, respectively. Thus, the "small n, large J" design does well (when $J = 100$ is viewed as large) in the case of the multi-site randomized trial.

But can the "small n, $J = 100$" design detect the variance of the treatment effect across sites? This parameter is important in gauging the generalizability of the treatment. Indeed, if ω_{11} is non-trivial, the main effect of treatment becomes misleading as a measure of effect at any specific site. In this case, a central F-distribution can be used to test the null hypothesis that $\omega_{11} = 0$ (see Raudenbush and Liu [30] for details).

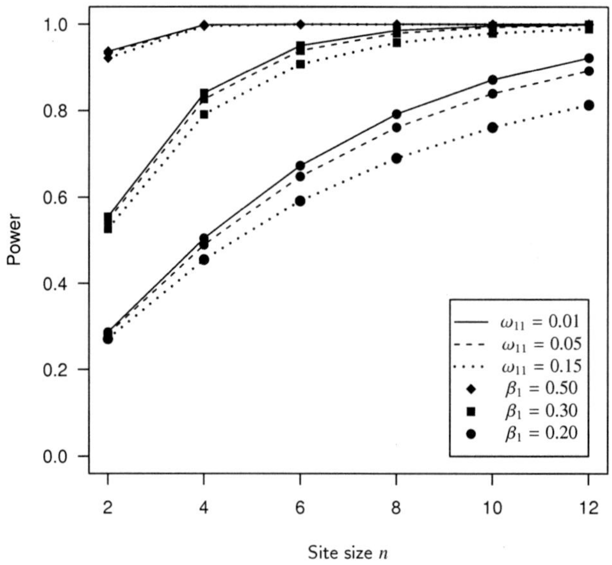

Fig. 5.2 Power to detect the main effect of treatment in a multi-site randomized trial for $J = 100$ sites with n participants at each site ($n/2$ in each treatment). Power is calculated at standardized effect sizes of 0.20, 0.30, and 0.50, and for effect size variances of 0.01, 0.05, and 0.15. Significance level is always 0.05.

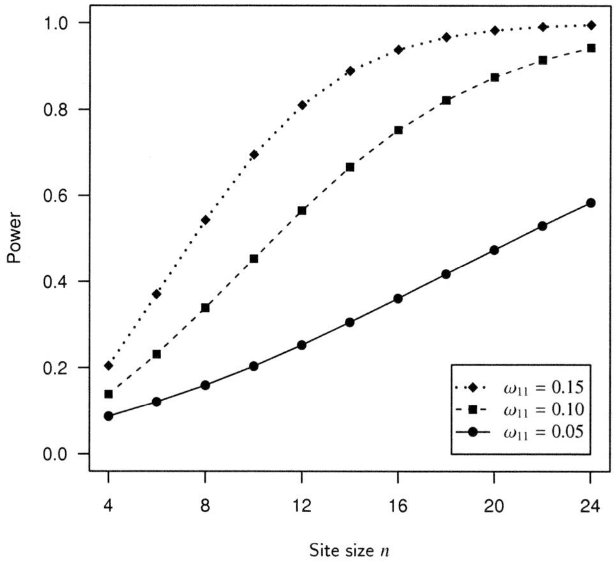

Fig. 5.3 Power to detect between-site variance at the treatment effect for $J = 100$ sites as a function of n, the number of participants per site ($n/2$ in each treatment). Between-site variances are 0.05, 0.10, and 0.15, and the significance level is always 0.05.

Figure 5.3 gives the results. The key conclusion is that the small n design (with $J = 100$) does not perform as well in detecting treatment-by-site variance as it does in detecting the main effect of treatment. While $n = 12$ and $n = 18$ are adequate to detect variances of $\omega_{11} = .15$ and $.10$, respectively, at power of .80, these variance values are quite large. Significantly more than 20 participants per site are required to detect a medium-sized variance of .05, not to mention a small variance of .01.

Conclusion

The results of power analysis for the cluster randomized trial and the multi-site randomized trial, while certainly not exhaustive, confirm what other practical experience has implied: Given moderately large J, small n designs can be very effective in detecting fixed effects of level-1 and level-2 predictors. These designs are much less adequate, however, in detecting variances of random coefficients. The exception occurs when the level-1 model fits well, as often occurs in studies of individual growth. But in two-level cross-sectional designs, the fit at level 1 is rarely good enough to allow high power for detecting the variances of random coefficients with small n as defined here, given J in the neighborhood of 100.

5.4 Validity of Statistical Inferences: Linear-Normal Case

Now-standard approaches to two-level data estimate variances and covariances via maximum likelihood (ML) or restricted maximum likelihood. Inferences about fixed and random regression coefficients are then conditional on these ML point estimates. Inferences about the variances are often based on the large-sample normal approximation to their sampling distributions. It is well known that this approach can work poorly when J is small, and especially when the data are greatly unbalanced (see Chapter 2 in this volume; see also Raudenbush and Bryk [28], Rubin [32], Seltzer [33]). These authors have recommended Bayesian methods and sensitivity analysis in the small J case. How well does the ML machinery work in the case of large J but small n?

The answer to this question appears to depend on the research focus. In particular, the validity of standard approaches depends on whether fixed regression coefficients, variance-covariance components, or the random coefficients themselves are of primary interest. We discuss each in turn.

5.4.1 Fixed Regression Coefficients

The short answer seems to be that inferences about the fixed regression coefficients proceeds comparatively smoothly in the case of small n but large J. The generalized least squares (GLS) estimator (5.7) sums over large J. Thus, imprecisions in estimating the weight matrices should have comparatively small effect. Moreover, the "small n, large J" case is ideal for the use of J-consistent robust standard errors [cf. 21]. These can be used as substitutes for model-based standard errors based on GLS and they also can gauge sensitivity of results to model assumptions and therefore signal the need to modify the hierarchical model.

To see how the robust standard errors are computed, we first rewrite the GLS estimator (5.6) in its more general form, based on the mixed model (5.1):

$$\underline{\hat{\boldsymbol{\beta}}} = \mathrm{Var}(\underline{\hat{\boldsymbol{\beta}}}) \sum_{j=1}^{J} \boldsymbol{X}_j' \boldsymbol{V}_j^{-1} \underline{\boldsymbol{y}}_j,$$

where $\boldsymbol{V}_j = \boldsymbol{Z}_j \boldsymbol{\Omega} \boldsymbol{Z}_j' + \sigma^2 \boldsymbol{I}_{n_j}$ and

$$\mathrm{Var}(\underline{\hat{\boldsymbol{\beta}}}) = \left(\sum_{j=1}^{J} \boldsymbol{X}_j' \boldsymbol{V}_j^{-1} \boldsymbol{X}_j \right)^{-1} . \tag{5.19}$$

Here the weight matrix for cluster j is $\boldsymbol{V}_j^{-1}$, evaluated at the ML estimates of $\boldsymbol{\Omega}$ and σ^2. Errors in these ML estimates will thus translate into errors in estimates of the weight matrix. However, when J is large, these errors will tend to be small, given that the GLS estimator is J-consistent.

Of course, the poor estimation of the weights, for example under-estimation of $\boldsymbol{\Omega}$, may distort the estimation of the standard errors (square roots of the diagonal elements of (5.19)). This would likely occur when the assumptions about the covariance structure are incorrect. To check on this possibility, we compute the robust variance

$$\text{Var}(\underline{\hat{\boldsymbol{\beta}}}) \left[\sum_{j=1}^{J} \boldsymbol{X}_j' \boldsymbol{V}_j^{-1} (\boldsymbol{y}_j - \boldsymbol{X}_j \hat{\boldsymbol{\beta}})(\boldsymbol{y}_j - \boldsymbol{X}_j \hat{\boldsymbol{\beta}})' \boldsymbol{V}_j^{-1} \boldsymbol{X}_j \right] \text{Var}(\underline{\hat{\boldsymbol{\beta}}}).$$

Given a total sample size of Jn (or sum of n_j in the unbalanced case), these robust estimators converge rapidly to the true variance when J is large and n is small. This convergence is not dependent on assumptions about which level-1 predictors have random coefficients. This last point is important in the "small n, large J" case because it is precisely in this case that one must impose constraints on the dimensionality of the random effects.

In conclusion, GLS seems to perform well in the "small n, large J" case. Moreover, the standard errors it produces are easily checked using robust standard errors, and the latter are especially useful in the "small n, large J" case.

5.4.2 Level-2 Variances

We noted earlier that while "small n, large J" tends to provide good power for detecting fixed effects of level-1 or level-2 predictors, power was more problematic when the aim was to detect heterogeneity of random coefficients. Similarly, threats to valid statistical inference arise when using ML to make inferences about such variance components.

The key problem is that the likelihood for level-2 variances can become quite positively skewed when n is small, even though J is fairly large. The skewness of the likelihood implies that the mode (i.e., the ML estimate) may poorly reflect the plausible values of the parameter. Moreover, in this case, a large-sample normal approximation to the likelihood will be poor.

These threats to valid inference are easily handled if the data are balanced. In this case, the F-distribution can supply the machinery to obtain accurate (asymmetric) confidence intervals. However, truly balanced data are rare in practice, often because of missing data, but possibly also because of cost considerations and because of the use of level-1 covariates that have different distributions in each cluster.

To illustrate the problems that can arise in making inferences about variances, we consider the simple case of a balanced, one-way random effects analysis of variance, i.e.,

$$\underline{y}_{ij} = \beta_{00} + \underline{\delta}_{0j} + \underline{\epsilon}_{ij},$$

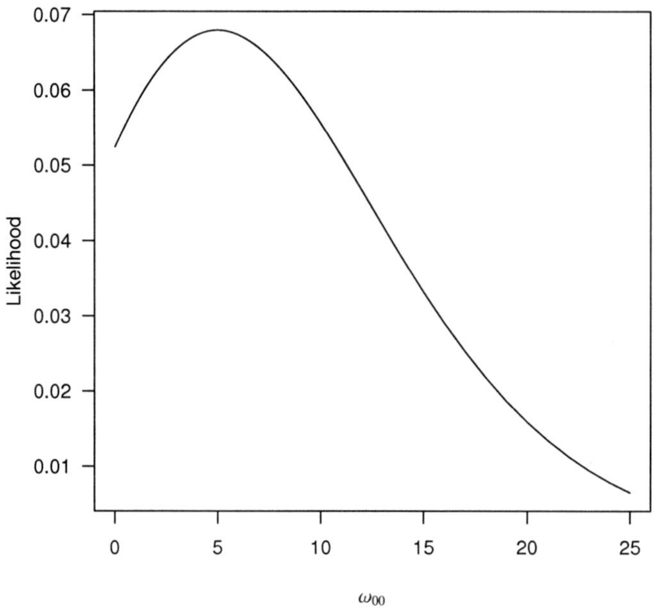

Fig. 5.4 Likelihood for ω_{00} in oneway ANOVA with $n = 2$, $J = 100$, and the mean square between equal to its expected value. True value of $\omega_{00} = 5.0$.

where $\underline{\delta}_{0j} \sim \mathcal{N}(0, \omega_{00})$ and $\underline{\epsilon}_{ij} \sim \mathcal{N}(0, \sigma^2)$. The likelihood is given by

$$\mathcal{L}(\beta_{00}, \omega_{00}, \sigma^2; \boldsymbol{y}) = (2\pi)^{-Jn/2}(\sigma^2)^{-J(n-1)/2}(n\omega_{00} + \sigma^2)^{-J/2} \times \exp\left[-\frac{1}{2}\left(\frac{SS_w}{\sigma^2} + \frac{SS_b}{n\omega_{00} + \sigma^2}\right)\right],$$

where SS_w and SS_b are the sums of squared deviations within and between clusters, respectively.

To illustrate the behavior of the likelihood, we set $\omega_{00} = 5$ and $\sigma^2 = 95$. Thus, the intra-cluster correlation coefficient is $\rho = .05$, a fairly typical value in several domains of multilevel research. For simplicity, we hold σ^2 equal to its true value. Figure 5.4 plots the likelihood for ω when $n = 2$ and $J = 100$ in a well-behaved case: the SS_w and SS_b are set to their expected values. We see that the likelihood, while globally skewed positively, is reasonably symmetric at its mode, the maximum likelihood estimator and, by construction, the true value of ω_{00}. Even in this case with $n = 2$, the modal value is reasonably representative of the plausible values of ω_{00}.

In reality, the SS_b will not be equal to its expected value. To see how the likelihood behaves in a somewhat less favorable setting, we set SS_b to one standard deviation below its expected value. Such a value could easily arise in practice. We plot the likelihood in this case in Fig. 5.5. Note the ML estimate

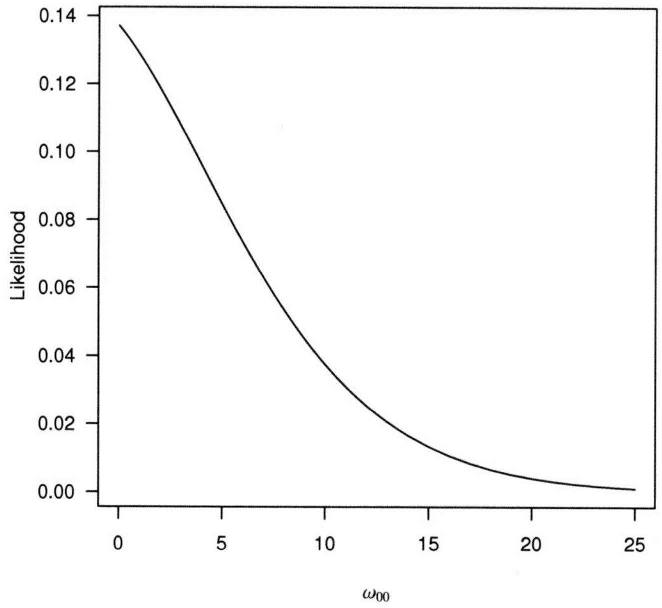

Fig. 5.5 Likelihood for ω_{00} in oneway ANOVA with $n = 2$, $J = 100$, and the mean square between equal to one standard deviation below its expected value. True value of $\omega_{00} = 5.0$.

is zero and the likelihood is very positively skewed. The modal value poorly represents the plausible values of ω_{00}.

How much do things improve when we increase n? Increasing n brings a fairly rapid improvement. Figure 5.6 plots the likelihood under the same conditions as in Fig. 5.5 but with n increased to 10. Fairly small increases in n create significant improvement in the shape of the likelihood.

I conclude that, when ρ is small and n is very small, likelihood-based inference about level-2 variance components can mislead the unwary. Small increases in n can improve things. These results have implications for inferences about random coefficients, the topic to which we now turn.

5.4.3 Inferences Concerning Random Effects

Obtaining good estimates of cluster-specific random coefficients or random effects is often, but not always, difficult in the "small n, large J" setting. The adequacy of these estimates depends strongly on the reliability $\boldsymbol{\Lambda}_j$. As we have seen, small n tends to work against large $\boldsymbol{\Lambda}_j$, but a good fit at level-1 (and therefore a small σ^2) or good leverage among the level-1 predictors can push $\boldsymbol{\Lambda}_j$ toward $\boldsymbol{I}_r$, even when n is small.

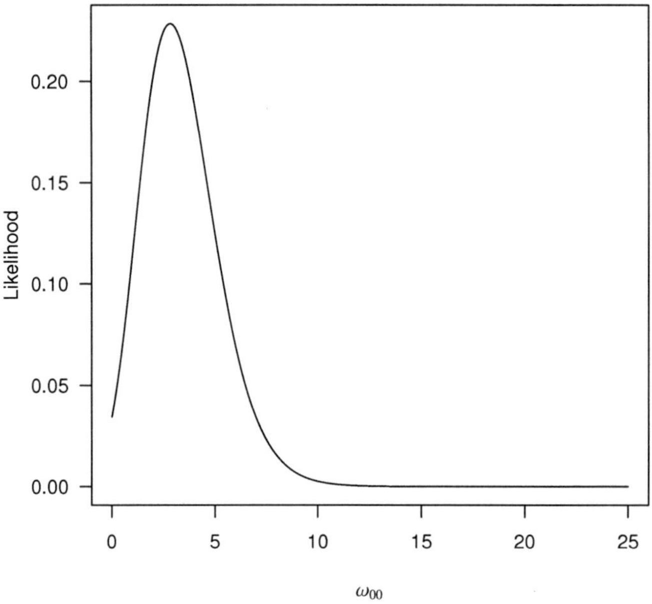

Fig. 5.6 Likelihood for ω_{00} in oneway ANOVA with $n = 10$, $J = 100$, and the mean square between equal to one standard deviation below its expected value. True value of $\omega_{00} = 5.0$.

In the previous section, we saw that the likelihood for level-2 variances (elements ω_{qq}) can be highly skewed when n is small even if J is moderately large. Empirical Bayes inferences for random coefficients, which condition on the MLE of variance components, may then "over-shrink" the point estimate toward zero and lead to under-estimates of the posterior variance.

A distinction arises in drawing inferences about random coefficients ($\underline{\boldsymbol{\beta}}_j$) as opposed to random effects ($\underline{\boldsymbol{\delta}}_j$). The validity of the latter depend more on large J than does the validity of the former.

5.4.4 Random Coefficients

Based on the two-level normal linear model of (5.4), the random coefficient vector, $\underline{\boldsymbol{\beta}}_j$, given the data $\boldsymbol{y}$ and the parameters σ^2 and $\boldsymbol{\Omega}$ is distributed as $\mathcal{N}(\boldsymbol{\beta}_j^*, \boldsymbol{V}_j^*)$ with

$$\boldsymbol{\beta}_j^* = \boldsymbol{C}_j^{-1}(\boldsymbol{y}_j - \boldsymbol{H}_j\hat{\gamma}) + \boldsymbol{H}_j\hat{\gamma}, \tag{5.20a}$$

$$\boldsymbol{V}_j^* = \sigma^2\left(\boldsymbol{C}_j^{-1} + \boldsymbol{C}_j^{-1}\boldsymbol{U}_j'\boldsymbol{H}_j \operatorname{Var}(\underline{\hat{\gamma}})\boldsymbol{H}_j'\boldsymbol{U}_j\boldsymbol{C}_j^{-1}\right), \tag{5.20b}$$

where $\boldsymbol{C}_j = \boldsymbol{U}_j'\boldsymbol{U}_j + \sigma^2\boldsymbol{\Omega}^{-1}$ and $\underline{\hat{\gamma}}$ is the generalized least squares estimator. A popular approach to empirical Bayes inference is based on the posterior distribution (5.20a) with restricted MLE substituted for the unknown $\boldsymbol{\Lambda}$, σ^2. A nice feature of this approach is that the posterior mean will exist even when $\boldsymbol{U}_j$ is less than full rank. However, when $\boldsymbol{U}_j$ is full rank, so that unit-specific least squares estimate $\hat{\boldsymbol{\beta}}_j$ exists, the posterior mean has the illuminating form (5.12), and the posterior variance matrix becomes

$$\boldsymbol{\Omega}(\boldsymbol{I}_r - \boldsymbol{\Lambda}_j) + (\boldsymbol{I}_r - \boldsymbol{\Lambda}_j)\boldsymbol{H}_j \operatorname{Var}(\hat{\underline{\gamma}})\boldsymbol{H}_j'(\boldsymbol{I}_r - \boldsymbol{\Lambda}_j)'. \tag{5.21}$$

The second term in (5.21) reflects uncertainty about γ. Thus, as J increases without bound (holding n_j constant), this term disappears, and (5.21) converges to $\boldsymbol{\Omega}(\boldsymbol{I}_r - \boldsymbol{\Lambda}_j)$. Note that as n_j increases without bound (holding J constant), $\boldsymbol{\Lambda}_j$ converges to $\boldsymbol{I}_r$ and (5.21) converges to the null matrix. Thus, large J is not essential in estimating $\underline{\boldsymbol{\beta}}_j$ using empirical Bayes. As we have seen, large n_j is not the only way to push $\boldsymbol{\Lambda}_j$ toward $\boldsymbol{I}_r$ inasmuch as small σ^2 or large level-1 dispersion of $\boldsymbol{U}_{ij}$ can also accomplish this.

5.4.5 Random Effects

While large J is not essential to ensure consistency of estimation of the random coefficients, large J is necessary to ensure consistency in estimating random effects, which are the discrepancies between the random coefficients $\underline{\boldsymbol{\beta}}_j$ and their expected values $\boldsymbol{H}_j\gamma$. To see this, note that the conditional distribution of the random effects vector $\underline{\boldsymbol{\delta}}_j$, given the data $\boldsymbol{y}$ and the parameters σ^2 and $\boldsymbol{\Omega}$, is distributed as $\mathcal{N}(\boldsymbol{\delta}_j^*, \boldsymbol{D}_j^*)$ where, in the full-rank case,

$$\boldsymbol{\delta}_j^* = \boldsymbol{\Lambda}_j(\hat{\boldsymbol{\beta}}_j - \boldsymbol{H}_j\hat{\gamma}), \tag{5.22a}$$

$$\boldsymbol{D}_j^* = \boldsymbol{\Omega}(\boldsymbol{I}_r - \boldsymbol{\Lambda}_j) + \boldsymbol{\Lambda}_j\boldsymbol{H}_j \operatorname{Var}(\hat{\underline{\gamma}})\boldsymbol{H}_j'\boldsymbol{\Lambda}_j'. \tag{5.22b}$$

As J increases without bound (holding n_j constant), the second term in (5.22b) disappears, and (5.22b) converges to $\boldsymbol{\Omega}(\boldsymbol{I}_r - \boldsymbol{\Lambda}_j)$. However, as n_j increases without bound (holding J constant), $\boldsymbol{\Lambda}_j$ converges to $\boldsymbol{I}_r$ and (5.22b) converges to $\operatorname{Var}(\boldsymbol{H}_j\hat{\underline{\gamma}})$. Thus, large J is essential for the consistency of the empirical Bayes estimator of $\underline{\boldsymbol{\delta}}_j$.

5.4.6 Over-Shrinkage and Under-Estimation of Uncertainty

If the likelihood for a level-2 variance is seriously skewed, as can occur when n_j is very small (even if J is moderate in size; see Fig. 5.5), $\hat{\boldsymbol{\Lambda}}_j$ based on the restricted MLE (or the unrestricted MLE) will be pulled toward the null matrix. This will pull empirical Bayes random effects estimates toward zero as it pulls the empirical Bayes random coefficients estimates toward their predicted values $\boldsymbol{H}_j\hat{\gamma}$. It will also tend to produce negatively biased estimates of the posterior variance (note that as $\boldsymbol{\Omega}$ decreases, $\boldsymbol{C}_j^{-1}$ also diminishes). In this setting, a Bayesian approach [see 34] or a better approximation to the posterior variance [19] will be helpful.

5.5 Validity of Statistical Inferences: Non-Linear Link Functions

We now consider the case in which $\underline{y}_{ij}$, given the random effect vector $\underline{\boldsymbol{\delta}}_j$, belongs to a one-parameter exponential family with canonical link function $\underline{\eta}_{ij} = \boldsymbol{X}'_{ij}\boldsymbol{\beta} + \boldsymbol{Z}'_{ij}\underline{\boldsymbol{\delta}}_j$ with $\underline{\boldsymbol{\delta}}_j \sim \mathcal{N}(\boldsymbol{\emptyset}, \boldsymbol{\Omega})$. Also, we can collect the elements $\underline{\eta}_{ij}$ into the vector $\underline{\boldsymbol{\eta}}_j = \boldsymbol{X}_j\boldsymbol{\beta} + \boldsymbol{Z}_j\underline{\boldsymbol{\delta}}_j$. The two-level linear models we have been discussing till now are a special case of this generalized linear mixed model with identity link and with $\underline{y}_{ij} \mid \boldsymbol{\delta}_j$ normally distributed. Among many examples of this model are binary y_{ij} with a logit link and a Bernoulli sampling model, counted y_{ij} (so that y_{ij} is a non-negative integer) with log link and Poisson sampling model, and continuous $y_{ij} > 0$ with reciprocal link and gamma sampling model.

A non-linear link function means that the conditional expectation of $\underline{\boldsymbol{y}}$ given the random effects induces an association between the estimate of the level-2 variance and the estimate of the fixed effects, with implications for inference in the "small n, large J" case. It also makes the problem of computing maximum likelihood estimates significantly more challenging than in the normal theory model with a linear link function.

5.5.1 Relation Between the Mean and the Variance

In our discussion of linear models with normal random effects at each level, we have seen that the "small n, large J" case can sometimes lead to problems in estimation of level-2 covariance components. And, because empirical Bayes bases inferences on ML estimates (or restricted ML estimates) of these variance components, these problems of variance estimation will tend also to affect the validity of inferences about random coefficients and random effects. However, these problems in variance estimation do not cause difficulty in making valid inferences about fixed effects. First, point estimates of the fixed effects are not sensitive to mis-estimation of variance components in the "small n, large J" case because, as J increases, the point estimate of the fixed effects is uncorrelated with point estimates of variance components. Second, robust standard errors are available in the "small n, large J", and these do not depend on the assumed variance-covariance structure.

When the level-1 link function is non-linear and the level-1 distribution is non-normal, the asymptotic orthogonality of the fixed effects estimates and the covariance estimates no longer holds. For example, in the case of a scalar random effect at level 2, we tend to see that larger estimates of the level-2 variance are associated with point estimates of the fixed effects that are farther from zero. This problem arises because the expected $\boldsymbol{y}$ is a non-linear function of the random effects. For example, in the logit linear model, we have the mixed model

$$\begin{aligned} E(\underline{y}_{ij} \mid \boldsymbol{\beta}, \boldsymbol{\delta}_j) &= \left[1 + \exp\{-(\boldsymbol{X}_{ij}'\boldsymbol{\beta} + \boldsymbol{Z}_{ij}'\boldsymbol{\delta}_j)\}\right]^{-1} \\ &= \mu_{ij}(\boldsymbol{\delta}_j) \\ &\approx \mu_{ij}(\boldsymbol{\emptyset}) + \mu_{ij}(\boldsymbol{\emptyset})\big(1 - \mu_{ij}(\boldsymbol{\emptyset})\big)\boldsymbol{Z}_{ij}'\boldsymbol{\delta}_j \\ &\quad + \mu_{ij}(\boldsymbol{\emptyset})\big(1 - \mu_{ij}(\boldsymbol{\emptyset})\big)\big(0.5 - \mu_{ij}(\boldsymbol{\emptyset})\big)\boldsymbol{Z}_{ij}'\boldsymbol{\delta}_j\boldsymbol{\delta}_j'\boldsymbol{Z}_{ij}. \end{aligned} \tag{5.23}$$

To the second order, the marginal expectation of $\underline{y}_{ij}$ is

$$E(\underline{y}_{ij}) \approx \mu_{ij}(\boldsymbol{\emptyset}) + \mu_{ij}(\boldsymbol{\emptyset})\big(1 - \mu_{ij}(\boldsymbol{\emptyset})\big)\big(0.5 - \mu_{ij}(\boldsymbol{\emptyset})\big)\boldsymbol{Z}_{ij}'\boldsymbol{\Omega}\boldsymbol{Z}_{ij}.$$

Clearly, the marginal expectation of $\underline{y}$ depends upon $\boldsymbol{\Omega}$ and, therefore, beliefs about $\boldsymbol{\beta}$ must depend upon $\boldsymbol{\Omega}$ as well. This dependence does not arise in the population-average model [40] wherein

$$E(\underline{y}_{ij}) = \left[1 + \exp\{-\boldsymbol{X}_{ij}'\boldsymbol{\beta}_{\text{pop.av}}\}\right]^{-1},$$

leading Heagerty and Zeger [15] to recommend population-average inference with robust standard errors as a robust alternative to "unit-specific" models of the type reflected in (5.23). The unit-specific coefficients $\boldsymbol{\beta}$ are not, of course, the same as the population-average coefficients $\boldsymbol{\beta}_{\text{pop.av}}$. The unit-specific coefficients define the expected change in the log odds that $\underline{y}_{ij} = 1$ given an increase in the corresponding $\boldsymbol{X}_{ij}$, *holding the random effect* $\underline{\delta}_j$ *constant.* In contrast, $\boldsymbol{\beta}_{\text{pop.av}}$ gives the expected change in the log odds $\underline{y}_{ij} = 1$ associated with a unit change in the corresponding $\boldsymbol{X}$, averaging over the distribution of the random effects. The distinction between unit-specific and population-average inference does not arise in linear models.

5.5.2 Estimation and Computation

Obtaining MLEs for hierarchical models is a two-step problem. The first step is to find the likelihood; this requires integration of the random effects from the joint distribution of the data and the random effects. The second step is to maximize the likelihood. For hierarchical linear models with normal random effects at each level, the first step is easy because the integration problem is soluble analytically. Maximization then proceeds using now-standard approaches such as the Expectation-Maximization (EM) algorithm [9, 10] or Fisher scoring [22].

When the level-1 link function is non-linear (and the level-1 sampling model is non-normal), the integration problem is much more challenging. And it tends to be especially challenging when n_j is small and when the level-2 variances are large. This logic becomes clear if we represent the likelihood as a Laplace transform.

The Likelihood

Let $f(y_j \mid \delta_j) = \exp\{\ell_j(\delta_j)\}$ denote the level-1 model. We use the binary outcome case with scalar random effect for illustrative purposes, in which case

$$\ell_j(\delta_j) = \sum_{i=1}^{n_j} \left[y_{ij}\eta_{ij} + \log(1 - \mu_{ij}) \right],$$

where

$$\begin{aligned} \underline{\eta}_{ij} &= \boldsymbol{X}'_{ij}\boldsymbol{\beta} + \underline{\delta}_j, \quad \underline{\delta}_j \sim \mathcal{N}(0, \omega), \\ \underline{\mu}_{ij} &= [1 + \exp\{-\underline{\eta}_{ij}\}]^{-1}. \end{aligned}$$

Defining the parameters as $\boldsymbol{\theta} = [\boldsymbol{\beta}, \omega]$, the likelihood of $\boldsymbol{\theta}$ at $\underline{\boldsymbol{y}} = \boldsymbol{y}$ is given by

$$\mathcal{L}(\boldsymbol{\theta}; \boldsymbol{y}) = \prod_{j=1}^{J} \left[(2\pi)^{-n_j/2} \omega^{-1/2} \int \exp\{\ell_j(\delta_j) - \tfrac{1}{2}\delta_j^2/\omega\} \, \mathrm{d}\delta_j \right]. \tag{5.24}$$

Laplace Transform

We now expand the integrand in an infinite Taylor series about its maximizer $\hat{\delta}_j$. We remove the j subscript for simplicity because the integral must be computed for every j. Thus, we have

$$\begin{aligned} h(\delta) &= \ell(\delta) - \tfrac{1}{2}\delta^2/\omega \\ &= [\ell(\hat{\delta}) - \tfrac{1}{2}\hat{\delta}^2/\omega] + [\ell^{(1)}(\hat{\delta}) - \hat{\delta}/\omega](\delta - \hat{\delta}) + \tfrac{1}{2}[\ell^{(2)} - \omega^{-1}](\delta - \hat{\delta})^2 + S, \end{aligned}$$

where

$$\begin{aligned} S &= \sum_{k=3}^{\infty} T_k, \\ T_k &= \frac{1}{k!}[\ell^{(k)}(\hat{\delta})](\delta - \hat{\delta})^k, \\ \ell^{(k)} &= \left. \frac{\mathrm{d}^k \ell(\delta)}{\mathrm{d}\delta^k} \right|_{\delta = \hat{\delta}}. \end{aligned}$$

Because $\hat{\delta}$ maximizes $h(\delta)$, the second term in the series vanishes. Substituting the Taylor series into the integral (5.24) thus yields a useful form of the likelihood

$$\begin{aligned} \mathcal{L}(\boldsymbol{\theta}; \boldsymbol{y}) = \prod_{j=1}^{J} \Big[&(2\pi)^{-n_j/2} \omega^{-1/2} \exp\{\ell_j(\hat{\delta}_j) - \tfrac{1}{2}\hat{\delta}_j^2/\omega\} \\ &\times E_{\mathcal{N}(0,\psi_j)}[\exp\{\underline{S}_j\}] \Big], \end{aligned} \tag{5.25}$$

where the expectation is taken over a normal distribution with mean 0 and variance $\psi_j = -\left[\ell_j^{(2)}(\hat{\delta}_j) - \omega^{-1}\right]^{-1}$.

Equation (5.25) creates a framework for approximating the likelihood and for evaluating the accuracy of the approximation. To examine how "small n, large J" affects accuracy of progressively better approximations, we note that

$$\psi_j = \left(\sum_{i=1}^{n_j} w_{ij} + \omega^{-1} \right)^{-1} = \omega(1 - \lambda_j),$$

where $w_{ij} = \mu_{ij}(\hat{\delta}_j)(1 - \mu_{ij}(\hat{\delta}_j))$ and $\lambda_j = \sum_{i=1}^{n_j} w_{ij} / (\sum_{i=1}^{n_j} w_{ij} + \omega^{-1})$. We highlight the form involving λ_j for continuity with the theme running throughout this chapter. We can view λ_j as the reliability of the iteratively reweighted least squares estimator of δ_j. As the cluster size n_j increases, λ_j converges to 1 and ψ_j approaches zero. Indeed, $\psi_j = O(n_j^{-1})$. Approximations increase in accuracy as more terms in $\exp\{S\}$ are absorbed in the expectation. Following Raudenbush et al. [31], we define the following Laplace approximations:

$$L_1 = \exp\{S\} = 1 + O(n^{-1}), \tag{5.26a}$$
$$L_2 = \exp\{S\} = L_1 + T_4 + T_3^2/2 + O(n^{-2}), \tag{5.26b}$$
$$L_3 = \exp\{S\} = L_2 + T_6 + 2T_3 T_5 + T_4^2/2 + 3T_3^2 T_4 + O(n^{-3}). \tag{5.26c}$$

Define $\zeta^k \triangleq E_{\mathcal{N}(0,\psi)}(\underline{\delta} - \hat{\delta})^k$. The asymptotic order of the approximations can be found by noting that $\zeta^k = O(n^{-k/2})$ for k even and $\ell^{(k)}(\hat{\delta}) = O(n)$. Thus,

$$E(\underline{T}_k) = \begin{cases} 0 & \text{for } k \text{ odd}, \\ \dfrac{\zeta^k\, \ell^{(k)}(\hat{\delta})}{k!} = O(n^{-k/2})\, O(n) = O(n^{-k/2+1}) & \text{for } k \text{ even}, \end{cases}$$

$$E(\underline{T}_k\, \underline{T}_m) = \begin{cases} 0 & \text{for } k+m \text{ odd}, \\ \dfrac{\zeta^{k+m}\, \ell^{(k)}(\hat{\delta})\, \ell^{(m)}(\hat{\delta})}{k!\, m!} = O(n^{-(k+m)/2})\, O(n^2) & \\ \qquad = O(n^{-(k+m-4)/2}) & \text{for } k+m \text{ even}, \end{cases}$$

$$E(\underline{T}_k\, \underline{T}_m\, \underline{T}_p) = \begin{cases} 0 & \text{for } k+m+p \text{ odd}, \\ \dfrac{\zeta^{k+m+p}\, \ell^{(k)}(\hat{\delta})\, \ell^{(m)}(\hat{\delta})\, \ell^{(p)}(\hat{\delta})}{k!\, m!\, p!} & \\ \qquad = O(n^{-(k+m+p-6)/2}) & \text{for } k+m+p \text{ even}, \end{cases}$$

$$E(\underline{T}_k\, \underline{T}_m\, \underline{T}_p\, \underline{T}_q) = \begin{cases} 0 & \text{for } k+m+p+q \text{ odd}, \\ \dfrac{\zeta^{k+m+p+q}\, \ell^{(k)}(\hat{\delta})\, \ell^{(m)}(\hat{\delta})\, \ell^{(p)}(\hat{\delta})\, \ell^{(q)}(\hat{\delta})}{k!\, m!\, p!\, q!} & \\ \qquad = O(n^{-(k+m+p+q-8)/2}) & \text{for } k+m+p+q \text{ even}. \end{cases}$$

Table 5.1 Error of approximation: Laplace versus adaptive Gauss-Hermite quadrature.

Order of Laplace	Number of quadrature points	Error
L_1	1 point	$O(n^{-1})$
L_2	4 points	$O(n^{-2})$
L_3	7 points	$O(n^{-3})$
L_4	10 points	$O(n^{-4})$

Equation (5.26) shows that the accuracy of the approximations depends on powers of the within-cluster sample size. This shows that for very small n, the approximations will tend to be poor (especially if ω is large). Eventually, of course, the rapidly increasing factorial denominators in the terms of S will dominate, ensuring convergence of higher-order approximations. These higher-order approximations are, however, tedious to derive.

The most common approach to approximation involves Gauss-Hermite quadrature [cf. 16]. This approach is useful when the dimension of $\boldsymbol{\Omega}$ is small. Even in the scalar case, comparatively large numbers of quadrature points will be needed if ω is large and n is small. Adaptive Gauss-Hermite quadrature [24], which centers the integrand around $\hat{\delta}_j$ (rather than around 0), will provide equal accuracy with many fewer quadrature points, but will still encounter difficulty when the dimension of the random effect is large.

Yosef [39] has clarified the relationship between the accuracy of the Laplace approximations and approximations to the likelihood based on adaptive Gauss-Hermite quadrature. We display these in Table 5.1. These are valid, however, only for scalar random effects.

Models with high-dimensional random effects are rarely feasible when n is small in the case of binary data, however. Laplace approximations of order L_2 work well in these high-dimensional cases assuming $n > 20$ [31].

References

1. R. C. Barnett, R. T. Brennan, S. W. Raudenbush, and N. L. Marshall. Gender and the relationship between marital role-quality and psychological distress: A study of dual-earner couples. *Journal of Personality and Social Psychology*, 64: 794–806, 1993.
2. U. Böckenholt. Hierarchical modeling of paired comparison data. *Psychological Methods*, 6:49–66, 2001.
3. G. Bond, L. Miller, R. Krumweid, and R. Ward. Assertive case management in three CMHs: A controlled study. *Hospital and Community Psychiatry*, 9: 411–418, 1988.
4. A. S. Bryk, V. Lee, and P. Holland. *Catholic Schools and the Common Good.* Harvard University Press, Cambridge, MA, 1993.

5. A. S. Bryk and S. W. Raudenbush. *Hierarchical Linear Models: Applications and Data Analysis Methods.* Sage, Newbury Park, CA, 1992.
6. B. Burns and A. Santos. Assertive community treatment: An update of randomized trials. *Psychiatric Services*, 47:669–675, 1995.
7. W. G. Cochran. Analysis of covariance: Its nature and uses. *Biometrics*, 13: 261–281, 1957.
8. J. Coleman, T. Hoffer, and S. Kilgore. *High School Achievement: Public, Catholic, and Other Private Schools Compared.* Basic, New York, 1982.
9. A. P. Dempster, N. M. Laird, and D. B. Rubin. Maximum likelihood from incomplete data via the *EM* algorithm. *Journal of the Royal Statistical Society, Series B*, 39:1–38, 1977. (with discussion)
10. A. P. Dempster, D. B. Rubin, and R. K. Tsutakawa. Estimation in covariance components models. *Journal of the American Statistical Association*, 76:341–353, 1981.
11. A. Donner, K. S. Brown, and P. Brasher. A methodological review of non-therapeutic intervention trials employing cluster randomization, 1979–1989. *International Journal of Epidemiology*, 19:795–800, 1990.
12. S. C. Duncan, T. E. Duncan, and A. Alpert. Alcohol use among African American and white siblings: A multilevel latent growth modeling approach. *Journal of Gender, Culture, and Health*, 3:209–225, 1998.
13. J. Finn and C. Achilles. Answers and questions about class size: A statewide experiment. *American Educational Research Journal*, 27:557–577, 1990.
14. C. Garner and S. W. Raudenbush. Neighborhood effects on educational attainment: A multilevel analysis of the influence of pupil ability, family, school, and neighborhood. *Sociology of Education*, 64:251–262, 1991.
15. P. J. Heagerty and S. L. Zeger. Marginalized multilevel models and likelihood inference. *Statistical Science*, 15:1–26, 2000. (with discussion)
16. D. Hedeker and R. D. Gibbons. A random-effects ordinal regression model for multilevel analysis. *Biometrics*, 50:933–944, 1994.
17. R. Ingersoll and S. Bobbitt. *Teacher Supply, Teacher Qualifications, and Teacher Turnover: 1990–91.* US National Center for Educational Statistics, Washington, DC, 1995.
18. B. A. Jones. Collaboration: The case for indigenous community-based organization support of drop-out prevention programming and implementation. *Journal of Negro Education*, 61:496–508, 1992.
19. R. E. Kass and D. Steffey. Approximate Bayesian inference in conditionally independent hierarchical models (parametric Empirical Bayes models). *Journal of the American Statistical Association*, 84:717–726, 1989.
20. M. Kuo, B. Mohler, S. W. Raudenbush, and F. J. Earls. Assessing exposure to violence using multiple informants: Application of hierarchical linear model. *The Journal of Child Psychology and Psychiatry*, 41:1049–1056, 2000.
21. K.-Y. Liang and S. L. Zeger. Longitudinal data analysis using generalized linear models. *Biometrika*, 73:13–22, 1986.
22. N. T. Longford. A fast scoring algorithm for maximum likelihood estimation in unbalanced mixed models with nested random effects. *Biometrika*, 74:817–827, 1987.

23. C. N. Morris. Parametric empirical Bayes inference: Theory and applications. *Journal of the American Statistical Association*, 78:47–65, 1983. (with discussion)
24. J. C. Pinheiro and D. M. Bates. Approximations to the log-likelihood function in the nonlinear mixed-effects model. *Journal of Computational and Graphical Statistics*, 4:12–35, 1995.
25. R. Plomin, R. N. Emde, J. M. Braungart, J. Campos, R. Corley, D. W. Fulker, J. Kagan, J. S. Reznick, J. Robinson, C. Zahn-Waxler, and J. DeFries. Genetic change and continuity from fourteen to twenty months: The MacArthur longitudinal twin study. *Child Development*, 64:1354–1376, 1993.
26. S. W. Raudenbush. Statistical analysis and optimal design for cluster randomized trials. *Psychological Methods*, 2:173–185, 1997.
27. S. W. Raudenbush, R. T. Brennan, and R. C. Barnett. A multivariate hierarchical model for studying psychological change within married couples. *Journal of Family Psychology*, 9:161–174, 1995.
28. S. W. Raudenbush and A. S. Bryk. Empirical Bayes meta-analysis. *Journal of Educational Statistics*, 10:75–98, 1985.
29. S. W. Raudenbush and A. S. Bryk. *Hierarchical Linear Models: Applications and Data Analysis Methods*, 2nd edition. Sage, Thousand Oaks, CA, 2002.
30. S. W. Raudenbush and X. Liu. Statistical power and optimal design for multisite randomized trials. *Psychological Methods*, 5:199–213, 2000.
31. S. W. Raudenbush, M.-L. Yang, and M. Yosef. Maximum likelihood for generalized linear models with nested random effects via high-order, multivariate Laplace approximation. *Journal of Computational and Graphical Statistics*, 9: 141–157, 2000.
32. D. B. Rubin. Estimation in parallel randomized experiments. *Journal of Educational Statistics*, 6:337–401, 1981.
33. M. H. Seltzer. Sensitivity analysis for fixed effects in the hierarchical model: A Gibbs sampling approach. *Journal of Educational Statistics*, 18:207–235, 1993.
34. M. H. Seltzer, W. H. Wong, and A. S. Bryk. Bayesian analysis in applications of hierarchical models: Issues and methods. *Journal of Educational and Behavioral Statistics*, 21:131–167, 1996.
35. Z. Shun. Another look at the salamander mating data: A modified Laplace approximation approach. *Journal of American Statistical Association*, 92:341–349, 1997.
36. R. Slavin. *Cooperative Learning.* Longman, New York, 1983.
37. J. B. Smith, V. E. Lee, and F. M. Newmann. *Instruction and Achievement in Chicago Elementary Schools.* Consortium on Chicago School Research, Chicago, 2001.
38. T. A. B. Snijders and R. J. Bosker. Standard errors and sample sizes for two-level research. *Journal of Educational Statistics*, 18:237–259, 1993.
39. M. Yosef. *A Comparison of Alternative Approximations to ML Estimation for Hierarchical Generalized Linear Models: The Logistic-Normal Model Case.* PhD thesis, Michigan State University, College of Education, East Lansing, 2001.
40. S. L. Zeger, K.-Y. Liang, and P. S. Albert. Models for longitudinal data: A generalized estimating equation approach. *Biometrics*, 44:1049–1060, 1988.

6

Multilevel Models for Ordinal and Nominal Variables

Donald Hedeker

University of Illinois at Chicago

6.1 Introduction

Reflecting the usefulness of multilevel analysis and the importance of categorical outcomes in many areas of research, generalization of multilevel models for categorical outcomes has been an active area of statistical research. For dichotomous response data, several approaches adopting either a logistic or probit regression model and various methods for incorporating and estimating the influence of the random effects have been developed [9, 21, 34, 37, 103, 115]. Several review articles [31, 39, 76, 90] have discussed and compared some of these models and their estimation procedures. Also, Snijders and Bosker [99, Chapter 14] provide a practical summary of the multilevel logistic regression model and the various procedures for estimating its parameters. As these sources indicate, the multilevel logistic regression model is a very popular choice for analysis of dichotomous data.

Extending the methods for dichotomous responses to ordinal response data has also been actively pursued [4, 29, 30, 44, 48, 58, 106, 113]. Again, developments have been mainly in terms of logistic and probit regression models, and many of these are reviewed in Agresti and Natarajan [5]. Because the proportional odds model described by McCullagh [71], which is based on the logistic regression formulation, is a common choice for analysis of ordinal data, many of the multilevel models for ordinal data are generalizations of this model. The proportional odds model characterizes the ordinal responses in C categories in terms of $C-1$ cumulative category comparisons, specifically, $C-1$ cumulative logits (i.e., log odds) of the ordinal responses. In the proportional odds model, the covariate effects are assumed to be the same across these cumulative logits, or proportional across the cumulative odds. As noted by Peterson and Harrell [77], however, examples of non-proportional odds are

J. de Leeuw, E. Meijer (eds.), *Handbook of Multilevel Analysis*,
© Springer 2008

not difficult to find. To overcome this limitation, Hedeker and Mermelstein [52] described an extension of the multilevel ordinal logistic regression model to allow for non-proportional odds for a set of regressors.

For nominal responses, there have been developments in terms of multilevel models as well. An early example is the model for nominal educational test data described by Bock [14]. This model includes a random effect for the level-2 subjects and fixed item parameters for the level-1 item responses nested within subjects. While Bock's model is a full-information maximum likelihood approach, using Gauss-Hermite quadrature to integrate over the random-effects distribution, it doesn't include covariates or multiple random effects. As a result, its usefulness for multilevel modeling is very limited. More general regression models of multilevel nominal data have been considered by Daniels and Gatsonis [25], Revelt and Train [88], Bhat [13], Skrondal and Rabe-Hesketh [97], and in Goldstein [38, Chapter 4]. In these models, it is common to adopt a reference cell approach in which one of the categories is chosen as the reference cell and parameters are characterized in terms of the remaining $C-1$ comparisons to this reference cell. Alternatively, Hedeker [47] adopts the approach in Bock's model, which allows any set of $C-1$ comparisons across the nominal response categories. Hartzel et al. [43] synthesizes some of the work in this area, describing a general mixed-effects model for both clustered ordinal and nominal responses, and Agresti et al. [3] describe a variety of social science applications of multilevel modeling of categorical responses.

This chapter describes multilevel models for categorical data that accommodate multiple random effects and allow for a general form for model covariates. Although only 2-level models will be considered here, 3-level generalizations are possible [35, 63, 83, 107]. For ordinal outcomes, proportional odds, partial proportional odds, and related survival analysis models for discrete- or grouped-time survival data are described. For nominal response data, models using both reference cell and more general category comparisons are described. Connections with item response theory (IRT) models are also made. A full maximum likelihood solution is outlined for parameter estimation. In this solution, multi-dimensional quadrature is used to numerically integrate over the distribution of random effects, and an iterative Fisher scoring algorithm is used to solve the likelihood equations. To illustrate application of the various multilevel models for categorical responses, several analyses of a longitudinal psychiatric dataset are described.

6.2 Multilevel Logistic Regression Model

Before considering models for ordinal and nominal responses, the multilevel model for dichotomous responses will be described. This is useful because both the ordinal and nominal models can be viewed as different ways of generalizing the dichotomous response model. To set the notation, let j denote the level-2 units (clusters) and let i denote the level-1 units (nested observations). Assume that there are $j = 1, \ldots, N$ level-2 units and $i = 1, \ldots, n_j$ level-1 units nested within each level-2 unit. The total number of level-1 observations across level-2 units is given by $n = \sum_{j=1}^{N} n_j$. Let $\underline{Y}_{ij}$ be the value of the dichotomous outcome variable, coded 0 or 1, associated with level-1 unit i nested within level-2 unit j. The logistic regression model is written in terms of the log odds (i.e., the logit) of the probability of a response, denoted $\underline{p}_{ij} = \Pr(\underline{Y}_{ij} = 1)$. Augmenting the standard logistic regression model with a single random effect yields

$$\log \left[\frac{\underline{p}_{ij}}{1 - \underline{p}_{ij}} \right] = \boldsymbol{x}'_{ij}\boldsymbol{\beta} + \underline{\delta}_j ,$$

where $\boldsymbol{x}_{ij}$ is the $s \times 1$ covariate vector (includes a 1 for the intercept), $\boldsymbol{\beta}$ is the $s \times 1$ vector of unknown regression parameters, and $\underline{\delta}_j$ is the random cluster effect (one for each level-2 cluster). These are assumed to be distributed in the population as $\mathcal{N}(0, \sigma_\delta^2)$. For convenience and computational simplicity, in models for categorical outcomes the random effects are typically expressed in standardized form. For this, $\underline{\delta}_j = \sigma_\delta \, \underline{\theta}_j$ and the model is given as

$$\log \left[\frac{\underline{p}_{ij}}{1 - \underline{p}_{ij}} \right] = \boldsymbol{x}'_{ij}\boldsymbol{\beta} + \sigma_\delta \, \underline{\theta}_j .$$

Notice that the random-effects variance term (i.e., the population standard deviation σ_δ) is now explicitly included in the regression model. Thus, it and the regression coefficients are on the same scale, namely in terms of the log-odds of a response.

The model can be easily extended to include multiple random effects. For this, denote $\boldsymbol{z}_{ij}$ as the $r \times 1$ vector of random-effect variables (a column of ones is usually included for the random intercept). The vector of random effects $\underline{\boldsymbol{\delta}}_j$ is assumed to follow a multivariate normal distribution with mean vector $\boldsymbol{\emptyset}$ and variance-covariance matrix $\boldsymbol{\Omega}$. To standardize the multiple random effects, $\underline{\boldsymbol{\delta}}_j = \boldsymbol{T}\underline{\boldsymbol{\theta}}_j$, where $\boldsymbol{T}\boldsymbol{T}' = \boldsymbol{\Omega}$ is the Cholesky decomposition of $\boldsymbol{\Omega}$. The model is now written as

$$\log \left[\frac{\underline{p}_{ij}}{1 - \underline{p}_{ij}} \right] = \boldsymbol{x}'_{ij}\boldsymbol{\beta} + \boldsymbol{z}'_{ij}\boldsymbol{T}\underline{\boldsymbol{\theta}}_j . \tag{6.1}$$

As a result of the transformation, the Cholesky factor $\boldsymbol{T}$ is usually estimated instead of the variance-covariance matrix $\boldsymbol{\Omega}$. As the Cholesky factor is essentially the matrix square root of the variance-covariance matrix, this allows more stable estimation of near-zero variance terms.

6.2.1 Threshold Concept

Dichotomous regression models are often motivated and described using the "threshold concept" [15]. This is also termed a latent variable model for dichotomous variables [65]. For this, it is assumed that a continuous latent variable $\underline{y}$ underlies the observed dichotomous response $\underline{Y}$. A threshold, denoted γ, then determines if the dichotomous response $\underline{Y}$ equals 0 ($\underline{y}_{ij} \leq \gamma$) or 1 ($\underline{y}_{ij} > \gamma$). Without loss of generality, it is common to fix the location of the underlying latent variable by setting the threshold equal to zero (i.e., $\gamma = 0$). Figure 6.1 illustrates this concept assuming that the continuous latent variable $\underline{y}$ follows either a normal or logistic probability density function (pdf).

As noted by McCullagh and Nelder [72], the assumption of a continuous latent distribution, while providing a useful motivating concept, is not a strict model requirement. In terms of the continuous latent variable $\underline{y}$, the model is written as

$$\underline{y}_{ij} = \boldsymbol{x}'_{ij}\boldsymbol{\beta} + \boldsymbol{z}'_{ij}\boldsymbol{T}\underline{\boldsymbol{\theta}}_j + \underline{\epsilon}_{ij} \,.$$

Note the inclusion of the errors $\underline{\epsilon}_{ij}$ in this representation of the model. In the logistic regression formulation, the errors $\underline{\epsilon}_{ij}$ are assumed to follow a standard logistic distribution with mean 0 and variance $\pi^2/3$ [2, 65]. The scale of the errors is fixed because $\underline{y}$ is not observed, and so the the scale is not separately identified. Thus, although the above model appears to be the

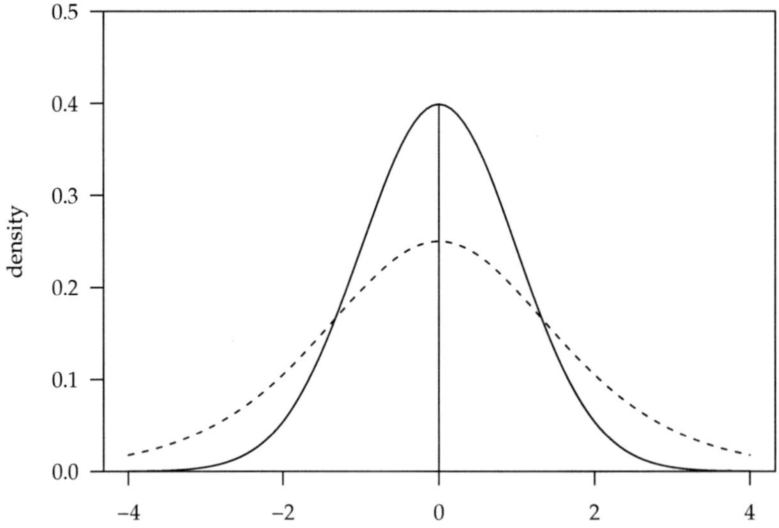

Fig. 6.1 Threshold concept for a dichotomous response (solid = normal, dashed = logistic).

same as an ordinary multilevel regression model for continuous outcomes, it is one in which the error variance is fixed and not estimated. This has certain consequences that will be discussed later.

Because the errors are assumed to follow a logistic distribution and the random effects a normal distribution, this model and models closely related to it are often referred to as logistic/normal or logit/normit models, especially in the latent trait model literature [11]. If the errors are assumed to follow a normal distribution, then the resulting model is a multilevel probit regression or normal/normal model. In the probit model, the errors have mean 0 and variance 1 (i.e., the variance of the standard normal distribution).

6.2.2 Multilevel Representation

For a multilevel representation of a simple model with only one level-1 covariate x_{ij} and one level-2 covariate x_j, the level-1 model is written in terms of the logit as

$$\log\left[\frac{\underline{p}_{ij}}{1-\underline{p}_{ij}}\right] = \underline{\beta}_{0j} + \underline{\beta}_{1j} x_{ij} ,$$

or in terms of the latent response variable as

$$\underline{y}_{ij} = \underline{\beta}_{0j} + \underline{\beta}_{1j} x_{ij} + \underline{\epsilon}_{ij} . \tag{6.2}$$

The level-2 model is then (assuming x_{ij} is a random-effects variable)

$$\underline{\beta}_{0j} = \beta_0 + \beta_2 x_j + \underline{\delta}_{0j} , \tag{6.3a}$$

$$\underline{\beta}_{1j} = \beta_1 + \beta_3 x_j + \underline{\delta}_{1j} . \tag{6.3b}$$

Notice that it's easiest, and in agreement with the normal-theory (continuous) multilevel model, to write the level-2 model in terms of the unstandardized random effects, which are distributed in the population as $\underline{\boldsymbol{\delta}}_j \sim \mathcal{N}(\boldsymbol{\emptyset}, \boldsymbol{\Omega})$. For models with multiple variables at either level 1 or level 2, the above level-1 and level-2 submodels are generalized in an obvious way.

Because the level-1 variance is fixed, the model operates somewhat differently than the more standard normal-theory multilevel model for continuous outcomes. For example, in an ordinary multilevel model, the level-1 variance term is typically reduced as level-1 covariates x_{ij} are added to the model. However, this cannot happen in the above model because the level-1 variance is fixed. As noted by Snijders and Bosker [99], what happens instead (as level-1 covariates are added) is that the random-effect variance terms tend to become larger as do the other regression coefficients, the latter become larger in absolute value.

6.2.3 Logistic and Probit Response Functions

The logistic model can also be written as

$$\underline{p}_{ij} = \Psi(\boldsymbol{x}'_{ij}\boldsymbol{\beta} + \boldsymbol{z}'_{ij}\boldsymbol{T}\underline{\boldsymbol{\theta}}_j),$$

where $\Psi(\eta)$ is the logistic cumulative distribution function (cdf), namely

$$\Psi(\eta) = \frac{\exp(\eta)}{1+\exp(\eta)} = \frac{1}{1+\exp(-\eta)}.$$

The cdf is also termed the response function of the model. A mathematical nicety of the logistic distribution is that the probability density function (pdf) is related to the cdf in a simple way, namely $\psi(\eta) = \Psi(\eta)[1 - \Psi(\eta)]$.

As mentioned, the probit model, which is based on the standard normal distribution, is often proposed as an alternative to the logistic model. For the probit model, the normal cdf $\Phi(\eta)$ and pdf $\phi(\eta)$ replace their logistic counterparts, and because the standard normal distribution has variance equal to 1, $\underline{\epsilon}_{ij} \sim \mathcal{N}(0, 1)$. As a result, in the probit model the underlying latent variable vector $\underline{\boldsymbol{y}}_j$ is distributed normally in the population with mean $\boldsymbol{X}_j\boldsymbol{\beta}$ and variance covariance matrix $\boldsymbol{Z}_j\boldsymbol{T}\boldsymbol{T}'\boldsymbol{Z}'_j + \boldsymbol{I}$. The latter, when converted to a correlation matrix, yields tetrachoric correlations for the underlying latent variable vector $\underline{\boldsymbol{y}}$ (and polychoric correlations for ordinal outcomes, discussed below). For this reason, in some areas, for example familial studies, the probit formulation is preferred to its logistic counterpart.

As can be seen in the earlier figure, both the logistic and normal distributions are symmetric around zero and differ primarily in terms of their scale; the standard normal has standard deviation equal to 1, whereas the standard logistic has standard deviation equal to $\pi/\sqrt{3}$. As a result, the two typically give very similar results and conclusions, though the logistic regression parameters (and associated standard errors) are approximately $\pi/\sqrt{3}$ times as large because of the scale difference between the two distributions. An alternative response function, which provides connections with proportional hazards survival analysis models (see Allison [7] and Section 6.3.2), is the complementary log-log response function $1 - \exp[-\exp(\eta)]$. Unlike the logistic and normal, the distribution that underlies the complementary log-log response function is asymmetric and has variance equal to $\pi^2/6$. Its pdf is given by $\exp(\eta)[1-p(\eta)]$. As Doksum and Gasko [26] note, large amounts of high-quality data are often necessary for response function selection to be relevant. Since these response functions often provide similar fits and conclusions, McCullagh [71] suggests that the response function choice should be based primarily on ease of interpretation.

6.3 Multilevel Proportional Odds Model

Let the C ordered response categories be coded as $c = 1, 2, \ldots, C$. Ordinal response models often utilize cumulative comparisons of the ordinal outcome. The cumulative probabilities for the C categories of the ordinal outcome $\underline{Y}$ are defined as $\underline{P}_{ijc} = \Pr(\underline{Y}_{ij} \leq c) = \sum_{k=1}^{c} \underline{p}_{ijk}$. The multilevel logistic model for the cumulative probabilities is given in terms of the cumulative logits as

$$\log\left[\frac{\underline{P}_{ijc}}{1-\underline{P}_{ijc}}\right] = \gamma_c - \left[\boldsymbol{x}'_{ij}\boldsymbol{\beta} + \boldsymbol{z}'_{ij}\boldsymbol{T}\underline{\boldsymbol{\theta}}_j\right] \qquad (c = 1, \ldots, C-1), \tag{6.4}$$

with $C-1$ strictly increasing model thresholds γ_c (i.e., $\gamma_1 < \gamma_2 < \cdots < \gamma_{C-1}$).

The relationship between the latent continuous variable $\underline{y}$ and an ordinal outcome with three categories is depicted in Fig. 6.2. In this case, the ordinal outcome $\underline{Y}_{ij} = c$ if $\gamma_{c-1} \leq \underline{y}_{ij} < \gamma_c$ for the latent variable (with $\gamma_0 = -\infty$ and $\gamma_C = \infty$). As in the dichotomous case, it is common to set a threshold to zero to set the location of the latent variable. Typically, this is done in terms of the first threshold (i.e., $\gamma_1 = 0$). In Fig. 6.2, setting $\gamma_1 = 0$ implies that $\gamma_2 = 2$.

At first glance, it may appear that the parameterization of the model in (6.4) is not consistent with the dichotomous model in (6.1). To see the

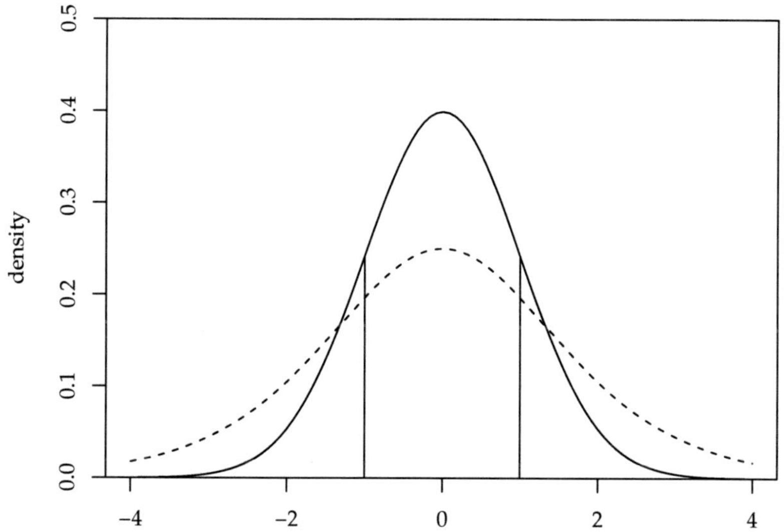

Fig. 6.2 Threshold concept for an ordinal response with 3 categories (solid = normal, dashed = logistic).

connection, notice that for a dichotomous outcome (coded 0 and 1), the model is written as

$$\log\left[\frac{\underline{P}_{ij0}}{1-\underline{P}_{ij0}}\right] = 0 - \left[\boldsymbol{x}'_{ij}\boldsymbol{\beta} + \boldsymbol{z}'_{ij}\boldsymbol{T}\underline{\boldsymbol{\theta}}_j\right],$$

and since for a dichotomous outcome $\underline{P}_{ij0} = \underline{p}_{ij0}$ and $1-\underline{P}_{ij0} = \underline{p}_{ij1}$,

$$\log\left[\frac{1-\underline{P}_{ij0}}{\underline{P}_{ij0}}\right] = \log\left[\frac{\underline{p}_{ij1}}{1-\underline{p}_{ij1}}\right] = \boldsymbol{x}'_{ij}\boldsymbol{\beta} + \boldsymbol{z}'_{ij}\boldsymbol{T}\underline{\boldsymbol{\theta}}_j,$$

which is the same as before. Also, in terms of the underlying latent variable $\underline{y}$, the multilevel representation of the ordinal model is identical to the dichotomous version presented earlier in (6.2). If the multilevel model is written in terms of the observed response variable $\underline{Y}$, then the level-1 model is written instead as

$$\log\left[\frac{\underline{P}_{ijc}}{1-\underline{P}_{ijc}}\right] = \gamma_c - \left[\underline{\beta}_{0j} + \underline{\beta}_{1j}x_{ij}\right]$$

for the case of a model with one level-1 covariate. Because the level-2 model does not really depend on the response function or variable, it would be the same as given above for the dichotomous model in (6.3a) and (6.3b).

Since the regression coefficients $\boldsymbol{\beta}$ do not carry the c subscript, they do not vary across categories. Thus, the relationship between the explanatory variables and the cumulative logits does not depend on c. McCullagh [71] calls this assumption of identical odds ratios across the $C-1$ cut-offs the proportional odds assumption. As written above, a positive coefficient for a regressor indicates that as values of the regressor increase so do the odds that the response is greater than or equal to c. Although this is a natural way of writing the model, because it means that for a positive β as x increases so does the value of $\underline{Y}$, it is not the only way of writing the model. In particular, the model is sometimes written as

$$\log\left[\frac{\underline{P}_{ijc}}{1-\underline{P}_{ijc}}\right] = \gamma_c + \boldsymbol{x}'_{ij}\boldsymbol{\beta} + \boldsymbol{z}'_{ij}\boldsymbol{T}\underline{\boldsymbol{\theta}}_j \qquad (c = 1,\ldots,C-1),$$

in which case the regression parameters $\boldsymbol{\beta}$ are identical but of opposite sign. This alternate specification is commonly used in survival analysis models (see Section 6.3.2).

6.3.1 Partial Proportional Odds

As noted by Peterson and Harrell [77], violation of the proportional odds assumption is not uncommon. Thus, they described a (fixed-effects) partial proportional odds model in which covariates are allowed to have differential effects on the $C-1$ cumulative logits. Similarly, Terza [109] developed a similar extension of the (fixed-effects) ordinal probit model. Hedeker and Mermelstein [52, 53] utilize this extension within the context of a multilevel ordinal regression model. For this, the model for the $C-1$ cumulative logits can be written as

$$\log\left[\frac{\underline{P}_{ijc}}{1-\underline{P}_{ijc}}\right] = \gamma_c - \left[(\boldsymbol{x}^*_{ij})'\boldsymbol{\beta}_c + \boldsymbol{x}'_{ij}\boldsymbol{\beta} + \boldsymbol{z}'_{ij}\boldsymbol{T}\underline{\boldsymbol{\theta}}_i\right] \qquad (c = 1,\ldots,C-1),$$

where $\boldsymbol{x}^*_{ij}$ is a $h \times 1$ vector containing the values of observation ij on the set of h covariates for which proportional odds is not assumed. In this model, $\boldsymbol{\beta}_c$ is a $h \times 1$ vector of regression coefficients associated with these h covariates. Because $\boldsymbol{\beta}_c$ carries the c subscript, the effects of these h covariates are allowed to vary across the $C-1$ cumulative logits. In many areas of research, this extended model is useful. For example, suppose that in a alchohol reduction study there are three response categories (abstinence, mild use, heavy use) and suppose that an intervention designed to reduce drinking is not successful in increasing the proportion of individuals in the abstinence category but is successful in moving individuals from heavy to mild use. In this case, the (covariate) effect of intervention group would not be observed on the first cumulative logit, but would be observed on the second cumulative logit. This extended model has been utilized in several articles [32, 114, 117], and a similar Bayesian hierarchical model is described in Ishwaran [57].

In general, this extension of the proportional odds model is not problematic; however, one caveat should be mentioned. For the explanatory variables without proportional odds, the effects on the cumulative log odds, namely $(\boldsymbol{x}^*_{ij})'\boldsymbol{\beta}_c$, result in $C-1$ non-parallel regression lines. These regression lines inevitably cross for some values of $\boldsymbol{x}^*$, leading to negative fitted values for the response probabilities. For $\boldsymbol{x}^*$ variables contrasting two levels of an explanatory variable (e.g., gender coded as 0 or 1), this crossing of regression lines occurs outside the range of admissible values (i.e., < 0 or > 1). However, if the explanatory variable is continuous, this crossing can occur within the range of the data, and so allowing for non-proportional odds can be problematic. A solution to this dilemma is sometimes possible if the variable has, say, m levels with a reasonable number of observations at each of these m levels. In this case, $m-1$ dummy-coded variables can be created and substituted into the model in place of the continuous variable. Alternatively, one might consider a nominal response model using Helmert contrasts [15] for the outcome variable. This approach, described in Section 6.4, is akin to the sequential logit models for nested or hierarchical response scales that are described in McCullagh and Nelder [72].

6.3.2 Survival Analysis Models

Several authors have noted the connection between survival analysis models and binary and ordinal regression models for survival data that are discrete or grouped within time intervals (for practical introductions, see Allison [6], Allison [7], D'Agostino et al. [24], and Singer and Willett [95]). This connection has been utilized in the context of categorical multilevel or mixed-effects regression models by many authors as well [42, 54, 94, 106, 108]. For this, assume that time (of assessment) can take on only discrete positive values $c = 1, 2, \ldots, C$.[1] For each level-1 unit, observation continues until time $\underline{Y}_{ij}$ at which point either an event occurs ($\underline{d}_{ij} = 1$) or the observation is censored ($\underline{d}_{ij} = 0$), where censoring indicates being observed at c but not at $c + 1$. Define P_{ijc} to be the probability of failure, up to and including time interval c, that is,

$$P_{ijc} = \Pr(\underline{Y}_{ij} \leq c),$$

and so the probability of survival beyond time interval c is simply $1 - P_{ijc}$.

Because $1 - P_{ijc}$ represents the survivor function, McCullagh [71] proposed the following grouped-time version of the continuous-time proportional hazards model

$$\log[-\log(1 - P_{ijc})] = \gamma_c + \boldsymbol{x}'_{ij}\boldsymbol{\beta}. \tag{6.5}$$

This is the aforementioned complementary log-log response function, which can be re-expressed in terms of the cumulative failure probability, $P_{ijc} = 1 - \exp(-\exp(\gamma_c + \boldsymbol{x}'_{ij}\boldsymbol{\beta}))$. In this model, $\boldsymbol{x}_{ij}$ includes covariates that vary either at level 1 or 2; however, they do not vary with time (i.e., they do not vary across the ordered response categories). They may, however, represent the average of a variable across time or the value of the covariate at the time of the event.

The covariate effects in this model are identical to those in the grouped-time version of the proportional hazards model described by Prentice and Gloeckler [79]. As such, the $\boldsymbol{\beta}$ coefficients are also identical to the coefficients in the underlying continuous-time proportional hazards model. Furthermore, as noted by Allison [6], the regression coefficients of the model are invariant to interval length. Augmenting the coefficients $\boldsymbol{\beta}$, the threshold terms γ_c represent the logarithm of the integrated baseline hazard (i.e., when $\boldsymbol{x} = \boldsymbol{\emptyset}$). While the above model is the same as that described in McCullagh [71], it is written so that the covariate effects are of the same sign as the Cox proportional hazards model. A positive coefficient for a regressor then reflects increasing hazard (i.e., lower values of $\underline{Y}$) with greater values of the regressor. Adding (standardized) random effects, we get

$$\log[-\log(1 - \underline{P}_{ijc})] = \gamma_c + \boldsymbol{x}'_{ij}\boldsymbol{\beta} + \boldsymbol{z}'_{ij}\boldsymbol{T}\underline{\boldsymbol{\theta}}_j\,. \tag{6.6}$$

This model is thus a multilevel ordinal regression model with a complementary log-log response function instead of the logistic. Though the logistic model

[1] To make the connection to ordinal models more direct, time is denoted here as c; however, more commonly it is denoted as t in the survival analysis literature.

has also been proposed for analysis of grouped- and/or discrete-time survival data, its regression coefficients are not invariant to time interval length and it requires the intervals to be of equal length [6]. As a result, the complementary log-log response function is generally preferred.

In the ordinal treatment, survival time is represented by the ordered outcome $\underline{Y}_{ij}$, which is designated as being censored or not. Alternatively, each survival time can be represented as a set of dichotomous dummy codes indicating whether or not the observation failed in each time interval that was experienced [6, 24, 95]. Specifically, each survival time $\underline{Y}_{ij}$ is represented as a vector with all zeros except for its last element, which is equal to $\underline{d}_{ij}$ (i.e., $= 0$ if censored and $= 1$ for an event). The length of the vector for observation ij equals the observed value of $\underline{Y}_{ij}$ (assuming that the survival times are coded as $1, 2, \ldots, C$). These multiple time indicators are then treated as distinct observations in a dichotomous regression model. In a multilevel model, a given cluster's response vector $\underline{\boldsymbol{Y}}_j$ is then of size $(\sum_{i=1}^{n_j} \underline{Y}_{ij}) \times 1$. This method has been called the pooling of repeated observations method by Cupples et al. [23]. It is particularly useful for handling time-dependent covariates and fitting non-proportional hazards models because the covariate values can change across time. See Singer and Willett [96] for a detailed treatment of this method.

For this dichotomous approach, define $\underline{\lambda}_{ijc}$ to be the probability of failure in time interval c, conditional on survival prior to c,

$$\underline{\lambda}_{ijc} = \Pr(\underline{Y}_{ij} = c \mid \underline{Y}_{ij} \geq c).$$

Similarly, $1 - \underline{\lambda}_{ijc}$ is the probability of survival beyond time interval c, conditional on survival prior to c. The multilevel proportional hazards model is then written as

$$\log[-\log(1 - \underline{\lambda}_{ijc})] = \boldsymbol{x}'_{ijc}\boldsymbol{\beta} + \boldsymbol{z}'_{ij}\boldsymbol{T}\underline{\boldsymbol{\theta}}_j\,, \qquad (6.7)$$

where now the covariates $\boldsymbol{x}$ can vary across time and so are denoted as $\boldsymbol{x}_{ijc}$. The first elements of $\boldsymbol{x}$ are usually time-point dummy codes. Because the covariate vector $\boldsymbol{x}$ now varies with c, this approach automatically allows for time-dependent covariates, and relaxing the proportional hazards assumption only involves including interactions of covariates with the time-point dummy codes.

Under the complementary log-log link function, the two approaches characterized by (6.6) and (6.7) yield identical results for the parameters that do not depend on c [28, 59]. Comparing these two approaches, notice that for the ordinal approach, each observation consists of only two pieces of data: the (ordinal) time of the event and whether it was censored or not. Alternatively, in the dichotomous approach, each survival time is represented as a vector of dichotomous indicators, where the size of the vector depends upon the timing of the event (or censoring). Thus, the ordinal approach can be easier to implement and offers savings in terms of the dataset size, especially as the number of time-points gets large, while the dichotomous approach is superior in its treatment of time-dependent covariates and relaxing of the proportional hazards assumption.

6.3.3 Estimation

For the ordinal models presented, the probability of a response in category c for a given level-2 unit j, conditional on the random effects $\boldsymbol{\theta}$, is equal to

$$\Pr(Y_{ij} = c \mid \boldsymbol{\theta}) = P_{ijc} - P_{ij,c-1} \,,$$

where $P_{ijc} = 1/[1+\exp(-\eta_{ijc})]$ under the logistic response function (formulas for other response functions are given in Section 6.2.3). Note that because $\gamma_0 = -\infty$ and $\gamma_C = \infty$, $P_{ij0} = 0$ and $P_{ijC} = 1$. Here, η_{ijc} denotes the response model, for example,

$$\eta_{ijc} = \gamma_c - \left[(\boldsymbol{x}^*_{ij})'\boldsymbol{\beta}_c + \boldsymbol{x}'_{ij}\boldsymbol{\beta} + \boldsymbol{z}'_{ij}\boldsymbol{T}\boldsymbol{\theta}_i \right] ,$$

or one of the other variants of η_{ijc} presented. In what follows, we'll consider the general model allowing for non-proportional odds, since the more restrictive proportional odds model is just a special case (i.e., when $\boldsymbol{\beta}_c = 0$).

Let $\boldsymbol{Y}_j$ denote the vector of ordinal responses from level-2 unit j (for the n_j level-1 units nested within). The probability of any pattern $\boldsymbol{Y}_j$ conditional on $\boldsymbol{\theta}$ is equal to the product of the probabilities of the level-1 responses,

$$\ell(\boldsymbol{Y}_j \mid \boldsymbol{\theta}) = \prod_{i=1}^{n_j} \prod_{c=1}^{C} (P_{ijc} - P_{ij,c-1})^{\mathrm{y}_{ijc}}, \qquad (6.8)$$

where $\mathrm{y}_{ijc} = 1$ if $Y_{ij} = c$ and 0 otherwise (i.e., for each ij-th observation, $\mathrm{y}_{ijc} = 1$ for only one of the C categories). For the ordinal representation of the survival model, where right-censoring is present, the above likelihood is generalized to

$$\ell(\boldsymbol{Y}_j \mid \boldsymbol{\theta}) = \prod_{i=1}^{n_j} \prod_{c=1}^{C} \left[(P_{ijc} - P_{ij,c-1})^{d_{ij}} \, (1 - P_{ijc})^{1-d_{ij}} \right]^{\mathrm{y}_{ijc}} , \qquad (6.9)$$

where $d_{ij} = 1$ if Y_{ij} represents an event, or $d_{ij} = 0$ if Y_{ij} represents a censored observation. Notice that (6.9) is equivalent to (6.8) when $d_{ij} = 1$ for all observations. With right-censoring, because there is essentially one additional response category (for those censored at the last category C), it is $\gamma_{C+1} = \infty$ and so $P_{ij,C+1} = 1$. In this case, parameters γ_c and $\boldsymbol{\beta}_c$ with $c = 1, \ldots, C$ are estimable, otherwise c only goes to $C - 1$.

The marginal density of $\underline{\boldsymbol{Y}}_j$ in the population is expressed as the following integral of the likelihood, $\ell(\cdot)$, weighted by the prior density $g(\cdot)$,

$$h(\boldsymbol{Y}_j) = \int_{\boldsymbol{\theta}} \ell(\boldsymbol{Y}_j \mid \boldsymbol{\theta})\, g(\boldsymbol{\theta})\, \mathrm{d}\boldsymbol{\theta}, \tag{6.10}$$

where $g(\boldsymbol{\theta})$ represents the multivariate standard normal density. The marginal log-likelihood from the N level-2 units, $\log L = \sum_j^N \log h(\boldsymbol{Y}_j)$, is then maximized to yield maximum likelihood estimates. For this, denote the conditional likelihood as ℓ_j and the marginal density as h_j. Differentiating first with respect to the parameters that vary with c, let $\boldsymbol{\alpha}_k$ represent a particular threshold γ_k or regression vector $\boldsymbol{\beta}_k^*$, where $k = 1, \ldots, C$ if right-censoring occurs, otherwise $k = 1, \ldots, C-1$. Then

$$\frac{\partial \log L}{\partial \boldsymbol{\alpha}_k} = \sum_{j=1}^{N} h_j^{-1} \frac{\partial h_j}{\partial \boldsymbol{\alpha}_k},$$

with

$$\frac{\partial h_j}{\partial \boldsymbol{\alpha}_k} = \int_{\boldsymbol{\theta}} \sum_{i=1}^{n_j} \sum_{c=1}^{C} \mathrm{y}_{ijc} \left[d_{ij} \frac{(\partial P_{ijc}) a_{ck} - (\partial P_{ij,c-1}) a_{c-1,k}}{P_{ijc} - P_{ij,c-1}} \right.$$
$$\left. - (1 - d_{ij}) \frac{(\partial P_{ijc}) a_{ck}}{1 - P_{ijc}} \right] \times \ell_j\, g(\boldsymbol{\theta})\, \frac{\partial \eta_{ijk}}{\partial \boldsymbol{\alpha}_k}\, \mathrm{d}\boldsymbol{\theta}, \tag{6.11}$$

where $\partial \eta_{ijk} / \partial \boldsymbol{\alpha}_k = 1$ and $-\boldsymbol{x}_{ij}^*$ for the thresholds and regression coefficients, respectively, and $a_{ck} = 1$ if $c = k$ (and $= 0$ if $c \neq k$). Also, ∂P_{ijc} represents the pdf of the response function; various forms of this are given in Section 6.2.3.

For the parameters that do not vary with c, let $\boldsymbol{\zeta}$ represent an arbitrary parameter vector; then for $\boldsymbol{\beta}$ and the vector $\mathrm{v}(\boldsymbol{T})$, which contains the unique elements of the Cholesky factor $\boldsymbol{T}$, we get

$$\frac{\partial \log L}{\partial \boldsymbol{\zeta}} = \sum_{j=1}^{N} h_j^{-1} \int_{\boldsymbol{\theta}} \sum_{i=1}^{n_j} \sum_{c=1}^{C} \mathrm{y}_{ijc} \left[d_{ij} \frac{\partial P_{ijc} - \partial P_{ij,c-1}}{P_{ijc} - P_{ij,c-1}} - (1 - d_{ij}) \frac{\partial P_{ijc}}{1 - P_{ijc}} \right]$$
$$\times \ell_j\, g(\boldsymbol{\theta})\, \frac{\partial \eta_{ijc}}{\partial \boldsymbol{\zeta}}\, \mathrm{d}\boldsymbol{\theta}, \tag{6.12}$$

where

$$\frac{\partial \eta_{ijc}}{\partial \boldsymbol{\beta}} = -\boldsymbol{x}_{ij}, \qquad \frac{\partial \eta_{ijc}}{\partial (\mathrm{v}(\boldsymbol{T}))} = -\mathrm{J}_r (\boldsymbol{\theta} \otimes \boldsymbol{z}_{ij}),$$

and J_r is the elimination matrix of Magnus [69], which eliminates the elements above the main diagonal. If $\boldsymbol{T}$ is an $r \times 1$ vector of independent variance terms (e.g., if $\boldsymbol{z}_{ij}$ is an $r \times 1$ vector of level-1 or level-2 grouping variables, see Section 6.7), then $\partial \eta_{ijc} / \partial \boldsymbol{T} = \boldsymbol{z}_{ij} \theta$ in the equation above.

Fisher's method of scoring can be used to provide the solution to these likelihood equations. For this, provisional estimates for the vector of parameters $\boldsymbol{\Theta}$, on iteration ι are improved by

$$\boldsymbol{\Theta}_{\iota+1} = \boldsymbol{\Theta}_{\iota} - \left\{ E \left[\frac{\partial^2 \log L}{\partial \boldsymbol{\Theta}_{\iota} \, \partial \boldsymbol{\Theta}'_{\iota}} \right] \right\}^{-1} \frac{\partial \log L}{\partial \boldsymbol{\Theta}_{\iota}} , \tag{6.13}$$

where, following Bock and Lieberman [17], the information matrix, or minus the expectation of the matrix of second derivatives, is given by

$$-E \left[\frac{\partial^2 \log L}{\partial \boldsymbol{\Theta}_{\iota} \, \partial \boldsymbol{\Theta}'_{\iota}} \right] = E \left[\sum_{j=1}^{N} h_j^{-2} \frac{\partial h_j}{\partial \boldsymbol{\Theta}_{\iota}} \left(\frac{\partial h_j}{\partial \boldsymbol{\Theta}_{\iota}} \right)' \right] .$$

Its estimator is obtained using the estimated parameter values and, at convergence, the large-sample variance covariance matrix of the parameter estimates is gotten as the inverse of the information matrix. The form on the right-hand side of the above equation is sometimes called the "outer product of the gradients." It was proposed in the econometric literature by Berndt et al. [12], and is often referred to as the BHHH method.

6.4 Multilevel Nominal Response Models

Let $\underline{Y}_{ij}$ now denote a nominal variable associated with level-2 unit j and level-1 unit i. Adding random effects to the fixed-effects multinomial logistic regression model (see Agresti [2] and Long [65]), we get that the probability that $\underline{Y}_{ij} = c$ (a response occurs in category c) for a given level-2 unit j is given by

$$\underline{p}_{ijc} = \Pr(\underline{Y}_{ij} = c) = \frac{\exp(\underline{\eta}_{ijc})}{1 + \sum_{h=2}^{C} \exp(\underline{\eta}_{ijh})} \qquad \text{for } c = 2, 3, \ldots, C, \tag{6.14a}$$

$$\underline{p}_{ij1} = \Pr(\underline{Y}_{ij} = 1) = \frac{1}{1 + \sum_{h=2}^{C} \exp(\underline{\eta}_{ijh})} , \tag{6.14b}$$

where the multinomial logit $\underline{\eta}_{ijc} = \boldsymbol{x}'_{ij}\boldsymbol{\beta}_c + \boldsymbol{z}'_{ij}\boldsymbol{T}_c\,\underline{\boldsymbol{\theta}}_j$. Comparing this to the logit for ordered responses, we see that all of the covariate effects $\boldsymbol{\beta}_c$ vary across categories ($c = 2, 3, \ldots, C$). Similarly for the random-effect variance term $\boldsymbol{T}_c$. As written above, an important distinction between the model for ordinal and nominal responses is that the former uses cumulative comparisons of the categories, whereas the latter uses comparisons to a reference category.

This model generalizes Bock's model for educational test data [14] by including covariates $\boldsymbol{x}_{ij}$ and by allowing a general random-effects design vector $\boldsymbol{z}_{ij}$ including the possibility of multiple random effects $\underline{\boldsymbol{\theta}}_j$. As discussed by

Bock [14], the model has a plausible interpretation. Namely, each nominal category is assumed to be related to an underlying latent "response tendency" for that category. The category c associated with the response variable $\underline{Y}_{ij}$ is then the category for which the response tendency is maximal. Notice that this assumption of C latent variables differs from the ordinal model where only one underlying latent variable is assumed. Bock [15] refers to the former as the extremal concept and the latter as the aforementioned threshold concept, and notes that both were introduced into psychophysics by Thurstone [111]. The two are equivalent only for the dichotomous case (i.e., when there are only two response categories).

The model as written above allows estimation of any pairwise comparisons among the C response categories. As characterized in Bock [14], it is benefical to write the nominal model to allow for any possible set of $C-1$ contrasts. For this, the category probabilities are written as

$$\underline{p}_{ijc} = \frac{\exp(\underline{\eta}_{ijc})}{\sum_{h=1}^{C} \exp(\underline{\eta}_{ijh})} \qquad \text{for } c = 1, 2, \ldots, C, \tag{6.15}$$

where now

$$\underline{\eta}_{ijc} = \boldsymbol{x}'_{ij}\boldsymbol{\Gamma}\boldsymbol{d}_c + (\boldsymbol{z}'_{ij} \otimes \underline{\boldsymbol{\theta}}'_j)\mathrm{J}'_{r^*}\boldsymbol{\Lambda}\boldsymbol{d}_c\,. \tag{6.16}$$

Here, $\boldsymbol{D}$ is the $(C-1) \times C$ matrix containing the contrast coefficients for the $C-1$ contrasts between the C logits and $\boldsymbol{d}_c$ is the c-th column vector of this matrix. The $s \times (C-1)$ parameter matrix $\boldsymbol{\Gamma}$ contains the regression coefficients associated with the s covariates for each of the $C-1$ contrasts. Similarly, $\boldsymbol{\Lambda}$ contains the random-effect variance parameters for each of the $C-1$ contrasts. Specifically,

$$\boldsymbol{\Lambda} = [\, \mathrm{v}(\boldsymbol{T}_1) \quad \mathrm{v}(\boldsymbol{T}_2) \quad \ldots \quad \mathrm{v}(\boldsymbol{T}_{C-1})\,],$$

where $\mathrm{v}(\boldsymbol{T}_c)$ is the $r^* \times 1$ vector ($r^* = r(r+1)/2$) of elements below and on the diagonal of the Cholesky (lower-triangular) factor $\boldsymbol{T}_c$ and J_{r^*} is the aforementioned elimination matrix of Magnus [69]. This latter matrix is necessary to ensure that the appropriate terms from the $1 \times r^2$ vector resulting from the Kronecker product $(\boldsymbol{z}'_{ij} \otimes \underline{\boldsymbol{\theta}}'_j)$ are multiplied with the $r^* \times 1$ vector resulting from $\boldsymbol{\Lambda}\boldsymbol{d}_c$. For the case of a random-intercepts model, the model simplifies to

$$\underline{\eta}_{ijc} = \boldsymbol{x}'_{ij}\boldsymbol{\Gamma}\boldsymbol{d}_c + \boldsymbol{\Lambda}\boldsymbol{d}_c\,\underline{\theta}_j\,,$$

with $\boldsymbol{\Lambda}$ as the $1 \times (C-1)$ vector $\boldsymbol{\Lambda} = [\,\sigma_1 \quad \sigma_2 \quad \ldots \quad \sigma_{C-1}\,]$.

Notice that if $\boldsymbol{D}$ equals

$$\boldsymbol{D} = \begin{pmatrix} 0\,1\,0\ldots0 \\ 0\,0\,1\ldots0 \\ \cdot\;\cdot\;\cdot\;\cdot\;\cdot\;\cdot\;\cdot \\ 0\,0\,0\ldots1 \end{pmatrix},$$

the model simplifies to the earlier representation in (6.14a) and (6.14b). The current formulation, however, allows for a great deal of flexibility in the types of comparisons across the C response categories. For example, if the categories are ordered, an alternative to the cumulative logit model of the previous section is to employ Helmert contrasts [15] within the nominal model. For this, with $C = 4$, the contrast matrix would be

$$\boldsymbol{D} = \begin{pmatrix} -1 & \frac{1}{3} & \frac{1}{3} & \frac{1}{3} \\ 0 & -1 & \frac{1}{2} & \frac{1}{2} \\ 0 & 0 & -1 & 1 \end{pmatrix} .$$

Helmert contrasts are similar to the category comparisons of continuation-ratio logit models, as described within a mixed-model formulation by Ten Have and Uttal [108]. However, the Helmert contrasts above are applied to the category logits, rather then the category probabilities as in continuation-ratio models.

6.4.1 Parameter Estimation

Estimation follows the procedure described for ordinal outcomes. Specifically, letting $\boldsymbol{Y}_j$ denote the vector of nominal responses from level-2 unit j (for the n_j level-1 units nested within), the probability of any $\boldsymbol{Y}_j$ conditional on the random effects $\boldsymbol{\theta}$ is equal to the product of the probabilities of the level-1 responses

$$\ell(\boldsymbol{Y}_j \mid \boldsymbol{\theta}) = \prod_{i=1}^{n_j} \prod_{c=1}^{C} (p_{ijc})^{\mathrm{y}_{ijc}} , \tag{6.17}$$

where $\mathrm{y}_{ijc} = 1$ if $Y_{ij} = c$, and 0 otherwise. The marginal density of the response vector $\boldsymbol{Y}_j$ is again given by (6.10). The marginal log-likelihood from the N level-2 units, $\log L = \sum_j^N \log h(\boldsymbol{Y}_j)$, is maximized to obtain maximum likelihood estimates of $\boldsymbol{\Gamma}$ and $\boldsymbol{\Lambda}$. Specifically, using $\boldsymbol{\Delta}$ to represent either parameter matrix,

$$\frac{\partial \log L}{\partial \boldsymbol{\Delta}'} = \sum_{j=1}^{N} h^{-1}(\boldsymbol{Y}_j) \int_{\boldsymbol{\theta}} \left[\sum_{i=1}^{n_j} \boldsymbol{D} \left(\mathbf{y}_{ij} - \boldsymbol{P}_{ij} \right) \otimes \partial \boldsymbol{\Delta} \right] \times \ell(\boldsymbol{Y}_j \mid \boldsymbol{\theta})\, g(\boldsymbol{\theta})\, d\boldsymbol{\theta}, \tag{6.18}$$

where

$$\partial \boldsymbol{\Gamma} = \boldsymbol{x}'_{ij} , \qquad \partial \boldsymbol{\Lambda} = \left[\mathrm{J}_{r^*} (\boldsymbol{\theta} \otimes \boldsymbol{z}_{ij}) \right]' ,$$

$\mathbf{y}_{ij}$ is the $C \times 1$ indicator vector, and $\boldsymbol{P}_{ij}$ is the $C \times 1$ vector obtained by applying (6.15) for each category. As in the ordinal case, Fisher's method of scoring can be used to provide the solution to these likelihood equations.

6.5 Computational Issues

In order to solve the above likelihood solutions for both the ordinal and nominal models, integration over the random-effects distribution must be performed. Additionally, the above likelihood solutions are only in terms of the regression parameters and variance-covariance parameters of the random-effects distribution. Often, estimation of the random effects is also of interest. These issues are described in great detail in Skrondal and Rabe-Hesketh [98]; here, we discuss some of the relevant points.

6.5.1 Integration over θ

Various approximations for evaluating the integral over the random-effects distribution have been proposed in the literature; several of these are compared in Chapter 9. Perhaps the most frequently used methods are based on first- or second-order Taylor expansions. Marginal quasi-likelihood (MQL) involves expansion around the fixed part of the model, whereas penalized or predictive quasi-likelihood (PQL) additionally includes the random part in its expansion [39]. Both of these are available in the MLwiN software program [84]. Unfortunately, several authors [19, 87, 90] have reported downwardly biased estimates using these procedures in certain situations, especially for the first-order expansions.

Raudenbush et al. [87] proposed an approach that uses a combination of a fully multivariate Taylor expansion and a Laplace approximation. Based on the results in Raudenbush et al. [87], this method yields accurate results and is computationally fast. Also, as opposed to the MQL and PQL approximations, the deviance obtained from this approximation can be used for likelihood-ratio tests. This approach has been incorporated into the HLM software program [86].

Numerical integration can also be used to perform the integration over the random-effects distribution. Specifically, if the assumed distribution is normal, Gauss-Hermite quadrature can approximate the above integral to any practical degree of accuracy [104]. Additionally, like the Laplace approximation, the numerical quadrature approach yields a deviance that can be readily used for likelihood-ratio tests. The integration is approximated by a summation on a specified number of quadrature points Q for each dimension of the integration. The solution via quadrature can involve summation over a large number of points, especially as the number of random effects is increased. For example, if there is only one random effect, the quadrature solution requires only one additional summation over Q points relative to the fixed-effects solution. For models with $r > 1$ random effects, however, the quadrature is performed over Q^r points, and so becomes computationally burdensome for $r > 5$ or so. Also, Lesaffre and Spiessens [61] present an example where the method only gives valid results for a high number of quadrature points. These authors advise practitioners to routinely examine results for the dependence on Q. To address these issues, several authors have described a method of adaptive quadrature that uses relatively few points per dimension (e.g., 3 or so), which are adapted to the location and dispersion of the distribution to be integrated [18, 64, 78, 80]. Simulations show that adaptive quadrature performs well in a wide variety of situations and typically outperforms ordinary quadrature [82]. Several software packages have implemented ordinary or adaptive Gauss-Hermite quadrature, including Egret® [22], gllamm [81], LIMDEP [40], MIXOR [49], MIXNO [46], Stata [101], and SAS PROC NLMIXED [93].

Another approach that is commonly used in econometrics and transportation research uses simulation methods to integrate over the random-effects distribution (see the introductory overview by Stern [102] and the excellent book by Train [112]). When used in conjunction with maximum likelihood estimation, it is called "maximum simulated likelihood" or "simulated maximum likelihood." The idea behind this approach is to draw a number of values from the random-effects distribution, calculate the likelihood for each of these draws, and average over the draws to obtain a solution. Thus, this method maximizes a simulated sample likelihood instead of an exact likelihood, but can be considerably faster than quadrature methods, especially as the number of random effects increases [41]. It is a very flexible and intuitive approach with many potential applications (see Drukker [27]). In particular, Bhat [13] and Glasgow [36] describe this estimation approach for multilevel models of nominal outcomes. In terms of software, LIMDEP [40] has included this estimation approach for several types of outcome variables, including nominal and ordinal, and Haan and Uhlendorff [41] describe a Stata routine for nominal data.

Bayesian approaches, such as the use of Gibbs sampling [33] and related methods [105], can also be used to integrate over the random-effects distribution. This approach is described in detail in Chapter 2. For nominal responses, Daniels and Gatsonis [25] use this approach in their multilevel polychotomous regression model. Similarly, Ishwaran [57] utilize Bayesian methods in modeling multilevel ordinal data. The freeware BUGS software program [100] can be used to facilitate estimation via Gibbs sampling. In this regard, Marshall and Spiegelhalter [70] provide an example of multilevel modeling using BUGS, including some syntax and discussion of the program.

6.5.2 Estimation of Random Effects and Probabilities

In many cases, it is useful to obtain estimates of the random effects and also to obtain fitted marginal probabilities. The random effects $\boldsymbol{\theta}_j$ can be estimated using empirical Bayes methods [16]. For the univariate case, this estimator $\hat{\theta}_j$ is given by the mean of the posterior distribution,

$$\hat{\theta}_j = E(\underline{\theta}_j \mid \boldsymbol{Y}_j) = \frac{1}{h(\boldsymbol{Y}_j)} \int_\theta \theta_j \, \ell(\cdot) \, g(\theta) \, \mathrm{d}\theta, \tag{6.19}$$

where $\ell(\cdot)$ is the conditional likelihood for the particular model (i.e., ordinal or nominal). The variance of the posterior distribution is obtained as

$$\mathrm{Var}(\hat{\underline{\theta}}_j \mid \boldsymbol{Y}_j) = \frac{1}{h(\boldsymbol{Y}_j)} \int_\theta (\theta_j - \hat{\theta}_j)^2 \, \ell(\cdot) \, g(\theta) \, \mathrm{d}\theta.$$

These quantities may be used, for example, to evaluate the response probabilities for particular level-2 units (e.g., person-specific trend estimates).

To obtain estimated marginal probabilities (e.g., the estimated response probabilities of the control group across time), an additional step is required for models with non-linear response functions (e.g., the models considered in this chapter). First, so-called "subject-specific" probabilities [75, 118] are estimated for specific values of covariates and random effects, say θ^*. These subject-specific estimates indicate, for example, the response probability for a subject with random effect level θ^* in the control group at a particular time-point. Denoting these subject-specific probabilities as $\hat{P}_{ss}$, marginal probabilities $\hat{P}_m$ can then be obtained by numerical quadrature, namely $\hat{P}_m = \int_\theta \hat{P}_{ss} \, g(\theta) \, \mathrm{d}\theta$, or by marginalizing the scale of the regression coefficients [51, p. 179]. Continuing with our example, the marginalized estimate would indicate the estimated response probability for the entire control group at a particular time-point. Both subject-specific and marginal estimates have their uses, since they are estimating different quantities, and several authors have characterized the differences between the two [45, 62, 75].

6.6 Intraclass Correlation

For a random-intercepts model (i.e., $\boldsymbol{z}_j = \boldsymbol{1}_{n_j}$), it is often of interest to express the level-2 variance in terms of an intraclass correlation. For this, one can make reference to the threshold concept and the underlying latent response tendency that determines the observed response. For the ordinal logistic model assuming normally distributed random effects, the estimated intraclass correlation equals $\hat{\sigma}^2/(\hat{\sigma}^2 + \pi^2/3)$, where the latter term in the denominator represents the variance of the underlying latent response tendency. As mentioned earlier, for the logistic model, this variable is assumed to be distributed as a standard logistic distribution with variance equal to $\pi^2/3$. For a probit model, this term is replaced by 1, the variance of the standard normal distribution.

For the nominal model, one can make reference to multiple underlying latent response tendencies, denoted as $\underline{y}_{ijc}$, and the associated regression model including level-1 residuals $\underline{\epsilon}_{ijc}$,

$$\underline{y}_{ijc} = \boldsymbol{x}'_{ij}\boldsymbol{\beta}_c + \boldsymbol{z}'_{ij}\boldsymbol{T}_c\underline{\boldsymbol{\theta}}_j + \underline{\epsilon}_{ijc}, \qquad c = 1, 2, \ldots, C.$$

As mentioned earlier, for a particular ij-th unit, the category c associated with the nominal response variable $\underline{Y}_{ij}$ is the one for which the latent $\underline{y}_{ijc}$ is maximal. Since, in the common reference cell formulation, $c = 1$ is the reference category, $\boldsymbol{T}_1 = \boldsymbol{\beta}_1 = 0$, and so the model can be rewritten as

$$\underline{y}_{ijc} = \boldsymbol{x}'_{ij}\boldsymbol{\beta}_c + \boldsymbol{z}'_{ij}\boldsymbol{T}_c\,\underline{\boldsymbol{\theta}}_j + (\underline{\epsilon}_{ijc} - \underline{\epsilon}_{ij1}), \qquad c = 2, \ldots, C,$$

for the latent response tendency of category c relative to the reference category. It can be shown that the level-1 residuals $\underline{\epsilon}_{ijc}$ for each category are distributed according to a type I extreme-value distribution [see 68, p. 60]. It can further be shown that the standard logistic distribution is obtained as the difference of two independent type I extreme-value variates [see 72, pp. 20 and 142]. As a result, the level-1 variance is given by $\pi^2/3$, which is the variance for a standard logistic distribution. The estimated intraclass correlations are thus calculated as $r_c = \hat{\sigma}_c^2/(\hat{\sigma}_c^2 + \pi^2/3)$, where $\hat{\sigma}_c^2$ is the estimated level-2 variance assuming normally distributed random intercepts. Notice that $C - 1$ intraclass correlations are estimated, one for each category c versus the reference category. As such, the cluster influence on the level-1 responses is allowed to vary across the nominal response categories.

6.7 Heterogeneous Variance Terms

Allowing for separate random-effect variance terms for groups of either i or j units is sometimes important. For example, in a twin study it is often necessary to allow the intra-twin correlation to differ between monozygotic and dizygotic twins. In this situation, subjects ($i = 1, 2$) are nested within twin pairs ($j = 1, \ldots, N$). To allow the level-2 variance to vary for these two twin-pair types, the random-effects design vector $\boldsymbol{z}_{ij}$ is specified as a 2×1 vector of dummy codes indicating monozygotic and dizygotic twin-pair status, respectively. $\boldsymbol{T}$ (or $\boldsymbol{T}_c$ in the nominal model) is then a 2×1 vector of independent random-effect standard deviations for monozygotics and dizygotics, and the cluster effect $\underline{\theta}_j$ is a scalar that is pre-multiplied by the vector $\boldsymbol{T}$. For example, for a random-intercepts proportional odds model, we would have

$$\log\left[\frac{\underline{P}_{ijc}}{1 - \underline{P}_{ijc}}\right] = \gamma_c - \left\{\boldsymbol{x}'_{ij}\boldsymbol{\beta} + [MZ_j \quad DZ_j]\begin{bmatrix}\sigma_{\delta\,(MZ)} \\ \sigma_{\delta\,(DZ)}\end{bmatrix}\underline{\theta}_j\right\},$$

where MZ_j and DZ_j are dummy codes indicating twin-pair status (i.e., if $MZ_j = 1$ then $DZ_j = 0$, and vice versa).

Notice, that if the probit formulation is used and the model has no covariates (i.e., only an intercept, $x_{ij} = 1$), the resulting intraclass correlations

$$ICC_{MZ} = \frac{\sigma^2_{\delta\,(MZ)}}{\sigma^2_{\delta\,(MZ)} + 1} \qquad \text{and} \qquad ICC_{DZ} = \frac{\sigma^2_{\delta\,(DZ)}}{\sigma^2_{\delta\,(DZ)} + 1}$$

are polychoric correlations (for ordinal responses) or tetrachoric correlations (for binary responses) for the within twin-pair data. Adding covariates then yields adjusted tetrachoric and polychoric correlations. Because estimation of polychoric and tetrachoric correlations is often important in twin and genetic studies, these models are typically formulated in terms of the probit link. Comparing models that allow homogeneous versus heterogeneous subgroup random-effects variance thus allows testing of whether the tetrachoric (or polychoric) correlations are equal across the subgroups.

The use of heterogeneous variance terms can also be found in some item response theory (IRT) models in the educational testing literature [14, 16, 92]. Here, item responses ($i = 1, 2, \ldots, m$) are nested within subjects ($j = 1, 2, \ldots, N$) and a separate random-effect standard deviation (i.e., an element of the $m \times 1$ vector $\boldsymbol{T}$) is estimated for each test item (i.e., each i unit). In the multilevel model this is accomplished by specifying $\boldsymbol{z}_{ij}$ as an $m \times 1$ vector of dummy codes indicating the repeated items. To see this, consider the popular two-parameter logistic model for dichotomous responses [66] that specifies the probability of a correct response to item i ($\underline{Y}_{ij} = 1$) as a function of the ability of subject j ($\underline{\theta}_j$),

$$\Pr(\underline{Y}_{ij} = 1) = \frac{1}{1 + \exp[-a_i(\underline{\theta}_j - b_i)]} ,$$

where a_i is the slope parameter for item i (i.e., item discrimination), and b_i is the threshold or difficulty parameter for item i (i.e., item difficulty). The distribution of ability in the population of subjects is assumed to be normal with mean 0 and variance 1 (i.e., the usual assumption for the random effects $\underline{\theta}_j$ in the multilevel model). As noted by Bock and Aitkin [16], it is convenient to let $c_i = -a_i b_i$ and write

$$\Pr(\underline{Y}_{ij} = 1) = \frac{1}{1 + \exp[-(c_i + a_i\,\underline{\theta}_j)]} ,$$

which can be recast in terms of the logit of the response as

$$\text{logit}_{ij} = \log\left[\frac{\underline{p}_{ij}}{1 - \underline{p}_{ij}}\right] = c_i + a_i\,\underline{\theta}_j\,.$$

As an example, suppose that there are four items. This model can be represented in matrix form as

$$\underbrace{\begin{pmatrix} \text{logit}_{1j} \\ \text{logit}_{2j} \\ \text{logit}_{3j} \\ \text{logit}_{4j} \end{pmatrix}}_{} = \underbrace{\begin{pmatrix} 1\,0\,0\,0 \\ 0\,1\,0\,0 \\ 0\,0\,1\,0 \\ 0\,0\,0\,1 \end{pmatrix}}_{\boldsymbol{X}_j} \underbrace{\begin{pmatrix} c_1 \\ c_2 \\ c_3 \\ c_4 \end{pmatrix}}_{\boldsymbol{c}} + \underbrace{\begin{pmatrix} 1\,0\,0\,0 \\ 0\,1\,0\,0 \\ 0\,0\,1\,0 \\ 0\,0\,0\,1 \end{pmatrix}}_{\boldsymbol{Z}_j} \underbrace{\begin{pmatrix} a_1 \\ a_2 \\ a_3 \\ a_4 \end{pmatrix}}_{\boldsymbol{a}} (\underline{\theta}_j),$$

showing that this IRT model is a multilevel model that allows the random-effects variance terms to vary across items (level 1). The usual IRT notation is a bit different than the multilevel notation, but $\boldsymbol{c}$ simply represents the fixed effects (i.e., $\boldsymbol{\beta}$) and $\boldsymbol{a}$ is the the random-effects standard deviation vector $\boldsymbol{T}' = [\,\sigma_{\delta 1} \quad \sigma_{\delta 2} \quad \sigma_{\delta 3} \quad \sigma_{\delta 4}\,]$.

The elements of the $\boldsymbol{T}$ vector can also be viewed as the (unscaled) factor loadings of the items on the (unidimensional) underlying ability variable ($\underline{\theta}$). A simpler IRT model that constrains these factor loadings to be equal is the one-parameter logistic model, the so-called Rasch model [116]. This constraint is achieved by setting $\boldsymbol{Z}_j = \boldsymbol{1}_{n_j}$ and $\boldsymbol{a} = a$ in the above model. Thus, the Rasch model is simply a random-intercepts logistic regression model with item indicators for $\boldsymbol{X}$.

Unlike traditional IRT models, the multilevel formulation of the model easily allows multiple covariates at either level (i.e., items or subjects). This and other advantages of casting IRT models as multilevel models are described in detail by Adams et al. [1] and Rijmen et al. [89]. In particular, this allows a model for examining whether item parameters vary by subject characteristics, and also for estimating ability in the presence of such item by subject interactions. Interactions between item parameters and subject characteristics, often termed item bias [20], is an area of active psychometric research. Also, although the above illustration is in terms of a dichotomous response model, the analogous multilevel ordinal and nominal models apply. For ordinal items responses, application of the cumulative logit multilevel models yields what Thissen and Steinberg [110] have termed "difference models," namely, the treatment of ordinal responses as developed by Samejima [92] within the IRT context. Similarly, in terms of nominal responses, the multilevel model yields the nominal IRT model developed by Bock [14].

6.8 Health Services Research Example

The McKinney Homeless Research Project (MHRP) study [55, 56] in San Diego, CA was designed to evaluate the effectiveness of using Section 8 certificates as a means of providing independent housing to the severely mentally ill homeless. Section 8 housing certificates were provided from the Department of Housing and Urban Development (HUD) to local housing authorities in San Diego. These housing certificates, which require clients to pay 30% of their income toward rent, are designed to make it possible for low-income individuals to choose and obtain independent housing in the community. Three hundred sixty-one clients took part in this longitudinal study employing a randomized factorial design. Clients were randomly assigned to one of two types of supportive case management (comprehensive vs. traditional) and to one of two levels of access to independent housing (using Section 8 certificates).

Eligibility for the project was restricted to individuals diagnosed with a severe and persistent mental illness who were either homeless or at high risk of becoming homeless at the start of the study. Individuals' housing status was classified at baseline and at 6-, 12-, and 24-month follow-ups.

In this illustration, focus will be on examining the effect of access to Section 8 certificates on repeated housing outcomes across time. Specifically, at each time-point each subjects' housing status was classified as either streets/shelters, community housing, or independent housing. This outcome can be thought of as ordinal with increasing categories indicating improved housing outcomes. The observed sample sizes and response proportions for these three outcome categories by group are presented in Table 6.1.

These observed proportions indicate a general decrease in street living and an increase in independent living across time for both groups. The increase in independent housing, however, appears to occur sooner for the Section 8 group relative to the control group. Regarding community living, across time this increases for the control group and decreases for the Section 8 group.

There is some attrition across time; attrition rates of 19.4% and 12.7% are observed at the final time-point for the control and Section 8 groups, respectively. Since estimation of model parameters is based on a full-likelihood approach, the missing data are assumed to be "ignorable" conditional on both the model covariates and the observed responses [60]. In longitudinal studies, ignorable nonresponse falls under the "missing at random" (MAR) assumption introduced by Rubin [91], in which the missingness depends only on observed data. In what follows, since the focus is on describing application of the various multilevel regression models, we will make the MAR assumption.

Table 6.1 Housing status across time by group: response proportions and sample sizes.

		Time-point			
Group	Status	Baseline	6-months	12-months	24-months
Control	Street	.555	.186	.089	.124
	Community	.339	.578	.582	.455
	Independent	.106	.236	.329	.421
	n	180	161	146	145
Section 8	Street	.442	.093	.121	.120
	Community	.414	.280	.146	.228
	Independent	.144	.627	.732	.652
	n	181	161	157	158

A further approach, however, that does not rely on the MAR assumption (e.g., a multilevel pattern-mixture model as described in Hedeker and Gibbons [50]) could be used. Missing data issues are described more fully in Chapter 10.

6.8.1 Ordinal Response Models

To prepare for the ordinal analyses, the observed cumulative logits across time for the two groups are plotted in Figs. 6.3 and 6.4. The first cumulative logit compares independent and community housing versus street living (i.e., categories 2 and 3 combined versus 1), while the second cumulative logit compares independent housing versus community housing and street living (i.e., category 3 versus 2 and 1 combined). For the proportional odds model to hold, these two plots should look the same, with the only difference being the scale difference on the y-axis. As can be seen, these plots do not look that similar. For example, the post-baseline group differences do not appear to be the same for the two cumulative logits. In particular, it appears that the Section 8 group does better more consistently in terms of the second cumulative logit (i.e., independent versus community and street housing). This would imply that the proportional odds model is not reasonable for these data.

To assess this more rigorously, two ordinal logistic multilevel models were fit to these data, the first assuming a proportional odds model and the second relaxing this assumption. For both analyses, the repeated housing status classifications were modeled in terms of time effects (6-, 12-, and 24-month follow-ups compared to baseline), a group effect (Section 8 versus control), and group by time interaction terms. The first analysis assumes these effects are the same across the two cumulative logits of the model, whereas the second

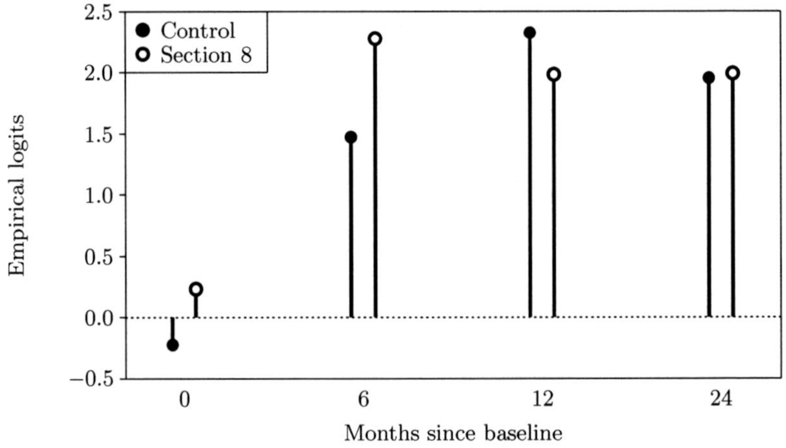

Fig. 6.3 First cumulative logit values across time by group.

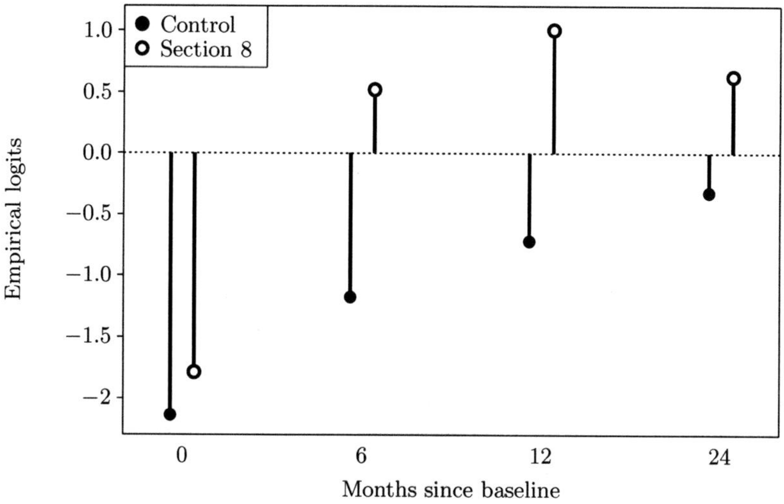

Fig. 6.4 Second cumulative logit values across time by group.

analysis estimates effects for each explanatory variable on each of the two cumulative logits. In terms of the multilevel part of the model, only a random subject effect was included in both analyses. Results from these analyses are given in Table 6.2.

The proportional odds model indicates significant time effects for all time-points relative to baseline, but only significant group by time interactions for the 6- and 12-month follow-ups. Marginally significant effects are obtained for the Section 8 effect and the Section 8 by t3 (24-months) interaction. Thus, the analysis indicates that the control group moves away from street living to independent living across time and that this improvement is more pronounced for Section 8 subjects at the 6- and 12-month follow-up. Because the Section 8 by t3 interaction is only marginally significant, the groups do not differ significantly in housing status at the 24-month follow-up as compared to baseline.

However, comparing log-likelihood values clearly rejects the proportional odds assumption (likelihood ratio $\chi^2_7 = 52.14$), indicating that the effects of the explanatory variables cannot be assumed identical across the two cumulative logits. Interestingly, none of the Section 8 by time interaction terms are significant in terms of the non-street logit (i.e., comparing categories 2 and 3 versus 1), while all of them are significant in terms of the independent logit (i.e., comparing category 3 versus 1 and 2 combined). Thus, as compared to baseline, Section 8 subjects are more likely to be in independent housing at all follow-up time-points, relative to the control group.

In terms of the random subject effect, it is clear that the data are correlated within subjects. Expressed as an intraclass correlation, the attributable

Table 6.2 Housing status across time: Ordinal logistic model estimates and standard errors (se).

	Proportional Odds		Non-Proportional Odds			
			Non-street[a]		Independent[b]	
Term	estimate	se	estimate	se	estimate	se
Intercept	−.220	.203	−.322	.218		
Threshold	**2.744**	.110			**2.377**	.279
t1 (6 month vs. base)	**1.736**	.233	**2.297**	.298	**1.079**	.358
t2 (12 month vs. base)	**2.315**	.268	**3.345**	.450	**1.645**	.336
t3 (24 month vs. base)	**2.499**	.247	**2.821**	.369	**2.145**	.339
Section 8 (yes=1, no=0)	*.497*	.280	*.592*	.305	.323	.401
Section 8 by t1	**1.408**	.334	.566	.478	**2.023**	.478
Section 8 by t2	**1.173**	.360	−.958	.582	**2.016**	.466
Section 8 by t3	*.638*	.331	−.366	.506	**1.073**	.472
Subject sd	**1.459**	.106	**1.457**	.112		
$-2\log L$	2274.39		2222.25			

bold indicates $p < .05$, *italic* indicates $.05 < p < .10$

[a] Logit comparing independent and community housing vs. street

[b] Logit comparing independent housing vs. community housing and street

variance at the subject level equals .39 for both models. Also, the Wald test is highly significant in terms of rejecting the null hypothesis that the (subject) population standard deviation equals zero. Strictly speaking, as noted by Raudenbush and Bryk [85] and others, this test is not to be relied upon, especially as the population variance is close to zero. In the present case, the actual significance test is not critical because it is more or less assumed that the population distribution of the subject effects will not have zero variance.

6.8.2 Nominal Response Models

For the initial set of analyses with nominal models, reference category contrasts were used and street/shelter was chosen as the reference category. Thus, the first comparison compares community to street responses, and the second compares independent to street responses. A second analysis using Helmert contrasts will be described later.

Corresponding observed logits for the reference-cell comparisons by group and time are given in Figs. 6.5 and 6.6. Comparing these plots, different patterns for the post-baseline group differences are suggested. It seems that the non-Section 8 group does better in terms of the community versus

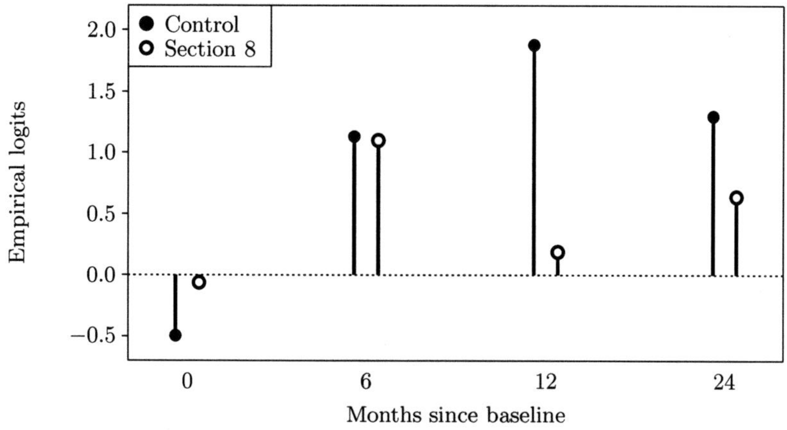

Fig. 6.5 First reference-cell logit values across time by group.

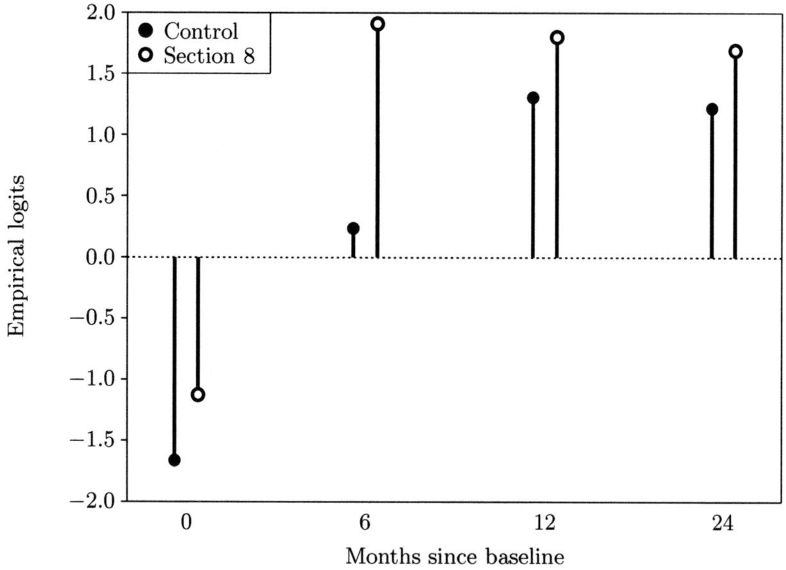

Fig. 6.6 Second reference-cell logit values across time by group.

street comparison, whereas the Section 8 group is improved for the independent versus street comparison. Further, the group differences appear to vary across time. The subsequent analyses will examine these visual impressions of the data.

To examine the sensitivity of the results to the normality assumption for the random effects, two multilevel nominal logistic regression models were fit to these data assuming the random effects were normally and uniformly

Table 6.3 Housing status (community versus street) across time: Nominal model estimates and standard errors (se).

	Normal prior		Uniform prior	
Term	estimate	se	estimate	se
Intercept	**−.452**	.192	**−.473**	.184
t1 (6 month vs. base)	**1.942**	.312	**1.850**	.309
t2 (12 month vs. base)	**2.820**	.466	**2.686**	.457
t3 (24 month vs. base)	**2.259**	.378	**2.143**	.375
Section 8 (yes=1, no=0)	*.521*	.268	*.471*	.258
Section 8 by t1	−.135	.490	−.220	.484
Section 8 by t2	**−1.917**	.611	**−1.938**	.600
Section 8 by t3	*−.952*	.535	*−.987*	.527
Subject sd	**.871**	.138	**.153**	.031
$-2\log L$	2218.73		2224.74	

bold indicates $p < .05$, *italic* indicates $.05 < p < .10$

distributed, respectively. Tables 6.3 and 6.4 list results for the two response category comparisons of community versus street and independent versus street, respectively. The time and group effects are the same as in the previous ordinal analyses.

Table 6.4 Housing status (independent versus street) across time: Nominal model estimates and standard errors (se).

	Normal prior		Uniform prior	
Term	estimate	se	estimate	se
Intercept	**−2.675**	.367	**−2.727**	.351
t1 (6 month vs. base)	**2.682**	.425	**2.540**	.422
t2 (12 month vs. base)	**4.088**	.559	**3.916**	.551
t3 (24 month vs. base)	**4.099**	.469	**3.973**	.462
Section 8 (yes=1, no=0)	.781	.491	.675	.460
Section 8 by t1	**2.003**	.614	**2.016**	.605
Section 8 by t2	.548	.694	.645	.676
Section 8 by t3	.304	.615	.334	.600
Subject sd	**2.334**	.196	**.490**	.040
$-2\log L$	2218.73		2224.74	

bold indicates $p < .05$, *italic* indicates $.05 < p < .10$

The results are very similar for the two multilevel models. Thus, the random-effects distributional form does not seem to play an important role for these data. Subjects in the control group increase both independent and community housing relative to street housing at all three follow-ups, as compared to baseline. Compared to controls, the increase in community versus street housing is less pronounced for Section 8 subjects at 12 months, but not statistically different at 6 months and only marginally different at 24 months. Conversely, as compared to controls, the increase in independent versus street housing is more pronounced for Section 8 subjects at 6 months, but not statistically different at 12 or 24 months. Thus, both groups reduce the degree of street housing, but do so in somewhat different ways. The control group subjects are shifted more toward community housing, whereas Section 8 subjects are more quickly shifted toward independent housing.

As in the ordinal case, the Wald tests are all significant for the inclusion of the random-effects variance terms. A likelihood-ratio test also clearly supports inclusion of the random subject effect (likelihood ratio $\chi_2^2 = 134.3$ and 128.3 for the normal and uniform distribution, respectively, as compared to the fixed-effects model, not shown). Expressed as intraclass correlations, $r_1 = .19$ and $r_2 = .62$ for community versus street and independent versus street, respectively. Thus, the subject influence is much more pronounced in terms of distinguishing independent versus street living, relative to community versus street living. This is borne out by contrasting models with separate versus a common random-effect variance across the two category contrasts (not shown), which yields a highly significant likelihood ratio $\chi_1^2 = 49.2$ favoring the model with separate variance terms.

An analysis was also done to examine if the random-effect variance terms varied significantly by treatment group. The deviance $(-2 \log L)$ for this model, assuming normally distributed random effects, equaled 2218.43, which was nearly identical to the value of 2218.73 (from Tables 6.3 and 6.4) for the model assuming homogeneous variances across groups. The control group and Section 8 group estimates of the subject standard deviations were respectively .771 (se = .182) and .966 (se = .214) for the community versus street comparison, and 2.228 (se = .299) and 2.432 (se = .266) for the independent versus street comparison. Thus, the homogeneity of variance assumption across treatment groups is clearly not rejected.

Finally, Table 6.5 lists the results obtained for an analysis assuming normally distributed random effects and using Helmert contrasts for the three response categories. From this analysis, it is interesting that none of the Section 8 by time interaction terms are observed to be statistically significant for the first Helmert contrast (i.e., comparing street to non-street housing). Thus, group assignment is not significantly related to housing when considering simply street versus non-street housing outcomes. However, the second Helmert contrast that contrasts the two types of non-street housing (i.e., independent

Table 6.5 Housing status across time: Nominal model estimates and standard errors (se) using Helmert contrasts.

	Independent & Community vs. Street		Independent vs. Community	
Term	estimate	se	estimate	se
Intercept	**−1.042**	.163	**−1.112**	.163
t1 (6 month vs. base)	**1.541**	.215	**.371**	.187
t2 (12 month vs. base)	**2.303**	.323	**.634**	.176
t3 (24 month vs. base)	**2.119**	.258	**.920**	.179
Section 8 (yes=1, no=0)	*.434*	.222	.130	.213
Section 8 by t1	*.623*	.330	**1.069**	.253
Section 8 by t2	−.457	.401	**1.233**	.256
Section 8 by t3	−.216	.345	*.628*	.255
Subject sd	**1.068**	.099	**.732**	.083
$-2 \log L = 2218.73$				

bold indicates $p < .05$, *italic* indicates $.05 < p < .10$

versus community) does reveal the benefical effect of the Section 8 certificate in terms of the positive group by time interaction terms. Again, the Section 8 group is more associated with independent housing, relative to community housing, than the non-Section 8 group. In many ways, the Helmert contrasts, with their intuitive interpretations, represent the best choice for the analysis of these data.

6.9 Discussion

Multilevel ordinal and multinomial logistic regression models are described for the analysis of categorical data. These models are useful for analysis of outcomes with more than two response categories. By and large, the models are seen as extensions of the multilevel logistic regression model. However, they generalize the model in different ways. The ordinal model uses cumulative dichotomizations of the categorical outcome. Alternatively, the nominal model typically uses dichotomizations that are based on selecting one category as the reference that the others are each compared to. This chapter has also described how other comparisons can be embedded within the nominal model.

For ordinal data, both proportional odds and non-proportional odds models are considered. Since, as noted by Peterson and Harrell [77], examples of non-proportional odds are not difficult to find; the latter model is especially attractive for analyzing ordinal outcomes. In the example presented, the non-proportional odds model provided more specific information about the effect of Section 8 certificates. Namely, as compared to baseline, these certificates were effective in increasing independent housing (versus community housing and street living combined) at all follow-up time-points. Interestingly, the same could not be said when comparing independent and community housing combined versus street living. Thus, the use of the non-proportional odds model was helpful in elucidating a more focused analysis of the effect of the Section 8 program.

For the nominal model, both reference cell and Helmert contrasts were applied in the analysis of these data. The former indicated an increase for community relative to street housing for the non-Section 8 group, and an increase for independent relative to street housing for the Section 8 group. Alternatively, the Helmert contrasts indicated that the groups did not differ in terms of non-street versus street housing, but did differ in terms of the type of non-street housing (i.e., the Section 8 group was more associated with independent housing). In either case, the nominal model makes an assumption that has been referred to as "independence of irrelevant alternatives" [10, 67, 68]. This is because the effect of an explanatory variable comparing two categories is the same regardless of the total number of categories considered. This assumption is generally reasonable when the categories are distinct and dissimilar, and unreasonable when the nominal categories are seen as substitutes for one another [8, 73]. Furthermore, McFadden [74] notes that the multinomial logistic regression model is relatively robust in many cases in which this assumption is implausible. In the present example, the outcome categories are fairly distinct and so the assumption would seem to be reasonable for these data. The possibility of relaxing this assumption, though, for a more general multilevel nominal regression model is discussed in detail in Train [112].

The example presented illustrated the usefulness of the multilevel approach for longitudinal categorical data. In particular, it showed the many possible models and category comparisons that are possible if the response variable has more than two categories. In terms of the multilevel part of the model, only random-intercepts models were considered in the data analysis. However, in describing model development, multiple random effects were allowed. An analysis of these data incorporating random subject intercepts and linear trends is discussed in Hedeker [46]. Additionally, the data had a relatively simple multilevel structure, in that there were only two levels, namely, repeated observations nested within subjects. Extensions of both the ordinal and nominal models for three and higher levels is possible in the MLwiN [84] and HLM [86] software programs.

Acknowledgements Thanks are due to Drs. Richard Hough and Michael Hurlburt for use of the longitudinal data, to Jan de Leeuw for organizing this collection, and especially to Erik Meijer for many helpful comments on previous drafts of this chapter. This work was supported by National Institutes of Mental Health Grant MH56146.

References

1. R. J. Adams, M. Wilson, and M. Wu. Multilevel item response models: An approach to errors in variable regression. *Journal of Educational and Behavioral Statistics*, 22:47–76, 1997.
2. A. Agresti. *Categorical Data Analysis*. Wiley, New York, 1990.
3. A. Agresti, J. G. Booth, J. P. Hobart, and B. Caffo. Random-effects modeling of categorical response data. *Sociological Methodology*, 30:27–80, 2000.
4. A. Agresti and J. B. Lang. A proportional odds model with subject-specific effects for repeated ordered categorical responses. *Biometrika*, 80:527–534, 1993.
5. A. Agresti and R. Natarajan. Modeling clustered ordered categorical data: A survey. *International Statistical Review*, 69:345–371, 2001.
6. P. D. Allison. Discrete-time methods for the analysis of event histories. *Sociological Methodology*, 13:61–98, 1982.
7. P. D. Allison. *Survival Analysis using the SAS System: A Practical Guide*. SAS Institute, Cary, NC, 1995.
8. T. Amemiya. Qualitative response models: A survey. *Journal of Economic Literature*, 19:483–536, 1981.
9. D. A. Anderson and M. Aitkin. Variance component models with binary response: Interviewer variability. *Journal of the Royal Statistical Society, Series B*, 47:203–210, 1985.
10. K. J. Arrow. *Social Choice and Individual Values*. Wiley, New York, 1951.
11. D. J. Bartholomew and M. Knott. *Latent Variable Models and Factor Analysis*, 2nd edition. Oxford University Press, New York, 1999.
12. E. K. Berndt, B. H. Hall, R. E. Hall, and J. A. Hausman. Estimation and inference in nonlinear structural models. *Annals of Economic and Social Measurement*, 3:653–665, 1974.
13. C. R. Bhat. Quasi-random maximum simulated likelihood estimation of the mixed multinomial logit model. *Transportation Research, Part B*, 35:677–693, 2001.
14. R. D. Bock. Estimating item parameters and latent ability when responses are scored in two or more nominal categories. *Psychometrika*, 37:29–51, 1972.
15. R. D. Bock. *Multivariate Statistical Methods in Behavioral Research*. McGraw-Hill, New York, 1975.
16. R. D. Bock and M. Aitkin. Marginal maximum likelihood estimation of item parameters: An application of the EM algorithm. *Psychometrika*, 46:443–459, 1981.

17. R. D. Bock and M. Lieberman. Fitting a response model for n dichotomously scored items. *Psychometrika*, 35:179–197, 1970.
18. R. D. Bock and S. Shilling. High-dimensional full-information item factor analysis. In M. Berkane, editor, *Latent Variable Modeling and Applications to Causality*, pages 163–176. Springer, New York, 1997.
19. N. E. Breslow and X. Lin. Bias correction in generalised linear mixed models with a single component of dispersion. *Biometrika*, 82:81–91, 1995.
20. G. Camilli and L. A. Shepard. *Methods for Identifying Biased Test Items*. Sage, Thousand Oaks, CA, 1994.
21. M. R. Conaway. Analysis of repeated categorical measurements with conditional likelihood methods. *Journal of the American Statistical Association*, 84: 53–61, 1989.
22. C. Corcoran, B. Coull, and A. Patel. *Egret for Windows User Manual*. Cytel Software Corporation, Cambridge, MA, 1999.
23. L. A. Cupples, R. B. D'Agostino, K. Anderson, and W. B. Kannel. Comparison of baseline and repeated measure covariate techniques in the Framingham Heart Study. *Statistics in Medicine*, 7:205–218, 1985.
24. R. B. D'Agostino, M.-L. Lee, A. J. Belanger, L. A. Cupples, K. Anderson, and W. B. Kannel. Relation of pooled logistic regression to time dependent Cox regression analysis: The Framingham Heart Study. *Statistics in Medicine*, 9: 1501–1515, 1990.
25. M. J. Daniels and C. Gatsonis. Hierarchical polytomous regression models with applications to health services research. *Statistics in Medicine*, 16:2311–2325, 1997.
26. K. A. Doksum and M. Gasko. On a correspondence between models in binary regression analysis and in survival analysis. *International Statistical Review*, 58:243–252, 1990.
27. D. M. Drukker. Maximum simulated likelihood: Introduction to a special issue. *The Stata Journal*, 6:153–155, 2006.
28. J. Engel. On the analysis of grouped extreme-value data with GLIM. *Applied Statistics*, 42:633–640, 1993.
29. F. Ezzet and J. Whitehead. A random effects model for ordinal responses from a crossover trial. *Statistics in Medicine*, 10:901–907, 1991.
30. A. Fielding, M. Yang, and H. Goldstein. Multilevel ordinal models for examination grades. *Statistical Modelling*, 3:127–153, 2003.
31. G. M. Fitzmaurice, N. M. Laird, and A. G. Rotnitzky. Regression models for discrete longitudinal responses. *Statistical Science*, 8:284–309, 1993.
32. S. A. Freels, R. B. Warnecke, T. P. Johnson, and B. R. Flay. Evaluation of the effects of a smoking cessation intervention using the multilevel thresholds of change model. *Evaluation Review*, 26:40–58, 2002.
33. S. Geman and D. Geman. Stochastic relaxation, Gibbs distributions and the Bayesian restoration of images. *IEEE Transactions on Pattern Analysis and Machine Intelligence*, 6:721–741, 1984.
34. R. D. Gibbons and R. D. Bock. Trend in correlated proportions. *Psychometrika*, 52:113–124, 1987.

35. R. D. Gibbons and D. Hedeker. Random-effects probit and logistic regression models for three-level data. *Biometrics*, 53:1527–1537, 1997.
36. G. Glasgow. Mixed logit models for multiparty elections. *Political Analysis*, 9: 116–136, 2001.
37. H. Goldstein. Nonlinear multilevel models, with an application to discrete response data. *Biometrika*, 78:45–51, 1991.
38. H. Goldstein. *Multilevel Statistical Models*, 3rd edition. Edward Arnold, London, 2003.
39. H. Goldstein and J. Rasbash. Improved approximations for multilevel models with binary responses. *Journal of the Royal Statistical Society, Series A*, 159: 505–513, 1996.
40. W. H. Greene. *LIMDEP Version 8.0 User's Manual*, 4th edition. Econometric Software, Plainview, NY, 2002.
41. P. Haan and A. Uhlendorff. Estimation of multinomial logit models with unobserved heterogeneity using maximum simulated likelihood. *The Stata Journal*, 6:229–245, 2006.
42. A. Han and J. A. Hausman. Flexible parametric estimation of duration and competing risk models. *Journal of Applied Econometrics*, 5:1–28, 1990.
43. J. Hartzel, A. Agresti, and B. Caffo. Multinomial logit random effects models. *Statistical Modelling*, 1:81–102, 2001.
44. D. A. Harville and R. W. Mee. A mixed-model procedure for analyzing ordered categorical data. *Biometrics*, 40:393–408, 1984.
45. P. J. Heagerty and S. L. Zeger. Marginalized multilevel models and likelihood inference. *Statisical Science*, 15:1–26, 2000. (with discussion)
46. D. Hedeker. MIXNO: A computer program for mixed-effects nominal logistic regression. *Journal of Statistical Software*, 4(5):1–92, 1999.
47. D. Hedeker. A mixed-effects multinomial logistic regression model. *Statistics in Medicine*, 21:1433–1446, 2003.
48. D. Hedeker and R. D. Gibbons. A random-effects ordinal regression model for multilevel analysis. *Biometrics*, 50:933–944, 1994.
49. D. Hedeker and R. D. Gibbons. MIXOR: A computer program for mixed-effects ordinal regression analysis. *Computer Methods and Programs in Biomedicine*, 49:157–176, 1996.
50. D. Hedeker and R. D. Gibbons. Application of random-effects pattern-mixture models for missing data in longitudinal studies. *Psychological Methods*, 2:64–78, 1997.
51. D. Hedeker and R. D. Gibbons. *Longitudinal Data Analysis*. Wiley, New York, 2006.
52. D. Hedeker and R. J. Mermelstein. A multilevel thresholds of change model for analysis of stages of change data. *Multivariate Behavioral Research*, 33: 427–455, 1998.
53. D. Hedeker and R. J. Mermelstein. Analysis of longitudinal substance use outcomes using random-effects regression models. *Addiction*, 95(Supplement 3):S381–S394, 2000.

54. D. Hedeker, O. Siddiqui, and F. B. Hu. Random-effects regression analysis of correlated grouped-time survival data. *Statistical Methods in Medical Research*, 9:161–179, 2000.
55. R. L. Hough, S. Harmon, H. Tarke, S. Yamashiro, R. Quinlivan, P. Landau-Cox, M. S. Hurlburt, P. A. Wood, R. Milone, V. Renker, A. Crowell, and E. Morris. Supported independent housing: Implementation issues and solutions in the San Diego McKinney homeless demonstration research project. In W. R. Breakey and J. W. Thompson, editors, *Mentally Ill and Homeless: Special Programs for Special Needs*, pages 95–117. Harwood, New York, 1997.
56. M. S. Hurlburt, P. A. Wood, and R. L. Hough. Providing independent housing for the homeless mentally ill: A novel approach to evaluating long-term longitudinal housing patterns. *Journal of Community Psychology*, 24:291–310, 1996.
57. H. Ishwaran. Univariate and multirater ordinal cumulative link regression with covariate specific cutpoints. *Canadian Journal of Statistics*, 28:715–730, 2000.
58. J. Jansen. On the statistical analysis of ordinal data when extravariation is present. *Applied Statistics*, 39:75–84, 1990.
59. E. Läärä and J. N. S. Matthews. The equivalence of two models for ordinal data. *Biometrika*, 72:206–207, 1985.
60. N. M. Laird. Missing data in longitudinal studies. *Statistics in Medicine*, 7: 305–315, 1988.
61. E. Lesaffre and B. Spiessens. On the effect of the number of quadrature points in a logistic random-effects model: An example. *Applied Statistics*, 50:325–335, 2001.
62. J. K. Lindsey and P. Lambert. On the appropriateness of marginal models for repeated measurements in clinical trials. *Statistics in Medicine*, 17:447–469, 1998.
63. L. C. Liu and D. Hedeker. A mixed-effects regression model for longitudinal multivariate ordinal data. *Biometrics*, 62:261–268, 2006.
64. Q. Liu and D. A. Pierce. A note on Gauss-Hermite quadrature. *Biometrika*, 81:624–629, 1994.
65. J. S. Long. *Regression Models for Categorical and Limited Dependent Variables.* Sage, Thousand Oaks, CA, 1997.
66. F. M. Lord. *Applications of Item Response Theory to Practical Testing Problems.* Erlbaum, Hillside, NJ, 1980.
67. R. D. Luce. *Individual Choice Behavior.* Wiley, New York, 1959.
68. G. S. Maddala. *Limited-Dependent and Qualitative Variables in Econometrics.* Cambridge University Press, Cambridge, UK, 1983.
69. J. R. Magnus. *Linear Structures.* Charles Griffin, London, 1988.
70. E. C. Marshall and D. Spiegelhalter. Institutional performance. In A. H. Leyland and H. Goldstein, editors, *Multilevel Modelling of Health Statistics*, pages 127–142. Wiley, New York, 2001.
71. P. McCullagh. Regression models for ordinal data. *Journal of the Royal Statistical Society, Series B*, 42:109–142, 1980. (with discussion)
72. P. McCullagh and J. A. Nelder. *Generalized Linear Models*, 2nd edition. Chapman & Hall, London, 1989.

73. D. McFadden. Conditional logit analysis of qualitative choice behavior. In P. Zarembka, editor, *Frontiers in Econometrics.* Academic Press, New York, 1973.
74. D. McFadden. Qualitative response models. In W. Hildenbrand, editor, *Advances in Econometrics*, pages 1–37. Cambridge University Press, Cambridge, UK, 1980.
75. J. M. Neuhaus, J. D. Kalbfleisch, and W. W. Hauck. A comparison of cluster-specific and population-averaged approaches for analyzing correlated binary data. *International Statistical Review*, 59:25–35, 1991.
76. J. F. Pendergast, S. J. Gange, M. A. Newton, M. J. Lindstrom, M. Palta, and M. R. Fisher. A survey of methods for analyzing clustered binary response data. *International Statistical Review*, 64:89–118, 1996.
77. B. Peterson and F. E. Harrell. Partial proportional odds models for ordinal response variables. *Applied Statistics*, 39:205–217, 1990.
78. J. C. Pinheiro and D. M. Bates. Approximations to the log-likelihood function in the nonlinear mixed-effects model. *Journal of Computational and Graphical Statistics*, 4:12–35, 1995.
79. R. L. Prentice and L. A. Gloeckler. Regression analysis of grouped survival data with application to breast cancer data. *Biometrics*, 34:57–67, 1978.
80. S. Rabe-Hesketh, A. Skrondal, and A. Pickles. Reliable estimation of generalized linear mixed models using adaptive quadrature. *The Stata Journal*, 2: 1–21, 2002.
81. S. Rabe-Hesketh, A. Skrondal, and A. Pickles. GLLAMM manual. Working Paper 160, U.C. Berkeley Division of Biostatistics, Berkeley, CA, 2004. (Downloadable from `http://www.bepress.com/ucbbiostat/paper160/`)
82. S. Rabe-Hesketh, A. Skrondal, and A. Pickles. Maximum likelihood estimation of limited and discrete dependent variable models with nested random effects. *Journal of Econometrics*, 128:301–323, 2005.
83. R. Raman and D. Hedeker. A mixed-effects regression model for three-level ordinal response data. *Statistics in Medicine*, 24:3331–3345, 2005.
84. J. Rasbash, F. Steele, W. J. Browne, and B. Prosser. *A User's Guide to MLwiN. Version 2.0.* Centre for Multilevel Modelling, University of Bristol, Bristol, UK, 2005.
85. S. W. Raudenbush and A. S. Bryk. *Hierarchical Linear Models: Applications and Data Analysis Methods*, 2nd edition. Sage, Thousand Oaks, CA, 2002.
86. S. W. Raudenbush, A. S. Bryk, Y. F. Cheong, and R. Congdon. *HLM 6: Hierarchical Linear and Nonlinear Modeling.* Scientific Software International, Chicago, 2004.
87. S. W. Raudenbush, M.-L. Yang, and M. Yosef. Maximum likelihood for generalized linear models with nested random effects via high-order, multivariate Laplace approximation. *Journal of Computational and Graphical Statistics*, 9: 141–157, 2000.
88. D. Revelt and K. Train. Mixed logit with repeated choices: Household's choices of appliance efficiency level. *Review of Economics and Statistics*, 80:647–657, 1998.

89. F. Rijmen, F. Tuerlinckx, P. De Boeck, and P. Kuppens. A nonlinear mixed model framework for item response theory. *Psychological Methods*, 8:185–205, 2003.
90. G. Rodríguez and N. Goldman. An assessment of estimation procedures for multilevel models with binary responses. *Journal of the Royal Statistical Society, Series A*, 158:73–89, 1995.
91. D. B. Rubin. Inference and missing data. *Biometrika*, 63:581–592, 1976. (with discussion)
92. F. Samejima. Estimation of latent ability using a response pattern of graded scores. *Psychometrika Monograph No. 17*, 1969.
93. SAS/Stat. *SAS/Stat User's Guide, version 9.1.* SAS Institute, Cary, NC, 2004.
94. T. H. Scheike and T. K. Jensen. A discrete survival model with random effects: An application to time to pregnancy. *Biometrics*, 53:318–329, 1997.
95. J. D. Singer and J. B. Willett. It's about time: Using discrete-time survival analysis to study duration and the timing of events. *Journal of Educational and Behavioral Statistics*, 18:155–195, 1993.
96. J. D. Singer and J. B. Willett. *Applied Longitudinal Data Analysis: Modeling Change and Event Occurrence.* Oxford University Press, Oxford, UK, 2003.
97. A. Skrondal and S. Rabe-Hesketh. Multilevel logistic regression for polytomous data and rankings. *Psychometrika*, 68:267–287, 2003.
98. A. Skrondal and S. Rabe-Hesketh. *Generalized Latent Variable Modeling: Multilevel, Longitudinal, and Structural Equation Models.* Chapman & Hall/CRC, Boca Raton, FL, 2004.
99. T. A. B. Snijders and R. J. Bosker. *Multilevel Analysis: An Introduction to Basic and Advanced Multilevel Modeling.* Sage, Thousand Oaks, CA, 1999.
100. D. J. Spiegelhalter, A. Thomas, N. G. Best, and W. R. Gilks. *BUGS: Bayesian Inference Using Gibbs Sampling, Version 0.50.* Technical Report, MRC Biostatistics Unit, Cambridge, UK, 1995.
101. StataCorp. *Stata Statistical Software: Release 9.* Stata Corporation, College Station, TX, 2005.
102. S. Stern. Simulation-based estimation. *Journal of Economic Literature*, 35: 2006–2039, 1997.
103. R. Stiratelli, N. M. Laird, and J. H. Ware. Random-effects models for serial observations with binary response. *Biometrics*, 40:961–971, 1984.
104. A. H. Stroud and D. Secrest. *Gaussian Quadrature Formulas.* Prentice Hall, Englewood Cliffs, NJ, 1966.
105. M. A. Tanner. *Tools for Statistical Inference: Methods for the Exploration of Posterior Distributions and Likelihood Functions*, 3rd edition. Springer, New York, 1996.
106. T. R. Ten Have. A mixed effects model for multivariate ordinal response data including correlated discrete failure times with ordinal responses. *Biometrics*, 52:473–491, 1996.
107. T. R. Ten Have, A. R. Kunselman, and L. A. Tran. A comparison of mixed effects logistic regression models for binary response data with two nested levels of clustering. *Statistics in Medicine*, 18:947–960, 1999.

108. T. R. Ten Have and D. H. Uttal. Subject-specific and population-averaged continuation ratio logit models for multiple discrete time survival profiles. *Applied Statistics*, 43:371–384, 1994.
109. J. Terza. Ordinal probit: A generalization. *Communications in Statistics: Theory and Methods*, 14:1–11, 1985.
110. D. Thissen and L. Steinberg. A taxonomy of item response models. *Psychometrika*, 51:567–577, 1986.
111. L. L. Thurstone. Psychophysical analysis. *American Journal of Psychology*, 38:368–389, 1927.
112. K. E. Train. *Discrete Choice Methods with Simulation.* Cambridge University Press, Cambridge, UK, 2003.
113. G. Tutz and W. Hennevogl. Random effects in ordinal regression models. *Computational Statistics & Data Analysis*, 22:537–557, 1996.
114. M. A. Wakefield, F. J. Chaloupka, N. J. Kaufman, C. T. Orleans, D. C. Barker, and E. E. Ruel. Effect of restrictions on smoking at home, at school, and in public places on teenage smoking: Cross sectional study. *British Medical Journal*, 321:333–337, 2001.
115. G. Y. Wong and W. M. Mason. The hierarchical logistic regression model for multilevel analysis. *Journal of the American Statistical Association*, 80: 513–524, 1985.
116. B. D. Wright. Solving measurement problems with the Rasch model. *Journal of Educational Measurement*, 14:97–116, 1977.
117. H. Xie, G. McHugo, A. Sengupta, D. Hedeker, and R. Drake. An application of the thresholds of change model to the analysis of mental health data. *Mental Health Services Research*, 3:107–114, 2001.
118. S. L. Zeger, K.-Y. Liang, and P. S. Albert. Models for longitudinal data: A generalized estimating equation approach. *Biometrics*, 44:1049–1060, 1988.

7

Multilevel and Related Models for Longitudinal Data

Anders Skrondal[1] and Sophia Rabe-Hesketh[2]

[1] Department of Statistics and Methodology Institute, London School of Economics & Division of Epidemiology, Norwegian Institute of Public Health.
[2] Graduate School of Education and Graduate Group in Biostatistics, University of California, Berkeley & Institute of Education, University of London.

7.1 Introduction

Longitudinal data, often called repeated measurements in medicine and panel data in the social sciences, arise when units provide responses on multiple occasions. Such data can be thought of as clustered or two-level data with occasions i at level 1 and units j at level 2.

One feature distinguishing longitudinal data from other types of clustered data is the chronological ordering of the responses, implying that level-1 units cannot be viewed as exchangeable. Another feature of longitudinal data is that they often consist of a large number of small clusters.

A typical aim in longitudinal analysis is to investigate the effects of covariates both on the overall level of the responses and on changes of the responses over time. An important merit of longitudinal designs is that they allow the separation of cross-sectional and longitudinal effects. They also allow the investigation of heterogeneity across units both in the overall level of the response and in the development over time. Heterogeneity not captured by observed covariates produces dependence among responses even after controlling for those covariates. This violates the typical assumptions of ordinary regression models and must be accommodated to avoid invalid inference.

It is useful to distinguish between longitudinal data with balanced and unbalanced occasions. The occasions are balanced if all units are measured at the same time points t_i, $i = 1, \ldots, n$, and unbalanced if units are measured at different time points, $t_{ij}, i = 1, \ldots, n_j$. In the case of balanced occasions, the data can also be viewed as single-level multivariate data where responses at different occasions are treated as different variables. One advantage of the univariate multilevel approach taken here is that unbalanced occasions and

J. de Leeuw, E. Meijer (eds.), *Handbook of Multilevel Analysis*,
© Springer 2008

missing data are accommodated without resorting to complete case analysis (sometimes called listwise deletion). We will use maximum likelihood estimation, which produces consistent estimates if responses are missing at random (MAR) as defined by Rubin [59]; see Chapter 10 [40] for other approaches in the case of MAR and Verbeke and Molenberghs [65] for approaches in the case of responses not missing at random (NMAR).

In this chapter we will consider both linear mixed models and generalized linear mixed models. A linear mixed model is written in Chapter 1, equation (1.4), as

$$\underline{\boldsymbol{y}}_j = \boldsymbol{X}_j\boldsymbol{\beta} + \boldsymbol{Z}_j\underline{\boldsymbol{\delta}}_j + \underline{\boldsymbol{\epsilon}}_j, \tag{7.1}$$

where $\underline{\boldsymbol{y}}_j$ is the vector of continuous responses for unit j. In this book the covariate matrices $\boldsymbol{X}_j$ and $\boldsymbol{Z}_j$ are treated as fixed. Extra assumptions are required when these matrices are treated as random; see, for instance, Rabe-Hesketh and Skrondal [54].

A generalized linear mixed model also accommodates non-continuous responses and can be written as

$$g(E(\underline{\boldsymbol{y}}_j \mid \underline{\boldsymbol{\delta}}_j)) = \boldsymbol{X}_j\boldsymbol{\beta} + \boldsymbol{Z}_j\underline{\boldsymbol{\delta}}_j \stackrel{\Delta}{=} \underline{\boldsymbol{\eta}}_j, \tag{7.2}$$

where $g(\cdot)$ is a link function and $\boldsymbol{\eta}_j$ is a vector of linear predictors. Conditional on the random effects $\underline{\boldsymbol{\delta}}_j$, the elements $\underline{y}_{ij}$ of $\underline{\boldsymbol{y}}_j$ have a distribution from the exponential family and are mutually independent. See Rabe-Hesketh and Skrondal [54] and Chapter 9 [58] for treatments of generalized linear mixed models.

For dichotomous and ordinal responses, generalized linear mixed models with logit and probit links can also be defined using a latent response formulation. A linear mixed model is in this case specified for an imagined continuous latent response $\underline{y}^*_{ij}$. The observed dichotomous or ordinal response $\underline{y}_{ij}$ with $S > 1$ categories results from partitioning $\underline{y}^*_{ij}$ into S segments using $S-1$ cut-points or thresholds; see Chapter 6 [31] for details.

We will use an example dataset to illustrate some of the ideas discussed in this chapter. The dataset comes from an American panel survey of 545 young males taken from the National Longitudinal Survey (Youth Sample) for the period 1980–1987. The data were previously analyzed by Vella and Verbeek [64] and can be downloaded from the web pages of Wooldridge [70] and Rabe-Hesketh and Skrondal [53]. The response variable is the natural logarithm of the hourly wage in US dollars and the following covariates will be used:

- `educ`: Years of schooling (x_{1j})
- `black`: Dummy variable for being black (x_{2j})
- `hisp`: Dummy variable for being Hispanic (x_{3j})
- `labex`: Labor market experience (in 2-year periods) (x_{4ij})

- `labexsq`: Labor market experience squared (x_{5ij})
- `married`: Dummy variable for being married (x_{6ij})
- `union`: Dummy variable for being a member of a union (x_{7ij})

The first three covariates are time-constant, whereas the next four are time-varying.

7.2 Models with Unit-Specific Intercepts

In longitudinal data it is usually impossible to capture all between-unit variability using observed covariates. If the remaining "unobserved heterogeneity" is ignored, it induces longitudinal dependence among the responses for the same unit (after controlling for the included covariates). A simple way of representing "unobserved heterogeneity" is by including unit-specific intercepts, which could be either random or fixed.

7.2.1 Random Intercept Models

Consider the response $\underline{y}_{ij}$ of unit j on occasion i ($i = 1, \ldots, n_j$). In a linear random intercept model, sometimes referred to as a one-way error component model, it is assumed that the unit-specific effects are realizations of a random variable $\underline{\delta}_j$,

$$\underline{y}_{ij} = \boldsymbol{x}'_{ij}\boldsymbol{\beta} + \underline{\delta}_j + \underline{\epsilon}_{ij},$$

where $\underline{\delta}_j$ and $\underline{\epsilon}_{ij}$ are independently distributed $\underline{\delta}_j \sim \mathcal{N}(0, \omega^2)$ and $\underline{\epsilon}_{ij} \sim \mathcal{N}(0, \sigma^2)$. The random intercept or "permanent component" $\underline{\delta}_j$ allows the level of the response to vary across units, whereas the "transitory component" $\underline{\epsilon}_{ij}$ varies over occasions within units. The model is a special case of a linear mixed model (7.1) with $\boldsymbol{Z}_j = \mathbf{1}_{n_j}$.

The variance-covariance matrix of the responses $\underline{\boldsymbol{y}}_j$, after controlling for $\boldsymbol{X}_j$, is given by

$$\mathrm{Cov}(\underline{\boldsymbol{y}}_j) = \mathrm{Cov}(\mathbf{1}_{n_j}\underline{\delta}_j + \underline{\boldsymbol{\epsilon}}_j) = \omega^2\mathbf{1}_{n_j}\mathbf{1}'_{n_j} + \sigma^2\boldsymbol{I}_{n_j},$$

with diagonal elements $\omega^2 + \sigma^2$ and off-diagonal elements ω^2. The residual intraclass correlation becomes

$$\mathrm{Corr}(\underline{y}_{ij}, \underline{y}_{i'j}) = \frac{\omega^2}{\omega^2 + \sigma^2}. \tag{7.3}$$

This covariance structure is also shown in panel A of Table 7.1. It is sometimes referred to as exchangeable since the joint distribution of the residuals for a given unit remains unchanged if the residuals are exchanged across occasions. The covariance structure also satisfies the sphericity property that the conditional variances $\mathrm{Var}(\underline{y}_{ij} - \underline{y}_{i'j})$ of all pairwise differences are equal. Note that

Table 7.1 Common dependence structures for longitudinal data ($\boldsymbol{\Psi}_j \triangleq \text{Cov}(\underline{\boldsymbol{y}}_j)$).

A. Random intercept structure:

$$\boldsymbol{\Psi}_j = \omega^2 \mathbf{1}_{n_j}\mathbf{1}'_{n_j} + \sigma^2 \boldsymbol{I}_{n_j} = \begin{bmatrix} \omega^2+\sigma^2 & & & \\ \omega^2 & \omega^2+\sigma^2 & & \\ \vdots & \vdots & \ddots & \\ \omega^2 & \omega^2 & \cdots & \omega^2+\sigma^2 \end{bmatrix}$$

B. Random coefficient structure:

$$\boldsymbol{\Psi}_j = \boldsymbol{Z}_j \boldsymbol{\Omega} \boldsymbol{Z}'_j + \sigma^2 \boldsymbol{I}_{n_j}$$

C. Autoregressive residual structure AR(1):

$$\boldsymbol{\Psi}_j = \frac{\sigma_u^2}{1-\alpha^2} \begin{bmatrix} 1 & & & \\ \alpha & 1 & & \\ \vdots & \vdots & \ddots & \\ \alpha^{n_j-1} & \alpha^{n_j-2} & \cdots & 1 \end{bmatrix}$$

D. Moving average residual structure MA(1):

$$\boldsymbol{\Psi}_j = \sigma_u^2 \begin{bmatrix} 1+a^2 & & & & \\ a & 1+a^2 & & & \\ 0 & a & 1+a^2 & & \\ \vdots & \vdots & \vdots & \ddots & \\ 0 & 0 & 0 & \cdots & 1+a^2 \end{bmatrix}$$

E. Autoregressive response structure AR(1):

$$\boldsymbol{\Psi}_j = \frac{\sigma_\epsilon^2}{1-\gamma^2} \begin{bmatrix} 1 & & & \\ \gamma & 1 & & \\ \vdots & \vdots & \ddots & \\ \gamma^{n_j-1} & \gamma^{n_j-2} & \cdots & 1 \end{bmatrix}$$

the covariances ω^2 are restricted to be nonnegative in the random intercept model. If this restriction is relaxed, the above covariance structure is often called compound symmetric. In the case of balanced occasions, we could also allow the variance of $\underline{\epsilon}_{ij}$ to take on a different value $\boldsymbol{\Sigma}_{ii}$ for each occasion.

Typically, the random intercept model is estimated by either maximum likelihood or restricted maximum likelihood [42]. The likelihood has a closed form, but iterative methods such as the EM algorithm, Newton-Raphson, Fisher scoring, or iterated generalized least squares (IGLS) must be used to estimate the parameters; see Chapter 1 [15].

Maximum likelihood estimates of the random intercept model for the wage panel data, obtained using Stata's [63] `xtmixed` command, are given in the first column of Table 7.2. As might be expected, more years of schooling, more labor market experience, being married, and being a union member are all associated with higher hourly wages, whereas being black decreases the wage compared with being white, and Hispanics' wages are similar to those of whites (controlling for the other covariates). The residual intraclass correlation is estimated as 0.47; 47% of the variance not explained by the covariates is therefore between individuals and 53% within individuals.

For generalized linear mixed models, the dependence among observed responses is generally difficult to express because the model-implied correlations and variances depend on the covariates. However, for a generalized linear random intercept model, obtained by substituting $\boldsymbol{Z}_j = \mathbf{1}_{n_j}$ in (7.2), with dichotomous or ordinal responses, the intraclass correlation of the latent responses is constant and given by (7.3) with σ^2 replaced by $\pi^2/3$ for logit models and 1 for probit models. An important interpretational issue in generalized linear mixed models concerns the distinction between conditional and marginal effects, which correspond to unit-specific and population-averaged effects in the longitudinal setting. We return to this in Section 7.6.

Generally, the marginal likelihood does not have a closed form for generalized linear mixed models, making estimation more difficult. Common approaches include penalized quasilikelihood [23], maximum likelihood using adaptive quadrature [56] and Markov Chain Monte Carlo (MCMC) [10]; see also Chapter 9 [58]. For dichotomous responses and counts, closed-form likelihoods can be achieved by specifying a conjugate distribution for the random intercepts, giving the beta-binomial and negative-binomial models, respectively [38].

Simulation studies [5, 26, 48, 69] suggest that inference for the random intercept model and similar models is relatively robust to violation of the normality assumption for the random intercept. However, to safeguard against distributional misspecification, the random intercept distribution can be left unspecified by using nonparametric maximum likelihood estimation [30, 34, 37]. The nonparametric maximum likelihood estimator of the random intercept distribution is discrete with estimated locations and masses, their number being determined to reach the largest maximized likelihood.

For the wage panel data, `gllamm` [53, 55] in Stata was used to estimate models with a discrete random effects distribution. The directional derivative [37] was used to determine whether the nonparametric maximum likelihood estimator (NPMLE) was achieved as described in Rabe-Hesketh et al. [52]. In the example, the NPMLE appears to have eight mass points whose estimated locations and masses are shown in Fig. 7.1. This estimated discrete distribution is quite symmetric apart from a tiny mass at 1.77, which appears to accommodate one outlying individual whose log wage exceeded the 99th

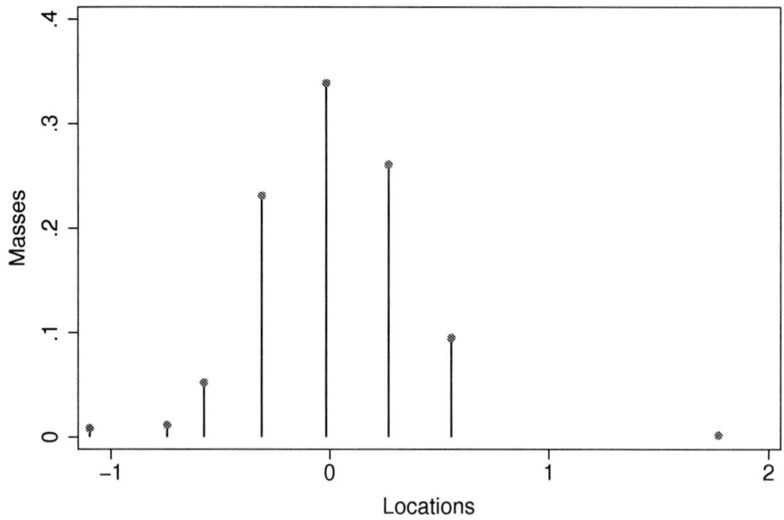

Fig. 7.1 Estimated discrete random intercept distribution from NPMLE.

percentile (across individuals and time) in 1981–1987. The standard deviation of the discrete distribution is very close to $\hat{\omega}_0$ for the conventional random intercept model, as are the estimates of the regression parameters $\boldsymbol{\beta}$ and variance parameter σ^2 given in the second column of Table 7.2.

As discussed for linear models in Chapter 3 [61], violation of the assumption that $\underline{\delta}_j$ has zero expectation can invalidate inference. Specifically, if $E(\underline{\delta}_j) = \boldsymbol{z}_j'\boldsymbol{\gamma}$, where $\mathbf{1}_{n_j}\boldsymbol{z}_j$ and $\boldsymbol{X}_j$ are nonorthogonal, the estimates of the regression coefficients $\boldsymbol{\beta}$ will be inconsistent. When the covariates $\boldsymbol{X}_j$ are treated as random variables $\underline{\boldsymbol{X}}_j$, this problem is referred to as endogeneity in econometrics because the covariates are correlated with the random effects.

The standard approach to handling endogeneity in econometrics is instrumental variables estimation [70]. In the present context, a simpler solution is to estimate the within-unit effects of $\boldsymbol{X}_j$, which can be achieved by also controlling for the cluster mean covariates $\bar{\boldsymbol{X}}_{\cdot j}$. An alternative for linear models is to use a fixed effects approach, which will be discussed next. Unfortunately, there are no easy fixes for violation of the assumption that $E(\underline{\epsilon}_{ij}) = 0$.

7.2.2 Fixed Intercept Models

A simple linear fixed intercept model or fixed effects model has the form

$$\underline{y}_{ij} = \boldsymbol{x}_{ij}'\boldsymbol{\beta} + \delta_j + \underline{\epsilon}_{ij}, \tag{7.4}$$

where δ_j are unit-specific intercepts or "fixed effects" and $\underline{\epsilon}_{ij}$ are identically and independently normally distributed residuals with $E(\underline{\epsilon}_{ij}) = 0$. Due to the

Table 7.2 Estimates for wage panel data.

	Intercept						Intercept and Slope						Autoregressive (1)			
	Random		NPML		Fixed		Both random		Random slope		Both fixed		AR Residual		AR Response	
	Est	(SE)	Est	(SE)	Est	(SE)	Est	(SE)	Est	(SE)	Est	(SE)	Est	(SE)	Est	(SE)
Fixed part																
β_0 [**cons**]	−0.11	(0.11)	−0.17	(0.10)			−0.16	(0.12)					−0.11	(0.11)	0.11	(0.06)
β_1 [**educ**]	0.10	(0.01)	0.11	(0.01)			0.11	(0.01)					0.10	(0.01)	0.05	(0.00)
β_2 [**black**]	−0.14	(0.05)	−0.14	(0.05)			−0.15	(0.04)					−0.14	(0.05)	−0.08	(0.02)
β_3 [**hisp**]	0.02	(0.04)	0.01	(0.03)			0.01	(0.04)					0.02	(0.04)	0.01	(0.02)
β_4 [**labex**]	0.22	(0.02)	0.22	(0.02)	0.23	(0.02)	0.21	(0.02)	0.22	(0.02)			0.23	(0.02)	0.03	(0.02)
β_5 [**labexsq**]	−0.01	(0.00)	−0.02	(0.00)	−0.02	(0.00)	−0.01	(0.00)	−0.01	(0.01)	−0.01	(0.01)	−0.02	(0.00)	0.00	(0.00)
β_6 [**married**]	0.06	(0.02)	0.07	(0.02)	0.05	(0.02)	0.07	(0.02)	0.05	(0.02)	0.04	(0.03)	0.06	(0.02)	0.05	(0.01)
β_7 [**union**]	0.11	(0.02)	0.10	(0.02)	0.08	(0.02)	0.11	(0.02)	0.08	(0.02)	0.04	(0.02)	0.09	(0.02)	0.07	(0.01)
γ [**lag**]															0.56	(0.01)
Random part																
ω_0	0.33	(0.01)	0.33	(–)			0.44	(0.02)					0.31	(0.01)		
ω_1							0.10	(0.01)	0.11	(0.01)						
ρ_{10}							−0.66	(0.04)								
σ	0.35	(0.00)	0.35	(0.00)	0.35	(0.00)	0.33	(0.00)	0.33	(0.00)			0.37	(0.00)	0.39	(0.00)
α													0.27	(0.01)		
Log-likelihood[a]	−2193.3		−2176.1				−2118.9						−2109.0			

[a] No log-likelihood given when estimates are based on transformed data or subset of data.

inclusion of fixed effects δ_j for each unit j, the mean structure of $\underline{\boldsymbol{y}}_j$ is saturated so that the regression coefficients $\boldsymbol{\beta}$ represent within-unit or longitudinal effects only. Unlike the random intercept model, the fixed intercept model no longer makes any assumptions regarding the cross-sectional component of the model, so that endogeneity bias can be avoided. The cost of this robustness is that regression parameters for time-constant covariates such as gender or treatment group cannot be estimated and all covariates must therefore be time-varying.

The fixed intercepts are rarely of interest in themselves and estimation can be involved when there are many units. An attractive alternative to estimating all parameters is to eliminate the fixed intercepts. This can be accomplished by transforming both the responses and covariates and then using ordinary least squares (OLS). In econometrics, two popular transformations are first-differencing: $\underline{y}_{ij} - \underline{y}_{i-1,j}$, $\boldsymbol{x}_{ij} - \boldsymbol{x}_{i-1,j}$, and cluster-mean centering: $\underline{y}_{ij} - \bar{\underline{y}}_{\cdot j}$, $\boldsymbol{x}_{ij} - \bar{\boldsymbol{x}}_{\cdot j}$. Both approaches yield consistent estimates of the remaining regression coefficients, but the latter, known as the fixed effects estimator, is more efficient if the residuals $\underline{\epsilon}_{ij}$ are mutually independent as assumed above [70]. Verbeke et al. [66] propose eliminating the intercepts by conditioning on the cluster mean responses and maximizing the resulting conditional likelihood. This can be implemented by premultiplying $\underline{\boldsymbol{y}}_j$ and $\boldsymbol{X}_j$ by a $n_j \times (n_j - 1)$ orthonormal contrast matrix. This approach yields identical estimates as the fixed effects estimator based on cluster mean centering, but has the advantage that the OLS standard error estimates need not be corrected for the loss of degrees of freedom.

Some insight can be gained [41] regarding the difference between fixed effects and random effects estimators of the regression coefficients by considering the generalized least squares (GLS) estimator for the latter. The GLS estimator is asymptotically equivalent to the maximum likelihood estimator but has a closed form. It can be shown that the GLS estimator is a matrix weighted average of the fixed effects (or within-unit) estimator and the between-unit estimator obtained by OLS estimation for the regression of the cluster-mean response on the cluster-mean covariates. If the random intercept model is correctly specified, the GLS estimator is more efficient since it uses cross-sectional information in addition to longitudinal information. However, if the cross-sectional component of the model is misspecified, the GLS estimator becomes inconsistent for the longitudinal effects in contrast to the fixed effects estimator. Thus, a difference between fixed effects and GLS estimates for $\boldsymbol{\beta}$ suggests that the random effects model is misspecified and is the basis for the popular Durbin-Wu-Hausman specification test [25] in this context.

Returning to the wage panel data, the fixed effects estimates of the coefficients of the time-varying covariates, obtained using Stata's `xtreg` command, are given in the third column of Table 7.2. The estimates are quite similar to the estimates for the random intercept model, suggesting that the cross-sectional component of the random intercept model is not severely misspecified.

In generalized linear models, except for linear Gaussian or log-linear Poisson models, inclusion of a fixed intercept for each unit leads to inconsistent estimates of the regression parameters $\boldsymbol{\beta}$, which is known as the incidental parameter problem [49]. For binary logistic models, the problem can be overcome by conditioning on the sum of the responses for each unit to eliminate the unit-specific intercepts, as mentioned above for linear models. In epidemiology, such a conditional maximum likelihood approach is used for matched case-control studies [7], in psychometrics for the Rasch measurement model [57], and in econometrics for panel data [8, 9]. In addition to the limitation of not permitting time-constant covariates, this approach also discards units with all responses equal to 0 or all equal to 1. Furthermore, conditional maximum likelihood estimation is impossible for some model types such as probit models.

7.3 Models with Unit-Specific Intercepts and Slopes

Sometimes units vary not just in the overall level of the response (controlling for covariates) but also in the effects of time-varying covariates on the response. A typical example is where the effect of time, i.e., the rate of change, varies between units. Such heterogeneity in the effects of covariates can be viewed as interactions between the included covariates and a categorical variable representing the units.

7.3.1 Continuous Random Coefficients

The random coefficient model [35] can be written as

$$\underline{y}_{ij} = \boldsymbol{x}'_{ij}\boldsymbol{\beta} + \boldsymbol{z}'_{ij}\underline{\boldsymbol{\delta}}_j + \underline{\epsilon}_{ij},$$

where $\boldsymbol{x}_{ij}$ denotes both time-varying and time-constant covariates with fixed coefficients $\boldsymbol{\beta}$ and $\boldsymbol{z}_{ij}$ denotes time-varying covariates with random coefficients $\underline{\boldsymbol{\delta}}_j \sim \mathcal{N}(\boldsymbol{\emptyset}, \boldsymbol{\Omega})$. Since the random coefficients have zero means, $\boldsymbol{x}_{ij}$ will typically contain all elements in $\boldsymbol{z}_{ij}$, with the corresponding fixed effects interpretable as the mean effects. The first element of these vectors is invariably equal to 1, corresponding to a fixed and random intercept, respectively. The random intercept model is thus the special case where $\boldsymbol{z}_{ij} = 1$. The random coefficient covariance structure of the vector $\underline{\boldsymbol{y}}_j$ is presented in panel B of Table 7.1.

A useful version of the random coefficient model for longitudinal data is a growth curve model where individuals are assumed to differ not only in their intercepts but also in other aspects of their trajectory over time, for example in the linear growth (or decline) of the response. These models include random coefficients for (functions of) time. For example, a linear growth curve model can be written as

$$\underline{y}_{ij} = \boldsymbol{x}'_{ij}\boldsymbol{\beta} + \underline{\delta}_{0j} + \underline{\delta}_{1j}t_{ij} + \underline{\epsilon}_{ij}, \tag{7.5}$$

where t_{ij}, the time at the i-th occasion for individual j, is one of the covariates in $\boldsymbol{x}_{ij}$. The random intercept $\underline{\delta}_{0j}$ and random slope $\underline{\delta}_{1j}$ represent unit-specific deviations from the mean intercept and slope, respectively. The random intercept and slope should not be specified as uncorrelated, because translation of the time scale t_{ij} changes the magnitude of the correlation [18, 39, 62].

In a linear growth curve model, the variance of the responses (controlling for the covariates) varies over occasions t_{ij},

$$\text{Var}(\underline{y}_{ij}) = \omega_0^2 + 2\omega_{10}t_{ij} + \omega_1^2 t_{ij}^2 + \sigma^2.$$

Note that the variance increases as a quadratic function of time if $t_{ij} \geq 0$ and $\omega_{10} \geq 0$. The covariance between two responses $\underline{y}_{ij}$ and $\underline{y}_{i'j}$ for a unit at different occasions i and i' becomes

$$\text{Cov}(\underline{y}_{ij}, \underline{y}_{i'j}) = \omega_0^2 + \omega_{10}(t_{ij} + t_{i'j}) + \omega_1^2 t_{ij}t_{i'j},$$

which depends on the time associated with the occasions.

For the wage panel data, we would expect wages to increase more rapidly for some individuals as they gain more labor market experience than for others. We therefore estimated a model with a random slope for `labex` in addition to the random intercept. Maximum likelihood estimates using `xtmixed` are given in the fourth column of Table 7.2. The fixed part estimates remain practically the same as for the random intercept model. There is a negative estimated correlation between the random intercept and random slope. To visualize the model, the bottom panel of Fig. 7.2 shows the fitted trajectories (obtained by plugging in empirical Bayes predictions of the random intercepts and slopes and setting `married` and `union` to zero) for the first 40 individuals. For comparison, the corresponding trajectories for the random intercept model are given in the top panel of the figure. These trajectories are nonlinear due to the quadratic term `labexsq` in the fixed part of the model.

For balanced occasions with associated times $t_{ij} = t_i$, the linear growth curve model can also be formulated as a two-factor model with fixed factor loadings,

$$\underline{y}_{ij} = \lambda_{0i}\,\underline{\beta}_{0j} + \lambda_{1i}\,\underline{\beta}_{1j} + \underline{\epsilon}_{ij}, \qquad \lambda_{0i} = 1, \quad \lambda_{1i} = t_i\,,$$

where

$$\underline{\beta}_{0j} = \beta_0 + \underline{\delta}_{0j}, \quad \underline{\beta}_{1j} = \beta_1 + \underline{\delta}_{1j}.$$

Note that the means of the factors cannot be set to zero here as is usually done in factor models. A path diagram of this model is shown in Fig. 7.3, where

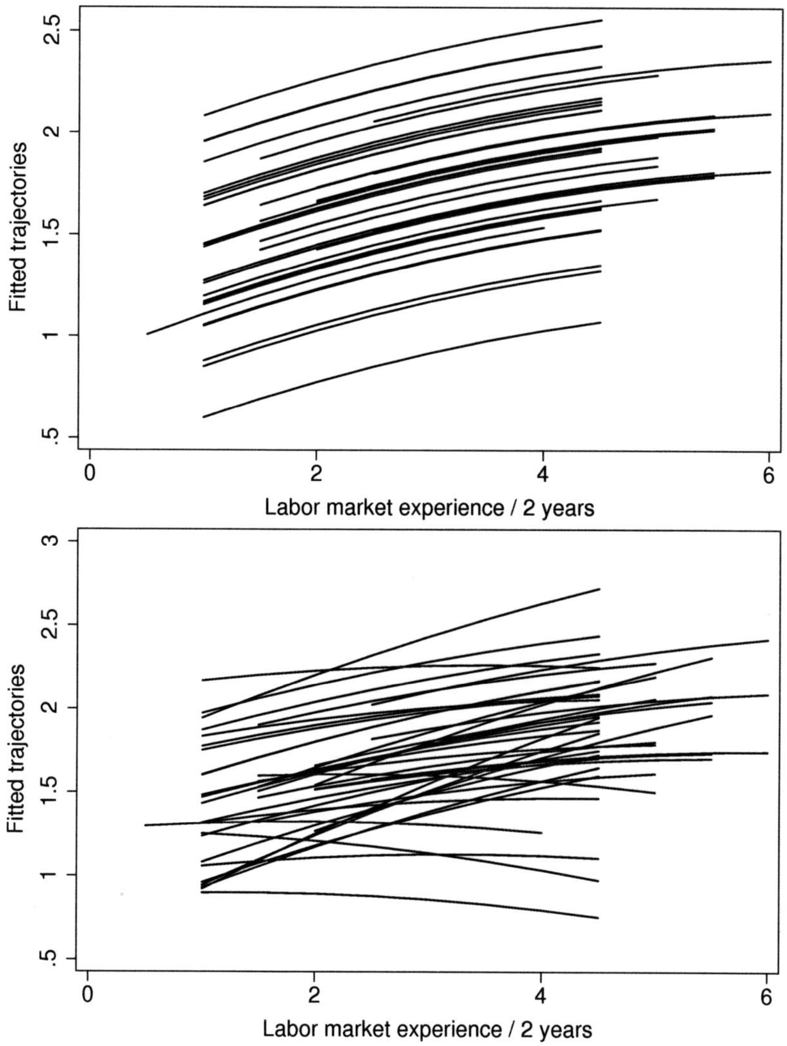

Fig. 7.2 Fitted trajectories for linear random intercept model (top) and random coefficient model (bottom). Empirical Bayes predictions are substituted for the random effects and `married` and `union` set to zero.

there are three occasions with times $t_1 = 0$, $t_2 = 1$, and $t_3 = 2$. Following the usual conventions, latent variables or random effects are represented by circles and observed variables by rectangles. Long arrows represent regressions and short arrows residual errors.

Meredith and Tisak [43] suggest using a two-factor model with free factor loadings λ_{1i} for $\underline{\beta}_{1j}$ (subject to identification restrictions, such as $\lambda_{11}=0$ and

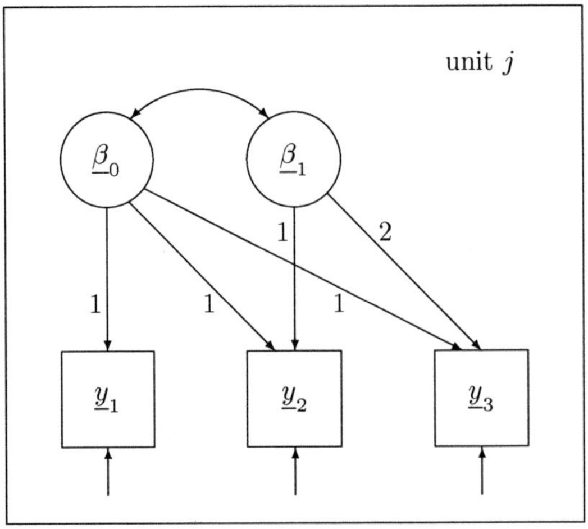

Fig. 7.3 Path diagram for growth curve model with balanced occasions.

$\lambda_{12}=1$) to model nonlinear growth. Estimation of this factor model requires balanced occasions, but modern software can handle missing data.

Generalized linear random coefficient models are defined analogously to the linear case. Maximum likelihood estimation using numerical integration becomes computationally more demanding as the number of random effects increases. Unfortunately, we can no longer exploit conjugacy to obtain closed-form likelihoods for counts and dichotomous responses.

7.3.2 Fixed Coefficients

Instead of considering the unit-specific intercepts and slopes as random, we can specify a model with fixed intercepts and slopes,

$$\underline{y}_{ij} = \boldsymbol{x}'_{ij}\boldsymbol{\beta} + \delta_{0j} + \delta_{1j} z_{ij} + \underline{\epsilon}_{ij}.$$

If the data are balanced, $z_{ij} = z_i$, and the differences $z_i - z_{i-1}$ are constant, then the δ_{0j} and δ_{1j} can be eliminated by double-differencing [70]. Alternatively, first-differencing can be used to turn the model into a fixed-intercepts model, which can subsequently be estimated by any of the methods discussed in Section 7.2.2. This approach was used to obtain the estimates for the wage panel data given in the sixth column of Table 7.2. The estimated regression coefficients for `married` and `union` are considerably closer to zero than in the random coefficient model. Wooldridge [70] also describes an approach for eliminating the intercepts and slopes in more general models with unbalanced $\boldsymbol{z}_{ij}$.

Verbeke et al. [66] suggest a hybrid approach, treating the intercepts as fixed but the slope(s) as random,

$$\underline{y}_{ij} = \boldsymbol{x}'_{ij}\boldsymbol{\beta} + \delta_{0j} + \underline{\delta}_{1j} z_{ij} + \underline{\epsilon}_{ij}.$$

The fixed intercepts are eliminated by forming contrasts using an orthonormal coefficient matrix as described in Section 7.2.2, corresponding to conditional maximum likelihood estimation. Estimates for the wage panel data using this approach are given in the fifth column of Table 7.2 and are quite similar to the estimates for the random coefficient model.

7.3.3 Discrete Random Coefficients

It is sometimes believed that the population consists of different subpopulations or classes of units characterized by different unknown patterns of development over time. Since class membership is not known, the parameters characterizing the development trajectory can be treated as discrete latent variables or random effects.

In a linear latent trajectory model or latent profile model [22] analogous to (7.5), the model for a unit in latent class c ($c = 1, \ldots, C$) is given by

$$\underline{y}_{ijc} = e_{0c} + e_{1c} t_{ij} + \underline{\epsilon}_{ijc}.$$

Each latent class is characterized by a pair of coefficients e_{0c} and e_{1c}, representing the intercept and slope of the latent trajectory. Other covariates can be included in the regression model above, so that the e_{0c} and e_{1c} describe the distinct patterns of deviations from the mean trajectory given the covariates. Alternatively, other covariates could be included in a multinomial logit model for the latent class membership probabilities, as is often done in conventional latent class models [14].

Interestingly, the number of classes cannot be increased indefinitely. If it is attempted to exceed the maximum possible number of classes, then estimated locations of some classes will either coincide or the probabilities of some classes tend to zero. The solution with the maximum number of classes then corresponds to the nonparametric maximum likelihood estimator [1]. An extension of the model would be to allow the variance of residuals $\underline{\epsilon}_{ijc}$ to differ between classes.

For balanced occasions, we do not have to assume that the latent trajectories are linear or have another particular shape but can, instead, specify an unstructured model with latent trajectory

$$\underline{y}_{ijc} = e_{ic} + \underline{\epsilon}_{ijc}, \qquad i = 1, \ldots, n,$$

for class c.

In the case of categorical responses, latent trajectory models are typically referred to as latent class growth models [47] and represent an application of mixture regression models [51, 68] to longitudinal data.

All these models assume that the responses on a unit are conditionally independent given latent class membership. Muthén and Shedden [46] relax this assumption for continuous responses in their growth mixture models by allowing the residuals $\underline{\epsilon}_{ijc}$ to be correlated conditional on latent class membership with covariance matrices differing between classes.

7.4 Models with Correlated Residual Errors

In the models considered so far, the residuals $\underline{\epsilon}_{ij}$ have been assumed to be mutually independent and the longitudinal dependence among the responses (given the covariates) has been accommodated by including either fixed or random unit-specific effects. In the case of random effects, the responses are conditionally independent given the random effects but marginally dependent with covariance structures for linear models given in Table 7.1.

These covariance structures may be overly restrictive, particularly for a random intercept model when there are a large number of occasions. For instance, the correlations between pairs of responses often decrease as the time lag increases, which is at odds with the constant correlations induced by the random intercept model. For such reasons, the residuals $\underline{\epsilon}_{ij}$ are sometimes allowed to be correlated. Caution should be exercised when combining a complex unit-level random part with a covariance structure for the residuals, as the resulting model may not be identified.

Allowing for dependence among the residuals can also be motivated as follows. Unit-specific intercepts and slopes accommodate the effects of only time-constant influences (not represented by the covariates). The independence assumption for the residuals then implies that time-varying random influences are immediate and do not persist over more than a single occasion. There is often no compelling reason to exclude a third type of random influence that is neither everlasting nor fleeting, but persists for an intermediate length of time, leading to serially correlated residual errors.

In the following subsections, we follow the treatment in Skrondal and Rabe-Hesketh [60]. We discuss the case of continuous responses, sometimes indicating how the models are modified for other response types. The models to be described can be generalized to dichotomous and ordinal responses using the latent response formulation.

7.4.1 Autoregressive Residuals

When occasions are equally spaced in time, a first-order autoregressive model AR(1) can be expressed as

$$\underline{\epsilon}_{ij} = \alpha \underline{\epsilon}_{i-1,j} + \underline{u}_{ij}, \tag{7.6}$$

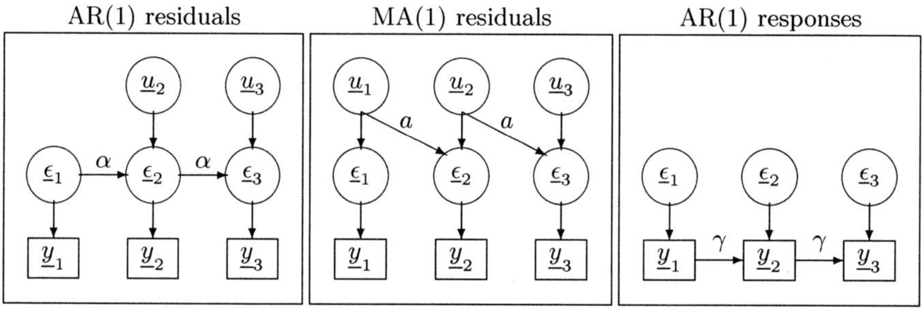

Fig. 7.4 Path diagrams for autoregressive responses and autoregressive and moving average residuals. Covariates and paths from covariates to responses omitted (Source: Skrondal and Rabe-Hesketh [60]).

where $\underline{\epsilon}_{i-1,j}$ is independently distributed from the "innovation errors" $\underline{u}_{ij}$, $\underline{u}_{ij} \sim \mathcal{N}(0, \sigma_u^2)$. This is illustrated in path diagram form in the first panel of Fig. 7.4. A "random walk" is obtained if the restriction $\alpha = 1$ is imposed.

Assuming that the process is weakly stationary, $|\alpha| < 1$, the covariance structure is as shown in panel C of Table 7.1. The correlations between responses at different occasions become

$$\text{Corr}(\underline{\epsilon}_{ij}, \underline{\epsilon}_{i+k,j}) = \alpha^k.$$

For non-equally spaced occasions, the correlation structure is often specified as

$$\text{Corr}(\underline{y}_{ij}, \underline{y}_{i+k,j}) = \alpha^{|t_{i+k}-t_i|},$$

where the correlation structure for unbalanced occasions is simply obtained by replacing t_i by t_{ij} [16]. In the case of balanced occasions, we can also specify a different parameter α_i for each occasion, giving an antedependence structure [21] for the residuals.

For the wage panel data, we estimated a random intercept model with AR(1) residuals by maximum likelihood using the `lme()` function in S-PLUS giving the estimates in column 7 of Table 7.2 (Stata's `xtregar` command can be used to estimate the model using the generalized least squares estimator proposed by Baltagi and Wu [4]). The autoregressive coefficient is estimated as $\hat{\alpha} = 0.27$ and the estimates of the regression parameters $\boldsymbol{\beta}$ are very similar to those given for the random intercept model in the first column. The random intercept model with AR(1) residuals has a considerably larger log-likelihood than the random intercept model with uncorrelated residuals. Introducing a random slope increases the log-likelihood to -2095.7 and reduces the estimated autoregressive coefficient to $\hat{\alpha} = 0.17$ (estimates not shown).

First-order autoregressive covariance structures are often as unrealistic as the random intercept structure since the correlations fall off too rapidly with increasing time lags. One possibility is to specify a higher-order autoregressive process of order k, AR(k),

$$\underline{\epsilon}_{ij} = \alpha_1 \underline{\epsilon}_{i-1,j} + \alpha_2 \underline{\epsilon}_{i-2,j} + \cdots + \alpha_k \underline{\epsilon}_{i-k,j} + \underline{u}_{ij}.$$

7.4.2 Moving Average Residuals

Random shocks disturb the response variable for some fixed number of periods before disappearing and can be modeled by moving averages [6]. A first-order moving average process MA(1) for the residuals can be specified as

$$\underline{\epsilon}_{ij} = \underline{u}_{ij} + a\,\underline{u}_{i-1,j}.$$

A path diagram for this model is given in the second panel of Fig. 7.4 and the covariance structure is presented in panel D of Table 7.1. The MA(1) process "forgets" what happened more than one period in the past, in contrast to the autoregressive processes.

The moving average model of order k, MA(k), is given as

$$\underline{\epsilon}_{ij} = \underline{u}_{ij} + a_1 \underline{u}_{i-1,j} + a_2 \underline{u}_{i-2,j} + \cdots + a_k \underline{u}_{i-k,j},$$

with "memory" extending k periods in the past.

7.5 Models with Lagged Responses

Lags of the response $\underline{y}_{ij}$ can be included as covariates in addition to $\boldsymbol{x}_{ij}$. Such models are usually called autoregressive models but are sometimes also referred to as transition models [17], Markov models [17], or conditional models [11].

When occasions are equally spaced in time, a first-order autoregressive model for the responses $\underline{y}_{ij}$ can be written as

$$\underline{y}_{ij} = \boldsymbol{x}'_{ij}\boldsymbol{\beta} + \gamma \underline{y}_{i-1,j} + \underline{\epsilon}_{ij}.$$

A path diagram for this model is shown under "AR(1) responses" in the third panel of Fig. 7.4. Assuming that the process is weakly stationary, $|\gamma| < 1$, the covariance structure is shown in panel E of Table 7.1. An extension of the autoregressive model is the antedependence model, which specifies a different parameter γ_i for each occasion.

A first-order autoregressive model for the responses was estimated for the wage panel data giving the estimates shown in the last column of Table 7.2. The regression coefficient of the lagged response is estimated as $\hat{\gamma} = 0.56$. As would be expected, many of the other regression coefficients change considerably due to controlling for the lagged response.

As for the residual autoregressive structure, the first-order autoregressive structure for responses is often deemed unrealistic, since the correlations fall off too rapidly with increasing time lags. Once again, this may be rectified by specifying a higher-order autoregressive process AR(k),

$$\underline{y}_{ij} = \boldsymbol{x}'_{ij}\boldsymbol{\beta} + \gamma_1 \underline{y}_{i-1,j} + \gamma_2 \underline{y}_{i-2,j} + \cdots + \gamma_k \underline{y}_{i-k,j} + \underline{\epsilon}_{ij}.$$

Apart from being of interest in its own right, the lagged response model is useful for distinguishing between different longitudinal models. Consider two simple models; a model with a lagged response and lagged covariate but independent residuals $\underline{\epsilon}_{ij}$

$$\underline{y}_{ij} = \gamma\, \underline{y}_{i-1,j} + \beta_1 x_{ij} + \beta_2 x_{i-1,j} + \underline{\epsilon}_{ij}, \tag{7.7}$$

and an autocorrelation model without lagged response or lagged covariate

$$\underline{y}_{ij} = \beta\, x_{ij} + \underline{\epsilon}_{ij},$$

but residuals $\underline{\epsilon}_{ij}$ having an AR(1) structure. Substituting first for $\underline{\epsilon}_{ij} = \alpha\underline{\epsilon}_{i-1,j} + \underline{u}_{ij}$ from (7.6), then for $\underline{\epsilon}_{i-1,j} = \underline{y}_{i-1,j} - \beta\, x_{i-1,j}$, and reexpressing, the autocorrelation model can alternatively be written as

$$\underline{y}_{ij} = \alpha\, \underline{y}_{i-1,j} + \beta\, x_{ij} - \alpha\beta\, x_{i-1,j} + \underline{u}_{ij}.$$

This model is equivalent to model (7.7) with the restriction $\beta_2 = -\gamma\beta_1$. This means that we can use (7.7) to discriminate between autocorrelated residuals and lagged responses.

Use of lagged response models should be conducted with caution. First, lags should be avoided if the lagged effects do not have a "causal" interpretation since the interpretation of $\boldsymbol{\beta}$ changes when $\underline{y}_{i-1,j}$ is included as an additional covariate. Second, the models require balanced data in the sense that all units are measured on the same occasions. It is problematic if the response for a unit is missing at an occasion. In practice, the entire unit is often discarded in this case. Third, lagged response models reduce the sample size. This is because the $\underline{y}_{ij}$ on the first occasions can only serve as covariates and cannot be regressed on lagged responses (which are missing). Alternatively, if the lagged responses are treated as endogenous, the sample size is not reduced, but an initial condition problem arises for the common situation where the process is ongoing when we start observing it [28]. Finally, if random effects are also included in the model, even the initial response (at the start of the process) becomes endogenous [28].

An advantage of lagged response models as compared to models with autoregressive residuals is that they can easily be used for response types other than the continuous. Heckman [29] discusses a very general framework for longitudinal modeling of dichotomous responses, for instance combining lagged responses with random effects.

7.6 Marginal Approaches

As is clear from the general form of generalized linear mixed models (including linear mixed models) in (7.2), the model linking the expectation to the covariates is specified conditional on the unit-specific random effects $\underline{\boldsymbol{\delta}}_j$. The regression coefficients $\boldsymbol{\beta}$ therefore have a conditional or unit-specific interpretation.

The marginal or population averaged expectations of the responses can be obtained by integrating the inverse link function of the linear predictor over the random effects distribution

$$E(\underline{y}_{ij}) = \int g^{-1}(\boldsymbol{x}'_{ij}\boldsymbol{\beta} + \boldsymbol{z}'_{ij}\boldsymbol{\delta}_j)\ \phi(\boldsymbol{\delta}_j; \boldsymbol{\emptyset}, \boldsymbol{\Omega})\ \mathrm{d}\boldsymbol{\delta}_j\,, \qquad (7.8)$$

where $\phi(\boldsymbol{\delta}_j; \boldsymbol{\emptyset}, \boldsymbol{\Omega})$ is the multivariate normal density of the random effects.

For linear mixed models, the link function $g(\cdot)$ is the identity and the population averaged expectation is simply the fixed part $\boldsymbol{x}'_{ij}\boldsymbol{\beta}$ of the model. Therefore, the regression coefficients $\boldsymbol{\beta}$ also have a population averaged interpretation in this case. In the linear case, it could therefore be argued that it does not matter whether the model is interpreted conditionally or marginally. However, in the marginal interpretation of the random part, only the covariance matrix $\boldsymbol{\Psi}_j$ of the total random part (as shown in Table 7.1) is interpreted, not the individual covariance matrices $\boldsymbol{\Omega}$ and $\boldsymbol{\Sigma}_j \stackrel{\Delta}{=} \mathrm{Cov}(\underline{\boldsymbol{\epsilon}}_j)$. Thus, Verbeke and Molenberghs [65] argue that the covariance matrices $\boldsymbol{\Omega}$ and $\boldsymbol{\Sigma}_j$ need not be positive semi-definite in this case as long as $\boldsymbol{\Psi}_j$ is positive semi-definite.

For link functions other than the identity, the expectation in (7.8) differs from the fixed part of the model. For generalized linear mixed models with probit links, we can derive a simple form for the population averaged expectation using the latent response formulation. The model can be specified as

$$\underline{y}^*_{ij} = \boldsymbol{x}'_{ij}\boldsymbol{\beta} + \boldsymbol{z}'_{ij}\underline{\boldsymbol{\delta}}_j + \underline{\epsilon}_{ij}, \qquad \underline{\boldsymbol{\delta}}_j \sim \mathcal{N}(\boldsymbol{\emptyset}, \boldsymbol{\Omega}), \quad \underline{\epsilon}_{ij} \sim \mathcal{N}(0, 1),$$

with $\underline{y}_{ij} = 1$ if $\underline{y}^*_{ij} > 0$ and $\underline{y}_{ij} = 0$ otherwise. The unit-specific model then becomes

$$E(\underline{y}_{ij} \mid \underline{\boldsymbol{\delta}}_j) = \Pr(\underline{y}_{ij} = 1 \mid \underline{\boldsymbol{\delta}}_j) = \Phi(\boldsymbol{x}'_{ij}\boldsymbol{\beta} + \boldsymbol{z}'_{ij}\underline{\boldsymbol{\delta}}_j),$$

where $\Phi(\cdot)$ is the standard normal cumulative distribution function, the inverse probit link. The corresponding marginal model is given by

$$\begin{aligned}
E(\underline{y}_{ij}) &= \Pr(\underline{y}_{ij} = 1) \\
&= \Pr(\underline{y}^*_{ij} > 0) \\
&= \Pr(\boldsymbol{x}'_{ij}\boldsymbol{\beta} + \boldsymbol{z}'_{ij}\underline{\boldsymbol{\delta}}_j + \underline{\epsilon}_{ij} > 0) \\
&= \Pr\big(-(\boldsymbol{z}'_{ij}\underline{\boldsymbol{\delta}}_j + \underline{\epsilon}_{ij}) \leq \boldsymbol{x}'_{ij}\boldsymbol{\beta}\big) \\
&= \Pr\left(\frac{\boldsymbol{z}'_{ij}\underline{\boldsymbol{\delta}}_j + \underline{\epsilon}_{ij}}{\sqrt{\boldsymbol{z}'_{ij}\boldsymbol{\Omega}\boldsymbol{z}_{ij} + 1}} \leq \frac{\boldsymbol{x}'_{ij}\boldsymbol{\beta}}{\sqrt{\boldsymbol{z}'_{ij}\boldsymbol{\Omega}\boldsymbol{z}_{ij} + 1}}\right) \\
&= \Phi\left(\frac{\boldsymbol{x}'_{ij}\boldsymbol{\beta}}{\sqrt{\boldsymbol{z}'_{ij}\boldsymbol{\Omega}\boldsymbol{z}_{ij} + 1}}\right),
\end{aligned} \tag{7.9}$$

where the denominator is greater than 1 if $\boldsymbol{\Omega} \neq \boldsymbol{\emptyset}$. For a random intercept probit model, the denominator is a constant $\sqrt{\omega^2 + 1}$ and the population averaged model has the same functional form as the unit-specific model but with attenuated regression coefficients $\boldsymbol{\beta}/\sqrt{\omega^2 + 1}$. This attenuation is shown graphically in Fig. 7.5, where the dashed curves represent unit-specific relationships for a random intercept probit model with a single covariate, whereas the flatter solid curve represents the population averaged relationship.

It can be seen from (7.9) that if any aspect of the random part of the model is altered, the regression coefficients must also be altered to obtain a

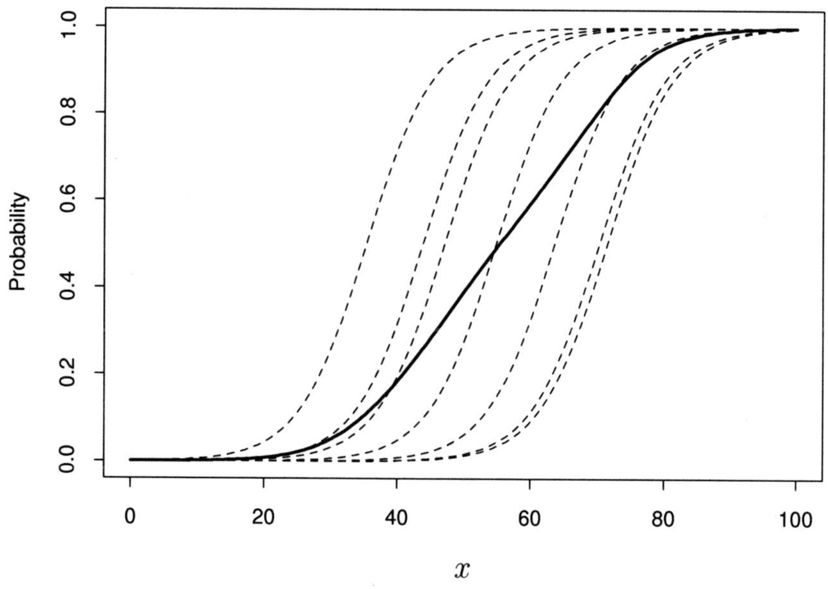

Fig. 7.5 Unit-specific versus population-averaged probit regression.

good fit to the empirical (marginal) relationship between the response and covariates. Therefore, estimates of the unit-specific regression parameters become inconsistent under misspecification of the random part except for linear mixed models.

Whether unit-specific or population averaged effects are of interest will depend on the context. For example, in public health, population averaged effects may be of interest, whereas unit-specific effects are important for the patient and clinician. An advantage of unit-specific effects is that they are more likely to be stable across populations, whereas marginal effects depend greatly on the between-unit heterogeneity, which will generally differ between populations.

If interest is focused on marginal effects and between-unit heterogeneity or longitudinal dependence are regarded as a nuisance, generalized estimating equations (GEE) [36, 71] can be used. The simplest version is to estimate the mean structure as if the responses were independent and then adjust standard errors for the dependence using the so-called sandwich estimator. The estimates of the population averaged regression parameters can be shown to be consistent, but if the responses are correlated, they are not efficient. To increase efficiency a "working correlation matrix" is therefore specified within a multivariate extension of the iteratively reweighted least squares algorithm for generalized linear models. Typically, one of the structures listed in Table 7.1 is used for the working correlation matrix of the residuals $\underline{y}_{ij} - g^{-1}(\boldsymbol{x}'_{ij}\boldsymbol{\beta})$, as well as unrestricted and independence correlation structures. The working correlation matrix is combined with the variance function of an appropriate generalized linear model, typically allowing for overdispersion if the responses are counts. It is important to note that, apart from continuous responses, the specified correlation structures generally cannot be derived from a statistical model. Thus, GEE is a multivariate quasi-likelihood approach with no proper likelihood.

There are also "proper" marginal statistical models with corresponding likelihoods. Examples include the Bahadur [2] and Dale [13] models, which parameterize dependence via marginal correlations and marginal bivariate odds ratios, respectively [19, 44]. See Molenberghs and Verbeke [45] for an overview of these models.

Heagerty and Zeger [27] introduce random effects models where the marginal mean is regressed on covariates as in GEE. In these models, the relationship between the conditional mean (given the random effects) and the covariates is found by solving the integral equation (7.8) linking the conditional and marginal means. Interestingly, the integral involved can be written as a unidimensional integral over the distribution of the sum of the terms in the random part of the model.

7.7 Concluding Remarks

It is straightforward to extend the longitudinal models discussed here to situations where units are clustered by including random effects varying at higher levels.

We have focused on linear and quadratic growth models, but nonlinear growth models can also be specified via linear mixed models using higher-order polynomials of time or splines [62]. Nonlinear mixed models [50] can be preferable if specific functional forms are suggested by substantive theory as in pharmacokinetics.

Useful books on modeling longitudinal data include Skrondal and Rabe-Hesketh [60], Hand and Crowder [24], Crowder and Hand [12], Vonesh and Chinchilli [67], Jones [33], Hsiao [32], Baltagi [3], Wooldridge [70], Lindsey [38], Verbeke and Molenberghs [65], Diggle et al. [17], and Fitzmaurice et al. [20].

References

1. M. Aitkin. A general maximum likelihood analysis of variance components in generalized linear models. *Biometrics*, 55:117–128, 1999.
2. R. R. Bahadur. A representation of the joint distribution of responses to n dichotomous items. In H. Solomon, editor, *Studies in Item Analysis and Prediction*, volume 6, pages 158–168. Stanford University Press, Stanford, CA, 1961.
3. B. H. Baltagi. *Econometric Analysis of Panel Data*, 2nd edition. Wiley, Chichester, 2001.
4. B. H. Baltagi and P. X. Wu. Unequally spaced panel data regressions with AR(1) disturbances. *Econometric Theory*, 15:814–823, 1999.
5. D. J. Bartholomew. The sensitivity of latent trait analysis to the choice of prior distribution. *British Journal of Mathematical and Statistical Psychology*, 41: 101–107, 1988.
6. G. E. P. Box, G. M. Jenkins, and G. C. Reinsel. *Time Series Analysis: Forecasting and Control*, 3rd edition. Prentice-Hall, Englewood Cliffs, NJ, 1994.
7. N. E. Breslow and N. Day. *Statistical Methods in Cancer Research. Vol. I — The Analysis of Case-Control Studies.* IARC, Lyon, 1980.
8. G. Chamberlain. Analysis of covariance with qualitative data. *Review of Economic Studies*, 47:225–238, 1980.
9. G. Chamberlain. Panel data. In Z. Griliches and M. D. Intriligator, editors, *Handbook of Econometrics*, volume 2, pages 1247–1318. North-Holland, Amsterdam, 1984.
10. D. G. Clayton. Generalized linear mixed models. In W. R. Gilks, S. Richardson, and D. J. Spiegelhalter, editors, *Markov Chain Monte Carlo in Practice*, pages 275–301. Chapman & Hall, London, 1996.

11. D. R. Cox. The analysis of multivariate binary data. *Applied Statistics*, 21: 113–120, 1972.
12. M. J. Crowder and D. J. Hand. *Analysis of Repeated Measures.* Chapman & Hall, London, 1990.
13. J. R. Dale. Global cross-ratio models for bivariate, discrete, ordered responses. *Biometrics*, 42:909–917, 1986.
14. C. M. Dayton and G. B. MacReady. Concomitant variable latent class models. *Journal of the American Statistical Association*, 83:173–178, 1988.
15. J. de Leeuw and E. Meijer. Introduction to multilevel analysis. In J. de Leeuw and E. Meijer, editors, *Handbook of Multilevel Analysis*, Chapter 1. Springer, New York, 2008. (this volume)
16. P. J. Diggle. An approach to analysis of repeated measures. *Biometrics*, 44: 959–971, 1988.
17. P. J. Diggle, P. J. Heagerty, K.-Y. Liang, and S. L. Zeger. *Analysis of Longitudinal Data*, 2nd edition. Oxford University Press, Oxford, UK, 2002.
18. R. C. Elston. On estimating time-response curves. *Biometrics*, 20:643–647, 1964.
19. G. M. Fitzmaurice, N. M. Laird, and A. G. Rotnitzky. Regression models for discrete longitudinal responses. *Statistical Science*, 8:284–309, 1993.
20. G. M. Fitzmaurice, N. M. Laird, and J. H. Ware. *Applied Longitudinal Analysis.* Wiley, New York, 2004.
21. K. R. Gabriel. Ante-dependence analysis of an ordered set of variables. *Annals of Mathematical Statistics*, 33:201–212, 1962.
22. W. A. Gibson. Three multivariate models: factor analysis, latent structure analysis and latent profile analysis. *Psychometrika*, 24:229–252, 1959.
23. H. Goldstein. Nonlinear multilevel models, with an application to discrete response data. *Biometrika*, 78:45–51, 1991.
24. D. J. Hand and M. J. Crowder. *Practical Longitudinal Data Analysis.* Chapman & Hall, London, 1996.
25. J. A. Hausman. Specification tests in econometrics. *Econometrica*, 46:1251–1271, 1978.
26. P. J. Heagerty and B. F. Kurland. Misspecified maximum likelihood estimates and generalized linear mixed models. *Biometrika*, 88:973–985, 2001.
27. P. J. Heagerty and S. L. Zeger. Marginalized multilevel models and likelihood inference. *Statistical Science*, 15:1–26, 2000. (with discussion)
28. J. J. Heckman. The incidental parameters problem and the problem of initial conditions in estimating a discrete stochastic process and some Monte Carlo evidence on their practical importance. In C. F. Manski and D. L. McFadden, editors, *Structural Analysis of Discrete Data with Econometric Applications*, pages 179–196. MIT Press, Cambridge, MA, 1981. (Downloadable from `http://elsa.berkeley.edu/~mcfadden/discrete.html`)
29. J. J. Heckman. Statistical models for discrete panel data. In C. F. Manski and D. L. McFadden, editors, *Structural Analysis of Discrete Data with Econometric Applications*, pages 114–178. MIT Press, Cambridge, MA, 1981. (Downloadable from `http://elsa.berkeley.edu/~mcfadden/discrete.html`)

30. J. J. Heckman and B. Singer. A method for minimizing the impact of distributional assumptions in econometric models for duration data. *Econometrica*, 52: 271–320, 1984.
31. D. Hedeker. Multilevel models for ordinal and nominal variables. In J. de Leeuw and E. Meijer, editors, *Handbook of Multilevel Analysis*, Chapter 6. Springer, New York, 2008. (this volume)
32. C. Hsiao. *Analysis of Panel Data*, 2nd edition. Cambridge University Press, Cambridge, UK, 2003.
33. R. H. Jones. *Longitudinal Data with Serial Correlation: A State-Space Approach.* Chapman & Hall, London, 1993.
34. N. M. Laird. Nonparametric maximum likelihood estimation of a mixing distribution. *Journal of the American Statistical Association*, 73:805–811, 1978.
35. N. M. Laird and J. H. Ware. Random-effects models for longitudinal data. *Biometrics*, 38:963–974, 1982.
36. K.-Y. Liang and S. L. Zeger. Longitudinal data analysis using generalized linear models. *Biometrika*, 73:13–22, 1986.
37. B. G. Lindsay and M. L. Lesperance. A review of semiparametric mixture models. *Journal of Statistical Planning and Inference*, 47:5–28, 1995.
38. J. K. Lindsey. *Models for Repeated Measurements*, 2nd edition. Oxford University Press, Oxford, UK, 1999.
39. N. T. Longford. *Random Coefficient Models.* Oxford University Press, Oxford, UK, 1993.
40. N. T. Longford. Missing data. In J. de Leeuw and E. Meijer, editors, *Handbook of Multilevel Analysis*, Chapter 10. Springer, New York, 2008. (this volume)
41. G. S. Maddala. The use of variance components models in pooling cross-section and time series data. *Econometrica*, 39:341–358, 1971.
42. C. E. McCulloch and S. R. Searle. *Generalized, Linear and Mixed Models.* Wiley, New York, 2001.
43. W. Meredith and J. Tisak. Latent curve analysis. *Psychometrika*, 55:107–122, 1990.
44. G. Molenberghs. Generalized estimating equations. In M. Aerts, H. Geys, G. Molenberghs, and L. M. Ryan, editors, *Topics in Modelling Clustered Data*, pages 47–75. Chapman & Hall/CRC, Boca Raton, FL, 2002.
45. G. Molenberghs and G. Verbeke. *Models for Discrete Longitudinal Data.* Springer, New York, 2005.
46. B. O. Muthén and K. Shedden. Finite mixture modeling with mixture outcomes using the EM algorithm. *Biometrics*, 55:463–469, 1999.
47. D. S. Nagin and K. C. Land. Age, criminal careers, and population heterogeneity: Specification and estimation of a nonparametric mixed Poisson model. *Criminology*, 31:327–362, 1993.
48. J. M. Neuhaus, W. W. Hauck, and J. D. Kalbfleisch. The effects of mixture distribution misspecification when fitting mixed-effects logistic models. *Biometrika*, 79:755–762, 1992.
49. J. Neyman and E. L. Scott. Consistent estimates based on partially consistent observations. *Econometrica*, 16:1–32, 1948.

50. J. C. Pinheiro and D. M. Bates. *Mixed-Effects Models in S and S-PLUS.* Springer, New York, 2000.
51. R. E. Quandt. A new approach to estimating switching regressions. *Journal of the American Statistical Association*, 67:306–310, 1972.
52. S. Rabe-Hesketh, A. Pickles, and A. Skrondal. Correcting for covariate measurement error in logistic regression using nonparametric maximum likelihood estimation. *Statistical Modelling*, 3:215–232, 2003.
53. S. Rabe-Hesketh and A. Skrondal. *Multilevel and Longitudinal Modeling Using Stata.* Stata Press, College Station, TX, 2005.
54. S. Rabe-Hesketh and A. Skrondal. Generalized linear mixed effects models. In G. Fitzmaurice, M. Davidian, G. Verbeke, and G. Molenberghs, editors, *Longitudinal Data Analysis: A Handbook of Modern Statistical Methods.* Chapman & Hall/CRC, Boca Raton, FL, 2007.
55. S. Rabe-Hesketh, A. Skrondal, and A. Pickles. GLLAMM manual. Working Paper 160, U.C. Berkeley Division of Biostatistics, Berkeley, CA, 2004. (Downloadable from `http://www.bepress.com/ucbbiostat/paper160/`)
56. S. Rabe-Hesketh, A. Skrondal, and A. Pickles. Maximum likelihood estimation of limited and discrete dependent variable models with nested random effects. *Journal of Econometrics*, 128:301–323, 2005.
57. G. Rasch. *Probabilistic Models for Some Intelligence and Attainment Tests.* Danmarks Pædagogiske Institut, Copenhagen, 1960.
58. G. Rodríguez. Multilevel generalized linear models. In J. de Leeuw and E. Meijer, editors, *Handbook of Multilevel Analysis*, Chapter 9. Springer, New York, 2008. (this volume)
59. D. B. Rubin. Inference and missing data. *Biometrika*, 63:581–592, 1976. (with discussion)
60. A. Skrondal and S. Rabe-Hesketh. *Generalized Latent Variable Modeling: Multilevel, Longitudinal, and Structural Equation Models.* Chapman & Hall/CRC, Boca Raton, FL, 2004.
61. T. A. B. Snijders and J. Berkhof. Diagnostic checks for multilevel models. In J. de Leeuw and E. Meijer, editors, *Handbook of Multilevel Analysis*, Chapter 3. Springer, New York, 2008. (this volume)
62. T. A. B. Snijders and R. J. Bosker. *Multilevel Analysis: An Introduction to Basic and Advanced Multilevel Modeling.* Sage, London, 1999.
63. StataCorp. *Stata Statistical Software: Release 9.0.* Stata Corporation, College Station, TX, 2005.
64. F. Vella and M. Verbeek. Whose wages do unions raise? A dynamic model of unionism and wage rate determination for young men. *Journal of Applied Econometrics*, 13:163–183, 1998.
65. G. Verbeke and G. Molenberghs. *Linear Mixed Models for Longitudinal Data.* Springer, New York, 2000.
66. G. Verbeke, B. Spiessens, and E. Lesaffre. Conditional linear mixed models. *American Statistician*, 55:25–34, 2001.
67. E. F. Vonesh and V. M. Chinchilli. *Linear and Nonlinear Models for the Analysis of Repeated Measurements.* Marcel Dekker, New York, 1997.

68. M. Wedel and W. A. Kamakura. *Market Segmentation: Conceptual and Methodological Foundations*, 2nd edition. Kluwer, Dordrecht, 2000.
69. M. Wedel and W. A. Kamakura. Factor analysis with (mixed) observed and latent variables. *Psychometrika*, 66:515–530, 2001.
70. J. M. Wooldridge. *Econometric Analysis of Cross Section and Panel Data.* MIT Press, Cambridge, MA, 2002.
71. S. L. Zeger and K.-Y. Liang. Longitudinal data analysis for discrete and continuous outcomes. *Biometrics*, 42:121–130, 1986.

8

Non-Hierarchical Multilevel Models

Jon Rasbash[1] and William J. Browne[2]

[1] University of Bristol, Graduate School of Education
[2] University of Bristol, Department of Clinical Veterinary Science

8.1 Introduction

In the models discussed in this book so far we have assumed that the structures of the populations from which the data have been drawn are hierarchical. This assumption is sometimes not justified. In this chapter two main types of non-hierarchical model are considered. Firstly, cross-classified models. The notion of cross-classification is probably reasonably familiar to most readers. Secondly, we consider multiple membership models, where lower level units are influenced by more than one higher-level unit from the same classification. For example, some pupils may attend more than one school. We also consider situations that contain a mixture of hierarchical, crossed and multiple membership relationships.

8.2 Cross-Classified Models

This section is divided into three parts. In the first part we look at situations that can give rise to a two way cross-classification and introduce some diagrams to describe the population structure, and discuss notation for constructing a statistical model. In the second part we discuss some of the possible estimation methods for estimating cross-classified models and give an example analysis of an educational data set. In the third part we then describe some more complex cross-classified structures and give an example analyses of a medical data set.

J. de Leeuw, E. Meijer (eds.), *Handbook of Multilevel Analysis*,
© Springer 2008

8.2.1 Two-Way Cross-Classifications: A Basic Model

Suppose, we have data on a large number of patients, attending many hospitals and we also know the neighbourhood in which the patient lives and that we regard patient, neighbourhood and hospital all as important sources of variation for the patient level outcome measure we wish to study. Now, typically hospitals will draw patients from many different neighbourhoods and the inhabitants of a neighbourhood will go to many hospitals. No pure hierarchy can be found and patients are said to be contained within a cross-classification of hospitals by neighbourhoods. This can be represented schematically, for the case of twelve patients contained within a cross-classification of three neighbourhoods by four hospitals as in Table 8.1.

In this example we have patients at level 1 and neighbourhood and hospital are cross-classified at level 2. The characteristic pattern of a cross-classification is shown: some rows contain multiple entries and some columns contain multiple entries. In a nested relationship, if the row classification is nested within the column classification, then all the entries across a row will fall under a single column and vice versa if the column classification is nested within the row classification. For example, if hospitals are nested within neighbourhoods, we might observe the pattern in Table 8.2.

Many studies follow this simple two-way crossed structure; here are a few examples:

- Education: students cross-classified by primary school and secondary school.
- Health: patients cross-classified by general practice and hospital.
- Survey data: individuals cross-classified by interviewer and area of residence.

Table 8.1 Patients cross-classified by hospital and neighbourhood.

	Neighbourhood 1	Neighbourhood 2	Neighbourhood 3
Hospital 1	XX	X	
Hospital 2	X	X	
Hospital 3		XX	X
Hospital 4		X	XXX

Table 8.2 Patients nested within hospitals within neighbourhoods.

	Neighbourhood 1	Neighbourhood 2	Neighbourhood 3
Hospital 1	XXX		
Hospital 2		XX	
Hospital 3			XXX
Hospital 4	XXXX		

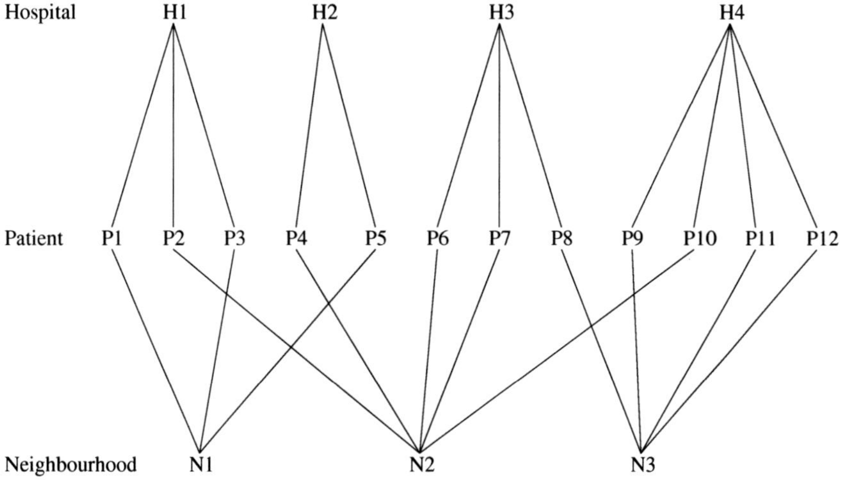

Fig. 8.1 Diagrams for crossed structure given in Table 8.1.

Diagrams for Representing the Relationship Between Classifications

We find two types of diagrams useful in expressing the nature of relationships between classifications. Firstly, unit diagrams where we draw every unit (patient, hospital and neighbourhood, in the case of our first example) and connect each lowest level unit (patient) to its parent units (hospital, neighbourhood). Such a representation of the data in Table 8.1 is shown in Fig. 8.1.

Note that we have two hierarchies present, patients within hospitals and patients within neighbourhoods. We have organised the topology of the diagram such that patients are nested within hospitals. However, when we come to add neighbourhoods to the diagram, we see that the connecting lines cross, indicating we have a cross-classification. Drawing the hierarchical structure shown in Table 8.2 gives the representation shown in Fig. 8.2.

Clearly, to draw such diagrams that include all units with large data sets is not practical, as there will be far too many nodes on the diagram to fit into a reasonable area. However, they can be used in schematic form to convey the structure of the relationship between classifications. But when we have four or five random classifications present (as commonly occur with social data), even schematic forms of these diagrams can become hard to read. There is a more minimal diagram, the classification diagram, which has one node for each classification. Nodes connected by an arrow indicate a nested relationship, nodes connected by a double arrow indicate a multiple-membership

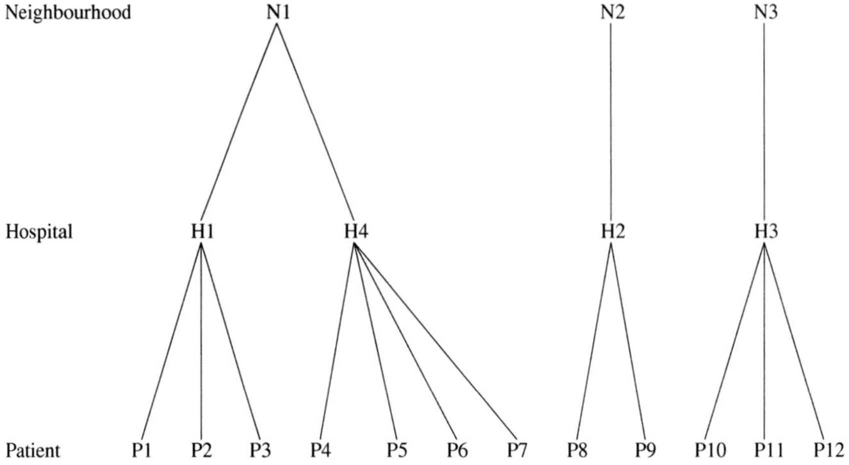

Fig. 8.2 Diagrams for completely nested structure given in Table 8.2.

relationship (examples are given later) and unconnected nodes indicate a crossed relationship. Thus the crossed structure in Fig. 8.1 and the completely nested structure of Fig. 8.2 are drawn as shown in Fig. 8.3.

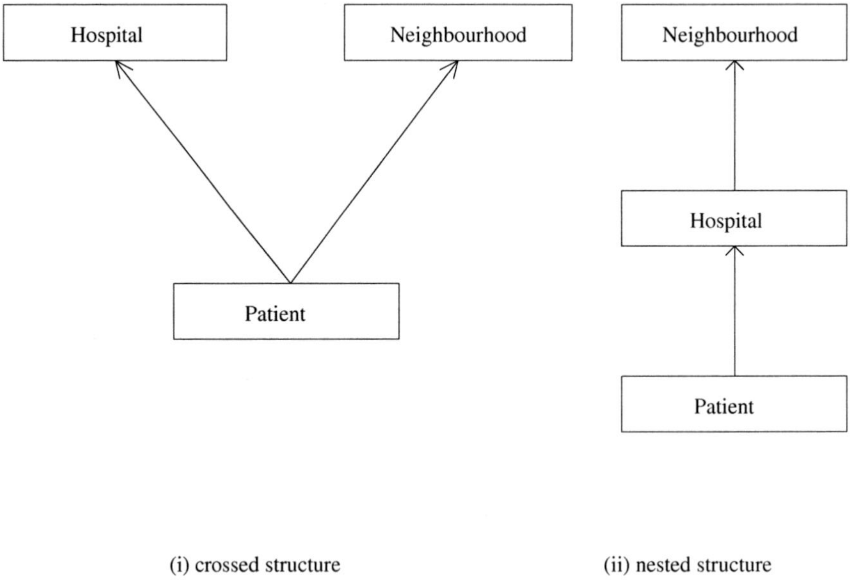

Fig. 8.3 Classification diagrams for nesting and crossing.

Some Notation for Constructing a Statistical Model

The matrix notation used in this book for describing hierarchical models, that is,

$$\underline{\boldsymbol{y}}_j = \boldsymbol{X}_j\boldsymbol{\beta} + \boldsymbol{Z}_j\underline{\boldsymbol{\delta}}_j + \underline{\boldsymbol{\epsilon}}_j \,,$$

does not readily extend to the case of cross-classifications. This is because this notation assumes a unique hierarchy where we write down the generic equation for the j-th level-2 unit. In a simple cross-classification we have two sets of level-2 units, for example, hospitals and neighbourhoods, so which classification is j indexing?

We can extend the basic scalar notation to handle cross-classified structures. Assume we have patients nested within a cross-classification of neighbourhoods by hospital, that is the case illustrated in Fig. 8.3(i). Suppose we want to estimate a simple variance components model giving estimates of the mean response and patient, hospital and neighbourhood level variation. In this case, we can write the model in scalar notation as

$$\underline{y}_{i(j_1,j_2)} = \beta_0 + \underline{\delta}_{j_1} + \underline{\delta}_{j_2} + \underline{\epsilon}_{i(j_1,j_2)} \,,$$

where β_0 estimates the mean response, j_1 indexes the hospital classification, j_2 indexes the neighbourhood classification, $\underline{\delta}_{j_1}$ is the random effect for hospital j_1, $\underline{\delta}_{j_2}$ is the random effect for neighbourhood j_2, $\underline{y}_{i(j_1,j_2)}$ is the response for the i-th patient from the cell in the cross-classification defined by hospital j_1 and neighbourhood j_2 and finally $\underline{\epsilon}_{i(j_1,j_2)}$ is the patient level residual for the i-th patient from the cell in the cross-classification defined by hospital j_1 and neighbourhood j_2.

Details of how this notation extends to represent more complex models and patterns of cross-classifications are given in Rasbash and Browne [14]. One problem with this notation is that as we fit models with more classifications and more complex patterns of crossing, the subscript notation that describes the patterns becomes very cumbersome and difficult to read. We therefore prefer an alternative notation introduced in Browne et al. [2].

We can write the same model as

$$\underline{y}_i = \beta_0 + \underline{\delta}^{(2)}_{hosp(i)} + \underline{\delta}^{(3)}_{nbhd(i)} + \underline{\epsilon}_i \,,$$

where i indexes the observation level, in this case patients, and $hosp(i)$ and $nbhd(i)$ are functions that return the unit number of the hospital and neighbourhood, respectively, that patient i belongs to. Thus, for the data structure drawn in Fig. 8.1, the values of $hosp(i)$ and $nbhd(i)$ are given in Table 8.3.

Therefore, the model for Patient 3 would be

$$\underline{y}_3 = \beta_0 + \underline{\delta}^{(2)}_1 + \underline{\delta}^{(3)}_1 + \underline{\epsilon}_3$$

Table 8.3 Indexing table for hospitals and neighbourhoods for patients given in Fig. 8.1.

i	$hosp(i)$	$nbhd(i)$
1	1	1
2	1	2
3	1	1
4	2	2
5	2	1
6	3	2
7	3	2
8	3	3
9	4	3
10	4	2
11	4	3
12	4	3

and for Patient 5 would be

$$\underline{y}_5 = \beta_0 + \underline{\delta}_2^{(2)} + \underline{\delta}_1^{(3)} + \underline{\epsilon}_5.$$

We number the classifications from 2 upwards as we use classification number 1 to represent the identity classification that applies to the observation level (like level 1 in a hierarchical model). This classification simply returns the row numbers in the data matrix. As can be seen, random effects require bracketed superscripting with their classification number to avoid ambiguity.

This simplified notation has the advantage that the subscripting notation does not increase in complexity as we add more classifications. This simplification is achieved because the notation makes no attempt to describe the patterns of crossing and nesting present. This is useful information and we therefore advocate the use of this notation in conjunction with the classification diagrams, as shown in Fig. 8.3, which display these patterns explicitly.

8.2.2 Estimation Algorithms

We will describe three estimation algorithms for fitting cross-classified models in detail and mention other alternatives. Each of these three methods has advantages and disadvantages in terms of speed, memory usage and bias and these will be discussed later. All three methods have been implemented in versions of the MLwiN software package [16] and all results in this chapter are produced by this package.

An IGLS Algorithm for Estimating Cross-Classified Models

The iterative generalized least squares estimates for a multilevel model are those estimates which simultaneously satisfy both of the following equations:

$$\hat{\boldsymbol{\beta}} = (\boldsymbol{X}'\boldsymbol{V}^{-1}\boldsymbol{X})^{-1}\boldsymbol{X}'\boldsymbol{V}^{-1}\boldsymbol{y},$$
$$\hat{\boldsymbol{\theta}} = \left(\boldsymbol{Z}^{*\prime}(\boldsymbol{V}^{*})^{-1}\boldsymbol{Z}^{*}\right)^{-1}\boldsymbol{Z}^{*\prime}(\boldsymbol{V}^{*})^{-1}\boldsymbol{y}^{*},$$

where $\hat{\boldsymbol{\beta}}$ are the estimated fixed coefficients and $\hat{\boldsymbol{\theta}}$ is a vector containing the estimated variances and covariances of the sets of random effects in the model. $\boldsymbol{V} = \mathrm{Cov}(\underline{\boldsymbol{y}} \mid \boldsymbol{X}, \boldsymbol{\beta})$ and an estimate of $\boldsymbol{V}$ is constructed from the elements of $\hat{\boldsymbol{\theta}}$ and $\boldsymbol{Z}$. $\boldsymbol{y}^*$ is the vector of elements of $(\boldsymbol{y} - \boldsymbol{X}\boldsymbol{\beta})(\boldsymbol{y} - \boldsymbol{X}\boldsymbol{\beta})'$ and therefore has length N^2 (N is the length of the data set). $\boldsymbol{V}^*$ has the form $\boldsymbol{V}^* = \boldsymbol{V} \otimes \boldsymbol{V}$ and $\boldsymbol{Z}^*$ is the design matrix linking $\underline{\boldsymbol{y}}^*$ to $\boldsymbol{V}$ in the regression of $\underline{\boldsymbol{y}}^*$ on $\boldsymbol{Z}^*$. See Goldstein [7] for more details. Some of these matrices are massive, for example, $(\boldsymbol{V}^*)^{-1}$ is dimensioned $N^2 \times N^2$, making a direct software implementation of these estimating equations extremely resource intensive both in terms of CPU time and memory consumed. However, in hierarchical models, $\boldsymbol{V}$ and $\boldsymbol{V}^*$ have a block-diagonal structure which can be exploited by customised algorithms [see 9] which allow efficient computation.

The problem presented by cross-classified models is that $\boldsymbol{V}$ (and therefore $\boldsymbol{V}^*$) no longer has the block-diagonal structure which the efficient algorithm requires.

Structure of $\boldsymbol{V}$ for Cross-Classified Models

Let's take a look at the structure of $\boldsymbol{V}$, the covariance matrix of $\underline{\boldsymbol{y}}$, for cross-classified models and see how we can adapt the standard IGLS algorithm to handle cross-classifications.

The basic two-level cross-classified model (with hospitals + neighbourhoods) can be written as:

$$\underline{y}_i = \boldsymbol{x}_i\boldsymbol{\beta} + \underline{\delta}^{(2)}_{hosp(i)} + \underline{\delta}^{(3)}_{nbhd(i)} + \underline{\epsilon}_i\,,$$
$$\underline{\delta}^{(2)}_{hosp(i)} \sim \mathcal{N}(0, \sigma^2_{\delta(2)}), \quad \underline{\delta}^{(3)}_{nbhd(i)} \sim \mathcal{N}(0, \sigma^2_{\delta(3)}), \quad \underline{\epsilon}_i \sim \mathcal{N}(0, \sigma^2_{\epsilon}).$$

The variance of our response is now

$$\mathrm{Var}(\underline{y}_i) = \mathrm{Var}\big(\underline{\delta}^{(2)}_{hosp(i)} + \underline{\delta}^{(3)}_{nbhd(i)} + \underline{\epsilon}_i\big) = \sigma^2_{\delta(2)} + \sigma^2_{\delta(3)} + \sigma^2_{\epsilon}.$$

The covariance between individuals a and b is

$$\mathrm{Cov}(\underline{y}_a, \underline{y}_b) = \mathrm{Cov}\big(\underline{\delta}^{(2)}_{hosp(a)} + \underline{\delta}^{(3)}_{nbhd(a)} + \underline{\epsilon}_a,\, \underline{\delta}^{(2)}_{hosp(b)} + \underline{\delta}^{(3)}_{nbhd(b)} + \underline{\epsilon}_b\big),$$

which simplifies to $\sigma^2_{\delta(2)}$ for two individuals from the same hospital but different neighbourhoods, to $\sigma^2_{\delta(3)}$ for two individuals from the same neighbourhood but different hospitals, to $\sigma^2_{\delta(2)} + \sigma^2_{\delta(3)}$ for two individuals from the same neighbourhood and the same hospital and to zero for two individuals who are from both different neighbourhoods and different hospitals. If we take a toy

Table 8.4 Indexing table for hospitals and neighbourhoods for 5 patients.

i	$hosp(i)$	$nbhd(i)$
1	1	1
2	1	2
3	1	1
4	2	2
5	2	1

example of five patients in two hospitals and introduce a cross-classification with two neighbourhoods, as shown in Table 8.4, this generates a 5×5 covariance matrix for the responses of the five patients with the following structure:

$$\boldsymbol{V} = \begin{pmatrix} h+n+p & h & h+n & 0 & n \\ h & h+n+p & h & n & 0 \\ h+n & h & h+n+p & 0 & n \\ 0 & n & 0 & h+n+p & h \\ n & 0 & n & h & h+n+p \end{pmatrix},$$

where $h = \sigma^2_{\delta(2)}$, $n = \sigma^2_{\delta(3)}$ and $p = \sigma^2_{\epsilon}$.

Here, the data is sorted patient within hospital. This allows us to split the covariance matrix into two components: A component for patients within hospitals which has a block-diagonal structure ($\boldsymbol{P}$) and a component for neighbourhoods ($\boldsymbol{Q}$): $\boldsymbol{V} = \boldsymbol{P} + \boldsymbol{Q}$, where

$$\boldsymbol{P} = \begin{pmatrix} h+p & h & h & 0 & 0 \\ h & h+p & h & 0 & 0 \\ h & h & h+p & 0 & 0 \\ 0 & 0 & 0 & h+p & h \\ 0 & 0 & 0 & h & h+p \end{pmatrix}$$

and

$$\boldsymbol{Q} = \begin{pmatrix} n & 0 & n & 0 & n \\ 0 & n & 0 & n & 0 \\ n & 0 & n & 0 & n \\ 0 & n & 0 & n & 0 \\ n & 0 & n & 0 & n \end{pmatrix}.$$

Given that the $\boldsymbol{V}$ matrix is sorted according to patient within hospital ($\boldsymbol{P}$), $\boldsymbol{P} + \boldsymbol{Q}$ cannot be simultaneously expressed in a single $\boldsymbol{V}$ matrix as block diagonal. Splitting the structure of $\boldsymbol{V}$ into a hierarchical, block-diagonal part that the IGLS algorithm can handle in an efficient way and a non-hierarchical, non-block-diagonal part forms the basis of a relatively efficient algorithm for handling cross-classified models.

If we take the dummy variable indicator matrix of neighbourhoods ($\boldsymbol{Z}$), then we have $\boldsymbol{Q} = \boldsymbol{Z}\boldsymbol{Z}'n$:

$$\boldsymbol{Z} = \begin{pmatrix} 1\,0 \\ 0\,1 \\ 1\,0 \\ 0\,1 \\ 1\,0 \end{pmatrix}, \quad \boldsymbol{Z}\boldsymbol{Z}'n = \begin{pmatrix} 1\,0\,1\,0\,1 \\ 0\,1\,0\,1\,0 \\ 1\,0\,1\,0\,1 \\ 0\,1\,0\,1\,0 \\ 1\,0\,1\,0\,1 \end{pmatrix} n.$$

We can define a "pseudo-unit" that spans the entire data set, in our toy example, all five points, and declare this pseudo-unit to be level 3 in the model (removing the neighbourhood level from the model). We can now form the three-level hierarchical model

$$\underline{y}_i = \beta_0 + \underline{\delta}^{(2)}_{hosp(i)} + \underline{\delta}^{(3)}_{punit(i),1} z_{i1} + \underline{\delta}^{(3)}_{punit(i),2} z_{i2} + \underline{\epsilon}_i\,,$$

$$\begin{pmatrix} \underline{\delta}^{(3)}_{punit(i),1} \\ \underline{\delta}^{(3)}_{punit(i),2} \end{pmatrix} \sim \mathcal{N}(0, \Sigma_{\delta(3)}), \quad \Sigma_{\delta(3)} = \begin{pmatrix} \sigma^2_{\delta(3),1} & 0 \\ 0 & \sigma^2_{\delta(3),2} \end{pmatrix},$$

$$\underline{\delta}^{(2)}_{hosp(i)} \sim \mathcal{N}(0, \sigma^2_{\delta(2)}), \quad \underline{\epsilon}_i \sim \mathcal{N}(0, \sigma^2_{\epsilon}).$$

Here the level structure is patients within hospitals within the pseudo-unit level. z_{i1} and z_{i2} are the first and second element, respectively, of the i-th row of $\boldsymbol{Z}$. $\sigma^2_{\delta(3),1}$ and $\sigma^2_{\delta(3),2}$ are both estimates of the between-neighbourhood variation; therefore, we constrain them to be equal. Thus we can use the standard IGLS hierarchical algorithm to define and estimate the correct covariance structure for a cross-classified model. Now if we had 200 hospitals and 100 neighbourhoods, we would have to form 100 dummy variables for the neighbourhoods, allow them all to have variances at level 3 and constrain the variances to be equal. Details of this algorithm are given in Rasbash and Goldstein [15] and Bull et al. [3] and it will be referred to as the RG algorithm in later sections.

MCMC

The MCMC estimation methods (see Chapter 2 of this book for a fuller description) aim to generate samples from the joint posterior distribution of all unknown parameters. They then use these samples to calculate point and interval estimates for each individual parameter. The Gibbs sampler algorithm produces samples from the joint posterior by generating in turn from the conditional posterior distributions of groups of unknown parameters. In Chapter 2, the Gibbs sampling algorithm for a normally distributed response hierarchical model is given.

As we have seen in the notation section, we can describe our model as a set of additive terms, one for the fixed part of the model and one for each of the random classifications. The MCMC algorithm works on each of these terms separately and, consequently, the algorithm for a cross-classified model is no more complicated than for a hierarchical model. For illustration, we present the steps for the following cross-classified model based on the variance components hospitals by neighbourhoods model and refer the interested reader to Browne et al. [2] for more general algorithms. Note that if the response is dichotomous or a count, then, as in Chapter 2, we can use the Metropolis-Gibbs hybrid method discussed there.

The basic two-level cross-classified model (with hospitals + neighbourhoods) can be written as:

$$\underline{y}_i = \boldsymbol{x}_i \boldsymbol{\beta} + \underline{\delta}^{(2)}_{hosp(i)} + \underline{\delta}^{(3)}_{nbhd(i)} + \underline{\epsilon}_i \,,$$
$$\underline{\delta}^{(2)}_{hosp(i)} \sim \mathcal{N}(0, \sigma^2_{\delta(2)}), \quad \underline{\delta}^{(3)}_{nbhd(i)} \sim \mathcal{N}(0, \sigma^2_{\delta(3)}), \quad \underline{\epsilon}_i \sim \mathcal{N}(0, \sigma^2_{\epsilon}).$$

We can split our unknown parameters into six distinct sets: the fixed effects, $\boldsymbol{\beta}$; the hospital random effects, $\underline{\delta}^{(2)}_{hosp(i)}$; the neighbourhood random effects, $\underline{\delta}^{(3)}_{nbhd(i)}$; the hospital variance, $\sigma^2_{\delta(2)}$; the neighbourhood variance, $\sigma^2_{\delta(3)}$ and the residual variance, σ^2_{ϵ}.

Then we need to generate random draws from the conditional distribution of each of these six groups of unknowns. MCMC algorithms are generally used in a Bayesian context and, consequently, we need to define prior distributions for our unknown parameters. For generality, we will use a multivariate normal prior for the fixed effects, $\underline{\boldsymbol{\beta}} \sim \mathcal{N}_{p_f}(\boldsymbol{\mu}_p, \boldsymbol{S}_p)$, and scaled inverse (SI) χ^2 priors for the three variances. For the hospital variance, $\underline{\sigma}^2_{\delta(2)} \sim SI\chi^2(\nu_2, s^2_2)$; for the neighbourhood variance, $\underline{\sigma}^2_{\delta(3)} \sim SI\chi^2(\nu_3, s^2_3)$ and for the residual variance, $\underline{\sigma}^2_{\epsilon} \sim SI\chi^2(\nu_\epsilon, s^2_\epsilon)$. The steps are then as follows:

- In step 1 of the algorithm, the conditional posterior distribution in the Gibbs update for the fixed effects parameter vector $\underline{\boldsymbol{\beta}}$ is multivariate normal with dimension p_f (the number of fixed effects):

  $$p(\underline{\boldsymbol{\beta}} \mid \boldsymbol{y}, \boldsymbol{\delta}^{(2)}, \boldsymbol{\delta}^{(3)}, \sigma^2_{\delta(2)}, \sigma^2_{\delta(3)}, \sigma^2_{\epsilon}) \sim \mathcal{N}_{p_f}(\hat{\boldsymbol{\beta}}, \hat{\boldsymbol{D}}),$$

  where

  $$\hat{\boldsymbol{D}} = \left(\sum_{i=1}^{N} \frac{\boldsymbol{x}_i' \boldsymbol{x}_i}{\sigma^2_\epsilon} + \boldsymbol{S}_p^{-1} \right)^{-1},$$
  $$\hat{\boldsymbol{\beta}} = \hat{\boldsymbol{D}} \left(\sum_{i=1}^{N} \frac{\boldsymbol{x}_i' d_i}{\sigma^2_\epsilon} + \boldsymbol{S}_p^{-1} \boldsymbol{\mu}_p \right),$$

  and

  $$d_i = y_i - \delta^{(2)}_{hosp(i)} - \delta^{(3)}_{nbhd(i)}.$$

- In step 2, we update the hospital residuals, $\underline{\delta}_k^{(2)}$, using Gibbs sampling with a univariate normal full conditional distribution:

$$p(\underline{\delta}_k^{(2)} \mid \boldsymbol{y}, \boldsymbol{\beta}, \boldsymbol{\delta}^{(3)}, \sigma^2_{\delta(2)}, \sigma^2_{\delta(3)}, \sigma^2_\epsilon) \sim \mathcal{N}(\hat{\delta}_k^{(2)}, \hat{D}_k^{(2)}),$$

where

$$\hat{D}_k^{(2)} = \left(\frac{n_k^{(2)}}{\sigma^2_\epsilon} + \frac{1}{\sigma^2_{\delta(2)}} \right)^{-1},$$

$$\hat{\delta}_k^{(2)} = \hat{D}_k^{(2)} \left(\sum_{i,hosp(i)=k} \frac{d_i^{(2)}}{\sigma^2_\epsilon} \right),$$

and

$$d_i^{(2)} = y_i - \boldsymbol{x}_i \boldsymbol{\beta} - \delta^{(3)}_{nbhd(i)}.$$

- In step 3, we update the neighbourhood residuals, $\underline{\delta}_k^{(3)}$, using Gibbs sampling with a univariate normal full conditional distribution:

$$p(\underline{\delta}_k^{(3)} \mid \boldsymbol{y}, \boldsymbol{\beta}, \boldsymbol{\delta}^{(2)}, \sigma^2_{\delta(2)}, \sigma^2_{\delta(3)}, \sigma^2_\epsilon) \sim \mathcal{N}(\hat{\delta}_k^{(3)}, \hat{D}_k^{(3)}),$$

where

$$\hat{D}_k^{(3)} = \left(\frac{n_k^{(3)}}{\sigma^2_\epsilon} + \frac{1}{\sigma^2_{\delta(3)}} \right)^{-1},$$

$$\hat{\delta}_k^{(3)} = \hat{D}_k^{(3)} \left(\sum_{i,nbhd(i)=k} \frac{d_i^{(3)}}{\sigma^3_\epsilon} \right),$$

and

$$d_i^{(3)} = y_i - \boldsymbol{x}_i \boldsymbol{\beta} - \delta^{(2)}_{hosp(i)}.$$

Note that in the above two steps $n_k^{(c)}$ refers to the number of individuals in the k-th unit of classification c.

- In step 4, we update the hospital variance $\underline{\sigma}^2_{\delta(2)}$ using Gibbs sampling and a Gamma full conditional distribution for $1/\underline{\sigma}^2_{\delta(2)}$:

$$p(1/\underline{\sigma}^2_{\delta(2)} \mid \boldsymbol{y}, \boldsymbol{\beta}, \delta^{(2)}, \delta^{(3)}, \sigma^2_{\delta(3)}, \sigma^2_\epsilon)$$
$$\sim \text{Gamma}\left(\frac{n_2 + \nu_2}{2}, \frac{1}{2} \left[\sum_{j=1}^{n_2} \left(\delta_j^{(2)}\right)^2 + \nu_2 s_2^2 \right] \right).$$

- In step 5, we update the neighbourhood variance $\underline{\sigma}^2_{\delta(3)}$ using Gibbs sampling and a Gamma full conditional distribution for $1/\underline{\sigma}^2_{\delta(3)}$:

$$p(1/\underline{\sigma}^2_{\delta(3)} \mid \boldsymbol{y}, \boldsymbol{\beta}, \boldsymbol{\delta}^{(2)}, \boldsymbol{\delta}^{(3)}, \sigma^2_{\delta(2)}, \sigma^2_{\epsilon}) \sim \text{Gamma}\left(\frac{n_3+\nu_3}{2}, \frac{1}{2}\left[\sum_{j=1}^{n_3}\left(\delta_j^{(3)}\right)^2 + \nu_3 s_3^2\right]\right).$$

- In step 6, we update the observation level variance $\underline{\sigma}^2_{\epsilon}$ using Gibbs sampling and a Gamma full conditional distribution for $1/\underline{\sigma}^2_{\epsilon}$:

$$p(1/\underline{\sigma}^2_{\epsilon} \mid \boldsymbol{y}, \boldsymbol{\beta}, \boldsymbol{\delta}^{(2)}, \boldsymbol{\delta}^{(3)}, \sigma^2_{\delta(2)}, \sigma^2_{\delta(3)}) \sim \text{Gamma}\left(\frac{N+\nu_\epsilon}{2}, \frac{1}{2}\left[\sum_{i=1}^{N}\epsilon_i^2 + \nu_\epsilon s_\epsilon^2\right]\right).$$

The above six steps are repeatedly sampled from in sequence to produce correlated chains of parameter estimates from which point and interval estimates can be created as in Chapter 2.

AIP Method

The Alternating Imputation Prediction (AIP) method is a data augmentation algorithm for estimating cross-classified models with large numbers of random effects. Comprehensive details of this algorithm are given in Clayton and Rasbash [4]. We now give an overview.

Data augmentation has been reviewed by Schafer [19]. Tanner and Wong [20] introduced the idea of data augmentation as a stochastic version of the EM algorithm for maximum likelihood estimation in problems involving missing data. Following Tanner and Wong we have

I(mputation) step — Impute missing data by sampling the distribution of the missing data conditional upon the observed data and current values of the model parameters.

P(osterior) step — Sample parameter values from the complete data posterior distribution; these will be used for the next I-step.

In the context of random effect models, the random effects play the role of missing data. If the observed data are denoted by $\boldsymbol{y}$, the random effects by $\underline{\boldsymbol{\delta}}$ and the model parameters by $\boldsymbol{\theta}$, then the algorithm is specified (at step t) by

I-step — Draw a sample $\boldsymbol{\delta}^{(t)}$ from $p(\underline{\boldsymbol{\delta}} \mid \boldsymbol{y}, \boldsymbol{\theta} = \boldsymbol{\theta}^{(t-1)})$.
P-step — Draw a sample $\boldsymbol{\theta}^{(t)}$ from $p(\underline{\boldsymbol{\theta}} \mid \boldsymbol{y}, \boldsymbol{\delta} = \boldsymbol{\delta}^{(t)})$.

Repeated application of these two steps delivers a stochastic chain with equilibrium distribution $p(\underline{\boldsymbol{\delta}}, \underline{\boldsymbol{\theta}} \mid \boldsymbol{y})$ in a similar way to the MCMC algorithm. Now let's look at how we can adapt this method to fit a crossed random effects model when the only estimating engine we have at our disposal is one optimized for fitting nested random effects.

An m-way cross-classified model can be broken down into m sub-models, each of which is a 2-level hierarchical model. For example, patients nested within a cross-classification of neighbourhood by hospital can be broken down into a patient within hospital sub-model and a patient within neighbourhood sub-model.

Take the simple model

$$\underline{y}_i = \boldsymbol{x}_i\boldsymbol{\beta} + \underline{\delta}^{(2)}_{hosp(i)} + \underline{\delta}^{(3)}_{nbhd(i)} + \underline{\epsilon}_i\,,$$

where hospital and neighbourhood are cross-classified. This cross-classified model can be portioned into two hierarchical sub-models: patients within neighbourhoods (model N) and patients within hospitals (model H). An informal description of the AIP algorithm is:

1. Start by fitting model N using an estimation procedure for 2-level models.
2. Sample the model parameters from an approximation to their joint posterior distribution. That is, sample the fixed effects, the neighbourhood level variance and the patient level variance; denote these samples by $\boldsymbol{\beta}_{[0,3]}$, $\sigma^2_{\delta[0,3]}$ and $\sigma^2_{\epsilon[0,3]}$, respectively. Here $[0,3]$ labels a term as belonging to AIP iteration 0, for classification number 3, that is neighbourhood. This is the P-step for the neighbourhood classification.
3. Next sample a set of neighbourhood level random effects ($\boldsymbol{o}_{[0,3]}$) from $p(\underline{\boldsymbol{\delta}}_{[0,3]} \mid \boldsymbol{y}, \boldsymbol{\beta}_{[0,3]}, \sigma^2_{\delta[0,3]}, \sigma^2_{\epsilon[0,3]})$. This is the I-step for the neighbourhood classification.
4. Offset $\boldsymbol{o}_{[0,3]}$ from $\boldsymbol{y}$, that is form $\tilde{\boldsymbol{y}} = \boldsymbol{y} - \boldsymbol{o}_{[0,3]}$, re-sort the data according to hospitals and fit model H using the new offset response $\tilde{\boldsymbol{y}}$.
5. Next sample $\boldsymbol{\beta}_{[0,2]}$, $\sigma^2_{\delta[0,2]}$ and $\sigma^2_{\epsilon[0,2]}$, from this second model, H. This is the P-step for the hospital classification.
6. Sample a set of hospital level random effects ($\boldsymbol{o}_{[0,2]}$) from $p(\underline{\boldsymbol{\delta}}_{[0,2]} \mid \boldsymbol{y}, \boldsymbol{\beta}_{[0,2]}, \sigma^2_{\delta[0,2]}, \sigma^2_{\epsilon[0,2]})$. This is the I-step for the hospital classification.

This completes one iteration of the AIP algorithm; this is an Imputation-Posterior algorithm that Alternates between the neighbourhood and hospital classifications. We proceed by forming $\tilde{\boldsymbol{y}} = \boldsymbol{y} - \boldsymbol{o}_{[0,2]}$, that is offsetting the sampled hospital residuals from $\boldsymbol{y}$ and using that as a response in step 1. After T iterations, the procedure delivers the following two chains, which can be used for inference:

$$\begin{gathered}\{\boldsymbol{\beta}_{[0,2]}, \sigma^2_{\delta[0,2]}, \sigma^2_{\epsilon[0,2]}\}, \{\boldsymbol{\beta}_{[1,2]}, \sigma^2_{\delta[1,2]}, \sigma^2_{\epsilon[1,2]}\}, \ldots, \{\boldsymbol{\beta}_{[T,2]}, \sigma^2_{\delta[T,2]}, \sigma^2_{\epsilon[T,2]}\} \\ \{\boldsymbol{\beta}_{[0,3]}, \sigma^2_{\delta[0,3]}, \sigma^2_{\epsilon[0,3]}\}, \{\boldsymbol{\beta}_{[1,3]}, \sigma^2_{\delta[1,3]}, \sigma^2_{\epsilon[1,3]}\}, \ldots, \{\boldsymbol{\beta}_{[T,3]}, \sigma^2_{\delta[T,3]}, \sigma^2_{\epsilon[T,3]}\}.\end{gathered}$$

Note that we get two sets of estimates for both the fixed effects and the level-1 variance with the AIP algorithm and the empirical distributions of these quantities should be equal. In our use of AIP, we run the chains for between 100 and 400 iterations and judge convergence for a parameter by looking at the mean of the distribution of that parameter's chain. To converge to the posterior distribution to get accurate estimates of, say, extreme percentiles, would have required many more iterations. However, compared to the MCMC algorithm outlined above, each iteration imposes a heavy computational burden. To avoid this problem, Clayton and Rasbash [4] employed a Rao-Blackwellisation [6] to estimate characteristics of the posterior distribution from short chains. However, the accuracy of this method was not thoroughly investigated.

Other Methods

Raudenbush [17] considers an empirical Bayes approach to fitting cross-classified models based on the EM algorithm. He considers the specific case of two classifications where one of the classifications has many units whilst the other has far fewer and shows two educational examples to illustrate the method.

Two other recent approaches that can be used for fitting cross-classified models, in particular with non-normal responses are Gauss-Hermite quadrature within PQL estimation [13] and the HGLM model framework as described in Lee and Nelder [12].

Comparison of Estimation Methods

The RG method when it works is generally fairly quick to converge where all or all but one of the crossed classifications have small numbers of units. When there are multiple crossed classifications with large numbers of units then the speed of the RG algorithm deteriorates and memory usage is greatly increased, often exhausting the available memory. The AIP method does not have these memory problems but will be slower for structures that are almost hierarchical. Although this method works reasonably well, if the response is a binary variable and quasi-likelihood methods need to be used, then this method like the RG method is still affected by the bias that is inherent in quasi-likelihood methods for binary response multilevel models [see 9]. The MCMC methods have no bias problems, although there are still issues on which prior distributions to use for the variance parameters. They also, like the AIP methods, do not have any memory problems. They are, however, generally computationally a lot slower, as they are estimating the whole distribution and not simply the mode, although as the structure of the data becomes more complex, the speed difference is reduced.

An Example Analysis of a Two-Way Cross-Classification: Primary Schools Crossed with Secondary Schools

We will here consider fitting the RG method using the IGLS algorithm, the MCMC method based on Gibbs sampling [2] and the AIP method to an educational example from Fife in Scotland. Here we have as a response the exam results of 3,435 children at age 16. We know for each child both the primary school and secondary school that they attended and we are interested in partitioning the variance between these two sources and individual pupil level variation. The classification diagram is shown in Fig. 8.4. There are 148 primary schools that feed into 19 secondary schools in the dataset. Of the 148 primary schools, 59 are nested within a single secondary school, whilst another 62 have at most 3 pupils that do not go to the main secondary school, so we have an almost nested structure. This structure is particularly suited for the RG algorithm.

We will fit the following model to the dataset

$$\underline{y}_i = \beta_0 + \underline{\delta}^{(2)}_{sec(i)} + \underline{\delta}^{(3)}_{prim(i)} + \underline{\epsilon}_i \,,$$

$$\underline{\delta}^{(2)}_{sec(i)} \sim \mathcal{N}(0, \sigma^2_{\delta(2)}), \quad \underline{\delta}^{(3)}_{prim(i)} \sim \mathcal{N}(0, \sigma^2_{\delta(3)}), \quad \underline{\epsilon}_i \sim \mathcal{N}(0, \sigma^2_{\epsilon}).$$

The results are shown in Table 8.5.

From Table 8.5 we can see that in this example there is more variation between primary schools than between secondary schools. The MCMC estimates replicate the IGLS estimates with slightly greater higher level variances (mean versus mode estimates) due to the skewness of the posterior distribution. The AIP method gives very similar results to the IGLS method. A further discussion of these results is given in Goldstein [8].

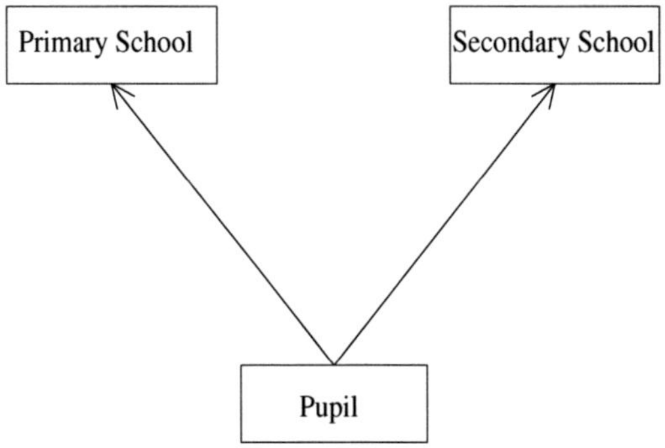

Fig. 8.4 Classification diagram for the Fife educational example.

Table 8.5 Point estimates for the Fife educational dataset.

Parameter		IGLS		MCMC		AIP	
Mean achievement	(β_0)	5.50	(0.17)	5.50	(0.18)	5.51	(0.19)
Secondary school variance	($\sigma^2_{\delta(2)}$)	0.35	(0.16)	0.41	(0.21)	0.34	(0.15)
Primary school variance	($\sigma^2_{\delta(3)}$)	1.12	(0.20)	1.15	(0.21)	1.11	(0.20)
Individual level variance	(σ^2_e)	8.10	(0.20)	8.12	(0.20)	8.11	(0.20)

8.2.3 Models for More Complex Population Structures

In this section we will consider expanding the simple two cross-classified structure to accommodate more classifications and more complex structures.

Example Scenarios

Let's take the situation described in the classification diagram drawn in Fig. 8.3(i) where patients lie within a cross-classification of hospitals by neighbourhoods. We may have information on the doctor that treated each patient and doctors may be nested within hospitals. The classification diagram for this structure is shown in Fig. 8.5.

A variance components model for this structure is written as

$$\underline{y}_i = \beta_0 + \underline{\delta}^{(2)}_{hosp(i)} + \underline{\delta}^{(3)}_{nbhd(i)} + \underline{\delta}^{(4)}_{doct(i)} + \underline{\epsilon}_i \,.$$

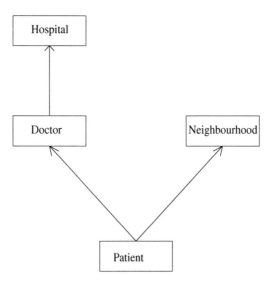

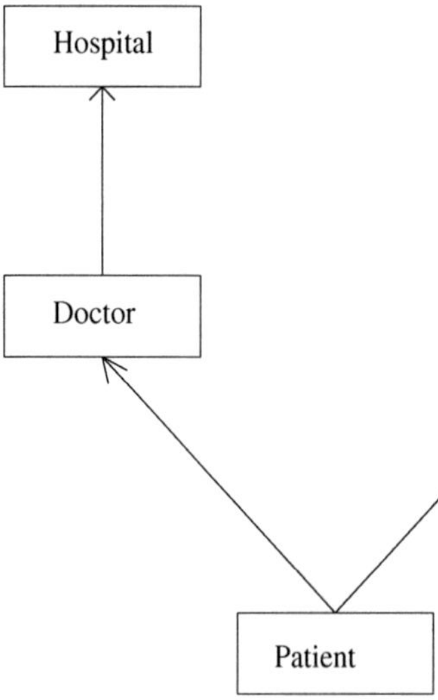

Fig. 8.5 Classification diagram for two crossed hierarchies (patients within doctors within hospitals) × (patients within neighbourhoods).

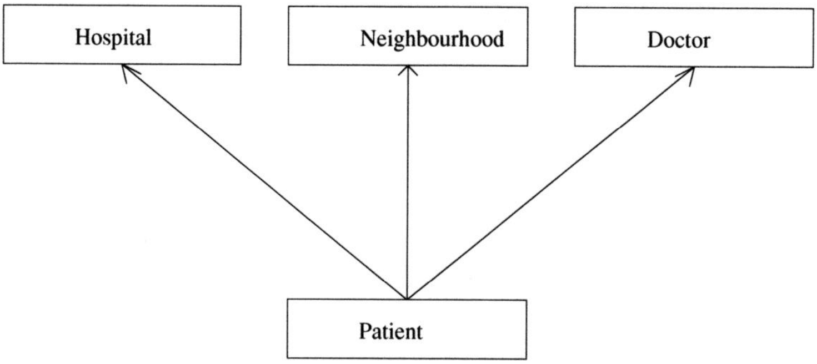

Fig. 8.6 Classification diagram for three crossed hierarchies (patients within hospitals) × (patients within doctors) × (patients within neighbourhoods).

If doctors work across hospitals and are therefore not nested within hospital, we then have a three way cross-classification which is drawn in Fig. 8.6.

Note that the variance components model for the structure in Fig. 8.6 is also described by the same equation. This is a reflection of the fact that the model notation for describing the random effects simply lists the classifications that are sources of variation for the response we are modelling. In the variance components model, we only have an intercept term which varies across all four classifications present. Suppose we had another explanatory variable, x_1, and we wished to allow its coefficient to vary across the doctor classifications; we would write this model as

$$\underline{y}_i = \beta_0 + \underline{\delta}^{(2)}_{hosp(i)} + \underline{\delta}^{(3)}_{nbhd(i)} + \underline{\delta}^{(4)}_{doct(i),0} + \beta_1 x_{1i} + \underline{\delta}^{(4)}_{doct(i),1} x_{1i} + \underline{\epsilon}_i \,,$$

or, alternatively, we can express the model as:

$$\begin{aligned} \underline{y}_i &= \underline{\beta}_{0i} + \underline{\beta}_{1i} x_{1i} + \underline{\epsilon}_i \,, \\ \underline{\beta}_{0i} &= \beta_0 + \underline{\delta}^{(2)}_{hosp(i)} + \underline{\delta}^{(3)}_{nbhd(i)} + \underline{\delta}^{(4)}_{doct(i),0} \,, \\ \underline{\beta}_{1i} &= \beta_1 + \underline{\delta}^{(4)}_{doct(i),1} . \end{aligned}$$

It may be that the scenario described in Fig. 8.6 is further complicated because hospitals, doctors and neighbourhoods are all nested within regions. In this case, the classification diagram becomes as in Fig. 8.7.

Extending the last model to incorporate a simple random effect for the region classification, we have

$$\begin{aligned} \underline{y}_i &= \underline{\beta}_{0i} + \underline{\beta}_{1i} x_{1i} + \underline{\epsilon}_i \,, \\ \underline{\beta}_{0i} &= \beta_0 + \underline{\delta}^{(2)}_{hosp(i)} + \underline{\delta}^{(3)}_{nbhd(i)} + \underline{\delta}^{(4)}_{doct(i),0} + \underline{\delta}^{(5)}_{reg(i)} \,, \\ \underline{\beta}_{1i} &= \beta_1 + \underline{\delta}^{(4)}_{doct(i),1} . \end{aligned}$$

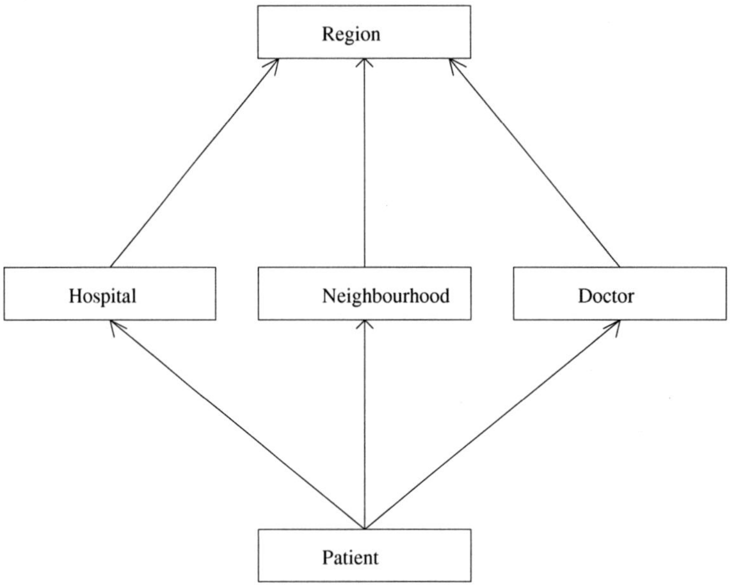

Fig. 8.7 Classification diagram for three crossed hierarchies nested within a higher-level classification.

These few example scenarios indicate how the classification diagrams and simplified notation can extend to describe patterns of crossings of arbitrary complexity.

An Example Analysis of a Complex Cross-Classified Structure: Artificial Insemination Data

We consider a data set concerning artificial insemination by donor. Detailed description of this data set and the substantive research questions addressed by modelling it within a cross-classified framework are given in Ecochard and Clayton [5]. The data was re-analysed in Clayton and Rasbash [4] as an example case study demonstrating the properties of the AIP algorithm for estimating cross-classified models.

The data consists of 1901 women who were inseminated by sperm donations from 279 donors. Each donor made multiple donations; there were 1328 donations in all. A single donation is used for multiple inseminations. Each woman receives a series of monthly inseminations, 1 insemination per ovulatory cycle. The data contain 12100 cycles within the 1901 women.

There are two crossed hierarchies, a hierarchy for donors and a hierarchy for women. Level 1 corresponds to measures made at each ovulatory cycle. The response we analyse is the binary variable indicating if conception occurs in a

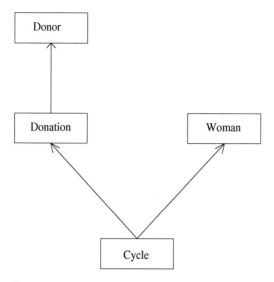

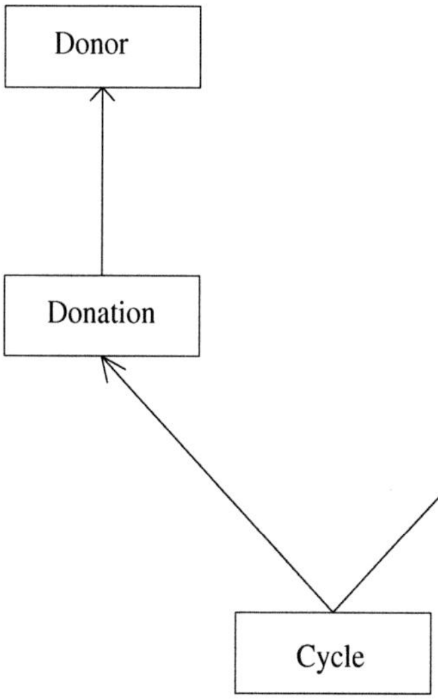

Fig. 8.8 Classification diagram for the artificial insemination example model.

given cycle. The hierarchy for women is cycles within women. The hierarchy for donors is cycles within donations within donors. Within a series of cycles, a woman may receive sperm from multiple donors/donations. The classification diagram for this structure is given in Fig. 8.8. The model fitted to the data is

$$\begin{aligned} \underline{y}_i &\sim \text{Bernoulli}(\underline{\pi}_i), \\ \text{logit}(\underline{\pi}_i) &= \beta_0 + azoo_i * \beta_1 + semenq_i * \beta_2 + (age > 35)_i * \beta_3 \\ &\quad + spermcount_i * \beta_4 + spermmot_i * \beta_5 + iearly_i * \beta_6 \\ &\quad + ilate_i * \beta_7 + \underline{\delta}^{(2)}_{woman(i)} + \underline{\delta}^{(3)}_{donation(i)} + \underline{\delta}^{(4)}_{donor(i)}, \end{aligned}$$

$$\underline{\delta}^{(2)}_{woman(i)} \sim \mathcal{N}(0, \sigma^2_{\delta(2)}), \quad \underline{\delta}^{(3)}_{donation(i)} \sim \mathcal{N}(0, \sigma^2_{\delta(3)}), \quad \underline{\delta}^{(4)}_{donor(i)} \sim \mathcal{N}(0, \sigma^2_{\delta(4)}).$$

Note that azoospermia (*azoo*) is a dichotomous variable indicating whether the fecundability of the woman is impaired (0 impaired, 1 not impaired). The results of fitting this model from the MCMC and AIP estimation procedures are given in Table 8.6. This model could not be fitted using the RG algorithm. This is because if the data is sorted according to women, then we need to fit 279 dummy variables for donors and 1328 dummy variables for donations. Alternatively, if we sort the data according to donations within donors, we have to fit 1901 dummy variables for women. Either way, the size of these data matrices cause problems of insufficient memory. Even if these memory problems can be worked around the numerical instability of the constraining procedure that attempts to constrain over a thousand separately estimated variances to be equal causes the adapted IGLS algorithm to fail to converge.

Table 8.6 Results for the Artificial Insemination example.

Parameter		MCMC		AIP	
Intercept	(β_0)	−3.92	(0.21)	−3.90	(0.21)
Azoospermia	(β_1)	0.21	(0.09)	0.22	(0.10)
Semen quality	(β_2)	0.18	(0.03)	0.18	(0.03)
Womens age > 35	(β_3)	−0.29	(0.12)	−0.27	(0.12)
Sperm count	(β_4)	0.002	(0.001)	0.002	(0.001)
Sperm motility	(β_5)	0.0002	(0.0001)	0.0002	(0.0001)
Insemination too early	(β_6)	−0.69	(0.17)	−0.67	(0.17)
Insemination too late	(β_7)	−0.27	(0.09)	−0.25	(0.09)
Women variance	($\sigma^2_{\delta(2)}$)	1.02	(0.15)	1.01	(0.11)
Donation variance	($\sigma^2_{\delta(3)}$)	0.36	(0.074)	0.34	(0.065)
Donor variance	($\sigma^2_{\delta(4)}$)	0.11	(0.06)	0.10	(0.06)

After inclusion of covariates, there is considerably more variation in the probability of a successful insemination attributable to the women hierarchy than the donor hierarchy. Both the AIP and MCMC methods give similar estimates for all parameters. The fixed effect estimates show that the probability of conception is increased with azoospermia and increased sperm quality, count and motility but decreased with the age of the woman and with inseminations that are too early or too late.

8.3 Multiple Membership Models

As we have seen from the previous section, allowing classifications to be crossed gives rise to a large family of additional model structures that can be estimated. The other main restriction of the basic multilevel model is the need for observations to belong to a unique classification unit, i.e., every pupil belongs to a particular class, every patient is treated at a particular hospital. Often however, over time a patient may be treated at several hospitals and depending on the response of interest all of these hospitals may have influence. In this section we will first introduce the idea of multiple membership and give some example scenarios where it may occur. We will then discuss the possible estimation procedures that can be used to fit multiple membership models and finish the chapter with a simulated example from the field of education.

8.3.1 A Basic Structure for Two-Level Multiple Memberships

Suppose we have data on a large number of patients that attend their local hospital and during the course of their hospital stay, they are treated by

Table 8.7 Table of patients that are seen by multiple nurses.

	Nurse 1	Nurse 2	Nurse 3
Patient 1	√		√
Patient 2	√		
Patient 3		√	√
Patient 4	√	√	

several nurses and we regard the nurses as an important factor on the patients outcome of interest. Now typically each patient will be seen by more than one nurse during their stay (although some will only see 1), but there are many nurses and so we will treat nurses as a random classification rather than as fixed effects. To illustrate this, Table 8.7 shows the nurses seen by the first 4 patients.

We can consider this structure in a unit diagram as shown in Fig. 8.9. Here each line in the diagram corresponds to a tick mark in the table. Again, as our data set gets larger, such unit diagrams become impractical, as there will be too many nodes and so we will resort to using the classification diagrams introduced earlier for cross-classified models. If we wish to include multiple membership classifications in such diagrams, we use the convention of a double arrow to represent multiple membership. This will lead to the classification diagram shown in Fig. 8.10 for the above patients and nurses example.

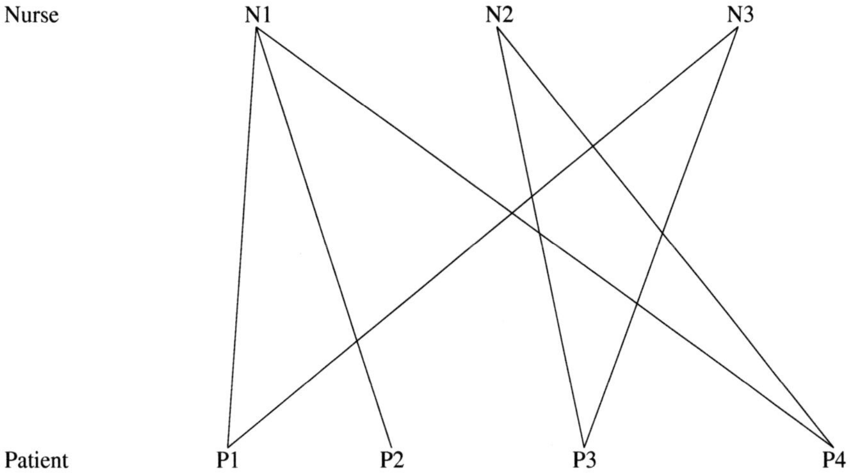

Fig. 8.9 Unit diagram for multiple membership patients within nurses example.

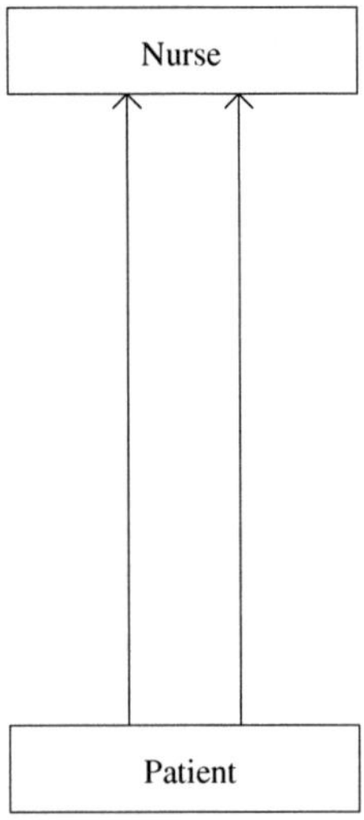

Fig. 8.10 Classification diagram for multiple membership patients within nurses example.

Example Scenarios

Many studies have multiple membership structure; here are a few examples:

- Education: Pupils change school/class over the course of their education and each school/class has an effect on their education.
- Health: Patients are seen by several doctors and nurses during the course of their treatment.
- Survey data: Over their lifetime, individuals move household and each household has a bearing on their lifestyle, health, salary etc.

Constructing a Statistical Model

Returning to our example of patients being seen by multiple nurses, we have Patient 1's response being affected by Nurse 1 and Nurse 3 while Patient 2 is only affected by Nurse 1. As we are treating nurse as a random classification, we would like each patient's response to have equal effect on the nurse classification variance, so we generally weight the random effects to sum to 1. For example, let's assume Patient 1 has been treated by Nurse 1 for 2 days and Nurse 3 for 1 day. Then we may give Nurse 1 a weight of $\frac{2}{3}$ and Nurse 3 a weight of $\frac{1}{3}$. Often we do not have information on the amount of time patients are seen by each nurse and so we commonly allocate equal weights (in this case $\frac{1}{2}$) to each nurse. We can then write down a general two-level multiple membership model as

$$\underline{y}_i = \boldsymbol{x}_i\boldsymbol{\beta} + \sum_{j \in nurse(i)} w_{i,j}^{(2)} \underline{\delta}_j^{(2)} + \underline{\epsilon}_i \,,$$

$$\underline{\delta}_j^{(2)} \sim \mathcal{N}(0, \sigma_{\delta(2)}^2), \quad \underline{\epsilon}_i \sim \mathcal{N}(0, \sigma_\epsilon^2),$$

where $nurse(i)$ is the set of nurses seen by patient i and $w_{i,j}^{(2)}$ is the weight given to nurse j for patient i. Here we assume that

$$\sum_{j \in nurse(i)} w_{i,j}^{(2)} = 1 \quad \forall i.$$

If we wish to write out this model for the first four patients from the example, we get

$$\underline{y}_1 = \boldsymbol{x}_1\boldsymbol{\beta} + \tfrac{1}{2}\underline{\delta}_1^{(2)} + \tfrac{1}{2}\underline{\delta}_3^{(2)} + \underline{\epsilon}_1 \,,$$

$$\underline{y}_2 = \boldsymbol{x}_2\boldsymbol{\beta} + \underline{\delta}_1^{(2)} + \underline{\epsilon}_2 \,,$$

$$\underline{y}_3 = \boldsymbol{x}_3\boldsymbol{\beta} + \tfrac{1}{2}\underline{\delta}_2^{(2)} + \tfrac{1}{2}\underline{\delta}_3^{(2)} + \underline{\epsilon}_3 \,,$$

$$\underline{y}_4 = \boldsymbol{x}_4\boldsymbol{\beta} + \tfrac{1}{2}\underline{\delta}_1^{(2)} + \tfrac{1}{2}\underline{\delta}_2^{(2)} + \underline{\epsilon}_4 \,.$$

8.3.2 Estimation Algorithms

There are two main algorithms for multiple membership models, an adaption of the Rasbash and Goldstein [15] algorithm described earlier and the MCMC method. The AIP method has not been extended to cater for multiple membership models.

An IGLS Algorithm for Multiple Membership Models

Earlier we described how to fit a cross-classified model by absorbing one of the cross-classifications into a set of dummy variables (the RG method). A slight modification is required to allow this technique to be used to fit multiple membership models. First let's consider a two-level hierarchical model for patients within nurses:

$$\underline{y}_i = \beta_0 + \underline{\delta}_{nurse(i)}^{(2)} + \underline{\epsilon}_i \,,$$

$$\underline{\delta}_{nurse(i)}^{(2)} \sim \mathcal{N}(0, \sigma_{\delta(2)}^2), \quad \underline{\epsilon}_i \sim \mathcal{N}(0, \sigma_\epsilon^2).$$

We can reparameterise this simple two-level model as

$$\underline{y}_i = \beta_0 + z_{i,1}\,\underline{\delta}_1^{(2)} + z_{i,2}\,\underline{\delta}_2^{(2)} + z_{i,3}\,\underline{\delta}_3^{(2)} + \cdots + z_{i,J}\,\underline{\delta}_J^{(2)} + \underline{\epsilon}_i\,,$$

$$\begin{pmatrix} \underline{\delta}_1^{(2)} \\ \underline{\delta}_2^{(2)} \\ \underline{\delta}_3^{(2)} \\ \vdots \\ \underline{\delta}_J^{(2)} \end{pmatrix} \sim \mathcal{N}(0, \Sigma_{\delta(2)}), \quad \Sigma_{\delta(2)} = \begin{pmatrix} \sigma^2_{\delta(2),1} & 0 & 0 & \dots & 0 \\ 0 & \sigma^2_{\delta(2),2} & 0 & \dots & 0 \\ 0 & 0 & \sigma^2_{\delta(2),3} & \cdots & 0 \\ \vdots & \vdots & \vdots & \ddots & \vdots \\ 0 & 0 & 0 & \dots & \sigma^2_{\delta(2),J} \end{pmatrix},$$

$$\underline{\epsilon}_i \sim \mathcal{N}(0, \sigma^2_\epsilon),$$

where $z_{i,j}$ is a dummy variable which is 1 if patient i is seen by nurse j, 0 otherwise and J is the total number of nurses. Also we add the constraint $\sigma^2_{\delta(2),1} = \sigma^2_{\delta(2),2} = \cdots = \sigma^2_{\delta(2),J}$. Now these two models will deliver the same estimates; however, the second formulation will take much longer to compute. The advantage of the second model formulation is that it is straightforward to extend it to the multiple membership case. Suppose patients are not nested within a single nurse but are multiple members of nurses with membership proportions, $\pi_{i,j}$. We can simply replace $z_{i,j}$ with $\pi_{i,j}$ in the second formulation and estimation can proceed in an identical fashion but will now deliver estimates for the multiple membership model.

MCMC

Once again we will use a Gibbs sampling algorithm that relies on updating groups of parameters in turn from their conditional posterior distributions. For illustration, we present the steps for the following simple multiple membership model based on the variance components model patients within nurses described earlier. We once again refer the interested reader to Browne et al. [2] for more general algorithms and note that if the response is dichotomous or a count, then, as in Chapter 2, we can use the Metropolis-Gibbs hybrid method discussed there.

The basic two-level multiple membership model (patients within nurses) can be written as:

$$\underline{y}_i = \boldsymbol{x}_i\underline{\boldsymbol{\beta}} + \sum_{j \in nurse(i)} w^{(2)}_{i,j}\,\underline{\delta}^{(2)}_j + \underline{\epsilon}_i\,,$$

$$\underline{\delta}^{(2)}_j \sim \mathcal{N}(0, \underline{\sigma}^2_{\delta(2)}), \quad \underline{\epsilon}_i \sim \mathcal{N}(0, \underline{\sigma}^2_\epsilon).$$

We can split our unknown parameters into four distinct sets: the fixed effects, $\underline{\boldsymbol{\beta}}$; the nurse random effects, $\underline{\delta}^{(2)}_j$; the nurse level variance, $\underline{\sigma}^2_{\delta(2)}$ and the patient level residual variance, $\underline{\sigma}^2_\epsilon$.

We then need to generate random draws from the conditional distribution of each of these four groups of unknowns. We will define prior distributions for our unknown parameters as follows: For generality, we will use a multivariate normal prior for the fixed effects, $\underline{\boldsymbol{\beta}} \sim \mathcal{N}_{p_f}(\boldsymbol{\mu}_p, \boldsymbol{S}_p)$, and scaled inverse χ^2 priors for the two variances. For the nurse level variance, $\underline{\sigma}^2_{\delta(2)} \sim SI\chi^2(\nu_2, s_2^2)$, and for the patient level variance, $\underline{\sigma}^2_{\epsilon} \sim SI\chi^2(\nu_\epsilon, s_\epsilon^2)$. The steps are then as follows:

- In step 1 of the algorithm, the conditional posterior distribution in the Gibbs update for the fixed effects parameter vector $\underline{\boldsymbol{\beta}}$ is multivariate normal with dimension p_f (the number of fixed effects):

$$p(\underline{\boldsymbol{\beta}} \mid \boldsymbol{y}, \boldsymbol{\delta}^{(2)}, \sigma^2_{\delta(2)}, \sigma^2_\epsilon) \sim \mathcal{N}_{p_f}(\hat{\boldsymbol{\beta}}, \hat{\boldsymbol{D}}),$$

  where

$$\hat{\boldsymbol{D}} = \left(\sum_{i=1}^{N} \frac{\boldsymbol{x}_i' \boldsymbol{x}_i}{\sigma^2_\epsilon} + \boldsymbol{S}_p^{-1} \right)^{-1},$$

$$\hat{\boldsymbol{\beta}} = \hat{\boldsymbol{D}} \left(\sum_{i=1}^{N} \frac{\boldsymbol{x}_i' d_i}{\sigma^2_\epsilon} + \boldsymbol{S}_p^{-1} \boldsymbol{\mu}_p \right),$$

  and

$$d_i = y_i - \sum_{j \in nurse(i)} w_{i,j}^{(2)} \delta_j^{(2)}.$$

- In step 2, we update the nurse residuals, $\underline{\delta}_k^{(2)}$, using Gibbs sampling with a univariate normal full conditional distribution:

$$p(\underline{\delta}_k^{(2)} \mid \boldsymbol{y}, \boldsymbol{\beta}, \sigma^2_{\delta(2)}, \sigma^2_\epsilon) \sim \mathcal{N}(\hat{\delta}_k^{(2)}, \hat{D}_k^{(2)}),$$

  where

$$\hat{D}_k^{(2)} = \left(\sum_{i,k \in nurse(i)} \frac{(w_{i,k}^{(2)})^2}{\sigma^2_\epsilon} + \frac{1}{\sigma^2_{\delta(2)}} \right)^{-1},$$

$$\hat{\delta}_k^{(2)} = \hat{D}_k^{(2)} \left(\sum_{i,k \in nurse(i)} \frac{w_{i,k}^{(2)} d_{i,k}^{(2)}}{\sigma^2_\epsilon} \right),$$

  and

$$d_{i,k}^{(2)} = y_i - \boldsymbol{x}_i \boldsymbol{\beta} - \sum_{j \in nurse(i), j \neq k} w_{i,j}^{(2)} \delta_j^{(2)}.$$

- In step 3, we update the nurse level variance $\underline{\sigma}^2_{\delta(2)}$ using Gibbs sampling and a Gamma full conditional distribution for $1/\underline{\sigma}^2_{\delta(2)}$:

$$p(1/\underline{\sigma}^2_{\delta(2)} \mid \boldsymbol{y}, \boldsymbol{\beta}, \delta^{(2)}, \sigma^2_\epsilon) \sim \text{Gamma}\left(\frac{n_2 + \nu_2}{2}, \frac{1}{2}\left[\sum_{j=1}^{n_2} \left(\delta_j^{(2)}\right)^2 + \nu_2 s_2^2\right]\right).$$

- In step 4, we update the patient level variance $\underline{\sigma}^2_\epsilon$ using Gibbs sampling and a Gamma full conditional distribution for $1/\underline{\sigma}^2_\epsilon$:

$$p(1/\underline{\sigma}^2_\epsilon \mid \boldsymbol{y}, \boldsymbol{\beta}, \delta^{(2)}, \sigma^2_{\delta(2)}) \sim \text{Gamma}\left(\frac{N + \nu_\epsilon}{2}, \frac{1}{2}\left[\sum_{i=1}^{N} \epsilon_i^2 + \nu_\epsilon s_\epsilon^2\right]\right).$$

The above four steps are repeatedly sampled from in sequence to produce correlated chains of parameter estimates from which point and interval estimates can be created as in Chapter 2.

Comparison of Estimation Methods

As in the comparison for cross-classified models, there are benefits for both methods. The RG method is fairly quick, but the number of level-2 units determines the size of some of the matrices involved and the number of constraints that the method has to apply. These dependencies lead to numerical instability or memory exhaustion in situations with more than a few hundred level-2 units. The MCMC methods, although again computationally slower, do not suffer from these memory problems.

An Example Analysis of a Two-Level Multiple Membership Model: Children Moving School

We consider a simulated data example based on the problem in education of adjusting for the fact that pupils move school during the course of their studies. We will consider a study with 4059 students from 65 schools taken from Rasbash et al. [16]. The actual data in the study has each child belonging to 1 school, but we will assume that over their education 10% of children moved school, so we will choose at random for 10% of the children a second school. We will assume that information about when the move occurred is unavailable and so for these children we will allocate equal weights of 0.5 to each school. Browne et al. [2] considered this as the basis for a simulation experiment by generating 1000 datasets with this structure to show the bias and coverage properties of the MCMC method. We will instead consider the true response on our modified structure. We have as a response the pupil's total (normalised) exam score in all GCSE exams taken at age 16 and as a

Table 8.8 Results for the multiple membership schools example.

Parameter		RG RIGLS		MCMC	
Intercept	(β_0)	0.002	(0.040)	0.003	(0.040)
LRT effect	(β_1)	0.565	(0.012)	0.565	(0.013)
School variance	$(\sigma^2_{\delta(2)})$	0.093	(0.018)	0.096	(0.020)
Pupil variance	(σ^2_{ϵ})	0.570	(0.013)	0.571	(0.013)

predictor the pupil's (standardised) score in a reading test taken at age 11. As we are interested in progress from age 11–16, it makes sense to consider the effect of all schools attended in this period. We will consider the following model:

$$\underline{normexam}_i = \beta_0 + \beta_1 standlrt_i + \sum_{j \in school(i)} w^{(2)}_{i,j} \underline{\delta}^{(2)}_j + \underline{\epsilon}_i \,,$$

$$\underline{\delta}^{(2)}_j \sim \mathcal{N}(0, \sigma^2_{\delta(2)}), \quad \underline{\epsilon}_i \sim \mathcal{N}(0, \sigma^2_{\epsilon}).$$

We fit this model using both the RG and MCMC methods and the results can be seen in Table 8.8. From the table, we can see that both methods give similar results. If we compare the results here with the results in Rasbash et al. [16], we see only slight changes to the estimates with the level-2 variance slightly decreased and the level-1 variance slightly increased. However, in cases where there is greater amounts of multiple membership the variance estimates can be altered if this multiple membership is ignored. For example, if we randomly assigned every pupil to a second school, the variances change to 0.088 and 0.609 at levels 1 and 2, respectively.

8.4 Combining Multiple Membership and Cross-Classified Structures in a Single Model

Consider two of our earlier examples in the field of education, firstly pupils in a crossing of primary schools and secondary schools and secondly pupils who are moving from school to school. We could assume that these two structures occur simultaneously and we will then end up with a model structure that contains both a multiple membership classification (secondary schools) and a second classification (primary schools) that is crossed with the first. This scenario can be represented by a classification diagram as in Fig. 8.11. Browne et al. [2] refer to models that contain both multiple memberships and cross classifications as multiple membership multiple classification (MMMC) models.

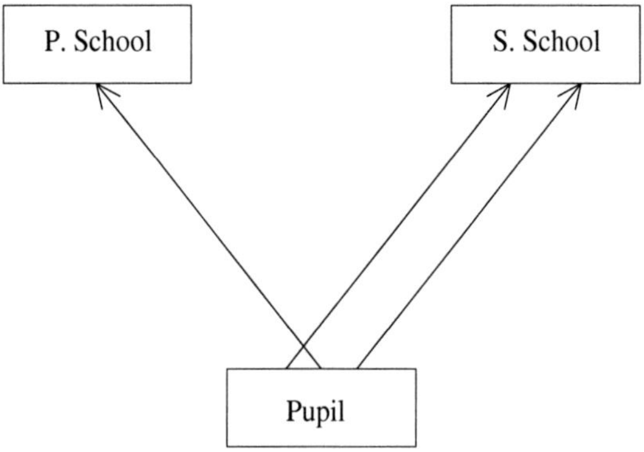

Fig. 8.11 Classification diagram for the primary schools/secondary schools multiple membership model.

8.4.1 Example Scenarios

Many studies have both cross-classified and multiple membership classifications in their structure. A few examples are the following:

- Education: Pupils can be affected by the crossing of the neighbourhood they live in and the school they attend. They could also change class over their period of education and so this multiple membership class classification will be crossed with the neighbourhood classification.
- Health: Patients are seen by several doctors during their treatment and may visit several hospitals. Doctors who are specialists may move from hospital to hospital and so are crossed with the hospitals.
- Survey Data: Individuals will belong to many households over the course of their lives and will reside in several properties. An entire household may move to a new property, so households can be crossed with properties and all the households/properties can have an effect on the individual. See Goldstein et al. [10] for more details.
- Spatial Data: Individuals will belong to a particular area but will also be affected by multiple neighbouring areas [11].

8.4.2 Constructing a Statistical Model

If we return to our example of pupils attending multiple secondary schools but coming from one primary school, we need to combine the multiple membership and cross-classified model structures into one model. As we are treating the secondary schools as a random classification, we would like each pupil to have an equal effect on the secondary school classification, so we will use weights that add to 1 when a pupil attends more than one secondary school. We will let *second*(i) be the list of secondary schools that child i has attended. We can then write down a general two-classification MMMC model as

$$y_i = \boldsymbol{x}_i \boldsymbol{\beta} + \sum_{j \in second(i)} w_{i,j}^{(2)} \delta_j^{(2)} + \delta_{prim(i)}^{(3)} + \epsilon_i \,,$$

$$\delta_j^{(2)} \sim \mathcal{N}(0, \sigma_{\delta(2)}^2), \quad \delta_{prim(i)}^{(3)} \sim \mathcal{N}(0, \sigma_{\delta(3)}^2), \quad \epsilon_i \sim \mathcal{N}(0, \sigma_\epsilon^2).$$

Here $w_{i,j}^{(2)}$ is the weight given to secondary school j for pupil i. We assume that $\sum_{j \in second(i)} w_{i,j}^{(2)} = 1, \forall i$. Both the RG algorithm and the MCMC method can be used to fit these models that combine both multiple membership and cross-classification.

8.4.3 An Example Analysis: Danish Poultry Farming

Rasbash and Browne [14] consider an example from veterinary epidemiology concerning the outbreaks of salmonella typhimurium in flocks of chickens in poultry farms in Denmark between 1995 and 1997. The response of interest is whether salmonella typhimurium is present in a flock and in the data collected 6.3% of flocks had the disease. At the observation level, each observation represents a flock of chickens. For each flock the response variable is whether or not there was an instance of salmonella in that flock. The basic data have a simple hierarchical structure, as each flock is kept in a house on a farm until slaughter. As flocks live for a short time before they are slaughtered, several flocks will stay in the same house each year. The hierarchy is as follows: 10,127 child flocks within 725 houses on 304 farms.

Each flock is created from a mixture of parent flocks (up to 6) of which there are 200 in Denmark and so we have a crossing between the child flock hierarchy and the multiple membership parent flock classification. The classification diagram can be seen in Fig. 8.12. We also know the exact makeup of each child flock (in terms of parent flocks) and so can use these as weights for each of the parent flocks. We are interested in assessing how much of the variability in salmonella incidence can be attributed to houses, farms and parent flocks.

There are also four hatcheries in which all the eggs from the parent flocks are hatched. We will therefore fit a variance components model that allows for different average rates of salmonella for each year with hatchery included in the fixed part as follows:

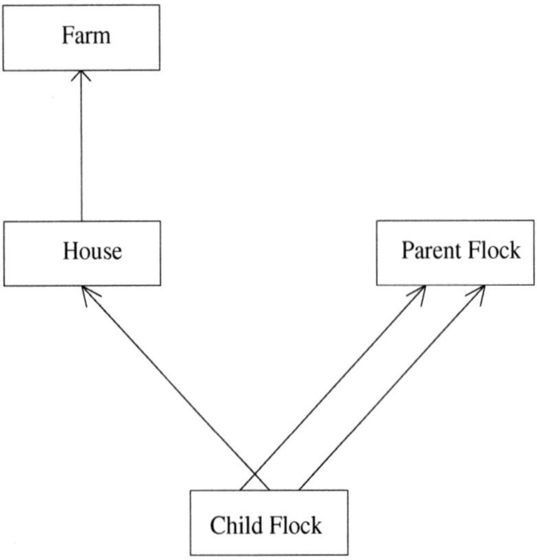

Fig. 8.12 Classification diagram for the Danish poultry model.

$$\underline{salmonella}_i \sim \text{Bernoulli}(\underline{\pi}_i), \tag{8.1a}$$

$$\begin{aligned} \text{logit}(\underline{\pi}_i) = \beta_0 + Y96 * \beta_1 + Y97 * \beta_2 + hatch2 * \beta_3 \\ + hatch3 * \beta_4 + hatch4 * \beta_5 + \underline{\delta}^{(2)}_{house(i)} + \underline{\delta}^{(3)}_{farm(i)} \\ + \sum_{j \in p.flock(i)} w^{(4)}_{i,j} \underline{\delta}^{(4)}_j , \end{aligned} \tag{8.1b}$$

$$\underline{\delta}^{(2)}_{house(i)} \sim \mathcal{N}(0, \sigma^2_{\delta(2)}), \quad \underline{\delta}^{(3)}_{farm(i)} \sim \mathcal{N}(0, \sigma^2_{\delta(3)}), \quad \underline{\delta}^{(4)}_j \sim \mathcal{N}(0, \sigma^2_{\delta(4)}). \tag{8.1c}$$

The results of fitting model (8.1) using both the Rasbash and Goldstein method with 1st-order MQL estimation and the MCMC method can be seen in Table 8.9. The quasi-likelihood methods are numerically rather unstable and we could not get either 2nd-order MQL or PQL to fit this model.

We can see here that there are large effects for the year the chickens were born, suggesting that salmonella was more prevalent in 1995 than the other years. The hatchery effects were also large, suggesting chickens produced in Hatcheries 1 and 3 had a larger incidence of salmonella. There is a large variability for the parent flock effects and for the farm effects which are of similar magnitude. There is less variability between houses within farms.

Method Comparison

The MCMC results were run for 50,000 iterations after a burn-in of 20,000 (This took 40 min on a 3.4 GHz PC), as we used arbitrary starting values and so the chain took a while to converge. From Table 8.9 we can see reasonable

Table 8.9 Results for the Danish poultry example.

Parameter		1st order MQL		MCMC	
Intercept	(β_0)	−1.862	(0.184)	−2.322	(0.213)
1996 effect	(β_1)	−1.004	(0.138)	−1.239	(0.162)
1997 effect	(β_2)	−0.852	(0.159)	−1.165	(0.187)
Hatchery 2 effect	(β_3)	−1.458	(0.222)	−1.733	(0.255)
Hatchery 3 effect	(β_4)	−0.250	(0.209)	−0.211	(0.252)
Hatchery 4 effect	(β_5)	−1.007	(0.353)	−1.062	(0.388)
House variance	($\sigma^2_{\delta(2)}$)	0.206	(0.096)	0.208	(0.108)
Farm variance	($\sigma^2_{\delta(3)}$)	0.639	(0.121)	0.927	(0.197)
Parent flock variance	($\sigma^2_{\delta(4)}$)	0.892	(0.184)	0.895	(0.179)

agreement between the two methods, although the fixed effects in MQL are all smaller, as is the farm level variance. This behaviour was shown in simulations on a nested three-level binary response data structure in Rodríguez and Goldman [18] with the improvements of the MCMC method shown in Browne and Draper [1] and so this suggests that the MCMC results should be more accurate. (See also Chapter 9 of this volume.)

8.4.4 Complex Random Effects

Model (8.1) is essentially another variance components model, but we could fit a model that has complex variation at one of the higher classifications. To illustrate this, we will modify the farm level variance to account for different variability between years at the farm level. That is, we replace the simple farm level random effects, $\underline{\delta}^{(3)}_{farm(i)}$ with three sets of effects, one for each year. Our expanded model is then as follows:

$$\underline{salmonella}_i \sim \text{Bernoulli}(\underline{\pi}_i), \tag{8.2a}$$

$$\begin{aligned}\text{logit}(\underline{\pi}_i) = \beta_0 &+ \mathit{Y96} * \beta_1 + \mathit{Y97} * \beta_2 + \mathit{hatch2} * \beta_3 \\ &+ \mathit{hatch3} * \beta_4 + \mathit{hatch4} * \beta_5 + \underline{\delta}^{(2)}_{house(i)} + \mathit{Y95} * \underline{\delta}^{(3)}_{farm(i),1} \\ &+ \mathit{Y96} * \underline{\delta}^{(3)}_{farm(i),2} + \mathit{Y97} * \underline{\delta}^{(3)}_{farm(i),3} + \sum_{j \in p.flock(i)} w^{(4)}_{i,j} \underline{\delta}^{(4)}_j ,\end{aligned} \tag{8.2b}$$

$$\begin{aligned}&\underline{\delta}^{(2)}_{house(i)} \sim \mathcal{N}(0, \sigma^2_{\delta(2)}), \quad \underline{\boldsymbol{\delta}}^{(3)}_{farm(i)} \sim \mathcal{N}_3(0, \Sigma_{\delta(3)}), \\ &\underline{\delta}^{(4)}_j \sim \mathcal{N}(0, \sigma^2_{\delta(4)}).\end{aligned} \tag{8.2c}$$

The parameter estimates for this extended model are given in Table 8.10. We see that the fixed effects estimates are fairly similar to model (8.1). It is interesting to see that all the covariances in the farm level variance matrix are positive. This suggests that after adjusting for other factors, if a farm has

Table 8.10 Estimates for the parameters in Model (8.2).

Parameter		MCMC estimates	
Intercept	(β_0)	−2.544	(0.240)
1996 effect	(β_1)	−1.149	(0.256)
1997 effect	(β_2)	−1.003	(0.293)
Hatchery 2 effect	(β_3)	−1.788	(0.265)
Hatchery 3 effect	(β_4)	−0.143	(0.252)
Hatchery 4 effect	(β_5)	−1.065	(0.383)
House variance	$(\sigma^2_{\delta(2)})$	0.271	(0.119)
Farm year95 variance	$(\Sigma_{\delta(3)}[1,1])$	1.416	(0.341)
Farm 95/96 covariance	$(\Sigma_{\delta(3)}[1,2])$	0.514	(0.262)
Farm 95/97 covariance	$(\Sigma_{\delta(3)}[1,3])$	0.415	(0.226)
Farm year96 variance	$(\Sigma_{\delta(3)}[2,2])$	1.239	(0.463)
Farm 96/97 covariance	$(\Sigma_{\delta(3)}[2,3])$	0.750	(0.321)
Farm year97 variance	$(\Sigma_{\delta(3)}[3,3])$	1.017	(0.482)
Parent flock variance	$(\sigma^2_{\delta(4)})$	0.878	(0.180)

an incidence of salmonella in 1995 then it is more likely to have an incidence again in 1996 and in 1997. In fact, the corresponding correlation estimates are 0.39, 0.35 and 0.67, respectively, showing that in particular there is a strong correlation between salmonella infection in farms in 1996 and 1997. On the other hand, these correlations are clearly not equal to 1, which is the value implied by model (8.1). Hence, this also shows the importance of allowing complex random effects and the strength of the versatile model specification approach presented here. The numerical instabilities of the quasi-likelihood methods mean that comparative estimates could not be calculated for this model.

8.5 Consequences of Ignoring Non-Hierarchical Structures

Analysing only hierarchical components of populations which have additional non-nested structures has two potentially negative consequences. First, the model is under-specified because there are sources of variation that have not been included in the model. This under-specification can lead to an underestimation of the standard errors of the parameters and therefore to incorrect inferences. Second, the variance components obtained from the simple hierarchical model, or sets of separate hierarchical models, cannot be trusted. They may change substantially if the additional non-nested structures are included in a single model. For example, we may wish to know about the relative importance of general practices and hospitals on the variation in some patient level outcome. If patients are cross-classified by hospital and general practice,

Table 8.11 Effects of ignoring a cross-classified structure.

Parameter	Model I		Model II		Model III	
Intercept	5.97	(0.07)	6.02	(0.07)	5.98	(0.07)
VRQ effect	0.16	(0.003)	0.16	(0.003)	0.16	(0.003)
Primary school variance	0.28	(0.06)			0.27	(0.06)
Secondary school variance			0.05	(0.02)	0.01	(0.02)
Pupil variance	4.25	(0.10)	4.48	(0.11)	4.25	(0.10)

we need to fit the full cross-classified model including patients, general practices and hospitals in order to address this question. Looking at two separate hierarchical analyses, one of patients within hospital, the other of patients within general practices, is not sufficient.

A numerical example of this is shown in Table 8.11, which shows results for three models fitted using the RG method to the educational attainment data from Fife in Scotland, where pupils are contained within a cross-classification of primary schools by secondary schools. Model I fits pupils within primary schools and ignores secondary school, Model II fits pupils within secondary schools and ignores primary school and Model III fits the cross-classification. The response is an attainment score at age 16, the explanatory variable VRQ is a verbal reasoning measure taken at age 11. When one side of the cross-classification is ignored, the released variance is split between the classification left in the model and the pupil level variance, inflating both estimates. This has the most drastic effect when the primary school hierarchy is ignored; in this case (Model II), the inflated estimate of the between secondary school variance is 2.5 times its standard error as opposed to 0.5 times its standard error in the full model.

References

1. W. J. Browne and D. Draper. A comparison of Bayesian and likelihood-based methods for fitting multilevel models. *Bayesian Analysis*, 1:473–549, 2006. (with discussion)
2. W. J. Browne, H. Goldstein, and J. Rasbash. Multiple membership multiple classification (MMMC) models. *Statistical Modelling*, 1:103–124, 2001.
3. J. M. Bull, G. D. Riley, J. Rasbash, and H. Goldstein. Parallel implementation of a multilevel modelling package. *Computational Statistics & Data Analysis*, 31:457–474, 1999.
4. D. G. Clayton and J. Rasbash. Estimation in large crossed random-effect models by data augmentation. *Journal of the Royal Statistical Society, Series A*, 162: 425–436, 1999.
5. R. Ecochard and D. G. Clayton. Multilevel modelling of conception in artificial insemination by donor. *Statistics in Medicine*, 17:1137–1156, 1998.

6. A. E. Gelfand and A. F. M. Smith. Sampling-based approaches to calculating marginal densities. *Journal of the American Statistical Association*, 85:398–409, 1990.
7. H. Goldstein. Multilevel mixed linear model analysis using iterative generalized least squares. *Biometrika*, 73:43–56, 1986.
8. H. Goldstein. *Multilevel Statistical Models*, 3rd edition. Edward Arnold, London, 2003.
9. H. Goldstein and J. Rasbash. Improved approximations for multilevel models with binary responses. *Journal of the Royal Statistical Society, Series A*, 159: 505–513, 1996.
10. H. Goldstein, J. Rasbash, W. J. Browne, G. Woodhouse, and M. Poulain. Multilevel modelling in the study of dynamic household structures. *European Journal of Population*, 16:373–387, 2000.
11. A. B. Lawson, W. J. Browne, and C. Vidal-Rodeiro. *Disease Mapping using WinBUGS and MLwiN*. Wiley, London, 2003.
12. Y. Lee and J. A. Nelder. Hierarchical generalised linear models: A synthesis of generalised linear models, random-effect models and structured dispersions. *Biometrika*, 88:987–1006, 2001.
13. J. X. Pan and R. Thompson. Generalized linear mixed models: An improved estimating procedure. In J. G. Bethlehem and P. G. M. van der Heijden, editors, *COMPSTAT: Proceedings in Computational Statistics, 2000*, pages 373–378. Physica, Heidelberg, 2000.
14. J. Rasbash and W. J. Browne. Modelling non-hierarchical structures. In A. H. Leyland and H. Goldstein, editors, *Multilevel Modelling of Health Statistics*, pages 93–106. Wiley, New York, 2001.
15. J. Rasbash and H. Goldstein. Efficient analysis of mixed hierarchical and crossed random structures using a multilevel model. *Journal of Behavioural Statistics*, 19:337–350, 1994.
16. J. Rasbash, F. Steele, W. J. Browne, and B. Prosser. *A User's Guide to MLwiN. Version 2.0.* Centre for Multilevel Modelling, University of Bristol, Bristol, UK, 2005.
17. S. W. Raudenbush. A crossed random effects model for unbalanced data with applications in cross-sectional and longitudinal research. *Journal of Educational Statistics*, 18:321–350, 1993.
18. G. Rodríguez and N. Goldman. An assessment of estimation procedures for multilevel models with binary responses. *Journal of the Royal Statistical Society, Series A*, 158:73–89, 1995.
19. J. L. Schafer. *Analysis of Incomplete Multivariate Data.* Chapman & Hall, London, 1997.
20. M. A. Tanner and W. H. Wong. The calculation of posterior distributions by data augmentation. *Journal of the American Statistical Association*, 82:528–550, 1987. (with discussion)

9

Multilevel Generalized Linear Models

Germán Rodríguez

Office of Population Research, Princeton University

9.1 Introduction

Two of the most influential papers in applied statistics published in the last few decades are Nelder and Wedderburn [65], introducing generalized linear models (GLMs), and Cox [20], the seminal paper introducing life tables with regression, better known as proportional hazard models. As we will see, these two developments are closely related. Nelder and Wedderburn's unique contribution was to provide a unified conceptual framework for studying a large range of statistical models, including not only classical linear models, but also logit and probit models for binary data, log-linear Poisson models for count data, and others. The unification was not only conceptual, but also led to common estimation procedures in the form of an iteratively re-weighted least squares (IRLS) algorithm. The first implementation of these procedures appeared in the highly successful program GLIM [3], which for many statisticians became synonymous with GLMs.

In this chapter we follow Wong and Mason [94], Longford [54, 56], Goldstein [30], Breslow and Clayton [11], and others in exploring extensions of GLMs to include random effects in a multilevel setting. Chapter 1 in this handbook has described multilevel models for continuous outcomes, while Chapter 6 has focused on multilevel models for categorical outcomes. Here we adopt a unified approach that views the general linear mixed model and many of the random-effects models for categorical data discussed in earlier chapters as special cases of the Multilevel Generalized Linear Model (MGLM). This approach has conceptual merit in emphasizing the similarities among these models, and provides a common framework to study and evaluate estimation methods. Alas, we do not have a single estimation procedure that can be applied to all MGLMs with the same measure of success that IRLS achieved

J. de Leeuw, E. Meijer (eds.), *Handbook of Multilevel Analysis*,
© Springer 2008

for GLMs. Instead, we must choose between quick but sometimes biased approximations, and more accurate but often compute-intensive maximum likelihood and Bayesian approaches. Part of our task in this chapter is to describe and illustrate the alternatives.

Section 9.2 develops the modeling framework. We introduce generalized linear models (GLMs) as an extension of linear models, and proceed to an analogous derivation of multilevel generalized linear models (MGLMs) as an extension of multilevel linear models. The ideas discussed apply more generally to generalized linear mixed models (GLMMs) and our notation reflects this broader applicability, but we tend to focus the narrative on the multilevel case. We review survival models, note their close connection with GLMs, and describe a natural extension to the multilevel case. We draw an important distinction between conditional and marginal models that is significant in the generalized linear case. Finally, we introduce non-linear mixed models and contrast them with MGLMs.

Section 9.3 is devoted to a discussion of estimation procedures. It turns out that calculation of the likelihood function for MGLMs involves intractable integrals. We discuss several alternatives and assess their performance in realistic situations, referring to some of our earlier work using simulated data and a case study [81, 82] and introducing new results. We review a range of approximate estimation procedures that, unfortunately, can be severely biased when random effects are substantial. We describe maximum likelihood estimation using Gauss-Hermite quadrature, a method that appears to work remarkably well, but is limited to relatively low-dimensional models. We also discuss Bayesian estimation procedures focusing on the Gibbs sampler, a Markov Chain Monte Carlo (MCMC) method that can be applied to more complex models involving high-dimensional integrals, albeit not without difficulty. We close this section with a brief discussion of other approaches to estimation, an active area of current research.

Section 9.4 is devoted to an application of MGLMs to the study of infant and child mortality in Kenya, using data from a national survey conducted in 1998. We use a three-level piece-wise exponential survival model that allows for clustering of infant and child deaths at both the family and community levels, and fit it to data using the equivalent MGLM with Poisson errors and log link. We compare estimates that ignore clustering, and estimates obtained by approximate quasi-likelihood and by full maximum likelihood. The discussion emphasizes interpretation of the results, particularly the family and community random parameters. Finally, we show how the model can be used to estimate measures of intra-family and intra-community correlation in infant and child deaths.

Section 9.5 is a brief discussion and summary of our conclusions.

9.2 Extending Multilevel Models

9.2.1 Generalized Linear Models

Consider briefly the general linear model. We usually view the outcome y_i for the i-th individual as a realization of a random variable (r.v.) $\underline{y}_i$ that depends on a vector $\boldsymbol{x}_i$ of predictors or explanatory variables through the equation

$$\underline{y}_i = \boldsymbol{x}_i'\boldsymbol{\beta} + \underline{\epsilon}_i, \tag{9.1}$$

where $\boldsymbol{\beta}$ is a vector of regression coefficients and $\underline{\epsilon}_i$ is an error term having a normal distribution with mean 0 and variance σ^2.

It will facilitate further generalization if we write this model in a slightly different way, noting that $\underline{y}_i$ has a normal distribution with mean μ_i and variance σ^2, which we write

$$\underline{y}_i \sim \mathcal{N}(\mu_i, \sigma^2), \tag{9.2}$$

and the expected value satisfies the linear model

$$\mu_i = \boldsymbol{x}_i'\boldsymbol{\beta}. \tag{9.3}$$

This approach draws a clear distinction between the stochastic structure of the data, specified in the first equation, and the systematic component, specified in the second.

The Exponential Family

Nelder and Wedderburn [65] generalize this model in two master strokes. First, they assume that the distribution of $\underline{y}_i$ is in an *exponential family* that includes as special cases many of the distributions we encounter in applied work, such as the normal, binomial, Poisson, gamma, and inverse Gaussian. The family may be written as

$$f(y_i) = \exp\left\{\frac{y_i\theta_i - b(\theta_i)}{a_i(\phi)} + c(y_i, \phi)\right\}, \tag{9.4}$$

where θ_i and ϕ are unknown parameters and $a_i(\cdot)$, $b(\cdot)$, and $c(\cdot)$ are known functions. Usually $a_i(\phi) = \phi/p_i$, where p_i is a known prior weight, and this will be assumed in the applications that follow. In this family, the mean is $E(\underline{y}_i) = b'(\theta_i)$ and the variance is $\text{Var}(\underline{y}_i) = b''(\theta_i)a_i(\phi)$. In applied work, we often express the variance as a function of the mean, say $\text{Var}(\underline{y}_i) = \phi V(\mu_i)$.

All the distributions mentioned above can be obtained from this general expression by suitable choice of parameters and functions. For example if we set $\theta_i = \mu_i$, $b(\theta_i) = \frac{1}{2}\theta_i^2$, $\phi = \sigma^2$, and $a_i(\phi) = \phi$, we obtain a normal distribution with mean μ_i and variance σ^2. In this case, the variance function is the identity. The Poisson distribution with mean μ_i has $\theta_i = \log \mu_i$, $b(\theta_i) = e^{\theta_i}$, $a_i(\phi) = \phi$, and $\phi = 1$, and the variance equals the mean. McCullagh and Nelder [59] show how you can obtain other special cases from the general formula.

The Link Function

The second aspect of the generalization is that instead of modeling the expected value of the outcome as a linear function of the covariates, we model a *transformation* of the expected value. Specifically, we introduce a one-to-one continuous differentiable transformation of the mean $\eta_i = g(\mu_i)$ and assume that the transformed mean follows a linear model, so that

$$g(\mu_i) = \eta_i = \boldsymbol{x}_i'\boldsymbol{\beta}. \tag{9.5}$$

The function $g(\cdot)$ is called the *link function*, and connects the mean with the linear predictor $\boldsymbol{x}_i'\boldsymbol{\beta}$ and thus the explanatory variables. The simplest possible link function is the identity, which leads to modeling the mean itself. Other transformations in common use are the logit, probit, log, inverse, and square root.

A key feature of GLMs is that the model for the transformed mean η_i is simple and has a familiar linear structure. Because the link function is one-to-one, we can always invert it to obtain a a model for the mean

$$\mu_i = g^{-1}(\boldsymbol{x}_i'\boldsymbol{\beta}), \tag{9.6}$$

but this model is usually more complicated. In particular, interpretation of the parameters is straightforward in the transformed scale, but may be rather involved in the original scale. Notable exceptions are models with log and logit links, where exponentiated coefficients may be interpreted as multiplicative effects on an expected count or an odds ratio, respectively. An example will follow in Section 9.4.

Link functions can often be motivated as a way to handle range restrictions on the mean μ_i. With count data, for example, a linear model is not attractive because the mean μ_i must be non-negative, but the linear predictor $\boldsymbol{x}_i'\boldsymbol{\beta}$ may yield positive or negative values. Modeling the log of the mean instead solves the problem. The link function can also make the assumption of linearity more plausible. With count data, for example, one often finds that effects are relative rather than absolute; an additive model in the log scale is equivalent to a multiplicative model in the original scale, and can thus represent relative effects. A link function that maps the mean μ_i into the parameter θ_i in the exponential family is said to be a canonical link. The canonical links for the Poisson and Bernoulli distributions are the log and the logit, respectively.

Estimation and Testing

An important practical feature of GLMs is that they can all be fit to data using the same algorithm, a form of iteratively reweighted least squares (IRLS). The algorithm may be motivated by considering a linearized form of the model, a fact that has motivated the adoption of similar strategies for MGLMs. Write the model as

$$\underline{\boldsymbol{y}} = \boldsymbol{\mu}(\boldsymbol{X\beta}) + \underline{\boldsymbol{\epsilon}}, \tag{9.7}$$

where $\boldsymbol{\mu}(\cdot)$ is the inverse link function applied element-wise to the linear predictor $\boldsymbol{\eta} = \boldsymbol{X\beta}$, and $\underline{\boldsymbol{\epsilon}}$ is a vector of independent heteroscedastic error terms with mean $\boldsymbol{\emptyset}$ and (diagonal) variance-covariance matrix $\phi\,\boldsymbol{V}(\boldsymbol{\mu})$. Expanding the link using a first-order Taylor series about a trial parameter value $\boldsymbol{\beta}_0$ and rearranging terms leads to the approximating linear model

$$\underline{\boldsymbol{y}}^* \approx \boldsymbol{X\beta} + \underline{\boldsymbol{\epsilon}}^*, \tag{9.8}$$

where $\underline{\boldsymbol{y}}^* = \boldsymbol{D}^{-1}(\underline{\boldsymbol{y}} - \boldsymbol{\mu}(\boldsymbol{X\beta}_0)) + \boldsymbol{X\beta}_0$ is a working response, $\underline{\boldsymbol{\epsilon}}^* = \boldsymbol{D}^{-1}\underline{\boldsymbol{\epsilon}}$ is a new error term with variance $\phi\boldsymbol{W}$, where $\boldsymbol{W} = \boldsymbol{D}^{-1}\boldsymbol{V}(\boldsymbol{\mu})\boldsymbol{D}^{-1}$ is a diagonal matrix of weights, and $\boldsymbol{D} = \partial\boldsymbol{\mu}/\partial\boldsymbol{\eta}$ is a diagonal matrix of derivatives of the link function with respect to the linear predictor. This approximating linear model may be fit using weighted least squares to obtain an improved estimate of $\boldsymbol{\beta}$, which can then be used to obtain a better approximating model, and so on to convergence. McCullagh and Nelder [59] show that this method is equivalent to Fisher scoring and leads to maximum likelihood estimates.

Under standard regularity conditions, the large sample distribution of the estimator $\underline{\hat{\boldsymbol{\beta}}}$ is approximately normal with mean equal to the true parameter value $\boldsymbol{\beta}$ and variance-covariance matrix $\phi\,(\boldsymbol{X}'\boldsymbol{W}\boldsymbol{X})^{-1}$. This result provides large-sample standard errors and a basis for Wald tests. Likelihood ratio tests are often preferable, and in the context of GLMs they are usually calculated by reference to a statistic known as the *deviance*. This statistic is constructed by considering a likelihood ratio test that compares the model of interest with a saturated model that has a separate parameter for each observation. The deviance is the product of the scale parameter ϕ and the usual likelihood ratio chi-squared statistic $-2\log\lambda$. A test comparing two nested models can then be computed as the difference of their scaled deviances.

9.2.2 Multilevel Generalized Linear Models

We now consider a similar extension for multilevel linear models. In previous chapters we have written the general linear mixed model in a form analogous to (9.1),

$$\underline{y}_i = \boldsymbol{x}_i'\boldsymbol{\beta} + \boldsymbol{z}_i'\underline{\boldsymbol{\delta}} + \underline{\epsilon}_i, \tag{9.9}$$

where $\underline{y}_i$ is the r.v. representing the outcome for the i-th individual, $\boldsymbol{x}_i$ is the i-th row of the model matrix for the fixed effects $\boldsymbol{\beta}$, $\boldsymbol{z}_i$ is the i-th row of the model matrix for the random effects $\underline{\boldsymbol{\delta}}$, and $\underline{\epsilon}_i$ is the individual error term. We assume that the random effects $\underline{\boldsymbol{\delta}}$ have a $\mathcal{N}(\boldsymbol{\emptyset}, \boldsymbol{\Omega})$ distribution, and the error terms are independent and identically distributed (i.i.d.) $\mathcal{N}(0, \sigma^2)$ r.v.'s.

We could write this model more compactly in terms of vectors, with $\underline{\boldsymbol{\epsilon}} \sim \mathcal{N}(\boldsymbol{\emptyset}, \sigma^2 \boldsymbol{I})$, but that would not be very productive for the generalization that follows. Instead, we will reformulate the model in terms of the *conditional* distribution of the outcomes $\underline{y}_i$ given the random effects $\boldsymbol{\delta}$, which we write as

$$\underline{y}_i \mid \boldsymbol{\delta} \sim \mathcal{N}(\mu_i, \sigma^2). \tag{9.10}$$

In words, we assume that given the random effects the outcomes are independent normally distributed r.v.'s with mean μ_i and variance σ^2. The conditional mean, in turn, follows the linear model

$$\mu_i = \boldsymbol{x}_i' \boldsymbol{\beta} + \boldsymbol{z}_i' \boldsymbol{\delta}, \tag{9.11}$$

depending on unknown coefficients $\boldsymbol{\beta}$ and given values $\boldsymbol{\delta}$ of the random effects. The essence of this approach is the recognition that *given* the random effects, the outcomes are independent and follow a linear model.

The stage is now set for the generalization. We retain the key assumption of conditional independence. However, instead of assuming that the conditional distribution of the outcomes $\underline{y}_i$ given the random effects $\underline{\boldsymbol{\delta}}$ is normal, we assume that the distribution is in the exponential family (9.4). This extends the general linear mixed model to situations where the conditional distribution of the responses is binomial, Poisson, gamma, or inverse Gaussian.

The second element of the generalization is the introduction of a link function. We assume that a *transformation* of the conditional mean, rather than the mean itself, follows a linear model, so that

$$g(\mu_i) = \boldsymbol{x}_i' \boldsymbol{\beta} + \boldsymbol{z}_i' \boldsymbol{\delta}. \tag{9.12}$$

The link function can be the identity, log, logit, probit, or any other one-to-one continuous differentiable transformation. This final extension leads to multilevel logit and probit models, multilevel log-linear models for count data, and many other applications.

By focusing on the conditional distribution of the outcomes given the random effects, we can apply without changes the entire conceptual apparatus of generalized linear models. In particular, random and fixed effects can be interpreted in a unified way, the interpretation is simple in the transformed scale because fixed and random effects enter linearly, and can often be translated meaningfully back to the original scale. We will return to these issues in Section 9.4.

9.2.3 Survival Models

Let us now consider models for time-to-event or survival data, which are closely related to GLMs. There is now an extensive literature on survival models; excellent texts include Kalbfleisch and Prentice [43], Cox and Oakes [22], and Therneau and Grambsch [89].

Hazards and Survival

In a standard hazard model, we assume that the survival experiences of different individuals are independent and that the hazard for individual i, or instantaneous risk of occurrence of the event at time t given that it has not occurred earlier, is given by

$$\lambda(t, \boldsymbol{x}_i) = \lambda_0(t) \exp\{\boldsymbol{x}_i'\boldsymbol{\beta}\}, \tag{9.13}$$

where $\lambda_0(t)$ represents a baseline hazard at time t and $\exp\{\boldsymbol{x}'\boldsymbol{\beta}\}$ is a relative risk associated with covariate values $\boldsymbol{x}$. The special case where $\lambda_0(t) = \lambda_0$ is the exponential survival model of Feigl and Zelen [26]. The model is easily extended to time-varying covariates $\boldsymbol{x}(t)$ and time-varying effects $\boldsymbol{\beta}(t)$. Note that taking logs yields a model that is linear in the relative risk parameters.

The cumulative hazard is defined as $\Lambda(t, \boldsymbol{x}_i) = \int_0^t \lambda(u, \boldsymbol{x}_i)\, du$, which for time-fixed covariates is simply the baseline cumulative hazard times the relative risk for individual i. We will also need the survival function or probability of being alive at time t, which can be obtained from the cumulative hazard as $S(t \mid \boldsymbol{x}_i) = \exp\{-\Lambda(t, \boldsymbol{x}_i)\}$, and therefore for fixed covariates satisfies

$$S(t \mid \boldsymbol{x}_i) = S_0(t)^{\exp\{\boldsymbol{x}_i'\boldsymbol{\beta}\}}, \tag{9.14}$$

where $S_0(t)$ is the baseline survival function.

Estimation with Censored Data

A distinctive feature of survival models is that observations are often *censored*, in the sense that for some individuals the event of interest has not yet occurred at the time the data are analyzed. Estimation of censored-data hazard models under parametric assumptions for the baseline hazard relies on the standard survival likelihood, to which an individual who dies at t contributes $\lambda(t, \boldsymbol{x})S(t, \boldsymbol{x})$, the density at t, and an individual who is censored at t contributes $S(t, \boldsymbol{x})$, the probability of surviving to t. This likelihood can be derived under the key assumption that censoring is non-informative, so all we know about an individual who is censored at t is that it survived that long; see Kalbfleisch and Prentice [43]. Cox [20, 21] introduced a partial likelihood that allows estimation of the relative risk coefficients $\boldsymbol{\beta}$ without assumptions about the shape of the baseline hazard $\lambda_0(t)$.

Several authors have noted a close relationship between hazard models and GLMs, and a number of papers show how various survival models can be fit using standard GLM software; see Aitkin and Clayton [4] for the exponential, Weibull, and extreme value distributions, Bennet and Whitehead [8] for the logistic and log-logistic, and Clayton and Cuzick [17] and Whitehead [93] for estimation using Cox's partial likelihood. In this section we focus on connections with Poisson models with log link, which we use in our application in Section 9.4, and binomial models with logit and c-log-log links.

Piece-Wise Exponential Survival

A flexible semi-parametric approach to hazard models is to partition time (or duration of exposure) into J intervals $[\tau_{j-1}, \tau_j)$ for $j = 1, \ldots, J$ with cutpoints $0 = \tau_0 < \tau_1 < \cdots < \tau_J$, and assume that the baseline hazard is constant within each interval, so that $\lambda_0(t) = \lambda_{0j}$ for $t \in [\tau_{j-1}, \tau_j)$. Judicious choice of cutpoints leads to good approximations to a wide range of hazard functions, using more closely spaced boundaries where the hazard varies rapidly and wider intervals where the hazard changes more slowly.

Holford [40] and Laird and Olivier [47] noted that the piece-wise exponential model is equivalent to a Poisson regression model. With censored data, we observe t_i, the total time lived by the i-th individual, and d_i, a death indicator that takes the value 1 if the individual died and 0 otherwise. Imagine defining analogous measures for each duration interval, so t_{ij} is the time lived by the i-th individual in the j-th interval, and d_{ij} is a death indicator that takes the value 1 if individual i died in interval j and 0 otherwise. Then a piece-wise exponential hazard model can be fitted by treating the death indicators d_{ij} as if they were independent Poisson observations with means $\mu_{ij} = \lambda_{ij} t_{ij}$, where λ_{ij} is the hazard for individual i in interval j.

The proof is not hard and can be sketched as follows. The contribution of the i-th individual to the standard survival log-likelihood for censored data has the form $d_i \log \lambda(t_i, \boldsymbol{x}_i) - \Lambda(t_i, \boldsymbol{x}_i)$. Suppose t_i falls in interval $j(i)$ and write $\lambda_{ij(i)}$ as shorthand for $\lambda(t_i, \boldsymbol{x}_i)$. The cumulative or integrated hazard can be computed easily because the hazard is constant in each interval, so $\Lambda(t_i, \boldsymbol{x}_i) = \sum_j \lambda_{ij} t_{ij}$, where the sum is over all intervals up to $j(i)$. There is a slight lack of symmetry in that we have only one term on the death indicator and $j(i)$ terms on the exposure times, but we can easily add the terms for previous intervals, which have $d_{ij} = 0$ and thus are all zero, to obtain

$$\log L_i = \sum_{j=1}^{j(i)} \{d_{ij} \log \lambda_{ij} - \lambda_{ij} t_{ij}\}. \tag{9.15}$$

This equation coincides with the log-likelihood that we would obtain if we treated d_{ij} as having a Poisson distribution with mean $\mu_{ij} = \lambda_{ij} t_{ij}$ except for a term $d_{ij} \log(t_{ij})$, but this is a constant depending on the data and not the parameters, so it can be ignored.

It is important to note that we have not assumed that the d_{ij} have independent Poisson distributions, because clearly they do not. If individual i died in interval j, then it must have been alive in all prior intervals, so the indicators couldn't possibly be independent. Moreover, each indicator can only take the values 1 and 0, so it couldn't possibly have a Poisson distribution that assigns probability to values greater than 1. The result is more subtle; it is the likelihood functions that coincide. Given a realization of a piece-wise exponential process, we can find a realization of a set of independent Poisson r.v.'s that happens to have the same probability and thus leads to the same estimates. The practical implication is that one can fit a piece-wise exponential model in terms of the equivalent GLM.

Discrete Survival Models

In his original paper, Cox [20] proposed a discrete version of the proportional hazards model by working with the conditional odds of dying at each possible failure time t_j given survival up to that point. Specifically, he proposed the model

$$\frac{\lambda(t_j \mid \boldsymbol{x})}{1 - \lambda(t_j \mid \boldsymbol{x})} = \frac{\lambda_0(t_j)}{1 - \lambda_0(t_j)} \exp\{\boldsymbol{x}'\boldsymbol{\beta}\}, \tag{9.16}$$

where $\lambda_0(t_j)$ is the baseline conditional probability of dying at t_j given survival to that time and $\exp\{\boldsymbol{x}'\boldsymbol{\beta}\}$ is the relative risk. In this model, the conditional log-odds of dying are linear in the relative risk parameters $\boldsymbol{\beta}$.

Cox [20] extended his partial likelihood approach to estimate $\boldsymbol{\beta}$ while treating the baseline hazards $\lambda_0(t_j)$ as nuisance parameters that could be conditioned out of the likelihood. Allison [5] noted that one could estimate the complete model, including a separate parameter for each discrete time of death t_j, by running a logistic regression on a set of pseudo-observations, in a procedure analogous to that described above for piece-wise exponential models.

An alternative extension of hazard models to discrete data assumes that the survival functions satisfy (9.14) and then solves for the conditional hazard at time t_j, to obtain

$$\lambda(t_j \mid \boldsymbol{x}) = 1 - (1 - \lambda_0(t_j))^{\exp\{\boldsymbol{x}'\boldsymbol{\beta}\}}. \tag{9.17}$$

The transformation that makes the right-hand side a linear function of the parameters is the complementary log-log, and the model can be fitted using a GLM with binomial structure and complementary log-log link.

This model can also be obtained by grouping time in a continuous-time proportional hazards model, see Prentice and Gloeckler [72] and Kalbfleisch and Prentice [43] for details. In this approach, time is grouped into intervals $[\tau_{j-1}, \tau_j)$ as before, but all we observe is whether an individual survives or dies in an interval. This construction imposes some constraints on censoring: If an individual is censored inside an interval, we do not know whether he or she would have survived the interval, and therefore must censor the observation back at the beginning of the interval. Unlike the piece-wise exponential setup, we cannot use information about exposure during part of an interval. On the other hand, we do not need to assume that the hazard is constant in each interval.

9.2.4 Multilevel Survival Models

In the last several years, there has been considerable interest in extending survival models by introducing random effects. A classic demographic contribution is Vaupel et al. [91], which introduced a gamma-distributed random effect to represent unobserved heterogeneity of frailty in univariate survival models, see also Aalen [1], Hougaard [41, 42], and Manton et al. [58]. The idea of frailty can be used to represent association of kindred lifetimes in a multivariate setting, see Clayton [14], Clayton and Cuzick [18], and Oakes [68]; to account for association in recurrent events and event history data, see Clayton [15] and Rodríguez [79]; and leads naturally to two- and three-level survival models, see Guo and Rodríguez [37], Sastry [83], and Barber et al. [7].

The multilevel extension follows the same strategy as for MGLMs. We assume that given a vector of random effects $\underline{\boldsymbol{\delta}}$, the survival experiences of different individuals are independent and follow a hazard model with conditional hazard

$$\lambda(t, \boldsymbol{x}_i \mid \boldsymbol{\delta}) = \lambda_0(t) \exp\{\boldsymbol{x}_i'\boldsymbol{\beta} + \boldsymbol{z}_i'\boldsymbol{\delta}\}. \tag{9.18}$$

In this generalization, the hazard for individual i depends not only on the fixed effects $\boldsymbol{\beta}$ with model vector $\boldsymbol{x}_i$, but also on the random effects $\boldsymbol{\delta}$ with model vector $\boldsymbol{z}_i$. Once again, the random effects enter a linear predictor in exactly the same form as the fixed effects. Calculation of the conditional cumulative hazard and the conditional survival function follows along the same lines as in ordinary survival models.

We can also calculate unconditional or marginal survival probabilities by integrating out the random effects. Calculation of unconditional hazards requires special care because hazards, by definition, are conditional on survival to time t. The extent of dependence of kindred lifetimes can be expressed in terms of measures of intraclass correlation. Estimation of both discrete- and continuous-time multilevel survival models can proceed by working in terms of the equivalent MGLM with binomial or Poisson errors. We will revisit these issues in the context of our application in Section 9.4.

9.2.5 Conditional and Marginal Models

An alternative approach to the analysis of correlated data that is popular in longitudinal or repeated-measurement studies focuses on the marginal distribution of the responses, see Diggle et al. [25, Chapter 8]. These models assume that the outcomes have a distribution in the exponential family, and that a transformation of the *marginal* mean is a linear function of observed covariates with coefficients $\boldsymbol{\beta}$. The models are usually fit to data using generalized estimating equations (GEE) that take into account the dependence of the observations. The method is very similar to the IRLS algorithm used in GLMs, using the same working dependent variable and the same set of weights, but instead of using weighted least squares (WLS) with a diagonal weight matrix, it uses generalized least squares (GLS) with a more general weight matrix, where the non-diagonal elements reflect the correlation structure of the observations.

In the linear case, the marginal and conditional models coincide, in the sense that in both instances the mean is a linear function of the covariates with the same coefficients $\boldsymbol{\beta}$. This is no longer true in the more general case; except for variance-component probit models, where the conditional and marginal models differ only by a scaling of the coefficients, the two approaches lead to different models. The distinction is particularly important in the case of survival models, where it can give rise to interesting paradoxes, see Vaupel and Yashin [92]. Marginal models are useful when one is interested in making inferences about population averages, whereas conditional models have a subject-specific interpretation, see Neuhaus et al. [66] for a comparison. As will be shown in our application, one can always use a conditional model to compute marginal quantities of interest, so in this sense the MGLM approach is richer, see also Goldstein [31, 32].

9.2.6 Non-Linear Models

Generalized linear models and the extensions considered so far expand the statistician's toolkit beyond the assumption of normally distributed outcomes, while retaining the assumption of linear effects on a transformed scale. Non-linear models are different; they retain the assumption of normally distributed outcomes, but move beyond the assumption of linear effects to consider more general structures where the parameters enter non-linearly. These models often have a natural physical interpretation, may be more parsimonious than linear models, and can provide more reliable predictions outside the observed range of the data. Needless to say, non-linear models have also been extended to include random effects at various levels of aggregation, see Davidian and Giltinan [24] and Pinheiro and Bates [71, Part II]. In this chapter we focus on MGLMs, but note that the two approaches share common estimation problems and have adopted similar solutions.

9.3 Approaches to Estimation

Estimation of multilevel linear models for normally distributed outcomes using maximum likelihood or restricted maximum likelihood is very well understood. Excellent implementations are available in specialized multilevel packages, namely HLM and MLwiN, as well as in general-purpose statistical packages, including Stata, SAS, and R/S-Plus. When it comes to MGLMs, however, the picture gets more complicated.

Estimation by maximum likelihood requires the marginal distribution of the responses. We assume that the random effects have density $g(\boldsymbol{\delta})$, a multivariate normal density with a patterned covariance structure. We further assume that the conditional density of the outcomes given the random effects, $f(\boldsymbol{y} \mid \boldsymbol{\delta})$, is a product of densities in the exponential family. The product of the marginal and conditional densities gives us the joint density of the outcomes and the random effects. Calculation of the marginal density of the outcome is then a "simple matter" of integrating out the random effects:

$$f(\boldsymbol{y}) = \int f(\boldsymbol{y} \mid \boldsymbol{\delta})\, g(\boldsymbol{\delta})\, \mathrm{d}\boldsymbol{\delta}. \tag{9.19}$$

Unfortunately, this integral is intractable, with no general closed-form solution.

There are some special cases of interest with a single random effect whose distribution is conjugate with the distribution of the outcome, see Lee and Nelder [49] for a general approach. For example, if $\underline{y}$ and $\underline{\delta}$ are scalars, the marginal distribution of the random effect is gamma and the conditional distribution of the outcome given the random effect is Poisson, then the marginal distribution of the outcome is negative binomial, see Lawless [48]. For binary outcomes the beta-binomial combination is popular, see Crowder [23]. A difficulty with these approaches is that they do not extend easily to models involving multiple dependent random effects. The flexibility of the assumption of multivariate normality for the random effects is, in fact, unmatched. To retain this flexibility, we need a way to get around the intractability of (9.19). We now turn to a discussion of the three main approaches to estimation in current use, starting with a simulation study used to evaluate them.

9.3.1 A Simulation Study

To assess the performance of alternative estimation procedures, we will use data from a simulation study described in Rodríguez and Goldman [81]. The study was motivated by work on health care utilization in Guatemala, where exploratory analyses had suggested large family and community effects on the use of modern health care, yet more formal analyses using multilevel models, as implemented in then current software, had failed to confirm the existence of large effects. To resolve this disparity, we ran a number of simulations, using what we then considered small and large variance components. Subsequent work revealed that the actual effects were in fact much larger than the values used in our simulations, see Pebley et al. [69] and the case study in Rodríguez and Goldman [82].

We will focus here on a set of simulations using the actual structure of Guatemalan data on prenatal care, with 2449 births to 1558 mothers who were living in 161 communities. We created three composite explanatory variables summarizing characteristics of the pregnancy, mother, and community, and set their fixed-effect coefficients to 1. We added random effects representing unobserved characteristics of the mother and community, sampled from normal distributions with mean 0 and variance 1. Finally, we simulated a binary response following a 3-level random-intercept logit model. This procedure was used to generate 100 datasets that have been used by several authors and are freely available at `http://data.princeton.edu/multilevel`.

Table 9.1 summarizes the results of trying various estimation procedures on these datasets. The results for MQL-1 and MQL-2 appeared in Rodríguez and Goldman [81]. Goldstein and Rasbash [34] reported results for PQL-2 using the first 25 of our 100 datasets; we have extended the analysis to cover all 100 and added PQL-1. The results using quadrature methods and the Gibbs sampler are new. We will comment on these results as we describe the various procedures. For brevity, we omit presentation and discussion of standard errors.

Browne and Draper [13] have also analyzed the first 25 of our datasets, and went on to generate a further 500 samples with the same multilevel structure, as part of an interesting simulation study contrasting Bayesian and likelihood-based procedures. The comparison includes MQL and PQL as well as a Bayesian approach, but excludes maximum likelihood via quadrature procedures. Their implementation of Bayesian estimation combines the Metropolis algorithm with Gibbs sampling and tries two choices of diffuse priors for the variances of the random effects. The evaluation criteria include the bias of point estimates and also the coverage rates of interval estimates. Their results parallel ours and lead to essentially the same conclusions regarding the relative merits of these methods.

Table 9.1 Estimates for simulated data using the Guatemala structure.

Estimation Method	Fixed Parameters (β) Individual	Family	Community	Random Parameters (σ) Family	Community
True Value	1.000	1.000	1.000	1.000	1.000
MQL-1	0.738	0.744	0.771	0.100	0.732
MQL-2	0.853	0.859	0.909	0.273	0.763
PQL-1	0.808	0.806	0.831	0.432	0.781
PQL-2	0.933	0.940	0.993	0.732	0.924
ML-5	0.983	0.988	1.037	0.962	0.981
ML-20	0.983	0.990	1.039	0.973	0.979
Gibbs	0.971	0.978	1.022	0.922	0.953

9.3.2 Marginal and Penalized Quasi-Likelihood

Goldstein [30] and collaborators have proposed a general approach to the estimation of MLGMs that relies on a linearization strategy, and has led to four different approximations, known as first- and second-order maximum quasi-likelihood (MQL) and penalized quasi-likelihood (PQL).

MQL-1

To motivate these approximations, we write the MLGM model as

$$\underline{\boldsymbol{y}} = \boldsymbol{\mu}(\boldsymbol{X}\boldsymbol{\beta} + \boldsymbol{Z}\underline{\boldsymbol{\delta}}) + \underline{\boldsymbol{\epsilon}}, \tag{9.20}$$

where $\underline{\boldsymbol{\epsilon}}$ is a heteroscedastic error term with mean $\boldsymbol{\emptyset}$ and variance $\boldsymbol{V}(\boldsymbol{\mu})$ depending on the mean. Goldstein [30] approximates the inverse link $\boldsymbol{\mu}(\boldsymbol{\eta})$ using a first-order Taylor series expansion around trial values $\boldsymbol{\beta} = \boldsymbol{\beta}_0$ and $\underline{\boldsymbol{\delta}} = \boldsymbol{\emptyset}$, to obtain

$$\underline{\boldsymbol{y}} = \boldsymbol{\mu}(\boldsymbol{X}\boldsymbol{\beta}_0) + \boldsymbol{D}\boldsymbol{X}(\boldsymbol{\beta} - \boldsymbol{\beta}_0) + \boldsymbol{D}\boldsymbol{Z}\underline{\boldsymbol{\delta}} + \underline{\boldsymbol{\epsilon}}, \tag{9.21}$$

where $\boldsymbol{D} = \partial\boldsymbol{\mu}/\partial\boldsymbol{\eta}_0$ is a diagonal matrix of derivatives of the mean with respect to the linear predictor evaluated at $\boldsymbol{\eta} = \boldsymbol{\eta}_0$. Pre-multiplying both sides of the equation by $\boldsymbol{D}^{-1}$ and rearranging terms gives

$$\underline{\boldsymbol{y}}^* = \boldsymbol{X}\boldsymbol{\beta} + \boldsymbol{Z}\underline{\boldsymbol{\delta}} + \underline{\boldsymbol{\epsilon}}^*, \tag{9.22}$$

where $\underline{\boldsymbol{y}}^* = \boldsymbol{D}^{-1}(\underline{\boldsymbol{y}} - \boldsymbol{\mu}_0) + \boldsymbol{X}\boldsymbol{\beta}_0$ and $\underline{\boldsymbol{\epsilon}}^*$ is an error term with mean $\boldsymbol{\emptyset}$ and variance $\boldsymbol{D}^{-1}\boldsymbol{V}(\boldsymbol{\mu})\boldsymbol{D}^{-1}$. (The variance is simpler for logit and other models where the derivative of the link $\boldsymbol{D}$ coincides with the variance function $\boldsymbol{V}(\boldsymbol{\mu})$.)

Equation (9.22) has the structure of a linear mixed model, with mean $E(\underline{\boldsymbol{y}}^*) = \boldsymbol{X}\boldsymbol{\beta}$ and variance

$$\mathrm{Var}(\underline{\boldsymbol{y}}^*) = \boldsymbol{Z}\boldsymbol{\Omega}\boldsymbol{Z}' + \boldsymbol{D}^{-1}\boldsymbol{V}(\boldsymbol{\mu}_0)\boldsymbol{D}^{-1}, \tag{9.23}$$

which has been evaluated at μ_0. Fitting this model by ML or REML leads to an improved estimate of the fixed effects $\boldsymbol{\beta}$, which can then be used to compute a new approximating model. The procedure is iterated to convergence. This method is termed maximum quasi-likelihood (MQL) because the approximating linear mixed model matches the mean and variance of the target model. Interestingly, if there are no random effects, the method coincides exactly with the IRLS algorithm used in GLMs and therefore leads to maximum likelihood estimates.

Longford [54, 56] adopted a different approach that, somewhat surprisingly, leads to an equivalent algorithm. He approximates the conditional likelihood $f(\boldsymbol{y} \mid \boldsymbol{\delta})$ using a second-order Taylor series expansion about $\boldsymbol{\delta} = \boldsymbol{\emptyset}$. The random effects appear in this expansion only in a quadratic form, which can be combined with a similar quadratic form in the marginal density $g(\boldsymbol{\delta})$ of the random effects to carry out the required integration analytically. Longford goes on to derive a Fisher scoring algorithm that provides estimates of both fixed and random effects. This strategy was first implemented in the multilevel package VARCL [55], and turns out to be exactly equivalent to Goldstein's MQL-1 procedure. For further details, see Rodríguez and Goldman [81].

Unfortunately, the results in Table 9.1 show that first-order MQL estimates can be biased, underestimating the fixed effects (β's) by 23–26% and the random parameters (σ's) by 27% at the community and 90% at the family level. For related results, see Breslow and Clayton [11] and Breslow and Lin [12].

MQL-2

Goldstein [30, p. 50] also proposed a quadratic approximation based on a second-order Taylor series expansion. Specifically, he adds the second-order terms corresponding to each of the random effects in the model, but omits second-order terms on the fixed effects as well as mixed derivatives. The resulting squared terms are treated as additional random effects in the approximating linear model. Because these are really not separate terms, their means and variances are not estimated, but rather are calculated from the variances of the original random effects under the assumption of normality. The resulting constrained model is easily fit using MLwiN. We refer to this approximation as MQL-2.

Experience suggests that this method is more accurate than MQL-1, although it doesn't always converge. Table 9.1 shows that the bias is reduced to 9–15% for the fixed parameters, and 24% and 73% for the community and family random parameters, respectively; a notable improvement, although substantial bias remains.

PQL-1

Simulations show that MQL-1 and MQL-2 work better when the random effects are small, i.e., their variances are close to zero. This fact should not be surprising considering that the approximation is based on a Taylor series expansion about $\boldsymbol{\delta} = \boldsymbol{\emptyset}$. An alternative procedure would be to expand about $\boldsymbol{\delta} = \boldsymbol{\delta}_0$ with a non-zero pivot, and the obvious candidate is the empirical Bayes estimate of the random effects, defined as $E(\underline{\boldsymbol{\delta}} \mid \boldsymbol{y})$, evaluated at current parameter values. The expansion then becomes

$$\underline{\boldsymbol{y}} = \boldsymbol{\mu}(\boldsymbol{\eta}_0) + \boldsymbol{DX}(\boldsymbol{\beta} - \boldsymbol{\beta}_0) + \boldsymbol{DZ}(\underline{\boldsymbol{\delta}} - \boldsymbol{\delta}_0) + \underline{\boldsymbol{\epsilon}}, \tag{9.24}$$

and leads to an approximating multilevel linear model with the same form as (9.22), except that the working response is now $\underline{\boldsymbol{y}}^* = \boldsymbol{D}^{-1}(\underline{\boldsymbol{y}} - \boldsymbol{\mu}_0) + \boldsymbol{X\beta}_0 + \boldsymbol{Z\delta}_0$. This model can be estimated using ML or REML, and the resulting estimates of both fixed and random effects are used to obtain a new approximating model. The procedure is then iterated to convergence.

The same procedure has been derived by other authors using different approaches. Laird [46] and Stiratelli et al. [87] derive it from a Bayesian perspective as an approximation to a posterior distribution using a diffuse prior. Schall [84] starts from a MGLM and uses a linearized form of the link function applied to the data. Breslow and Clayton [11] derive the procedure using Laplace's method for integral approximation, and term it penalized quasi-likelihood or PQL by relating it to results of Green [36].

Our experience suggests that PQL-1 tends to perform better than MQL-1, is sometimes competitive with MQL-2, and is more likely to converge. For the simulated data, the PQL-1 estimates of the fixed effects are not quite as good as MQL-2, but the estimates of the random parameters are better, although the family standard deviation is still seriously biased.

PQL-2

Goldstein and Rasbash [34] have proposed an improved version of PQL, termed PQL-2, that extends the Taylor series to include second-order terms on the random effects, but no second-order terms on the fixed effects and no mixed derivatives. The resulting squared terms are treated exactly the same way as in MQL-2, as additional random effects whose variance is not estimated but rather calculated from the other parameters.

We have found PQL-2 to be the most accurate method in this series, although sometimes it fails to converge. The results in Table 9.1 show that PQL-2 has only a 1–7% bias for the fixed parameters, and underestimates the community random parameter by 8%, although there is still a 27% bias in the estimation of the family random parameter.

Bootstrapping

One way to reduce the bias in the approximate estimation procedures is by bootstrapping, see Kuk [45] and Goldstein [33], and the detailed discussion in Chapter 11. We used MLwiN to bootstrap MQL-1 and PQL-1 estimates in a case study involving three-level random-intercept logit models [82]. We found that the procedure was successful in correcting the bias of the estimates of both fixed and random parameters. However, the technique is extremely compute-intensive (more so than the MCMC methods discussed below), taking days to converge in one of our datasets and failing after 400 replicates in another. In both cases, however, we noted that the first few iterations achieved large bias corrections, suggesting that one could run a few bootstrap iterations as a diagnostic technique. For more details, see Rodríguez and Goldman [82, Fig. 3].

9.3.3 Gauss-Hermite Quadrature

A second approach to estimation of MGLMs is to calculate the integral (9.19) representing the marginal likelihood using numerical quadrature procedures. Previous work along these lines includes Anderson and Aitkin [6] and Hedeker and Gibbons [38, 39], see also Chapter 6 in this handbook. For an excellent introduction to numerical integration methods with applications to statistics, see Thisted [90, Chapter 5].

Table 9.1 shows the results of computing maximum likelihood estimates for our simulated data using 5-point and 20-point Gauss-Hermite quadrature. We find no evidence of bias in the estimation of the fixed effects, and only about a 2% bias in the estimation of the random parameters, well within the margin of error of our simulations. We now describe the method in some detail.

Quadrature Rules

Quadrature methods approximate an integral as a weighted sum of function values evaluated over a grid of points, so that

$$\int f(x)\,\mathrm{d}x \approx \sum_q w_q f(x_q). \tag{9.25}$$

Simple methods, such as the trapezoidal rule and Simpson's rule, evaluate the integral at equally spaced points and can integrate certain polynomials exactly; in general, k points lead to exact integration of polynomials of degree $k-1$ with appropriate choice of weights.

Gaussian quadrature rules choose not only the weights, but also the evaluation points or abscissæ, and can achieve higher precision with a fixed number of points. In particular, Gauss-Hermite quadrature (so called because the evaluation points are zeroes of the Hermite polynomials) can be used with integrals of the form $\int f(x)e^{-x^2}\, dx$, and works best when $f(x)$ can be well approximated by a polynomial. The abscissæ and weights for this rule may be found in Abramowitz and Stegun [2] or may be computed using the function `gauher` in Press et al. [73].

In our applications, we need to evaluate integrals of the form $\int f(z)\,\phi(z)\, dz$, where $\phi(\cdot)$ is the standard normal density. A simple change of variables leads to the approximation $\sum w_q f(z_q)$, where w_q is the Gauss-Hermite weight divided by $\sqrt{\pi}$ and z_q is the Gauss-Hermite abscissa times $\sqrt{2}$.

Two-Level Likelihood

Consider a two-level random-intercept model with n_j observations in cluster j. Let $\underline{\delta}_j \sim \mathcal{N}(0, \sigma^2)$ denote the cluster effect. We assume that given δ_j the n_j observations are independent and have a distribution in the exponential family $f(y_{ij} \mid \delta_j)$. We further assume that the conditional mean $E(\underline{y}_{ij} \mid \delta_j) = \mu_{ij}$ satisfies a generalized linear model with $g(\mu_{ij}) = \boldsymbol{x}'_{ij}\boldsymbol{\beta} + \delta_j$. We write $\underline{\delta}_j = \sigma \underline{z}_j$, so we only need to consider standard normal random effects.

Let $L_j(z_j) = \prod_i f(y_{ij} \mid z_j)$ denote the conditional likelihood for cluster j given the random effect. We can evaluate the marginal likelihood for the cluster using Q-point Gauss-Hermite quadrature as a simple weighted average

$$L_j = \sum_{q=1}^{Q} w_q L_{jq}, \tag{9.26}$$

where we have written L_{jq} as shorthand for $L_j(z_q)$, the likelihood for cluster j evaluated at the q-th quadrature point.

Two-Level Score

First and second derivatives of the likelihood can also be evaluated as weighted averages, but we usually work with the log-likelihood instead. Let $\boldsymbol{\theta}$ denote the model parameters, including $\boldsymbol{\beta}$ and σ (or better still $\log \sigma$, which avoids range restrictions and is usually better behaved).

Let $\boldsymbol{u}_j = \partial \log L_j / \partial \boldsymbol{\theta}$ denote the score vector for cluster j. Simple calculus shows that

$$\boldsymbol{u}_j = \sum_{q=1}^{Q} w^*_{jq} \boldsymbol{u}_{jq}, \tag{9.27}$$

where $\boldsymbol{u}_{jq}$ is the score corresponding to the log-likelihood for cluster j evaluated at the q-th quadrature point, and $w^*_{jq} = w_q L_{jq} / L_j$.

The new weight w_{jq}^* has an interesting interpretation. One can view the approximate likelihood (9.26) as a mixture model where cluster j comes from one of Q discrete classes with random effects z_q and prior probabilities w_q. The new weight w_{jq}^* is the posterior probability that the cluster came from class q given the data $\boldsymbol{y}_j$. Thus, the quadrature score is the posterior average of the scores evaluated at the quadrature points.

Two-Level Hessian

Let $\boldsymbol{H}_j = \partial^2 \log L_j / \partial\boldsymbol{\theta}\, \partial\boldsymbol{\theta}'$ denote the Hessian or matrix of second derivatives of the log-likelihood for cluster j. It can be shown that this matrix satisfies

$$\boldsymbol{H}_j = \sum_{q=1}^{Q} w_{jq}^* \boldsymbol{H}_{jq} + \sum_{q=1}^{Q} w_{jq}^* (\boldsymbol{u}_{jq} - \boldsymbol{u}_j)(\boldsymbol{u}_{jq} - \boldsymbol{u}_j)', \tag{9.28}$$

where $\boldsymbol{H}_{jq}$ is the Hessian for cluster j evaluated at the q-th quadrature point. Thus, the Hessian is the posterior average of the Hessians evaluated at the quadrature points plus the variance of the scores evaluated at the quadrature points.

This equation is formally identical to a well-known result for maximum likelihood estimation using the EM algorithm, which views the random effects $\underline{\delta}_j$ as missing data, and shows that the incomplete data information equals the expected complete data information minus the variance of the scores, which represents the missing information; see Louis [57].

Adaptive Quadrature

Liu and Pierce [53] proposed an extension of Gauss-Hermite quadrature where the variable of integration is transformed so the integrand is sampled in a more appropriate region. The starting point is the observation that the integrand in (9.19) is the product of the prior density of the random effect and the density of the data given the random effect, and is therefore proportional to the posterior distribution of the random effect. This, in turn, can be approximated using a Gaussian density. To fix ideas, consider a two-level variance-components model where the random effect has a $\mathcal{N}(0, \sigma^2)$ prior and write the contribution of a cluster to the likelihood as

$$\int f(\boldsymbol{y} \mid \delta)\, \phi(\delta; 0, \sigma^2)\, \mathrm{d}\delta = \int \left\{ \frac{f(\boldsymbol{y} \mid \delta)\phi(\delta; 0, \sigma^2)}{\phi(\delta; \mu, \gamma^2)} \right\} \phi(\delta; \mu, \gamma^2)\, \mathrm{d}\delta, \tag{9.29}$$

where $\phi(\delta; \mu, \gamma^2)$ denotes the normal density with mean μ and variance γ^2.

Liu and Pierce [53] choose μ and γ^2 to match the mode and the curvature at the mode of the posterior density. The integral on the right-hand side is then evaluated using Gaussian quadrature, following a change of variables from δ to $(\delta - \mu)/\gamma$. This has the effect of sampling the integrand in a more relevant range, and improves accuracy as long as the ratio in braces is better approximated by a low-order polynomial than the likelihood. The method with a single node is equivalent to the Laplace (or PQL-1) approximation to the integral, so this approach may be viewed as an extension of Laplace approximation.

Pinheiro and Bates [70] derived this algorithm, which they termed *adaptive* Gaussian quadrature, from an interesting perspective. They viewed Gaussian quadrature as a deterministic version of Monte Carlo integration and proposed adaptive quadrature as a deterministic version of importance sampling, which tends to be much more efficient than simple Monte Carlo integration, using a Gaussian density with the same mode and curvature as the posterior density as the importance distribution.

Rabe-Hesketh et al. [74] proposed a slightly different approach that simplifies the calculations required to place the nodes; instead of matching the mode and curvature, they use the posterior mean and variance of the random effects, which are calculated by building on work of Naylor and Smith [64]. Their approach, embodied in the `gllamm` command in Stata, was the first implementation of adaptive quadrature for multilevel models, and has now replaced Gauss-Hermite quadrature in other Stata procedures, including the official commands for random effects logit, probit, and Poisson models. Another implementation of adaptive methods may be found in R's `lme4`.

Although adaptive quadrature requires additional computational effort to place the abscissæ, it usually pays off by requiring many fewer quadrature points. In our original analysis of Guatemalan data reported in Pebley et al. [69], we used Gauss-Hermite quadrature with 20 nodes at each level, so each likelihood evaluation required going over a 400-point grid. Recently, we were able to replicate the results exactly using `gllamm` with the default 6 points per level. The `gllamm` code is slow because it is interpreted, but speed has improved as critical parts of the algorithm have been converted to internal code in Stata. For further details, see Rabe-Hesketh et al. [74].

Extension to More Dimensions

So far we have discussed a two-level model with a single random effect, but the quadrature approach can be extended to higher-dimensional models. Consider first a three-level random-intercept model. Because there is only one random effect at each level, the model can be estimated by recursive application of the method described so far. Specifically, the likelihood for a level-3 unit is computed as a weighted sum of level-3 likelihoods evaluated at the quadrature points. These are products of level-2 likelihoods, each computed using (9.26).

Consider next a two-level random-slope model where we have two random coefficients, say $\underline{\alpha}_j = \alpha + \underline{\delta}_{1j}$ and $\underline{\beta}_j = \beta + \underline{\delta}_{2j}$. Fitting this model requires evaluating a bivariate normal integral, but we can always transform to independence; in the simplest case by using the marginal distribution of $\underline{\delta}_{1j}$ and the conditional distribution of $\underline{\delta}_{2j} \mid \delta_{1j}$, which can then be standardized. Extension to higher-dimensional models follows along similar lines using a Cholesky decomposition.

Optimization Algorithms

The foregoing results can be used in a Newton-Raphson algorithm for maximizing the log-likelihood function. Our experience using the built-in function minimizers in S-Plus and R, as well as code in Press et al. [73], suggests that the extra expense of computing second derivatives is not always worthwhile. Instead, we provide first derivatives only, letting the algorithms compute numerical second derivatives, or use variable-metric methods such as DFP or BFGS that build an approximation to the Hessian in the course of iteration. However, we do use analytic results to evaluate the Hessian after convergence, in order to obtain more accurate standard errors.

The first statistical package to incorporate quadrature methods was Egret [19]. The latest version of Stata can fit two-level random-intercept logit and probit models using adaptive quadrature, and has a nice provision for checking the procedure by comparing results with different numbers of points. A more general implementation of quadrature methods may be found in the package aML [51], which can handle, at least in principle, several levels and multiple random effects.

The computational burden of Gauss-Hermite quadrature increases rapidly with the dimensionality of the problem. For an m-dimensional model using Q quadrature points for each random effect, each evaluation of the likelihood function is equivalent to Q^m evaluations of a GLM likelihood. Using 12 quadrature points, which seems a reasonable standard for general use, one can easily fit three-level random-intercept models and two-level models with two random coefficients, say an intercept and a slope, with each likelihood evaluation the equivalent of 144 GLM likelihoods. But using 12-point quadrature to evaluate the likelihood of a three-level model with two random coefficients at each level is equivalent to evaluating almost 21,000 GLM likelihoods. Obviously, the technique works best for relatively low-dimensional models.

9.3.4 Bayesian Estimation Using the Gibbs Sampler

Recent advances in Bayesian estimation avoid the need for numerical integration by taking repeated samples from the posterior distribution of the parameters of interest. In particular, use of the Gibbs sampler in the context of MGLMs was first proposed by Zeger and Karim [95], and has been discussed in greater detail by Clayton [16]. See also Chapter 2 in this Handbook and the Browne and Draper [13] evaluation cited earlier.

Gibbs Sampling

To apply this framework, we adopt a Bayesian perspective, treating all parameters as random variables and assigning prior (or hyperprior) distributions to the fixed-effect parameters $\underline{\boldsymbol{\beta}}$ and to the precisions $\underline{\boldsymbol{\tau}}$ (the reciprocals of the variances) of the random effects. To obtain Bayesian estimates that are roughly comparable to maximum likelihood estimates, many analysts use vague or non-informative priors. Fixed effects are typically assumed to come from normal distributions with mean zero and very large variances, and precisions are sampled from diffuse gamma or Pareto distributions, see Spiegelhalter et al. [86].

A popular method for sampling from the posterior distribution of the parameters given the data is the Gibbs sampler, a Markov chain Monte Carlo (MCMC) method that focuses on the so-called full conditional distributions of each parameter given all others, turning a complex multivariate problem into a series of simpler univariate ones. This approach has been combined with a general method for drawing samples from any log-concave distribution, called adaptive rejection sampling [29]. The combination is available in the software package BUGS [86]. Convergence diagnostics can be calculated using a set of R or S-Plus functions, see Best et al. [9].

Results for Simulated Data

We tried the Gibbs sampler on our simulated Guatemalan data. We used non-informative priors, treating all four fixed-effect parameters as i.i.d. normal variates with mean 0 and precision 10^{-6}. For the two random-effect parameters representing the precision of the family and community random effects, we used a $\Gamma(\epsilon, \epsilon)$ distribution with $\epsilon = 0.001$, so the mean is 1 and the variance is 1000. We then ran a naive Gibbs sampler with a burn-in of 200 iterations followed by a further 1000 iterations. We are very grateful to David Clayton for sharing with us a set of C functions for MCMC estimation of generalized linear mixed models and for adapting his driver program to handle our simulated data. These routines have now been incorporated in the R package `GLMMGibbs`, see Myles and Clayton [63].

The results in Table 9.1 are very encouraging, showing practically no bias in the estimation of the fixed effects, about a 5% bias in the estimation of the community effect, and an 8% bias for the family effect. We did some further work exploring the nature of the remaining bias and discovered that we could essentially eliminate it by either (1) using informative priors for the precisions of the random effects, or (2) using a much larger sample size, simulated by combining our original samples in groups of five. For additional simulation results, see Browne and Draper [13].

Experience with Real Data

Our experience applying MCMC methods to real data has been somewhat mixed. In a case study fitting a three-level random-intercept logit model to data on immunization from Guatemala, we found slow mixing and poor convergence, particularly for parameters representing the variances of the random effects. Deciding whether a run is adequate often requires a battery of diagnostic procedures; we have used tests due to Geweke [28] and Roberts [78], and have found very useful the `gibbsit` software of Raftery and Lewis [75, 76], which provides an estimate of the number of iterations required to estimate credible limits for each parameter with given probability of attaining a desired precision.

Fitting a similar model for prenatal care data characterized by heavier clustering, particularly at the family level, proved substantially more difficult, with estimates of the efficiency of our chains as low as 1%. Rather than running much longer chains, we heeded the advice of Gelman and Rubin [27] and ran multiple chains with different starting values. The S-Plus function `itsim` was very useful in checking the output from multiple chains before pooling them to produce final estimates. In the end, the MCMC approach required extensive computation and judging convergence proved something of an arcane art form. For more details, see Rodríguez and Goldman [82].

9.3.5 Other Approaches to Estimation

High-Order Laplace

Breslow and Lin [12] proposed a fourth-order Laplace approximation for two-level models with a single random effect per cluster, and Lin and Breslow [52] extended the result to multiple independent random effects per cluster. More recently, Raudenbush et al. [77] further extended this approach to high-order approximations for multiple dependent random effects. They report that the method is remarkably accurate and computationally fast, and validate it by comparison to Gauss-Hermite quadrature with up to 40 points, using real and simulated data. This promising strategy was first implemented for two-level models in version 5 of HLM, but has now been extended to three-level models in version 6.

Simulated Maximum Likelihood

Monte Carlo integration is not restricted to Bayesian models, but can also be used for simulating the likelihood; see Lerman and Manski [50] for an early application. Closely related approaches are the method of simulated moments (MSM) introduced by McFadden [61], and the method of simulated scores (MSS), see Keane [44]. These methods are often used by applied economists estimating complex structural models. A useful survey may be found in Gouriéroux and Monfort [35].

In the context of generalized linear mixed models, McCulloch [60] developed Monte Carlo variants of the Expectation-Maximization (EM) and Newton-Raphson algorithms, as well as simulated maximum likelihood (SML). Booth et al. [10] compare several stochastic alternatives to numerical integration, including simulated maximum likelihood using importance sampling. These methods are particularly appropriate for high-dimensional models where quadrature succumbs to the curse of dimensionality.

Recently, Ng et al. [67] evaluated several simulation-based approaches for maximum likelihood estimation in multilevel models with binary outcomes, including bias correction using Kuk's bootstrap (described earlier) and the Robbins-Monro stochastic approximation method, and estimation using simulated maximum likelihood (SML). They conclude that SML performs comparably with the other methods, but has the advantage of yielding variance estimates—which can be used to construct Wald tests and confidence regions—as well as the value of the likelihood at the maximum, which is useful for constructing likelihood ratio tests to compare nested models. They note that SML requires good starting values, confirming results in [60], but is otherwise less prone to computational problems than the other algorithms, and gives results similar to numerical integration.

9.4 Infant and Child Mortality in Kenya

Our illustration of MGLMs uses data from the 1998 Kenya Demographic and Health Survey (KDHS) to study infant and child mortality.

9.4.1 The Kenya Survey

The 1998 Kenya Demographic and Health Survey (KDHS) is a national survey conducted by the National Council for Population and Development (NCPD) in collaboration with the Central Bureau of Statistics (CBS) and Macro International, which provided technical assistance. The survey is national in scope but excluded seven districts accounting for less than 4% of the population. The sample was selected using a two-stage stratified design and relied on a sampling frame maintained by the CBS. Field work was conducted between February and July 1998, and achieved an overall response rate of 96.8% of households and 95.7% of women aged 15–49 who were eligible for an individual interview. The interview included a retrospective maternity history that collects data on date of birth, survival status, and age at death for all children each woman has given birth to.

We selected for analysis all births in the 10 years preceding the interview, but excluded 170 pairs of twins and one set of triplets, which have much higher mortality risks than singletons. The final sample consists of 10,878 births to 4,939 women who live in 530 communities, defined in terms of the ultimate area units used in the sample design. One objective of our analysis is to determine the extent to which infant and child deaths are clustered within families and within communities.

We must note at the outset a limitation of the data: The community is defined in terms of the respondent's residence at the time of the survey, but our analysis uses retrospective mortality data over a 10-year period. While this is far from ideal, we claim three extenuating circumstances. First, a large fraction of respondents have always lived in the place where they were interviewed, and 80.9% of all births in our sample were born while the mother resided in her current community. Second, migration would tend to attenuate the influence of the community, so our estimates can be considered lower bounds on the true effects. Third, as a sensitivity test we repeated our analysis using only births in the last five years, and discovered that our estimates were remarkably resilient to the choice of reference period.

9.4.2 A Three-Level Hazard Model

Let $\underline{\lambda}_{ijk}(t)$ denote the risk of dying at age t for the i-th child of the j-th mother in the k-th community. We assume that the hazard depends on age t, a set of observed child, family, and community covariates $\boldsymbol{x}_{ijk}$, and unobserved family and community random effects $\underline{\delta}_{jk}$ and $\underline{\delta}_k$ via a conditional proportional hazards model:

$$\underline{\lambda}_{ijk}(t) = \lambda_0(t)\exp\{\boldsymbol{x}'_{ijk}\boldsymbol{\beta} + \underline{\delta}_{jk} + \underline{\delta}_k\}, \qquad (9.30)$$

where $\lambda_0(t)$ is a baseline hazard, $\boldsymbol{\beta}$ is a vector of fixed parameters representing the effects of observed covariates, and the unobserved family and community effects are normally distributed, $\underline{\delta}_{jk} \sim \mathcal{N}(0, \sigma_2^2)$ and $\underline{\delta}_k \sim \mathcal{N}(0, \sigma_3^2)$.

Choice of Duration Categories

We assume that the baseline hazard is constant in intervals defined by cutpoints $0 = \tau_0 < \tau_1 < \cdots < \tau_D$, so that $\lambda_0(t) = \lambda_{0d}$ if $t \in [\tau_{d-1}, \tau_d)$. The choice of cutpoints is dictated by the shape of the hazard and constraints in data collection.

The KDHS recorded age at death in days, months, or years. Days are used for neonatal deaths (occurring in the first month of life), months are used mostly for infant deaths (occurring before age 1), and years are used predominantly for deaths at ages 2 or higher. We first tabulated events and exposure by single months up to age 1 and by single years thereafter. In calculating exposure for deaths at ages 2 and higher, we treated deaths as occurring at the midpoint of an interval constrained by the reported age at death in years and the date of interview. No such approximation is required for deaths at earlier ages or for survivors.

Following some exploratory work, we decided to use separate exposure categories for the first month of life, and then for ages 1–5, 6–11, 12–23, and 24–59 completed months, with more detail at ages where the hazard is changing rapidly. These five categories capture more than 90% of the variation in the hazard by duration (as measured by the deviance in a marginal Poisson model), and yield 48,094 pseudo-observations. For some preliminary analyses, we used only three categories: the first month of life, the rest of the first year, and older ages, which reduced the number of pseudo-observations to 30,456 and yielded very similar results.

Selection of Explanatory Variables

Our selection of variables has been guided by previous work in the field; see Mosley and Chen [62] for a conceptual framework. We included only one community-level variable, type of place of residence, classified as urban or rural. Residence is coded at the time of the survey, so the same caveat we discussed for community effects applies here.

Our only family-level variable is mother's education, which can be coded in terms of completed years or using dummy variables to mark achievements such as completing primary or secondary school. Our exploratory analysis indicated that the most efficient way to capture the educational effect was to use linear and quadratic terms. We found that mortality increased as one moved up from no education to complete primary, and decreased only when one went past secondary education, but this tendency became less noticeable after controlling for mother's age, which plays the role of a confounding factor: The children of very young mothers have higher mortality risks, but young women also tend to be more educated than older women, a fact that actually lowers their children's risk.

All remaining variables are defined at the individual level. Males are known to have higher mortality than females, so we included a dummy variable for sex. First- and high-order births are also at increased risk. We considered using dummy variables for first births and for births of order six and higher, but noticed that linear and quadratic terms did a better job of capturing what appeared to be a gradual increase in risk with birth order.

An important determinant of mortality is length of the preceding birth interval, which of course is defined only for births of order two or higher. Children born shortly after a previous birth are known to have much higher risks, either because of maternal depletion or because they have to compete with older siblings for scarce resources. To capture this effect, we used a linear spline defined as $30-i$ (where i is interval length) for intervals shorter than 30 months and 0 for first births and for longer intervals. The linear spline proved significantly better than a simple dummy for short intervals.

The final individual variable in our model is age of the mother at the time of birth of the child, which is known to have a U-shaped relationship with mortality, with higher hazards for the youngest and oldest mothers. We tried dummy variables for mothers aged < 20 and 40+ at the time of birth of the child, but discovered that linear and quadratic terms on age at birth did a better job.

As part of our exploratory work, we allowed all of these variables to interact with child's age. We found no evidence of non-proportional effects except possibly for mother's education, which appeared to have a larger effect beyond the first month of life. However, the reduction in deviance did not justify the additional number of parameters required, as judged by Akaike's information criterion, so we retained the simpler proportional hazards model.

Estimation Results

Table 9.2 shows the results of fitting our final model by first-order MQL, first-order PQL, and maximum likelihood via 12-point Gauss-Hermite quadrature (ML). We also include for comparison results from a marginal Poisson model that ignores clustering at the family and community levels. Unlike some of the results we have obtained for heavily clustered binary data, in this application all three methods yield similar estimates of the fixed effects. In fact, the results are very similar to the marginal model as well, except possibly for cohort and birth order. However, the marginal model underestimates standard errors by an average of 8%, and does a poor job estimating the precision of the urban effect. The estimates of the random parameters, reported here in terms of the standard deviation of the family and community effects, are unusual in that MQL and PQL lead to slightly larger values than Gauss-Hermite quadrature.

First-order MQL converged quickly and uneventfully. First-order PQL alternated between two sets of estimates of the random parameters, one of which had the family variance component set to zero. The other, reported in Table 9.2, yielded results similar to MQL. We tried second-order MQL and PQL, but both failed repeatedly from a variety of starting points. We also tried these procedures with the smaller sample of 30,456 pseudo-observations

Table 9.2 Parameter estimates for the multilevel model of infant and child survival in Kenya.

Variable	Term	GLM	MQL-1	PQL-1	ML
		Fixed Effects			
Constant	1	−4.189	−4.163	−4.164	−4.588
		(0.095)	(0.105)	(0.106)	(0.118)
Age	1–5	−1.669	−1.646	−1.647	−1.642
(months)		(0.089)	(0.090)	(0.090)	(0.089)
	6–11	−2.062	−2.005	−2.007	−1.998
		(0.096)	(0.097)	(0.097)	(0.097)
	12–23	−2.912	−2.830	−2.834	−2.822
		(0.105)	(0.104)	(0.105)	(0.106)
	24–59	−3.748	−3.641	−3.646	−3.632
		(0.108)	(0.106)	(0.108)	(0.109)
Sex	male	0.080	0.087	0.087	0.087
		(0.065)	(0.067)	(0.067)	(0.068)
Cohort	1993+	0.195	0.173	0.173	0.173
		(0.066)	(0.068)	(0.069)	(0.069)
Mother's	$a-25$	−0.060	−0.048	−0.048	−0.047
Age		(0.010)	(0.011)	(0.011)	(0.011)
	$(a-25)^2$	0.003	0.003	0.003	0.003
		(0.001)	(0.001)	(0.001)	(0.001)
Birth	$o-3$	0.079	0.046	0.047	0.043
Order		(0.035)	(0.038)	(0.038)	(0.039)
	$(o-3)^2$	0.005	0.004	0.004	0.004
		(0.004)	(0.005)	(0.005)	(0.005)
Birth	$(30-i)_+$	0.039	0.036	0.036	0.036
Interval		(0.006)	(0.006)	(0.006)	(0.006)
Mother's	$e-7$	−0.074	−0.066	−0.066	−0.068
Education		(0.014)	(0.015)	(0.015)	(0.015)
	$(e-7)^2$	−0.008	−0.007	−0.007	−0.007
		(0.002)	(0.003)	(0.003)	(0.003)
Residence	urban	0.022	−0.001	0.001	0.040
		(0.102)	(0.144)	(0.144)	(0.142)
		Random Effects			
Family	σ_2	–	0.732	0.696	0.613
	$\log\sigma_2$	–	−0.312	−0.363	−0.489
		–	(0.102)	(0.096)	(0.140)
Community	σ_3	–	0.747	0.745	0.680
	$\log\sigma_3$	–	−0.291	−0.294	−0.386
		–	(0.068)	(0.058)	(0.081)
Log-likelihood		−5688.86	–	–	−5602.12

Standard errors shown in parentheses.

using only three duration categories and obtained similar results. We believe that further exploration of the properties of MQL and PQL for Poisson data with moderate and large amounts of clustering would be useful. The ML estimates converged quickly. We verified our calculations for two-level models that included only the family or community effect by running Stata's `xtpois` procedure, which uses adaptive Gaussian quadrature for normal random effects, obtaining practically identical results.

Testing Random Parameters

A final technical point before we turn to the interpretation of the results concerns testing for family and community effects. In Table 9.2, we report standard errors for $\log \sigma$ rather than σ because normal approximations tend to work better in the unconstrained scale. One must be careful not to divide the estimate by its standard error, as this would test the hypothesis $H_0 : \sigma = 1$ rather than $H_0 : \sigma = 0$. Instead, we build 95% confidence intervals in the log scale and exponentiate to obtain intervals for σ. In our example, the confidence intervals are $(0.467, 0.807)$ for the family and $(0.580, 0.797)$ for the community σ, indicating large effects. Note that by construction these intervals cannot include zero, so they should not be used as formal tests.

Likelihood ratio tests are preferable, but are not without difficulties. Because the null hypothesis $H_0 : \sigma = 0$ is on the boundary of the parameter space, the likelihood ratio test does not have the usual large sample chi-squared distribution with degrees of freedom equal to the number of parameters set to zero, see Self and Liang [85] and Stram and Lee [88]. These authors suggest treating the test for $H_0 : \sigma_2 = \sigma_3 = 0$ as a 50-50 mixture of χ^2_1 and χ^2_2 rather than the nominal χ^2_2. Similarly, a test of $H_0 : \sigma_2 = 0$ or $H_0 : \sigma_3 = 0$ would be treated as an equal mixture of zero and a χ^2_1. Pinheiro and Bates [71] simulate likelihood ratio statistics in the context of linear mixed models and note that these adjustments are not always successful. A simpler approach is to use the nominal degrees of freedom, understanding that the test would then be conservative. In our application, twice the difference in log-likelihoods between the marginal and conditional models is 173.5, and the effect is highly significant no matter how we treat the test criterion.

9.4.3 Fixed Effects Estimates

The first thing to note in Table 9.2 is the remarkable decline in risk with age. Exponentiating the coefficients for durations 1–5 and 12–23 we see that by ages one to five completed months, the risk is 81% lower—and by age one completed year, it is 95% lower—than in the first month of life. Males in this sample have a 9% higher risk than females with the same characteristics, but this difference is not significant.

Children born in the period since January 1993, however, have 19% *higher* risk at any given age than children born in 1992 or earlier. We examined this result closely for possible artifacts, including sensitivity to the choice of duration categories, and found it to be robust. We also looked at survival to age 1 using logit models to compare births in the periods 1–4 and 5–9 years before the survey, with similar results. It seems clear that infant and child mortality increased in Kenya in the late 1990s, an unfortunate development that is probably related to the AIDS pandemic.

Mother's age at the time of birth of the child has a significant effect on survival. The left panel in Fig. 9.1 shows the expected U-shaped relationship. The risk reaches its minimum around age 32, at which point a 10-year difference in either direction increases the risk by as much as 40%, everything else being equal. Birth order, on the other hand, has no significant effect on survival, with sample estimates suggesting, if anything, a linear increase in risk with parity. The excess risk often observed for first-order births appears to have been captured by mother's age.

Short birth intervals have a strong negative effect on infant and child survival, as expected. The hazard increases 4% for each month that the interval falls short of 30, the arbitrary cutoff point in our linear spline. This translates into a 24% excess risk for children born two years after a sibling, compared to children born after an interval of two and a half years.

Mother's education, which ranges from 0 to 19 years with quartiles at 4, 7, and 8, has a large effect on infant and child mortality. The right panel in Fig. 9.1 shows the overall relationship: We see little if any effect of just a few years of primary education, but a large (and increasing) effect after that. Around the median, each year of education is associated with a 7% decline in risk.

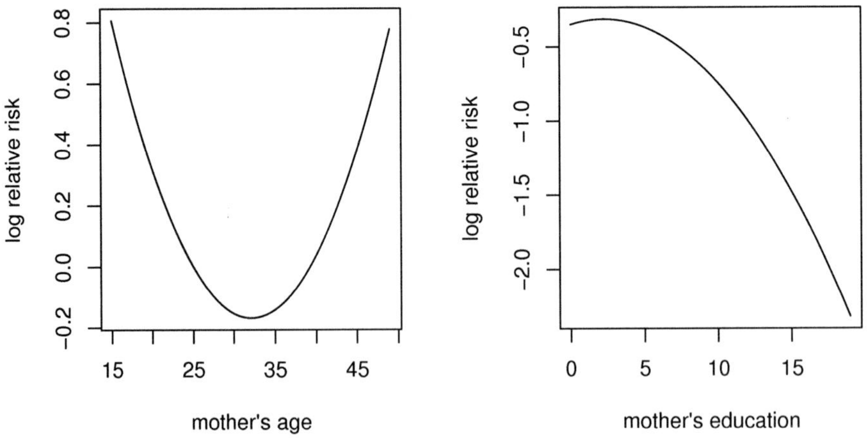

Fig. 9.1 The effects of mother's age and education on the log relative risk.

Finally, we find no significant effect of residence on child survival. Interestingly, urban residents in our sample have a 4% higher risk than their rural counterparts. We speculate that the erosion of the traditional differential that favors urban residence may be associated with higher prevalence of AIDS in the cities.

9.4.4 The Random Parameters

The most remarkable feature of our results concerns the large amount of clustering observed at both the family and community levels. In an analysis of family effects on infant and child mortality in Guatemala, Guo and Rodríguez [37] find much smaller family effects, and note that their results are in line with previous work in the area; see also Sastry [83].

Consider first the family random effect, which is estimated to have a standard deviation of 0.61. Because this effect is in the scale of the linear predictor, it can be interpreted exactly the same way as a fixed coefficient pertaining to an observed covariate. In particular, the children of a mother who is one standard deviation above the mean in a latent distribution of family frailty have 85% higher risk than the children of an average mother. In contrast, the children of a mother who is one standard deviation below the mean enjoy 45% lower risk than the children of the average mother. In both cases, the comparison is with children with identical observed characteristics who live in the same community.

The community random effect is, surprisingly, even larger, with a standard deviation of 0.68. Children who live in a community whose frailty is one standard deviation above the mean have almost double the risk—while those who live in communities one standard deviation below the mean have about half the risk—compared to children with the same observed characteristics who live in an average community. From a public health point of view, it would be interesting to identify communities with large estimated random effects, in search for an explanation of these findings.

One way to put these results in perspective is to look at the effect of observed characteristics other than age of the child. We computed the observed log relative risk, defined as the linear predictor omitting the constant, the dummy variables representing duration, and both random effects. The way we coded our covariates, this risk is zero for a third child, female, born before 1993, born at least two and a half years after the second birth, whose mother was 25 at the time of birth, had completed seven years of education, and lived in a rural area. For a similar male born after 1993 in a city, the log relative risk is 0.30. In our sample, log relative risks range from -2.04 to 2.16; selected percentiles are shown in Table 9.3.

Exponentiating these numbers, we find that children in the third quartile of relative risk have 61% higher risk than those in the first quartile. In contrast,

Table 9.3 Selected percentiles of log relative risk.

P	1	5	10	25	50	75	90	95	99
lrr	−0.78	−0.30	−0.13	0.15	0.38	0.63	0.87	1.02	1.31

the inter-quartile ranges in unobserved family and community characteristics translate into 2.3-fold and 2.5-fold increases in risk, respectively. Similarly, the range from the first to the 99th percentile in observed risk factors translates into an 8-fold increase in risk, whereas the equivalent ranges in the normal distributions representing family and community effects translate into 17-fold and 24-fold increases in risk, respectively. By this account, substantial relative risks associated with family and community frailty remain unobserved.

9.4.5 Survival Probabilities

We now translate our results into conditional and marginal probabilities of surviving to (or dying by) selected ages. This calculation can be done for selected values of the covariates, and helps present the results of hazard models in a less technical language.

Table 9.4 shows the conditional probabilities of infant and child death for our reference category and for children at the first and third quartile of observed risk factors and unobserved family and community effects. The underlying survival probabilities are all estimated as

$$S(t \mid \boldsymbol{x}_{ijk}, \delta_{jk}, \delta_k) = \exp\{-\Lambda_0(t)\exp\{\boldsymbol{x}'_{ijk}\hat{\beta} + \delta_{jk} + \delta_k\}\}, \qquad (9.31)$$

with the log relative risk $\boldsymbol{x}'_{ijk}\hat{\boldsymbol{\beta}}$ set to the observed quartiles 0.15 and 0.63, and the unobserved frailties set to the normal quartiles $\pm 0.67\hat{\sigma}_2$ and $\pm 0.67\hat{\sigma}_3$.

Table 9.4 Estimated infant and child mortality at first and third quartiles of observed and unobserved risk.

Risk Factor			Mortality	
Observed	Family	Community	Infant	Child
Q1	Q1	Q1	0.014	0.022
		Q3	0.034	0.053
	Q3	Q1	0.031	0.049
		Q3	0.075	0.118
Q3	Q1	Q1	0.022	0.035
		Q3	0.054	0.085
	Q3	Q1	0.049	0.078
		Q3	0.119	0.183
Baseline			0.028	0.044

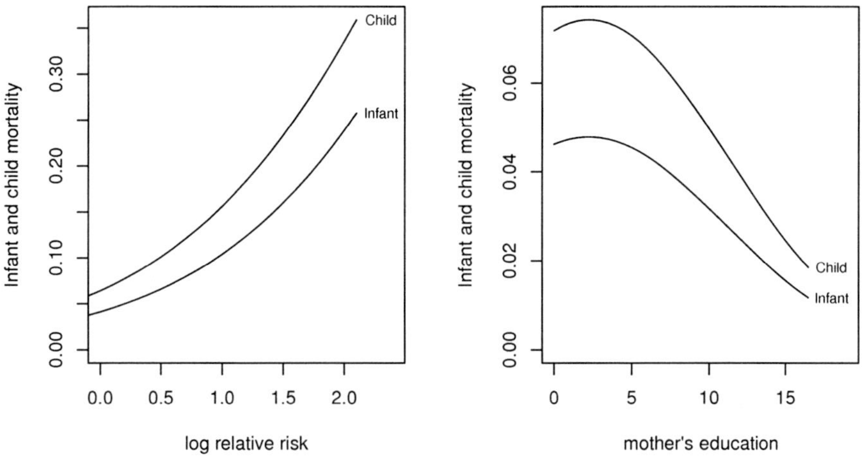

Fig. 9.2 Marginal probabilities of infant and child death by log relative risk and by mother's education.

As we move up the quartiles of observed and unobserved risk factors, the probability of an infant death increases from 14 to 119 per thousand, and the probability of a child death increases from 22 to 183 per thousand.

Figure 9.2 shows the marginal probabilities of infant and child death as a function of the log relative risk that combines all observed predictors (left panel), and as a function of mother's education with all other variables set to their reference values (right panel). The corresponding survival probabilities are estimated by evaluating the double integral

$$S(t \mid \boldsymbol{x}_{ijk}) = \int \int S(t \mid \boldsymbol{x}_{ijk}, \delta_{jk}, \delta_k) \, \mathrm{d}\delta_{jk} \, \mathrm{d}\delta_k \tag{9.32}$$

using 12-point Gauss-Hermite quadrature with conditional probabilities estimated using (9.31).

The marginal probability of infant death varies from 6 to 258 per thousand as a function of observed risk factors. The equivalent range for mortality up to age five is 9 to 359 per thousand. The effect of mother's education is fairly substantial. The probability that a child in our reference category will die before age one ranges from 47 per thousand if the mother has only a few years of education to 25 per thousand for high school graduates (and even less for the few women with higher education), after averaging over unobserved family and community attributes. Similarly, the probability of dying before age five declines from 75 to 39 per thousand, on average, as mother's education increases through upper primary and high school.

9.4.6 Intraclass Correlations

The variance parameters in random intercept models are closely related to measures of intraclass correlation. In a two-level linear model, the Pearson correlation between any two observations in the same cluster is $\rho = \sigma_2^2/(\sigma_2^2+\sigma_1^2)$. In a two-level logit model, the correlation is usually calculated by reference to the *latent* variable formulation of the model, setting $\sigma_1^2 = \pi^2/3$, the variance of the underlying standard logistic error, see Chapter 6. Rodríguez and Elo [80] show that the correlation of observed or *manifest* binary outcomes in two-level models can be quite different, and provide a Stata command `xtrho` that can be used to compute marginal and joint probabilities, and hence measures of correlation such as Person's r or Yule's Q, by numerical integration. Their ideas are easily extended to three-level survival, as shown below.

In the context of survival models, Oakes [68] has shown that the variance in a two-level model where frailty has a gamma distribution is closely related to Kendall's τ, a coefficient of ordinal association. No similar results have been obtained in general, but having fitted a multilevel survival model we can estimate any measure of association as a function of the estimated joint and marginal distributions. Because we followed children up to age 5 only, we are not in a position to estimate the correlation of lifetimes, but we can estimate correlation in survival up to ages one and five.

We calculate three marginal probabilities that are useful in constructing measures of intraclass correlation. First, we need the probability that a child with covariates $\boldsymbol{x}$ will live to age t, which is given by (9.32). Second, we need the probability that two children of the same mother both survive to age t. Because the survival experiences of these two children are independent given the family and community random effects, we can calculate the bivariate survival probability as

$$\begin{aligned} S_2(t, t \mid \boldsymbol{x}_{ijk}, \boldsymbol{x}_{i'jk}) \\ = \int\int S(t \mid \boldsymbol{x}_{ijk}, \delta_{jk}, \delta_k)\, S(t \mid \boldsymbol{x}_{i'jk}, \delta_{jk}, \delta_k)\, \mathrm{d}\delta_{jk}\, \mathrm{d}\delta_k, \quad (9.33) \end{aligned}$$

where the double integral is evaluated by Gauss-Hermite quadrature. We usually set $\boldsymbol{x}_{ijk} = \boldsymbol{x}_{i'jk}$, although only variables at levels 2 and 3 would need to be the same. Third, we need the probability that two children of different mothers who live in the same community will both survive to age t. Given the community random effect δ_k the survival experiences of these two children are independent, and the probability of surviving to age t can be calculated for each one by integrating out the corresponding family effect. The probability in question is then

$$S_3(t,t \mid \boldsymbol{x}_{ijk}, \boldsymbol{x}_{i'j'k}) = \int \left(\int S(t \mid \boldsymbol{x}_{ijk}, \delta_{jk}, \delta_k) \, \mathrm{d}\delta_{jk} \right.$$
$$\left. \times \int S(t \mid \boldsymbol{x}_{i'j'k}, \delta_{j'k}, \delta_k) \, \mathrm{d}\delta_{j'k} \right) \mathrm{d}\delta_k, \quad (9.34)$$

and can also be evaluated by Gauss-Hermite quadrature. We usually set $\boldsymbol{x}_{ijk} = \boldsymbol{x}_{i'j'k}$, although only variables at level 3 need be the same.

With these three probabilities in hand, we can now calculate any measure of correlation for binary outcomes. For example, the Pearson correlation between the indicators of survival to age t for two children of the same mother with observed covariates $\boldsymbol{x}$ is given by

$$\rho_2(t, \boldsymbol{x}) = \frac{S_2(t,t \mid \boldsymbol{x}, \boldsymbol{x}) - S(t \mid \boldsymbol{x})^2}{S(t \mid \boldsymbol{x})[1 - S(t \mid \boldsymbol{x})]}, \quad (9.35)$$

where $S_2(t,t \mid \boldsymbol{x}, \boldsymbol{x})$ is the joint survival probability from (9.33) and $S(t \mid \boldsymbol{x})$ is the marginal probability from (9.32). A similar expression applies to the correlation for children of different mothers living in the same community, but using (9.34) for the joint probability. These measures of intraclass correlation are a function of the marginal and joint probabilities of survival to age one or five, which in turn depend on the linear predictor as well as the variances of the random effects.

Figure 9.3 shows these correlations calculated over the entire range of observed relative risks in Kenya using the estimated values of σ_2 and σ_3 in

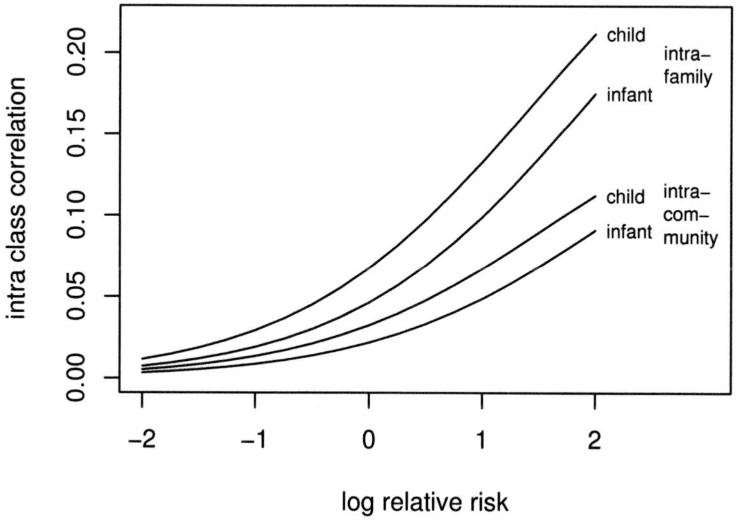

Fig. 9.3 Intra-family and intra-community correlations in infant and child mortality, by log relative risk.

Table 9.2. The intra-family correlations, which result from children sharing unobserved family and community characteristics, are always higher than the intra-community correlations, which result from sharing unobserved community characteristics only. The correlations are also higher for child than for infant mortality (or their complements, survival to ages five and one, respectively), and increase with the relative risk as measured from observed covariates. For our reference cell, the intra-family correlation is 0.05 for infant and 0.07 for child deaths, but these numbers increase to 0.18 and 0.21 at the highest levels of risk. The fact that the correlation between observed outcomes in the same family or community increases with the level of risk parallels the results obtained for two-level logit models in [80].

9.5 Summary and Conclusions

In this chapter we described generalizations of the multilevel model that go beyond normally distributed outcomes to cover a wide range of continuous and discrete responses, including binary, count, and survival data. The distinguishing feature of the generalization is the assumption that *conditional* on a set of random effects, the outcomes are independent and follow a standard generalized linear model. In this extension, a transformation of the conditional mean given a set of observed covariates and unobserved random effects follows a linear model. In a survival context, the conditional hazard has a similar structure. We contrasted this approach with models that focus on the *marginal* distribution of the outcomes, and with models that assume Gaussian outcomes but a non-linear structure of effects.

We reviewed the three main approaches to estimation, including marginal and penalized quasi-likelihood, maximum likelihood using Gauss-Hermite quadrature, and Bayesian estimation using the Gibbs sampler. We reported results of a simulation study showing that for heavily clustered binary responses quasi-likelihood estimates can be severely biased, while maximum likelihood estimates are approximately unbiased. Bayesian estimates showed a small bias that could be eliminated by using informative priors or larger samples. We also commented on a case study using binary data from Guatemala that leads to similar conclusions, but reveals some of the convergence problems that arise with bootstrapping and Bayesian estimates. Finally, we presented an application to survival data from Kenya where the approximate procedures fared better. On balance, there is a clear need for fast and accurate estimation procedures that can be applied to a wide variety of models and datasets.

Our analysis of infant and child mortality in Kenya illustrates the close connection between piece-wise exponential survival models and generalized linear models with Poisson errors and log link. We showed how the risk of death varies between birth and age five as a function of observed characteristics of the child, mother, and community, as well as unobserved random effects representing heterogeneity of frailty across families and communities. We found large effects on the hazard, and translated these into marginal and conditional probabilities of dying by age one and by age five. Finally, we developed measures of intra-family and intra-community correlation in infant and child deaths. The study illustrates how much more can be learned from a dataset by taking into account the group structure in the framework of multilevel generalized linear models.

Acknowledgements I am grateful to David Clayton for sharing his Gibbs sampling code and to Noreen Goldman and Erik Meijer for helpful comments on the manuscript. This work was supported by National Institute of Child Health and Human Development grant R01 HD35277.

References

1. O. O. Aalen. Heterogeneity in survival analysis. *Statistics in Medicine*, 7: 1121–1137, 1988.
2. M. Abramowitz and I. A. Stegun, editors. *Handbook of Mathematical Functions.* Number 55 in National Bureau of Standards Applied Mathematics Series. U.S. Government Printing Office, Washington, DC, 1964.
3. M. Aitkin, D. Anderson, B. Francis, and J. Hinde. *Statistical Modelling in GLIM.* Clarendon Press, Oxford, 1989.
4. M. Aitkin and D. G. Clayton. The fitting of exponential, Weibull and extreme value distributions to complex censored survival data using GLIM. *Applied Statistics*, 29:156–163, 1980.
5. P. D. Allison. Discrete-time methods for the analysis of event histories. *Sociological Methodology*, 13:61–98, 1982.
6. D. A. Anderson and M. Aitkin. Variance component models with binary response: Interviewer variability. *Journal of the Royal Statistical Society, Series B*, 47:203–210, 1985.
7. J. S. Barber, S. A. Murphy, W. G. Axinn, and J. Maples. Discrete-time multilevel hazard analysis. *Sociological Methodology*, 30:201–235, 2000.
8. S. Bennet and J. Whitehead. Fitting logistic and log-logistic regression models to censored data using GLIM. *GLIM Newsletter*, 4:12–19, 1981.
9. N. G. Best, M. K. Cowles, and S. K. Vines. *CODA: Convergence Diagnosis and Output Analysis Software for Gibbs Sampling Output, version 0.40.* Medical Research Council Biostatistics Unit, Cambridge, UK, 1997.
10. J. G. Booth, J. P. Hobert, and W. Jank. A survey of Monte Carlo algorithms for maximizing the likelihood of a two-stage hierarchical model. *Statistical Modelling*, 1:333–349, 2001.
11. N. E. Breslow and D. G. Clayton. Approximate inference in generalized linear mixed models. *Journal of the American Statistical Association*, 88:9–25, 1993.

12. N. E. Breslow and X. Lin. Bias correction in generalised linear mixed models with a single component of dispersion. *Biometrika*, 82:81–91, 1995.
13. W. J. Browne and D. Draper. A comparison of Bayesian and likelihood-based methods for fitting multilevel models. *Bayesian Analysis*, 1:473–549, 2006. (with discussion)
14. D. G. Clayton. A model for association in bivariate life tables and its application in epidemiological studies of familial tendency in chronic disease incidence. *Biometrika*, 65:141–151, 1978.
15. D. G. Clayton. The analysis of event history data: Review of progress and outstanding problems. *Statistics in Medicine*, 7:819–841, 1988.
16. D. G. Clayton. Generalized linear mixed models. In W. R. Gilks, S. Richardson, and D. J. Spiegelhalter, editors, *Markov Chain Monte Carlo in Practice*, pages 275–301. Chapman & Hall, London, 1996.
17. D. G. Clayton and J. Cuzick. The EM algorithm for Cox's regression model using GLIM. *Applied Statistics*, 34:148–156, 1985.
18. D. G. Clayton and J. Cuzick. Multivariate generalizations of the proportional hazards model. *Journal of the Royal Statistical Society, Series B*, 148:82–117, 1985. (with discussion)
19. C. Corcoran, B. Coull, and A. Patel. *Egret for Windows User Manual.* Cytel Software Corporation, Cambridge, MA, 1999.
20. D. R. Cox. Regression models and life tables. *Journal of the Royal Statistical Society, Series B*, 34:187–220, 1972. (with discussion)
21. D. R. Cox. Partial likelihood. *Biometrika*, 62:269–276, 1975.
22. D. R. Cox and D. Oakes. *Analysis of Survival Data.* Chapman & Hall, London, 1984.
23. M. J. Crowder. Beta-binomial Anova for proportions. *Applied Statistics*, 27: 34–37, 1978.
24. M. Davidian and D. M. Giltinan. *Nonlinear Models for Repeated Measurement Data.* Chapman & Hall, London, 1995.
25. P. J. Diggle, K.-Y. Liang, and S. L. Zeger. *Analysis of Longitudinal Data.* Oxford University Press, Oxford, UK, 1994.
26. P. Feigl and M. Zelen. Estimation of exponential survival probabilities with concomitant information. *Biometrics*, 21:826–838, 1967.
27. A. Gelman and D. B. Rubin. Inference from iterative simulation using multiple sequences. *Statistical Science*, 7:457–511, 1992. (with discussion)
28. J. Geweke. Evaluating the accuracy of sampling-based approaches to the calculation of posterior moments. In J. M. Bernardo, J. O. Berger, A. P. Dawid, and A. F. M. Smith, editors, *Bayesian Statistics 4*, pages 169–194. Oxford University Press, Oxford, UK, 1992.
29. W. R. Gilks and P. Wild. Adaptive rejection sampling for Gibbs sampling. *Applied Statistics*, 41:337–348, 1992.
30. H. Goldstein. Nonlinear multilevel models, with an application to discrete response data. *Biometrika*, 78:45–51, 1991.
31. H. Goldstein. Multilevel models and generalised estimating equations. *Multilevel Modelling Newsletter*, 5(2):2, 1993.

32. H. Goldstein. Multilevel unit specific and population average generalised linear models. *Multilevel Modelling Newsletter*, 7(3):4–5, 1995.
33. H. Goldstein. Consistent estimators for multilevel generalised linear models using an iterated bootstrap. *Multilevel Modelling Newsletter*, 8(1):3–6, 1996.
34. H. Goldstein and J. Rasbash. Improved approximations for multilevel models with binary responses. *Journal of the Royal Statistical Society, Series A*, 159: 505–513, 1996.
35. C. Gouriéroux and A. Monfort. *Simulation-Based Econometric Methods.* Oxford University Press, Oxford, UK, 1996.
36. P. J. Green. Penalized likelihood for general semi-parametric regression models. *International Statistical Review*, 55:245–259, 1987.
37. G. Guo and G. Rodríguez. Estimating a multivariate proportional hazards model for clustered data using the EM algorithm, with an application to child survival in Guatemala. *Journal of the American Statistical Association*, 87: 969–976, 1992.
38. D. Hedeker and R. D. Gibbons. A random-effects ordinal regression model for multilevel analysis. *Biometrics*, 50:933–944, 1994.
39. D. Hedeker and R. D. Gibbons. MIXOR: A computer program for mixed-effects ordinal regression analysis. *Computer Methods and Programs in Biomedicine*, 49:157–176, 1996.
40. T. R. Holford. The analysis of rates and survivorship using log-linear models. *Biometrics*, 36:299–306, 1980.
41. P. Hougaard. Life table methods for heterogeneous populations: Distributions describing the heterogeneity. *Biometrika*, 71:75–83, 1984.
42. P. Hougaard. Survival models for heterogeneous populations derived from stable distributions. *Biometrika*, 73:387–396, 1986.
43. J. D. Kalbfleisch and R. L. Prentice. *The Statistical Analysis of Failure Time Data*, 2nd edition. Wiley, New York, 2002.
44. M. P. Keane. Simulation estimation for panel data models with limited dependent variables. In G. S. Maddala, C. R. Rao, and H. D. Vinod, editors, *Handbook of Statistics*, volume 11, pages 545–571. North-Holland, Amsterdam, 1993.
45. A. Y. C. Kuk. Asymptotically unbiased estimation in generalized linear models with random effects. *Journal of the Royal Statistical Society, Series B*, 57: 395–407, 1995.
46. N. M. Laird. Empirical Bayes methods for two-way contingency tables. *Biometrika*, 65:581–590, 1978.
47. N. M. Laird and D. Olivier. Covariance analysis of censored survival data using log-linear analysis techniques. *Journal of the American Statistical Association*, 76:231–240, 1981.
48. J. F. Lawless. Regression methods for Poisson process data. *Journal of the American Statistical Association*, 82:808–815, 1987.
49. Y. Lee and J. A. Nelder. Hierarchical generalized linear models. *Journal of the Royal Statistical Society, Series B*, 58:619–678, 1996.
50. S. R. Lerman and C. F. Manski. On the use of simulated frequencies to approximate choice probabilities. In C. F. Manski and D. McFadden, editors, *Structural*

Analysis of Discrete Data with Econometric Applications, pages 305–319. MIT Press, Cambridge, MA, 1981.

51. L. A. Lillard and C. W. A. Panis. *aML: Multilevel Multiprocess Statistical Software, Version 2.0.* EconWare, Los Angeles, CA, 2003.
52. X. Lin and N. E. Breslow. Bias correction in generalized linear mixed models with multiple components of dispersion. *Journal of the American Statistical Association*, 91:1007–1016, 1996.
53. Q. Liu and D. A. Pierce. A note on Gauss-Hermite quadrature. *Biometrika*, 81: 624–629, 1994.
54. N. T. Longford. A quasi-likelihood adaptation for variance component analysis. In *American Statistical Association Proceedings of the Statistical Computing Section*, pages 137–142. 1988.
55. N. T. Longford. *VARCL: Software for Variance Component Analysis of Data with Nested Random Effects (Maximum Likelihood).* Educational Testing Service, Princeton, NJ, 1988.
56. N. T. Longford. Logistic regression with random coefficients. *Computational Statistics & Data Analysis*, 17:1–15, 1994.
57. T. A. Louis. Finding the observed information matrix when using the *EM* algorithm. *Journal of the Royal Statistical Society, Series B*, 44:226–233, 1982.
58. K. G. Manton, E. Stallard, and J. W. Vaupel. Alternative models for the heterogeneity of mortality risks among the aged. *Journal of the American Statistical Association*, 81:635–644, 1986.
59. P. McCullagh and J. A. Nelder. *Generalized Linear Models*, 2nd edition. Chapman & Hall, London, 1989.
60. C. E. McCulloch. Maximum likelihood algorithms for generalized linear mixed models. *Journal of the American Statistical Association*, 92:162–170, 1997.
61. D. McFadden. A method of simulated moments for estimation of discrete response models without numerical integration. *Econometrica*, 57:995–1026, 1989.
62. W. H. Mosley and L. C. Chen. An analytical framework for the study of child survival in developing countries. *Population and Development Review*, 10:25–45, 1984.
63. J. Myles and D. G. Clayton. *GLMMGibbs: An R Package for Estimating Bayesian Generalised Linear Mixed Models by Gibbs Sampling.* Comprehensive R Archive Network (devel section), 2001. URL http://cran.r-project.org
64. J. C. Naylor and A. F. M. Smith. Applications of a method for the efficient computation of posterior distributions. *Applied Statistics*, 31:214–225, 1980.
65. J. A. Nelder and R. Wedderburn. Generalized linear models. *Journal of the Royal Statistical Society, Series B*, 135:370–384, 1972.
66. J. M. Neuhaus, J. D. Kalbfleisch, and W. W. Hauck. A comparison of cluster-specific and population-averaged approaches to analyzing correlated binary data. *International Statistical Review*, 59:25–35, 1991.
67. E. S. W. Ng, J. R. Carpenter, H. Goldstein, and J.Rasbash. Estimation in generalised linear mixed models with binary outcomes by simulated maximum likelihood. *Statistical Modelling*, 6:23–42, 2006.

68. D. Oakes. A model for association in bivariate survival data. *Journal of the Royal Statistical Society, Series B*, 44:414–422, 1982.
69. A. R. Pebley, N. Goldman, and G. Rodríguez. Prenatal and delivery care and childhood immunization in Guatemala: Do family and community matter? *Demography*, 33:231–247, 1996.
70. J. C. Pinheiro and D. M. Bates. Approximations to the log-likelihood function in the nonlinear mixed-effects model. *Journal of Computational and Graphical Statistics*, 4:12–35, 1995.
71. J. C. Pinheiro and D. M. Bates. *Mixed-Effects Models in S and S-PLUS*. Springer, New York, 2000.
72. R. L. Prentice and L. A. Gloeckler. Regression analysis of grouped survival data with application to breast cancer data. *Biometrics*, 34:57–67, 1978.
73. W. H. Press, S. A. Teukolsky, W. T. Vetterling, and B. P. Flannery. *Numerical Recipes in C*, 2nd edition. Cambridge University Press, Cambridge, MA, 1992.
74. S. Rabe-Hesketh, A. Skrondal, and A. Pickles. Reliable estimation of generalized linear mixed models using adaptive quadrature. *The Stata Journal*, 2:1–21, 2002.
75. A. E. Raftery and S. M. Lewis. How many iterations in the Gibbs sampler? In J. M. Bernardo, J. O. Berger, A. P. Dawid, and A. F. M. Smith, editors, *Bayesian Statistics 4*, pages 763–773. Oxford University Press, Oxford, UK, 1992.
76. A. E. Raftery and S. M. Lewis. Implementing MCMC. In W. R. Gilks, S. Richardson, and D. J. Spiegelhalter, editors, *Markov Chain Monte Carlo in Practice*, pages 115–130. Chapman & Hall, London, 1996.
77. S. W. Raudenbush, M.-L. Yang, and M. Yosef. Maximum likelihood for generalized linear models with nested random effects via high-order, multivariate Laplace approximation. *Journal of Computational and Graphical Statistics*, 9: 141–157, 2000.
78. G. O. Roberts. Markov chain concepts related to sampling algorithms. In W. R. Gilks, S. Richardson, and D. J. Spiegelhalter, editors, *Markov Chain Monte Carlo in Practice*, pages 45–57. Chapman & Hall, London, 1996.
79. G. Rodríguez. Event history analysis. In S. Kotz, C. B. Read, and D. L. Banks, editors, *Encyclopedia of Statistical Sciences, Update Volume*, pages 222–230. Wiley, New York, 1997.
80. G. Rodríguez and I. Elo. Intra-class correlation in random-effects models for binary data. *The Stata Journal*, 3:32–46, 2003.
81. G. Rodríguez and N. Goldman. An assessment of estimation procedures for multilevel models with binary responses. *Journal of the Royal Statistical Society, Series A*, 158:73–89, 1995.
82. G. Rodríguez and N. Goldman. Improved estimation procedures for multilevel models with binary response: A case-study. *Journal of the Royal Statistical Society, Series A*, 164:339–355, 2001.
83. N. Sastry. A nested frailty model for survival data, with an application to the study of child survival in northeast Brazil. *Journal of the American Statistical Association*, 92:426–435, 1997.
84. R. Schall. Estimation in generalized linear models with random effects. *Biometrika*, 78:719–727, 1991.

85. S. G. Self and K.-Y. Liang. Asymptotic properties of maximum likelihood estimators and likelihood ratio tests under non-standard conditions. *Journal of the American Statistical Association*, 82:605–610, 1987.
86. D. J. Spiegelhalter, A. Thomas, N. G. Best, and W. R. Gilks. *BUGS: Bayesian Inference Using Gibbs Sampling.* Medical Research Council Biostatistics Unit, Cambridge, UK, 1996.
87. R. Stiratelli, N. M. Laird, and J. H. Ware. Random-effects models for serial observations with binary response. *Biometrics*, 40:961–971, 1984.
88. D. O. Stram and J. W. Lee. Variance components testing in the longitudinal mixed-effects model. *Biometrics*, 50:1171–1177, 1994.
89. T. M. Therneau and P. M. Grambsch. *Modeling Survival Data: Extending the Cox Model.* Springer, New York, 2000.
90. R. A. Thisted. *Elements of Statistical Computing: Numerical Computation.* Chapman & Hall, London, 1988.
91. J. Vaupel, K. G. Manton, and E. Stallard. The impact of heterogeneity in individual frailty on the dynamics of mortality. *Demography*, 16:439–454, 1979.
92. J. Vaupel and A. Yashin. Heterogeneity's ruses: Some surprising effects of selection on population dynamics. *American Statistician*, 39:176–185, 1985.
93. J. Whitehead. Fitting Cox's regression model to survival data using GLIM. *Applied Statistics*, 29:268–275, 1980.
94. G. Y. Wong and W. M. Mason. The hierarchical logistic regression model for multilevel analysis. *Journal of the American Statistical Association*, 80:513–524, 1985.
95. S. L. Zeger and M. R. Karim. Generalized linear models with random effects: A Gibbs sampling approach. *Journal of the American Statistical Association*, 86:79–86, 1991.

10

Missing Data

Nicholas T. Longford

SNTL and Universitat Pompeu Fabra

10.1 Background and Generalities

Nonresponse is a ubiquitous feature of large-scale studies that collect information from human subjects or their organizations, such as schools, households or businesses. The contacted sources of data (subjects, their parents, representatives of schools, and the like) may refuse to respond to some or all of the questionnaire items, may not have ready access to the requested information, and the record of the responses may be corrupted or lost altogether during its transfer and conversion to electronic format.

Concerns about missing values, and solutions commensurate with the computing facilities available at the time, can be traced back to Yates [30] and Healy and Westmacott [9]. These methods can be motivated as estimation of the missing values followed by an adjustment of the degrees of freedom due to the lost items of data. The main impetus for the modern approaches, based on computationally intensive methods, can be identified with Orchard and Woodbury [22], Rubin [23, 24] and Dempster et al. [4], and their application in a wide range of areas has been greatly promoted by Little and Rubin [11] and Rubin [25]. For a comprehensive review and discussion, see Rubin [26]. The early methodological developments were restricted to specific methods of analysis applied to data from small-scale experiments, such as the analysis of variance (ANOVA) of field experiments, and saw virtue in computational simplicity. In contrast, modern approaches pursue flexibility and versatility, aiming to deal with missing information by a module attached to the procedure that would have been employed had the data been complete. Indeed, this *complete-data analysis* is a key concept in these approaches.

As a consequence of missing values, less information is collected than was planned. At the same time, the representativeness of the units for which the

J. de Leeuw, E. Meijer (eds.), *Handbook of Multilevel Analysis*,
© Springer 2008

values of a variable have been recorded (the *responding* units) is questionable. Methods for addressing nonresponse can be divided into two categories: those that reduce the dataset (by deleting the records of some units) and those that make up the data so as to generate, structurally, a look-alike of the complete dataset. The latter are referred to as *imputation* methods. The aim of all of these methods is to adapt the complete-data analysis so that it would yield an estimator with good resampling properties.

In this chapter, we adopt the frequentist perspective, and so by "good properties" (efficiency) we understand small mean squared error (MSE) in repeated applications of the estimator on the datasets generated by hypothetical replications of the process that yielded the original (realized) dataset. This process is a superposition of sampling of the units (generation of the complete data) and nonresponse (deletion of a part of the data). An important prerequisite is that the complete-data analysis is efficient. That is, the quantities (parameters) of interest would be estimated efficiently had the data been complete.

10.1.1 Example

An example with computer generated data is summarized in Fig. 10.1. The complete data are pairs (x_i, y_i), $i = 1, \ldots, n = 20$, a random sample from a bivariate normal distribution. This dataset, a 20×2 array, is denoted by $\boldsymbol{A}$. The value of x_1 is not available to the analyst; the horizontal line drawn at y_1 indicates the uncertainty about x_1, although the value of x_1 is available to *us*, marked by a circle ○.

Suppose we wish to estimate the regression slope $\beta = \operatorname{Cov}(\underline{x}, \underline{y}) / \operatorname{Var}(\underline{x})$. The corresponding complete-data analysis is $\hat{\beta} = S_{xy}/S_{xx}$, where $S_{xx} = \sum_i (x_i - \bar{x})^2$ and $S_{xy} = \sum_i (x_i - \bar{x})(y_i - \bar{y})$ are the corrected totals of squares x^2 and cross products xy, respectively ($\bar{z}$ is the sample mean of z; z is either x or y). These two totals cannot be evaluated because the contribution of unit 1 to them is not available.

The obvious deletion method reduces the sample to the units $i = 2, \ldots n$ that have complete records. At this point we have to distinguish between the missing values in the realized dataset (one value of x missing), and missing values in hypothetical replications of the sampling and nonresponse processes. After all, these processes may yield a dataset with no missing values, or with more than just one missing value.

Denote the units with incomplete records by M, and the realized dataset by $\boldsymbol{A}_{-M}$. For $\boldsymbol{A}_{-M}$, the estimator $\hat{\beta}$ is well defined. We distinguish between the two estimators, $\hat{\beta}(\boldsymbol{A})$ and $\hat{\beta}(\boldsymbol{A}_{-M})$, by indicating the dataset used as the argument of $\hat{\beta}$; $\hat{\beta}(\boldsymbol{A})$ is efficient because the units $i = 1, \ldots, 20$ were obtained as a random sample (from a well-defined population). Without the

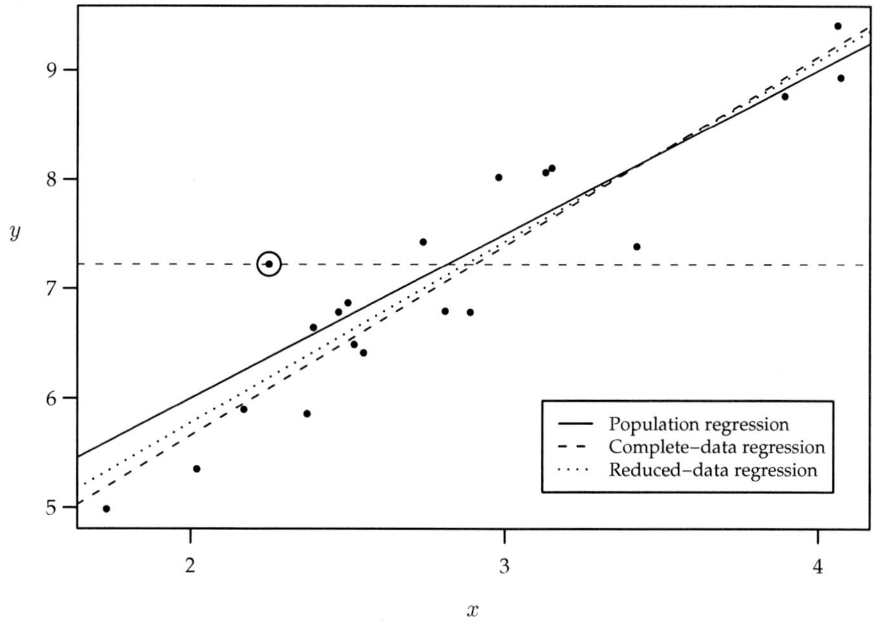

Fig. 10.1 A computer generated dataset with one missing value of x (in circle). The lines drawn are the "true" regression (solid), the complete-data regression (dots), and the regression based on the complete records (dashes).

missing unit(s), the resulting subsample may no longer be a random sample and $\hat{\beta}(\boldsymbol{A}_{-M})$ no longer efficient.

However, this should not stop us in our tracks. It can reasonably be argued that a few missing values are unlikely to alter the results radically. This we can check by deleting at random, or by a deliberate choice, another value of y, obtaining the dataset $\boldsymbol{A}_{-(M,k)}$ and comparing $\hat{\beta}(\boldsymbol{A}_{-M})$ with $\hat{\beta}(\boldsymbol{A}_{-(M,k)})$. Also, the normality of $\underline{x}$ provides some protection from extreme values of x. The value of x_1 can be further narrowed down by realizing that $\underline{y}_1 - \beta\underline{x}_1$ is also "constrained" by normality. On the other hand, the mechanism of missingness may target units with extreme residuals. Indeed, in Fig. 10.1 the complete- and incomplete-data analyses yield rather similar regressions. But recall, that our focus is on properties of estimators (random variables), not the values of their realizations, and so the proximity of the values of $\hat{\beta}(\boldsymbol{A})$ and $\hat{\beta}(\boldsymbol{A}_{-M})$ does not justify a dismissal of the issue.

10.1.2 Imputation Methods

Imputation methods generate a value for each missing item (in Section 10.1.1 for x_1), thus completing the dataset. Then the complete-data analysis can proceed, oblivious to the fact that the value of x_1 has been "manufactured" by the analyst. Since in this way we pretend to have more information than was recorded, we can expect the inferences to indicate higher precision than is warranted. The extent of this problem depends on how we impute for missing values, and how we exploit the information about the causes that give rise to the missing values.

Imputing for x_1 the sample mean $\hat{x}_1 = \bar{x}$ or the back-calculated value of x_1, $\hat{\hat{x}}_1 = \{y_1 - \bar{y}(\boldsymbol{A}_{-M})\} / \hat{\beta}(\boldsymbol{A}_{-M})$, are simple examples of imputation methods. Each of these imputed values is associated with a model for the mechanism of missingness. For $\hat{x}_1$, it is

$$\underline{x}_1 = \mu_x + \underline{\eta}_1 \tag{10.1}$$

(μ_x is the population mean of $\underline{x}$), and so we estimate x_1 by the fit of this model. For $\hat{\hat{x}}_1$, the model is

$$\underline{y}_1 = \alpha + \beta x_1 + \underline{\eta}_1 \,, \tag{10.2}$$

where $\underline{\eta}_1$ is a random draw from a centered normal distribution. Neither of these models is necessarily correct; the process of missingness may prefer to select extreme values of $\underline{x}$, or extreme values of the deviation $\underline{\eta}$. However, in some vague sense, the model in (10.2) is better informed because it makes use of the realized value of $\underline{y}_1$.

The complete-data analysis can be straightforwardly applied to the *completed dataset* $\hat{\boldsymbol{A}} = \{(\hat{x}_1, y_1), (x_2, y_2), \ldots, (x_n, y_n)\}$, but the estimator $\underline{\hat{\beta}}(\hat{\boldsymbol{A}})$ will not have the desired property of efficiency, and its sampling variance will not be estimated with honesty (without bias). Not only have we "invented" one value (in general, several values) for the missing data, our invention ($\hat{x}_1$ or $\hat{\hat{x}}_1$) looks too good. We have substituted $\eta_1 = 0$, that is, no deviation from the fit, and that could hardly be expected. This could be remedied by drawing a value of $\underline{\eta}_1$ at random from the distribution fitted for $\underline{\eta}$. In this way, we will not recover the value of x_1 with precision, but will mimic the assumed mechanism of missingness more closely.

The following example illustrates why estimating each missing value without bias is sometimes not very useful. Suppose the value of x is missing but we know that it is equal to -1, 0, or $+1$. Further, ± 1 are equally likely, but $x = 0$ is most likely, say, with probability 0.7. Imputing $\hat{x} = 0$ would seem to be reasonable. However, if the quantity of interest is x^2 and we would estimate it by $(\hat{x})^2 = 0$, with MSE equal to 0.3, we would fail to protect our inferences against the possibility that $x^2 = 1$. A better choice for x^2 is its expected value, $\widehat{x^2} = 0.3$, because, assuming that the conditional distribution of $\underline{x}$ is correct, we would estimate x^2 with smaller MSE, equal to 0.21. Note that in this way we break some rules of arithmetic; $(\hat{x})^2 \neq \widehat{x^2}$. In fact, if $\underline{\hat{x}}$ and $\underline{\widehat{\mathrm{Var}}(\hat{x})}$ are unbiased for x and $\mathrm{Var}(\underline{\hat{x}})$, respectively, then $(\underline{\hat{x}})^2 - \underline{\widehat{\mathrm{Var}}(\hat{x})}$ is unbiased for x^2.

10.2 Models for Missing Values

Having highlighted the importance of the mechanism of missingness, we now introduce a notation and associated terminology. To align it with the conventions in the literature [e.g., 11], we define the complete data as $(\boldsymbol{X}, \boldsymbol{Y})$ where $\boldsymbol{X}$ are the values of the variables which are never missing (such as variables set by design, or available prior to data collection), and some values of $\boldsymbol{Y}$ may be missing. The *indicator of missingness* is the array $\boldsymbol{R}$ of the same dimensions as $\boldsymbol{Y}$, defined as follows: the element of $\boldsymbol{R}$ for unit i and variable k is $R_{ik} = 1$ if Y_{ik} is recorded, and $R_{ik} = 0$ otherwise. The rows of $\boldsymbol{R}$ are denoted by $\boldsymbol{r}_i$ and columns by $\boldsymbol{R}_k$, and the analogous notation is used for $\boldsymbol{X}$ and $\boldsymbol{Y}$. For instance, the record of a subject on four variables may be $\boldsymbol{y}_i = (1, ?, ?, 5.37)$; '?' indicates a missing value. The corresponding vector $\boldsymbol{r}$ is $\boldsymbol{r}_i = (1, 0, 0, 1)$. We call $\boldsymbol{r}$ the *pattern of missingness.* A practical way of summarizing the extent of missing values is by tabulating $\boldsymbol{r}$. An example is given in Table 10.1 (the commas separating the elements of $\boldsymbol{r}$ are omitted to save space). A less complete summary enumerates only the number of missing values for each variable; this can be expressed as $n - \sum_i \boldsymbol{r}_i$, where n is the sample size. For Table 10.1, this summary is $(117, 97, 174, 115)$.

Thus the sampling and missingness (nonresponse) processes are described by the random array $(\underline{\boldsymbol{R}}, \underline{\boldsymbol{Y}}; \boldsymbol{X})$, or its joint distribution. By $\underline{\boldsymbol{Y}}_r$ we denote the recorded part of $\underline{\boldsymbol{Y}}$, that is, all $\underline{y}_{ik}$ for which $\underline{r}_{ik} = 1$. By $\underline{\boldsymbol{Y}}_m$ we denote the missing part of $\underline{\boldsymbol{Y}}$. Note that $\underline{\boldsymbol{Y}}_r$ and $\underline{\boldsymbol{Y}}_m$ may be "ragged" arrays, with gaps.

When the missingness process is simple random sampling we say that data are *missing completely at random* (MCAR); in this case, $\underline{\boldsymbol{R}}$ depends neither on $\boldsymbol{X}$ nor on $\boldsymbol{Y}$:

$$(\underline{\boldsymbol{R}}) \sim (\underline{\boldsymbol{R}} \mid \boldsymbol{Y}; \boldsymbol{X}),$$

where $\sim$ stands for "has the same distribution as". Simple random sampling is a very special process, unlikely to arise without either being promoted in some way or the cause of missingness being unrelated to the processes involved in generating the complete data.

A much more general mechanism of missingness arises by stratified random sampling. With it, MCAR applies within each subpopulation (stratum) defined by the categories of one or several completely recorded discrete variables. We can extend this definition to continuous variables by a limiting argument

Table 10.1 Tabulation of the pattern of missingness. An example.

	Pattern $\boldsymbol{r}$							
	(1111)	(1110)	(1101)	(1100)	(0011)	(0110)	(0100)	(0000)
Count	1233	17	87	24	43	11	9	54

(using finer and finer discretization). This mechanism of missingness is referred to as *missing at random* (MAR); $\underline{\boldsymbol{R}}$ depends on the complete data only through the recorded data:

$$(\underline{\boldsymbol{R}} \mid \boldsymbol{Y}; \boldsymbol{X}) \sim (\underline{\boldsymbol{R}} \mid \boldsymbol{Y}_r; \boldsymbol{X}).$$

Finally, mechanisms of missingness in which dependence on the missing values is essential (is present even after conditioning on all the recorded data) are called *missing not at random* (MNAR). Note that MAR and MNAR are qualified by the variables in $\boldsymbol{X}$ and $\boldsymbol{Y}$.

The following is an example of MNAR. A survey of alcohol consumption among the students of a college is to be conducted by telephone interviewing. Since students attend lectures on weekdays, they will be contacted in their homes on a Saturday morning and asked to recall their consumption the previous day. It is reasonable to anticipate that some of the students who have consumed alcohol in excess the night before will not be well disposed to respond. This is likely to be the case not only among students in general (so that this is not MCAR), but even within any conceivable category of students defined at the outset of the study. So this process of nonresponse is not a MAR either.

The central role of the MAR mechanism stems from the following characterization. When MAR applies, the missing part of a record, $\underline{\boldsymbol{Y}}_m$, is associated with the recorded part $\boldsymbol{Y}_r$ in the same way as the corresponding components in complete records. This enables us to establish, in principle, the marginal distribution of the missing part of a record. In practice, this marginal distribution is estimated, and estimation can be improved by using the records which are not complete but their recorded parts overlap with the missing part $\underline{\boldsymbol{Y}}_m$.

The joint distribution $(\underline{\boldsymbol{R}}, \underline{\boldsymbol{Y}}; \boldsymbol{X})$ can be expressed in terms of the conditional distributions $(\underline{\boldsymbol{R}} \mid \boldsymbol{Y}; \boldsymbol{X})$ and $(\underline{\boldsymbol{Y}} \mid \boldsymbol{R}; \boldsymbol{X})$ as

$$(\underline{\boldsymbol{R}}, \underline{\boldsymbol{Y}}; \boldsymbol{X}) \sim (\underline{\boldsymbol{R}} \mid \boldsymbol{Y}; \boldsymbol{X}) \times (\underline{\boldsymbol{Y}}; \boldsymbol{X}) \tag{10.3}$$

and

$$(\underline{\boldsymbol{R}}, \underline{\boldsymbol{Y}}; \boldsymbol{X}) \sim (\underline{\boldsymbol{Y}} \mid \boldsymbol{R}; \boldsymbol{X}) \times (\underline{\boldsymbol{R}}; \boldsymbol{X}), \tag{10.4}$$

respectively. The decomposition in (10.3) comprises the complete-data distribution $(\underline{\boldsymbol{Y}}; \boldsymbol{X})$ and the distribution of the deletion (selection) process $(\underline{\boldsymbol{R}} \mid \boldsymbol{Y}; \boldsymbol{X})$; it is referred to as the *selection model.* Its obvious appeal is in the correspondence with our description of the sampling and missingness processes. The decomposition in (10.4) corresponds to separate models for each pattern of missingness. It is referred to as the *pattern-mixture* model. Since a dataset may contain many patterns of missingness (up to 2^K for K variables), it may not be practical to associate each pattern $\boldsymbol{r}$ with a separate set of parameters describing the conditional distribution $(\underline{Y} \mid \boldsymbol{r})$. Also, no assumptions about the missing values could be supported empirically from the subset of data with the given pattern $\boldsymbol{r}$. For instance, the conditional distribution $(\underline{Y} \mid \boldsymbol{r} = (1,1,0,0); \boldsymbol{X})$ cannot inform us about the third and fourth components of $\boldsymbol{Y}$ without using the data on units with the other patterns. So, the use of pattern-mixture models is somewhat more complicated, but their flexibility can be used with advantage. In general, pattern-mixture and selection models yield different distributions for the missing values when particular assumptions (distributions and parameters) are specified for their components. The two kinds of models can be combined. For instance, instead of conditioning on each pattern separately, models can be formulated for sets of patterns, and (some of) these models may be selection models. All these models are informed by the associations among the (pairs of) variables, and so the units with complete records, $\boldsymbol{r} = (1, \ldots, 1)$, play a central role.

10.3 EM Algorithm and Multiple Imputation

We have so far considered a complete-data analysis $\underline{\hat{\beta}}$ as an estimator $\underline{\hat{\beta}}(\boldsymbol{A})$ that would be obtained had the complete data $\boldsymbol{A}$ been available. Deletion methods apply $\underline{\hat{\beta}}$ to a reduced dataset $\boldsymbol{A}_-$ which contains no missing values; imputation methods define values for the missing items, thus creating a completed dataset $\boldsymbol{A}_+$, and evaluate $\underline{\hat{\beta}}(\boldsymbol{A}_+)$. These estimators are deficient even when $\underline{\hat{\beta}}(\boldsymbol{A})$ is efficient and the deletion is "minimal" or the imputation "intelligent" (unbiased). When the dataset has few missing values, this problem can be dealt with by a caveat in the discussion of the analysis. When there are numerous missing values, their impact is no longer innocuous.

EM algorithm and multiple imputation are two general approaches to estimation with incomplete data. These approaches seek to adjust (efficient) complete-data estimators in such a way that the resulting (incomplete-data) estimator would also be efficient, with reference to replications of the sampling and missingness processes. In the EM algorithm, the computational procedure for the complete-data estimator is adjusted; with multiple imputation, additional "data" is generated to complete the observed dataset, but the complete-data estimator is applied without any alterations.

10.3.1 EM Algorithm

Let the complete-data likelihood for a parameter vector $\boldsymbol{\theta}$ be $L(\boldsymbol{\theta}; \boldsymbol{Y}, \boldsymbol{X})$. The likelihood for the observed data is

$$\int L(\boldsymbol{\theta}; \boldsymbol{Y}_r, \boldsymbol{Y}_m; \boldsymbol{X})\, p(\boldsymbol{R} \mid \boldsymbol{Y}_r; \boldsymbol{X})\; \mathrm{d}F(\boldsymbol{Y}_m; \boldsymbol{X}), \tag{10.5}$$

where p is the conditional probability of the pattern of missing data, and F is the distribution function of the missing data. This integral is in general difficult to evaluate or maximize. The EM algorithm avoids its direct maximization. EM is an iterative procedure, with iterations comprising two steps, E (estimation) and M (maximization). In the E-step, the expectation of the complete-data log-likelihood L is evaluated at the current estimate $\hat{\boldsymbol{\theta}}$. The expectation is taken over the conditional distribution of $\underline{\boldsymbol{Y}}_m$ given $\boldsymbol{Y}_r$ and $\boldsymbol{X}$. The M-step maximizes this expected log-likelihood, and the value of $\boldsymbol{\theta}$ at which the maximum is attained is adopted as the updated value of $\hat{\boldsymbol{\theta}}$. The iterations of EM are then repeated until the updating changes the value of $\hat{\boldsymbol{\theta}}$ only slightly.

A substantial simplification in (10.5) takes place when the mechanism of missingness can be eliminated from the likelihood. This happens when the sets of parameters characterizing the sampling and missingness processes are disjoint and functionally unrelated (*separated*) and MAR applies. When these two conditions hold the mechanism of missingness is said to be *ignorable.* Separation is usually satisfied, but MAR is the key assumption that makes the estimation task manageable. Under ignorable missingness

$$L(\boldsymbol{\theta}; \boldsymbol{Y}_r, \boldsymbol{X}) = \int L(\boldsymbol{\theta}; \boldsymbol{Y}_r, \boldsymbol{Y}_m; \boldsymbol{X})\, \mathrm{d}F(\boldsymbol{Y}_m; \boldsymbol{X}),$$

and so the conditional expectations in the E-step are much simpler.

The log-likelihood can often be expressed as a linear function of a small number of statistics (called *sufficient statistics*). In such a setting, the E-step estimates the contribution of the missing values to these statistics, and in the M-step the estimates of these statistics are used in place of their (unknown) complete-data values. The EM algorithm requires a value of $\hat{\boldsymbol{\theta}}$ for the first iteration. This has to be obtained separately, although this estimator need not have good properties; the estimator based on the complete records is usually satisfactory. Often the first few iterations of the EM algorithm move $\hat{\boldsymbol{\theta}}$ most of the way toward the maximum likelihood estimate, but then many further iterations are required to get very close to the solution.

For a proof of convergence of the EM algorithm and the associated regularity conditions, see Dempster et al. [4]. Convergence properties of the EM algorithm are derived by Wu [29]. The EM algorithm is particularly appealing when the M-step evaluates a simple estimator, because this evaluation will be conducted many times. The general idea of the EM algorithm has been extended in several directions. They include dealing with more complex problems, widening the scope of the EM algorithm, and acceleration of its convergence. See *Statistica Sinica* [6] for several innovations and Meng and van Dyk [21] for a review.

The estimated Hessian obtained from the last M-step estimates the complete-data Hessian which is related to the complete-data information. Its inverse estimates $\mathrm{Var}\{\hat{\underline{\theta}}(\boldsymbol{A})\}$; it underestimates $\mathrm{Var}\{\hat{\underline{\theta}}(\boldsymbol{A}_{-M})\}$. The difference $\mathrm{Var}\{\hat{\underline{\theta}}(\boldsymbol{A}_{-M})\} - \mathrm{Var}\{\hat{\underline{\theta}}(\boldsymbol{A})\}$ is the inflation of variance due to missing data. Louis [17], Meilijson [18] and Meng and Rubin [19] describe approaches to estimating the incomplete-data sampling variance (matrix) from the EM algorithm. These methods are not applicable or suitable in all settings. For example, there may be no shortlist of sufficient statistics. The second-order partial derivatives of the log-likelihood can always be approximated numerically by finite differences of the values of the log-likelihood, and these values can be approximated by simulations. This is a fall-back option when no computationally less demanding approach is suitable.

For univariate θ, the fraction of the information that is missing is defined as $\gamma = 1 - \mathrm{Var}\{\hat{\underline{\theta}}(\boldsymbol{A})\} / \mathrm{Var}\{\hat{\underline{\theta}}(\boldsymbol{A}_{-M})\}$. This definition is extended to multivariate $\boldsymbol{\theta}$ by considering estimation of the linear combinations $\boldsymbol{\theta}'\boldsymbol{c}$ for various vectors $\boldsymbol{c}$. The speed of convergence of the EM algorithm is closely related to the fraction γ, or to the largest fraction $\gamma(\boldsymbol{c})$ [29].

10.4 Multiple Imputation

With multiple imputation, a small number of alternative completions of the observed dataset are generated and the complete-data analysis is carried out on each completed dataset. The generation of these completions (*completed* datasets), say, $\boldsymbol{A}^{(1)}, \ldots, \boldsymbol{A}^{(K)}$ is usually the most complex task, but the remainder is straightforward. The complete-data analysis is applied on each completion, yielding estimates $\hat{\beta}^{(1)} = \hat{\beta}(\boldsymbol{A}^{(1)}), \ldots, \hat{\beta}^{(K)} = \hat{\beta}(\boldsymbol{A}^{(K)})$ and estimated complete-data sampling variances $\hat{s}^2(\boldsymbol{A}_1), \ldots, \hat{s}^2(\boldsymbol{A}_K)$. Although this requires K times as much computing, it entails little programming effort additional to that for conducting one complete-data analysis. The estimator for the recorded data is obtained by averaging the K results of the completed-data analyses:

$$\begin{aligned} \tilde{\underline{\beta}} &= \frac{1}{K} \sum_{k=1}^{K} \hat{\underline{\beta}}^{(k)}, \\ \tilde{\underline{s}}^2 &= \frac{1}{K} \sum_{k=1}^{K} \hat{\underline{s}}_k^2 + \frac{K+1}{K(K-1)} \sum_{k=1}^{K} \left(\hat{\underline{\beta}}^{(k)} - \tilde{\underline{\beta}}\right)^2 . \end{aligned} \tag{10.6}$$

The completions, sets of *plausible values* $\tilde{\boldsymbol{Y}}_k$, are generated by simulations from the model that relates the missing values to the observed data. For this, it is essential that missingness is ignorable. Otherwise, the details of the departure from ignorability have to be specified in detail. Suppose the missing values are related to the recorded values by the model

$$\underline{\boldsymbol{y}}_m = g(\boldsymbol{y}_r\,; \boldsymbol{\psi}) + \underline{\xi}\,, \tag{10.7}$$

where g is a function involving a set of parameters $\boldsymbol{\psi}$ and ξ is a random sample from a distribution (such as normal) with one or several unknown parameters $\boldsymbol{\omega}$. The parameter (vectors) $\boldsymbol{\psi}$ and $\boldsymbol{\omega}$ can be estimated from the complete records, although other records may also contribute (e.g., by using an EM algorithm). One set of plausible values is generated by drawing a set of plausible parameters $\underline{\tilde{\boldsymbol{\psi}}}$ and $\underline{\tilde{\boldsymbol{\omega}}}$ from the estimated sampling distribution of $\underline{\hat{\boldsymbol{\psi}}}$ and $\underline{\hat{\boldsymbol{\omega}}}$, followed by the "prediction" $\underline{\tilde{\boldsymbol{y}}}_m = g(\boldsymbol{y}_r\,;\underline{\tilde{\boldsymbol{\psi}}}) + \underline{\tilde{\xi}}$, where $\underline{\tilde{\xi}}$ is drawn at random from the estimated distribution of $\underline{\xi}$ (such as the centered normal distribution with variance $\tilde{\omega}$). This can be interpreted as a meticulous reflection of the uncertainty in both the estimated parameters and in the missing values. This is an important ingredient of the method, contributing to the "inheritance" of the good properties of the complete-data estimator.

Suppose the complete-data estimator $\underline{\hat{\theta}}(\boldsymbol{A})$ is unbiased and its sampling variance $s^2(\boldsymbol{A}) = \mathrm{Var}\{\underline{\hat{\theta}}(\boldsymbol{A})\}$ is estimated, by $\underline{\hat{s}}^2$, without bias and with sampling variance $\mathrm{Var}\{\underline{\hat{s}}^2(\boldsymbol{A})\}$ much smaller than s^4. Further, suppose the model for missing data (as in (10.7)) is correctly specified. Then, using infinitely many imputations, the estimator $\underline{\tilde{\beta}}$ is also unbiased, and its sampling variance is

$$\mathrm{Var}\big(\underline{\tilde{\beta}}\big) \doteq E\big(\underline{\hat{s}}^2\big) + B\,, \tag{10.8}$$

where $B = \mathrm{Var}_k\big(\underline{\hat{\beta}}^{(k)}\big)$ is the between-imputation variance; its finite-K estimator is

$$\underline{\hat{B}}_K = \frac{1}{K-1}\sum_{k=1}^{K}\Big(\underline{\hat{\beta}}^{(k)} - \underline{\tilde{\beta}}\Big)^2 .$$

For proof of (10.8), see Rubin [25].

In practice, only a finite number of imputations is used. The results about $\underline{\tilde{\beta}}$ remain valid, with an "approximation" caveat. The estimator $\underline{\tilde{\beta}}$ remains unbiased. The estimator of the sampling variance can be expressed as

$$\underline{\tilde{s}}^2 = \overline{\underline{\hat{s}}^2} + \left(1 + \frac{1}{K}\right)\underline{\hat{B}}_K\,,$$

where the bar $\overline{}$ denotes the average over the K imputations. As $K \longrightarrow +\infty$, $\underline{\hat{B}}_K$ converges to B, so B/K is the contribution to the sampling variance due to having used only K sets of imputations. The average $\overline{\underline{\hat{s}}^2}$ estimates the complete-data sampling variance $\mathrm{Var}\{\underline{\hat{\beta}}(\boldsymbol{A})\}$ and B can be interpreted as the inflation of the sampling variance caused by the missing values. It is useful to define the *fraction of information that is missing* as $\gamma = B/\{s^2(\boldsymbol{A}) + B\}$. When several parameters are estimated, their fractions γ may differ.

The number of imputations K has an impact on the precision of $\underline{\tilde{\beta}}$. By using an additional set of imputations, the sampling variance is reduced by $B/K - B/(K+1) = B/\{K(K+1)\}$, that is, by approximately $100\gamma/\{K(K+1)\}\%$. Thus, the value of γ is an important factor in choosing K. The gains in precision due to the first few sets of imputations are quite dramatic, but then the gains diminish. For instance, the sixth set of imputed values reduces the sampling variance by $B/30$, whereas infinitely many imputations after the fifth would have reduced it by $B/5$. The decision about setting K should be based on the anticipated fraction of the information missing. Nowadays, the amount of computing is not a serious concern, and so it should not enter into consideration about setting K. A more serious concern is the storage of the plausible values; this grows linearly with K. Of course, sets of plausible values do not have to be stored if they are immediately used in the completed-data analysis. However, then new sets of plausible values have to be generated for each analysis.

The model for missing data can rarely be established with any certainty. Rubin [26] and Schafer [27] recommend involving as many covariates as is practicable, so as to improve the chances of MAR being appropriate. If the model contains some unimportant or redundant terms the uncertainty about the missing values is greater than it could be otherwise, but this is less of a concern than unbiasedness in the estimation of the parameters associated with the missing values—that no important variable is omitted.

10.5 Missing Values in Multilevel Data

The previous sections summarized the general approach to dealing with missing values. Here we discuss issues specific to multilevel analysis. We consider the model

$$\underline{\boldsymbol{y}}_j = \boldsymbol{X}_j\boldsymbol{\beta} + \boldsymbol{Z}_j\underline{\boldsymbol{\delta}}_j + \underline{\boldsymbol{\epsilon}}_j\,, \tag{10.9}$$

where $\boldsymbol{\beta}$ is a set of regression parameters and $\underline{\boldsymbol{\delta}}_j$, $j = 1, \ldots, N_2$, and the $N = n_1 + \cdots + n_{N_2}$ elements of $\underline{\boldsymbol{\epsilon}}_j$ are mutually independent random samples from centered normal distributions with $p_z \times p_z$ variance matrix $\boldsymbol{\Omega}$ and variance σ^2, respectively; p_z is the number of columns in $\boldsymbol{Z}$. This model has the form of an analysis of covariance (ANCOVA) model; the only difference from the standard setting is in the status of the terms $\underline{\boldsymbol{\delta}}_j$. In ANCOVA, they are parameters (subject to an identification constraint, such as $\boldsymbol{\delta}_1 + \cdots + \boldsymbol{\delta}_{N_2} = \boldsymbol{0}_{p_z}$). In our model they are random variables described by their variance matrix $\boldsymbol{\Omega}$. We say that two such models are *paired*. We denote $\boldsymbol{X} = (\boldsymbol{X}_1', \ldots, \boldsymbol{X}_{N_2}')'$, $\boldsymbol{Z} = (\boldsymbol{Z}_1', \ldots, \boldsymbol{Z}_{N_2}')'$ and $\underline{\boldsymbol{y}} = (\underline{\boldsymbol{y}}_1', \ldots, \underline{\boldsymbol{y}}_{N_2}')'$, so that, for instance, $E(\underline{\boldsymbol{y}}) = \boldsymbol{X}\boldsymbol{\beta}$. Further, $\boldsymbol{V}_j = \mathrm{Var}(\underline{\boldsymbol{y}}_j) = \sigma^2\boldsymbol{I}_{n_j} + \boldsymbol{Z}_j\boldsymbol{\Omega}\boldsymbol{Z}_j'$. It is practical to define $\boldsymbol{\Psi} = \sigma^{-2}\boldsymbol{\Omega}$, so that σ^2 can be factored out of $\mathrm{Var}(\underline{\boldsymbol{y}}_j)$: $\boldsymbol{V}_j = \sigma^2\boldsymbol{W}_j$, with $\boldsymbol{W}_j = \boldsymbol{I}_{n_j} + \boldsymbol{Z}_j\boldsymbol{\Psi}\boldsymbol{Z}_j'$. The log-likelihood is $l = l_1 + \cdots + l_{N_2}$, where

$$-2\,l_j = n_j \log(2\pi\sigma^2) + \log\left(\det \boldsymbol{W}_j\right) + (\boldsymbol{y}_j - \boldsymbol{X}_j\boldsymbol{\beta})'\boldsymbol{W}_j^{-1}(\boldsymbol{y}_j - \boldsymbol{X}_j\boldsymbol{\beta})\,.$$

We assume that the parameters in $\boldsymbol{\Psi}$ are functionally related to neither $\boldsymbol{\beta}$ nor σ^2.

We will refer to the Fisher scoring algorithm [12, 13] as the complete-data analysis. The algorithm can be compactly described by the following equations:

$$
\begin{aligned}
\hat{\boldsymbol{\beta}} &= \left(\boldsymbol{X}'\hat{\boldsymbol{W}}^{-1}\boldsymbol{X}\right)^{-1}\boldsymbol{X}'\hat{\boldsymbol{W}}^{-1}\boldsymbol{y},\\
\hat{\sigma}^2 &= \frac{1}{N}(\boldsymbol{y}-\boldsymbol{X}\hat{\boldsymbol{\beta}})'\hat{\boldsymbol{W}}^{-1}(\boldsymbol{y}-\boldsymbol{X}\hat{\boldsymbol{\beta}}),\\
\frac{\partial l_j}{\partial \theta} &= \frac{1}{2\sigma^2}\sum_j (\boldsymbol{y}_j-\boldsymbol{X}_j\boldsymbol{\beta})'\boldsymbol{W}_j^{-1}\boldsymbol{Z}_j\frac{\partial \boldsymbol{\Psi}}{\partial \theta}\boldsymbol{Z}_j'\boldsymbol{W}_j^{-1}(\boldsymbol{y}_j-\boldsymbol{X}_j\boldsymbol{\beta})\\
&\quad - \frac{1}{2}\sum_j \operatorname{tr}\left(\boldsymbol{Z}_j'\boldsymbol{W}_j^{-1}\boldsymbol{Z}_j\frac{\partial \boldsymbol{\Psi}}{\partial \theta}\right),\\
-E\left(\frac{\partial^2 l_j}{\partial \theta_1\,\partial \theta_2}\right) &= \frac{1}{2}\sum_j \operatorname{tr}\left(\boldsymbol{Z}_j'\boldsymbol{W}_j^{-1}\boldsymbol{Z}_j\frac{\partial \boldsymbol{\Psi}}{\partial \theta_1}\boldsymbol{Z}_j'\boldsymbol{W}_j^{-1}\boldsymbol{Z}_j\frac{\partial \boldsymbol{\Psi}}{\partial \theta_2}\right),
\end{aligned}
$$

where $\boldsymbol{W} = \bigoplus_{j=1}^{N_2}\boldsymbol{W}_j$ is the block-diagonal matrix with blocks $\boldsymbol{W}_j$. ($\boldsymbol{V}$ can be defined similarly.) The first two equations are explicit, although they depend on the estimate of $\boldsymbol{W}_j$ which involves $\hat{\boldsymbol{\Psi}}$. In the third and fourth equations, θ, θ_1 and θ_2 are any parameters involved in $\boldsymbol{\Psi}$. The Hessian matrix $\boldsymbol{H}$ and the scoring vector $\boldsymbol{s}$ for the parameters $\boldsymbol{\theta}$ involved in $\boldsymbol{\Psi}$ are assembled from these equations, and an update of them is given by

$$\hat{\boldsymbol{\theta}}_{\text{new}} = \hat{\boldsymbol{\theta}}_{\text{old}} + \boldsymbol{H}^{-1}\boldsymbol{s}\,,$$

with $\boldsymbol{H}$ and $\boldsymbol{s}$ evaluated at the current (old) estimates of $\boldsymbol{\beta}$, σ^2 and $\boldsymbol{\Psi}$. The formulas for $\hat{\boldsymbol{\beta}}$ and $\hat{\sigma}^2$ have to be evaluated after each update, unless the update is very close to $\boldsymbol{\emptyset}$. Provisions have to be made to ensure that the estimated variance matrix $\hat{\boldsymbol{\Psi}}$ is non-negative definite. A practical proposition is to estimate a decomposition of $\boldsymbol{\Psi}$, such as the Cholesky or singular value decomposition; see Lindstrom and Bates [10] or Longford [16, Chap. 7].

10.5.1 EM Algorithm

We use the Fisher scoring algorithm as the M-step of the EM algorithm. Central to this is a description of the complete-data log-likelihood as a function of sufficient statistics. Noting that

$$
\begin{aligned}
\boldsymbol{W}_j^{-1} &= \boldsymbol{I}_{n_j} - \boldsymbol{Z}_j\boldsymbol{\Psi}\boldsymbol{G}_j^{-1}\boldsymbol{Z}_j'\,,\\
\det(\boldsymbol{W}_j) &= \det(\boldsymbol{G}_j)\,,
\end{aligned}
\tag{10.10}
$$

where $\boldsymbol{G}_j = \boldsymbol{I}_{p_z} + \boldsymbol{Z}_j'\boldsymbol{Z}_j\boldsymbol{\Psi}$, a convenient set of minimal sufficient statistics is

$$(\boldsymbol{X}_j\ \boldsymbol{y}_j)'\,\boldsymbol{Z}_j\,,$$
$$(\boldsymbol{X}\ \boldsymbol{y})'\,(\boldsymbol{X}\ \boldsymbol{y})\,.$$

These statistics are minimal sufficient also for the paired ANCOVA model.

The role of the E-step is to estimate these summaries. The building blocks of the summaries are the missing values themselves and products $u_1 u_2$, where at least one of u_1 and u_2 is a missing value of a variable in $\boldsymbol{X}$ or $\boldsymbol{y}$, on the same elementary unit ij. In summary, we require the conditional expectation and variance of each missing value and the conditional covariance of every pair of values missing for the same elementary unit. The conditioning is on the recorded data.

Some variables in $\boldsymbol{X}$ may be constant within clusters j ($X_{ij} \equiv X_j$); for a missing value on such a variable its conditional expectation and variance are still required, but it is only one value each, common to all the elementary units in the cluster. We split the task of evaluating the conditional moments of the missing values to the missing outcomes $\underline{\boldsymbol{y}}_m$ and the missing values of the covariates, $\underline{\boldsymbol{X}}_m$, according to the partitioning of the joint distribution of $\underline{\boldsymbol{X}}$ and $\underline{\boldsymbol{y}}$:

$$(\underline{\boldsymbol{y}}_r\,, \underline{\boldsymbol{X}}_r)(\underline{\boldsymbol{y}}_m \mid \boldsymbol{y}_r\,, \boldsymbol{X})(\underline{\boldsymbol{X}}_m \mid \boldsymbol{y}_r\,, \boldsymbol{X}_r).$$

Values Missing on $\underline{y}$ Only

We assume first that values are missing only on the outcome $\underline{y}$. Information about a missing value on $\underline{y}$ is contained in the values of the covariates for the unit and in the other units of the cluster, which share the same value of $\underline{\boldsymbol{\delta}}_j$. Assuming MAR given $\boldsymbol{X}$, the conditional distribution of a missing value $\underline{y}_{ij}$, given $\boldsymbol{y}_r$ and $\boldsymbol{X}$, is obtained from the joint distribution of $\underline{\boldsymbol{y}}_j$:

$$\underline{\boldsymbol{y}}_j \sim \mathcal{N}\left(\boldsymbol{X}_j\boldsymbol{\beta}\,, \sigma^2\boldsymbol{W}_j\right).$$

For notational simplicity, suppose the values $\underline{\boldsymbol{y}}_{m,j} = (\underline{y}_{1j}\,, \underline{y}_{2j}\,, \ldots, \underline{y}_{q_j j})'$ are missing, and the remainder of $\underline{\boldsymbol{y}}_j$ is recorded. Let $\boldsymbol{X}_j$, $\boldsymbol{Z}_j$ and $\boldsymbol{W}_j$ have the corresponding partitioning, $\boldsymbol{X}_j = (\boldsymbol{X}'_{m,j}\ \boldsymbol{X}'_{r,j})'$, $\boldsymbol{Z}_j = (\boldsymbol{Z}'_{m,j}\ \boldsymbol{Z}'_{r,j})'$ and

$$\boldsymbol{W}_j = \begin{pmatrix} \boldsymbol{W}_{r,j} & \boldsymbol{W}_{rm,j} \\ \boldsymbol{W}_{mr,j} & \boldsymbol{W}_{m,j} \end{pmatrix}.$$

The conditional distribution of $\underline{\boldsymbol{y}}_{m,j}$ given $\boldsymbol{y}_{r,j}$ is

$$\mathcal{N}\{\boldsymbol{X}_{m,j}\boldsymbol{\beta} + \boldsymbol{W}_{mr,j}\boldsymbol{W}_{r,j}^{-1}(\boldsymbol{y}_{r,j} - \boldsymbol{X}_{r,j}\boldsymbol{\beta}), \sigma^2(\boldsymbol{W}_{m,j} - \boldsymbol{W}_{mr,j}\boldsymbol{W}_{r,j}^{-1}\boldsymbol{W}_{rm,j})\}\,,$$

from which the required expectations are extracted straightforwardly. Since $\boldsymbol{W}_{r,j}$ and $\boldsymbol{W}_{m,j}$ have the same form as $\boldsymbol{W}_j$, they can be inverted either directly, or by using (10.10):

$$\boldsymbol{W}_{r,j}^{-1} = \boldsymbol{I}_{r_j} - \boldsymbol{Z}_{r,j}\boldsymbol{\Psi}\left(\boldsymbol{I}_{p_z} + \boldsymbol{Z}_{r,j}'\boldsymbol{Z}_{r,j}\boldsymbol{\Psi}\right)^{-1}\boldsymbol{Z}_{r,j}',$$

where $r_j = n_j - q_j$, and similarly for $\boldsymbol{W}_{m,j}$. Of course, this identity is useful only when the dimensions of $\boldsymbol{W}_{r,j}$ are much greater than those of $\boldsymbol{\Psi}$.

Missing Covariate Values

We separate the variables in $\underline{\boldsymbol{X}}$ into those that are never missing, $\boldsymbol{T}$, and those that contain some missing values, $\underline{\boldsymbol{U}}$; either of these matrices, or their cluster-level submatrices $\boldsymbol{T}_j$ and $\underline{\boldsymbol{U}}_j$ may be empty.

In estimating the contributions of the missing values of $\underline{y}$ we have relied on the model $(\underline{\boldsymbol{y}} \mid \boldsymbol{X})$, (10.9). Similarly, for estimating the contributions of $\underline{\boldsymbol{U}}$ we require a model for $\underline{\boldsymbol{U}}$. A practical proposition is to specify a distribution for $(\underline{\boldsymbol{U}} \mid \boldsymbol{T})$; we omit conditioning on $\boldsymbol{y}_r$ for convenience. The variables in $\underline{\boldsymbol{U}}$ can be categorical or continuous, and defined for elementary units or clusters, so several cases have to be distinguished. For the set of continuous variables in $\underline{\boldsymbol{U}}$, $\underline{\boldsymbol{U}}_{\text{con}}$, the natural model to consider is the multivariate multilevel model with no covariates:

$$\underline{\boldsymbol{U}}_{ij,\text{con}} = \boldsymbol{\mu}_{\text{con}} + \underline{\boldsymbol{\delta}}_{j,\text{con}} + \underline{\boldsymbol{\epsilon}}_{ij,\text{con}}, \tag{10.11}$$

with mutually independent centered normal random samples $\{\underline{\boldsymbol{\delta}}_{j,\text{con}}\}$ and $\{\underline{\boldsymbol{\epsilon}}_{ij,\text{con}}\}$. Dependence on the (cross-classified) categories of $\underline{\boldsymbol{U}}_{\text{cat}}$ can be introduced by allowing the mean to depend on the category. Such a variable can also be associated with cluster-level variation, or the components of $\underline{\boldsymbol{\epsilon}}_{\text{con}}$ may have category-specific variances. Note, however, that the number of categories grows very quickly with the number of variables, and so parsimony may be essential. If the normality assumption for some of the components of $\underline{\boldsymbol{U}}_{\text{con}}$ is not appropriate, the log-transformation may be applied. The E-step requires the expectations on the original scale; these are

$$\begin{aligned} E\{\exp(\underline{x})\} &= \exp\{E(\underline{x}) + \tfrac{1}{2}\operatorname{Var}(\underline{x})\}, \\ \operatorname{Var}\{\exp(\underline{x})\} &= \exp\{2\,E(\underline{x}) + \operatorname{Var}(\underline{x})\}\left[\exp\{\operatorname{Var}(\underline{x})\} - 1\right]. \end{aligned}$$

For some other non-linear transformations there are no analytical expressions. Although the delta method can be employed, one should be weary of applying approximations repeatedly (for several missing values) in iterations. The model in (10.11) can be expanded by replacing the vector of means $\boldsymbol{\mu}_{\text{con}}$ with a linear regression on $\boldsymbol{T}$ and could also include variables representing the categorical variables and their interactions.

For $\underline{\boldsymbol{U}}_{\text{cat}}$, the categorical versions of the multilevel models can be adopted, but fitting them is much more time-consuming (and ML estimation is only approximate), so they are not always practicable for the unrecorded categories. More practical alternatives assume that the within-cluster (multinomial) distributions are identical, or that the within-cluster probabilities satisfy a *normal* multilevel model with the dependence structure implied by multinomiality. That is, let $\underline{\boldsymbol{p}}_j$ be the $K \times 1$ vector of probabilities in cluster j. For each j, these probabilities add up to unity. We specify for them that

$$\underline{\boldsymbol{p}}_j = \boldsymbol{p} + \underline{\boldsymbol{\delta}}_j \,, \tag{10.12}$$

where $\underline{\boldsymbol{\delta}}_j$ is a random sample from a centered multivariate normal distribution, $\mathcal{N}(\boldsymbol{\emptyset}, \boldsymbol{\Omega})$. To satisfy the identity $\underline{\boldsymbol{p}}_j' \mathbf{1}_K = 1$ for each j, $\boldsymbol{\Omega}$ has to be such that $\mathbf{1}_K' \boldsymbol{\Omega} \mathbf{1}_K = 0$. For the vectors of counts $\underline{\boldsymbol{c}}_j$, the samples from the respective multinomial distributions with sample sizes n_j and vectors of probabilities $\boldsymbol{p}_j$, we have

$$\operatorname{Var}(\underline{\boldsymbol{c}}_j \mid \boldsymbol{p}_j) = n_j \left\{ \operatorname{diag}(\boldsymbol{p}_j) - \boldsymbol{p}_j \boldsymbol{p}_j' \right\} . \tag{10.13}$$

Equations (10.12) and (10.13) specify a two-level model which, with the admittedly invalid assumptions of normality, can be fitted by Fisher scoring. As an alternative, $\boldsymbol{\Omega}$ can be estimated by moment matching, as the variance matrix of the sample proportions $\underline{\boldsymbol{c}}_j / n_j$ in excess of that implied by (10.13).

In its generality, the EM algorithm appears rather extensive and requiring a substantial programming effort. Its implementation is practical when only a few patterns of missingness arise, so that a limited number of cases discussed above have to be dealt with. A distinct disadvantage of the general approach is that the already iterative and not particularly simple algorithm has to be interfered with. The compounding of two kinds of iterations, Fisher scoring and EM algorithm, generates no problems additional to those for either kind [20]; after a few EM iterations, one iteration of Fisher scoring within each EM iteration is usually sufficient.

10.5.2 Multiple Imputation

Multiple imputation overcomes some of the difficulties arising in the EM algorithm. First of all, the (Fisher scoring or another) complete-data algorithm is applied without any alterations. This is a great advantage for an analyst not acquainted with all the details of the algorithm, or when the algorithm is available only in a compiled form.

Like the EM, multiple imputation is based on a model for the missing values. A proposition practical in many settings is that of MAR, after conditioning on as many variables as is feasible. If we cannot condition on all the variables, those that are more closely associated (correlated) with the incompletely recorded variables should be preferred.

The plausible values can be generated in stages. For this, the incompletely recorded variables are classified into a small number of groups. The model for missing data for the first group relies on the information from that group and the completely recorded variables. At the second stage, missing data for the second group of variables are generated from a model which conditions on the first-group and completely recorded variables, and so on. At each stage the imputed values in the previous stages are conditioned on. It is advantageous to impute first for variables with few missing values which are useful for conditioning at later stages, and to deal with variables with a lot of missing values last.

When multiple imputation is organized in stages information about the missing values is not used fully; we trade efficiency for computational tractability or simplicity. However, efficiency is lost only due to handling missing-data information; only the component B in (10.8) is affected. An iterated version of this scheme makes better use of the information in incomplete records. The method described in van Buuren et al. [28] corresponds to the setting in which each incompletely recorded variable forms a group on its own.

10.5.3 Monotone Pattern of Missingness

In general, data can be missing with any conceivable pattern $\boldsymbol{r}$. However, some sets of variables may have a restricted pattern of missingness. An important example is when the only patterns occurring for a set of variables, suitably permuted, are such that $\boldsymbol{r}$ is non-increasing. That is, each $\boldsymbol{r}$ comprises a segment of ones, followed by a segment of zeros; either segment may be empty. This commonly arises in longitudinal studies, in which a subject may drop out at any stage, but no returns to the study occur after skipping a stage.

In this setting, multiple imputation by stages can be applied, with each variable included in a stage on its own. The plausible values for the first time-point are generated first; then the plausible values for the second time-point are generated, conditioning on the (recorded or imputed) values for the first time point, and so on for values at the second and subsequent time-points.

In this way, the plausible values are generated using univariate analyses (completing one variable at a time), and all information about the missing values is exploited. Variations on this theme include grouping the variables so that missingness is monotone with respect to these groups; if a subject has some values missing for variables in group k, all values are missing for variables in groups $k+1$, $k+2$, $\ldots$. Also, the monotone pattern may apply but for a small number of exceptions. In such a case, we may proceed with imputation for the remaining units, and make different arrangements for the few units that break the monotone pattern.

10.5.4 Sensitivity Analysis

All theoretical results about multiple imputation rely on the correctness of the assumption about the model for the missing values. Since this model, typically based on MAR, cannot be verified, we have to address the concern that the possibly inappropriate model for missing data has induced a bias of the incomplete-data estimator $\underline{\tilde{\theta}}$.

Exploring the entire range of alternative models for missing values is rarely feasible because of the vast variety of MNAR mechanisms that can be conceived. However, when estimating a specific parameter θ, we can define alternative models for missing data that stack the plausible values against the inference drawn from MAR-generated plausible values.

By way of an example, suppose the difference of the means of a variable for men and women in a given population is of interest. In the analysis of the study that collected the relevant data, MAR is assumed for the missing values. Suppose the hypothesis of no difference between men and women is rejected; the estimated mean for men is higher than for women. In the sensitivity analysis, we alter the plausible values, "disadvantaging" men's missing values. A practical way of doing this is to reduce each plausible value (replacement for a missing value) for men by c and increase each plausible value for women by c. Instead of adding or subtracting, for a variable with positive values, we can multiply or divide each plausible value by a given positive constant. This constant, or its additive version, describes the extent of departure from MAR. Of interest is the smallest value c for which substantially different conclusions are drawn. We can then speculate whether the MNAR process given by this constant is feasible. If it is not, we conclude that the result obtained assuming MAR holds for all feasible mechanisms of missingness. Otherwise, the result is inconclusive because of the uncertainty about this mechanism.

10.5.5 Generalized Linear Mixed Models

In generalized linear models, the (univariate) outcome $\underline{y}$ is related to the covariates $\boldsymbol{x}$ by the formula

$$E(\,\underline{y} \mid \boldsymbol{x}; \boldsymbol{\beta}) = f(\boldsymbol{x}\boldsymbol{\beta})\,,$$

where f is a strictly monotone function (its inverse is called the *link* function), and the conditional distribution of $\underline{y}$, given $\boldsymbol{x}\boldsymbol{\beta}$, is in a specified parametric family. The ordinary regression model corresponds to identity f and normal distribution of $\underline{y}$. The natural extension to generalized linear mixed models (GLMM) is by a reference to generalized ANCOVA (gANCOVA) models, that is, for the outcomes $\underline{\boldsymbol{y}}_j$ in cluster j

$$E(\,\underline{\boldsymbol{y}}_j \mid \boldsymbol{X}_j\,, \boldsymbol{Z}_j\,, \boldsymbol{\beta}, \boldsymbol{\delta}_j\,) = f\left(\boldsymbol{X}_j\boldsymbol{\beta} + \boldsymbol{Z}_j\boldsymbol{\delta}_j\right)\,,$$

where $\boldsymbol{X}_j$ and $\boldsymbol{Z}_j$ are the regression and variation design matrices for cluster j, $\boldsymbol{\beta}$ is a vector of (average) regression parameters, $\boldsymbol{\delta}_j$ is the vector of deviations of the regression specific to cluster j from the average regression, and the function f is applied elementwise. In GLMM, $\underline{\boldsymbol{\delta}}_j \sim \mathcal{N}(\boldsymbol{\emptyset}, \boldsymbol{\Omega})$, iid, for some variance matrix $\boldsymbol{\Omega}$, whereas in the paired ANCOVA model $\boldsymbol{\delta}_j$ are parameters, subject to an identification constraint. Note that the qualifier "average" is appropriate only on the linear scale ($\boldsymbol{X}_j\boldsymbol{\beta}$ and $\boldsymbol{Z}_j\boldsymbol{\delta}_j$), not on the scale of outcomes $\underline{\boldsymbol{y}}_j$, unless the link function f^{-1} is linear.

We can associate a specific GLMM with a (normal) random coefficient model, by replacing the link with identity, and the distributional assumption with normality. The algorithms for fitting a GLMM are rather complex. One class of them is based on analytical approximations to the log-likelihood [1], and another on numerical approximations, using Gaussian quadrature [14]. The implementation of Markov chain Monte Carlo (MCMC) algorithms [7] has revolutionized model fitting, although some computational issues associated with MCMC are still awaiting practical resolution. For a more extensive treatment of these models and algorithms, see Chapters 6 and 9.

None of the algorithms for fitting GLMM admits a description in terms of a set of sufficient statistics much smaller than the number of observations. As a consequence, implementing the EM algorithm with either of them is extremely difficult. Multiple imputation, on the other hand, is much better suited with GLMM because the model for missing data, and the process of generating plausible values is unaffected by the complete-data model and the complete-data algorithm is used without any alteration.

The advantages of multiple imputation are not specific to GLMM; they pertain to all complex complete-data analyses, including random coefficient models with multiple layers of nesting and cross-classifications, as well as their non-linear extensions.

10.6 Other Applications of EM and MI

The range of problems that can be formulated as analysis of incomplete data is much wider than the stereotype setting in which, contrary to the plan, some observations were not recorded. We can declare some data as missing even if we never had any intention to collect its values. An important generic example is that of the measurement error in the covariates. We illustrate this application first on a simple example, and then proceed to general cases.

Suppose we are interested in the simple regression

$$\underline{y} = b_0 + b_1 x + \underline{\epsilon}\,,$$

with the usual assumptions of independence, normality and homoscedasticity of $\underline{\epsilon}$, $\underline{\epsilon} \sim \mathcal{N}(0, \sigma^2)$, iid. Instead of the covariate x we observe only its corrupted version, $\underline{u} = x + \underline{\xi}$, where $\underline{\xi}$ is $\mathcal{N}(0, \tau^2)$, iid, independent of x and $\underline{y}$. This simple (measurement error) model/process can be more appropriately named the *corruption* model/process, since the deviation of $\underline{u}$ from x may be due to causes other than measurement. Suppose first that τ^2 is known.

In the complete-data analysis, we evaluate

$$\hat{\underline{b}}_1 = \frac{\sum_{i=1}^{n}(x_i - \bar{x})(\underline{y}_i - \underline{\bar{y}})}{\sum_{i=1}^{n}(x_i - \bar{x})^2} \tag{10.14}$$

($\bar{x}$ is the sample mean of the x_i and $\underline{\bar{y}}$ the sample mean of the $\underline{y}_i$). Since x_i are not observed, we estimate their contributions to the numerator and denominator in (10.14) by functions of $\underline{u}_i$. This is easy to accomplish by moment matching. In the following identities, E_ξ indicates expectation over the deviations $\underline{\xi}_i$. We have

$$E_\xi\{(\underline{u}_i - \underline{\bar{u}})(y_i - \bar{y})\} = (x_i - \bar{x})(y_i - \bar{y})$$

but

$$E_\xi\{(\underline{u}_i - \underline{\bar{u}})^2\} = (x_i - \bar{x})^2 + \frac{n-1}{n}\tau^2 .$$

So, the numerator in (10.14) can be estimated naively, replacing each x_i with $\underline{u}_i$, whereas the denominator has to be adjusted; its unbiased estimator is

$$\sum_{i=1}^{n}(\underline{u}_i - \underline{\bar{u}})^2 - (n-1)\tau^2 .$$

The combination of the E- and M-steps yields the estimator

$$\tilde{\underline{b}}_1 = \frac{\sum_{i=1}^{n}(\underline{u}_i - \underline{\bar{u}})(\underline{y}_i - \bar{y})}{\sum_{i=1}^{n}(\underline{u}_i - \underline{\bar{u}})^2 - (n-1)\tau^2} .$$

This slope estimator is steeper than the naive estimator of b_1; this phenomenon is generally referred to as *attenuation*. The estimator $\tilde{\underline{b}}_1$ can be derived without any reference to the EM algorithm. In any case, this EM algorithm is very unusual; only one application of the E- and M-steps is required because the E-step is independent of the M-step.

When τ^2 is not known, its estimator can be used. For this, the design may have to be expanded and a random sample of the values of x_i observed independently, by the same corrupted process, twice or several times. With such observations, τ^2 can be estimated from the within-x variation. For an application in educational testing, see Longford [15].

Multiple imputation entails simulating plausible values of x_i, followed by their substitution in the complete-data analysis (10.14). From

$$\begin{pmatrix} \underline{x} \\ \underline{u} \end{pmatrix} \sim \mathcal{N}\left\{ \begin{pmatrix} \mu_x \\ \mu_x \end{pmatrix}, \begin{pmatrix} \sigma_x^2 & \sigma_x^2 \\ \sigma_x^2 & \sigma_x^2 + \sigma_\xi^2 \end{pmatrix} \right\}$$

we have

$$(\underline{x}_i \mid u_i) \sim \mathcal{N}\left(\frac{\sigma_x^2}{\sigma_x^2 + \sigma_\xi^2} u_i + \frac{\sigma_\xi^2}{\sigma_x^2 + \sigma_\xi^2}\mu_x , \frac{\sigma_x^2\sigma_\xi^2}{\sigma_x^2 + \sigma_\xi^2} \right) .$$

This can be interpreted as shrinkage toward the mean μ_x or preferring to err on the side of the overall mean. The reward for this strategy is smaller mean squared error on average.

Since the complete-data analysis is very simple, involving a short list of linear and quadratic statistics, the EM algorithm is much more practical. However, for generalized linear models and their multilevel extensions, multiple imputation is much more versatile because simulation of the missing values is unaffected by the complexity of the complete-data analysis.

An alternative approach to handling measurement error, competing in versatility with multiple imputation, is the simulation-extrapolation method (SIMEX) of Carroll et al. [2]. With this method, corrupted values of the covariates are generated with a range of levels of the corruption greater than the realized level. The complete-data estimator is evaluated for each dataset completed with the simulated corrupted values, and inferences are made by extrapolating the values of the estimates to zero corruption. SIMEX relies on a good method of extrapolation and an appropriate choice of the levels of corruption. Difficulties arise with complex complete-data analyses when little is known about the dependence of the estimators on the extent of corruption and when several variables are subject to corruption. In comparison, simulation of the plausible values in multiple imputation is based solely on the assumed process of corruption, and is oblivious to the complexity of the complete-data analysis.

Misclassification can be regarded as a corruption process. The corrupted version of a dichotomous variable $\underline{x}$ is another dichotomous variable $\underline{u}$; $\underline{x} \neq \underline{u}$ with the misclassification probability. With multiple imputation, several draws are made from the plausible distribution of $\underline{x}_m$, which is established from the estimated conditional distribution of $\underline{x}_m$ given x_r and u. This distribution usually depends on the probability of misclassification. The plausible distribution is based on a plausible (randomly drawn) value of this probability. When the misclassification probability depends on some covariates, and the parameters of this dependence are estimated, the uncertainty about these parameters also has to be reflected in the plausible values of $\underline{x}$.

10.6.1 Random Coefficients as the Missing Data

Analysis with random coefficient models can be naturally formulated as a missing information problem. If the random coefficients were known the (ordinary least squares) analysis would be straightforward. The EM algorithm with this approach was developed by Dempster et al. [5]. This EM algorithm converges slowly when the fraction of the missing information is substantial, and such cases are encountered frequently. With direct maximization algorithms, such as Fisher scoring and iteratively reweighted least squares [8], such problems are much less acute. However, direct algorithms are difficult to implement for very complex models. The programming task can be reduced by a judicious assignment of a set of coefficients as missing data, selecting a simpler complete-data analysis. The E-step of the EM algorithm would estimate the contributions of the "missing" random coefficients to the sufficient statistics. In most settings, EM algorithm is not feasible, and the missing data have to be represented by multiple imputation. Since the complete-data algorithm is iterative, multiple imputation may have to be applied in each iteration. This complicates the assessment of convergence somewhat. Also, there is no proof that the good properties of the multiple-imputation estimator are maintained at the converging iteration. See Clayton and Rasbash [3] for a study of such an algorithm.

10.7 Summary

In most of statistical enterprise, inferences are made about specified (or implied) populations. Multilevel analysis, as many other generic methods of analysis, assume that the analysed dataset is representative of the studied population. Good representation is often eroded by selective missingness, and so methods for dealing with incomplete data should be in the toolkit of every statistical analyst. This imperative is even stronger in studying human populations because human subjects are often poorly motivated, easily distracted while responding, and do not cooperate with study protocols perfectly.

Although several kinds of data incompleteness can be handled by multilevel analysis without having to make special arrangements, invisible bias may be incurred when the analysed dataset is treated as complete. This chapter discussed two general approaches to dealing with missing values—the EM algorithm and multiple imputation. Both approaches consider an efficient complete-data analysis (typically, by maximizing a likelihood). In the EM algorithm, this analysis is adjusted, and applied iteratively. In multiple imputation, the complete-data analysis is used without any alteration, but multiple sets of replacements for the missing values have to be generated. Multiple imputation is more versatile, applicable with complex complete-data analyses in which EM would be very difficult to implement.

Methods for missing data are applicable in a much wider range of problems. Many complex problems could be simplified if some additional information (data) were available. If such data is regarded as missing a general approach to dealing with missing information can be invoked. The chapter discussed measurement error and complex random coefficient models as examples in which secondary applications of missing-data methods can be applied, leading to a reduction in the computational (programming) effort and enabling us to exploit available algorithms constructed for simpler problems.

References

1. N. E. Breslow and D. G. Clayton. Approximate inference in generalized linear mixed models. *Journal of the American Statistical Association*, 88:9–25, 1993.
2. R. J. Carroll, D. Ruppert, and L. A. Stefanski. *Measurement Error in Nonlinear Models.* Chapman & Hall, London, 1995.
3. D. G. Clayton and J. Rasbash. Estimation in large crossed random-effect models by data augmentation. *Journal of the Royal Statistical Society, Series A*, 162: 425–448, 1999. (with discussion)
4. A. P. Dempster, N. M. Laird, and D. B. Rubin. Maximum likelihood from incomplete data via the *EM* algorithm. *Journal of the Royal Statistical Society, Series B*, 39:1–38, 1977. (with discussion)
5. A. P. Dempster, D. B. Rubin, and R. K. Tsutakawa. Estimation in covariance components models. *Journal of the American Statistical Association*, 76:341–353, 1981.
6. EM and related algorithms (special issue). *Statistica Sinica*, 5(1):1–107, 1995.
7. A. Gelman, J. B. Carlin, H. S. Stern, and D. B. Rubin. *Bayesian Data Analysis.* Chapman & Hall, London, 1995.
8. H. Goldstein. *Multilevel Statistical Models*, 3rd edition. Edward Arnold, London, 2003.
9. M. J. R. Healy and M. Westmacott. Missing values in experiments analyzed on automatic computers. *Applied Statistics*, 5:203–206, 1956.
10. M. J. Lindstrom and D. M. Bates. Newton-Raphson and EM algorithms for linear mixed-effects models for repeated-measures data. *Journal of the American Statistical Association*, 83:1014–1022, 1988.
11. R. J. A. Little and D. B. Rubin. *Statistical Analysis with Missing Data.* Wiley, New York, 1987.
12. N. T. Longford. A fast scoring algorithm for maximum likelihood estimation in unbalanced mixed models with nested random effects. *Biometrika*, 74:817–827, 1987.
13. N. T. Longford. *Random Coefficient Models.* Oxford University Press, Oxford, UK, 1993.
14. N. T. Longford. Logistic regression with random coefficients. *Computational Statistics & Data Analysis*, 17:1–15, 1994.
15. N. T. Longford. Reliability of essay rating and score adjustment. *Journal of Educational and Behavioral Statistics*, 19:171–201, 1994.
16. N. T. Longford. *Missing Data and Small-Area Estimation. Modern Analytical Equipment for the Survey Statistician.* Springer, New York, 2005.
17. T. A. Louis. Finding the observed information matrix when using the *EM* algorithm. *Journal of the Royal Statistical Society, Series B*, 44:226–233, 1982.
18. I. Meilijson. A fast improvement to the EM algorithm on its own terms. *Journal of the Royal Statistical Society, Series B*, 51:127–138, 1989.
19. X.-L. Meng and D. B. Rubin. Using EM to obtain asymptotic variance-covariance matrices: The SEM algorithm. *Journal of the American Statistical Association*, 86:899–909, 1991.
20. X.-L. Meng and D. B. Rubin. Maximum likelihood estimation via the ECM algorithm: A general framework. *Biometrika*, 80:267–278, 1993.

21. X.-L. Meng and D. van Dyk. The EM algorithm — an old folk-song sung to a fast new tune. *Journal of the Royal Statistical Society, Series B*, 59:511–567, 1997. (with discussion)
22. T. Orchard and M. A. Woodbury. A missing information principle: Theory and applications. In L. M. Le Cam, J. Neyman, and E. L. Scott, editors, *Proceedings of the Sixth Berkeley Symposium on Mathematical Statistics and Probability*, volume 1, pages 697–715. University of California Press, Berkeley, 1972.
23. D. B. Rubin. Inference and missing data. *Biometrika*, 63:581–592, 1976. (with discussion)
24. D. B. Rubin. Noniterative least squares estimates, standard errors and F-tests for analyses of variance with missing data. *Journal of the Royal Statistical Society, Series B*, 38:270–274, 1976.
25. D. B. Rubin. *Multiple Imputation for Nonresponse in Surveys.* Wiley, New York, 1987.
26. D. B. Rubin. Multiple imputation after 18+ years. *Journal of the American Statistical Association*, 91:473–489, 1996.
27. J. L. Schafer. *Analysis of Incomplete Multivariate Data.* Chapman & Hall, London, 1997.
28. S. van Buuren, H. C. Boshuisen, and D. L. Knook. Multiple imputation of missing blood pressure covariates in survival analysis. *Statistics in Medicine*, 18:681–694, 1999.
29. C. F. J. Wu. On the convergence properties of the EM algorithm. *Annals of Statistics*, 11:95–103, 1983.
30. F. Yates. The analysis of replicated experiments when the field results are incomplete. *Empirical Journal of Experimental Agriculture*, 1:129–142, 1933.

11

Resampling Multilevel Models

Rien van der Leeden[1], Erik Meijer[2], and Frank M. T. A. Busing[1]

[1] Leiden University, Department of Psychology
[2] University of Groningen, Faculty of Economics, and RAND Corporation

11.1 Introduction

Estimation in (linear) multilevel models usually relies on maximum likelihood (ML) methods. The various computer programs for multilevel analysis employ versions of full information (FIML) and restricted maximum likelihood (REML) methods. Two vital assumptions underlying ML theory are that (a) the residuals are i.i.d. with a distribution from a specified class, usually the multivariate normal, and (b) the sample size is (sufficiently) large. More specifically, the attractive properties of FIML estimators—consistency, (asymptotic) efficiency and (asymptotic) normality—are derived from the supposition that the sample size goes to infinity. In practice, however, these assumptions will frequently be met only approximately, which may lead to severely biased estimators and incorrect standard errors [7].

Resampling methods can be used to obtain consistent estimators of bias and standard errors, and to obtain confidence intervals and bias-corrected estimates of model parameters. A number of general resampling approaches are found throughout the literature, of which we mention the bootstrap and the jackknife, permutation, and cross-validation. Bayesian Markov chain Monte Carlo methods [e.g., 16, 23] and simulation-based estimators for mixed nonlinear models [e.g., 27, 61, 63] are also closely related to these resampling methods. Particularly, bootstrap and jackknife procedures have proven to be methods that yield satisfactory results in small sample situations under minimal assumptions. In this chapter, we discuss resampling of multilevel data by means of bootstrap and jackknife procedures. In cases where the assumptions underlying ML methods for estimating multilevel models are violated, bootstrap and jackknife estimation may provide useful alternatives.

J. de Leeuw, E. Meijer (eds.), *Handbook of Multilevel Analysis*,
© Springer 2008

The application of the bootstrap and the jackknife to multilevel models is not straightforward. For the bootstrap, there are several possibilities, depending on the nature of the data and the assumptions one is willing to make. Each of them, however, has its own associated problems. In this chapter, we discuss three different approaches, which are derived from general principles of bootstrap theory and apply concepts adapted from bootstrapping regression models.

The application of the jackknife to multilevel analysis is based on a version of the delete-m jackknife approach [57, Section 2.3]. In this procedure, subsamples are obtained from the original sample by successively removing mutually exclusive groups of size m. For the application to multilevel analysis, the delete-m jackknife has been adapted for groups of unequal size.

Bootstrap and jackknife estimation in the context of multilevel analysis have been studied by several authors and for various models and situations. Laird and Louis [33, 34] discuss empirical Bayes confidence intervals based on bootstrap samples and Moulton and Zeger [45] study bootstrapping a model for repeated measurements. Bellmann et al. [3] use a parametric bootstrap for a panel data model that is essentially a multilevel model and Booth [4] similarly uses a parametric bootstrap for generalized linear mixed models. Goldstein [24] presents an iterated bootstrap based on the results of Kuk [32]. A theoretical analysis of nonparametric bootstrapping of balanced two-level models without covariates is given in Davison and Hinkley [14, pp. 100–102]. Our discussion largely follows the lines of the systematic development of resampling methods for multilevel models in Busing et al. [8, 9, 10, 12], Van der Leeden et al. [68], and Meijer et al. [42, 43].

In this chapter, we focus on `FIML` estimation for multilevel linear models with two levels. The ideas, however, are directly applicable to `REML` estimators and generalize straightforwardly to models with three or more levels.

In Section 11.2, we define the model upon which we center our discussion and we elaborate on the consequences of violating the assumptions of `ML` estimation in multilevel models. In Section 11.3, we briefly discuss the general ideas of the bootstrap and the jackknife, and the specific issues involved in the application of the bootstrap to regression models. Section 11.4 provides an extensive discussion of the three methods for bootstrap implementation, as well as a number of approaches to construct confidence intervals. In Section 11.5, the application of the jackknife to multilevel models is discussed. Section 11.6 briefly discusses the availability of resampling options in existing software for multilevel analysis. In Section 11.7, we discuss some results of evaluation studies of the various resampling approaches and in Section 11.8, we briefly discuss application of the presented approaches to other types of multilevel models, and we mention various possible extensions and alternatives to the (bootstrap) resampling methods presented.

11.2 Model, ML Estimation, and Assumptions

For our discussion of resampling methods for multilevel models, we consider a version of the two-level mixed linear model. Suppose data are obtained from n individuals nested within J groups, with group j containing n_j individuals. For each group j, the model is given by

$$\underline{\boldsymbol{y}}_j = \boldsymbol{X}_j \boldsymbol{\beta} + \boldsymbol{Z}_j \underline{\boldsymbol{\delta}}_j + \underline{\boldsymbol{\epsilon}}_j, \tag{11.1a}$$

with

$$\begin{pmatrix} \underline{\boldsymbol{\epsilon}}_j \\ \underline{\boldsymbol{\delta}}_j \end{pmatrix} \sim \mathcal{N}\left(\begin{pmatrix} \boldsymbol{\emptyset} \\ \boldsymbol{\emptyset} \end{pmatrix}, \begin{pmatrix} \sigma^2 \boldsymbol{I}_{n_j} & \boldsymbol{\emptyset} \\ \boldsymbol{\emptyset} & \boldsymbol{\Omega} \end{pmatrix} \right) \tag{11.1b}$$

and $(\underline{\boldsymbol{\epsilon}}_j, \underline{\boldsymbol{\delta}}_j) \perp (\underline{\boldsymbol{\epsilon}}_\ell, \underline{\boldsymbol{\delta}}_\ell)$ for all $j \neq \ell$. Under these assumptions,

$$\underline{\boldsymbol{y}}_j \sim \mathcal{N}(\boldsymbol{X}_j \boldsymbol{\beta}, \boldsymbol{Z}_j \boldsymbol{\Omega} \boldsymbol{Z}_j' + \sigma^2 \boldsymbol{I}_{n_j}). \tag{11.2}$$

In some situations, however, we assume that the explanatory variables are not fixed constants, but random variables with unspecified distributions, so that we should write $\underline{\boldsymbol{X}}_j$ and $\underline{\boldsymbol{Z}}_j$ instead of $\boldsymbol{X}_j$ and $\boldsymbol{Z}_j$. In that case, (11.2) should be viewed as the conditional distribution of $\underline{\boldsymbol{y}}_j$ given $\underline{\boldsymbol{X}}_j = \boldsymbol{X}_j$ and $\underline{\boldsymbol{Z}}_j = \boldsymbol{Z}_j$.

We tend to think of the model as being derived from a two-level slopes-as-outcomes model, which is a special case of the mixed linear model presented here, but we do not need that in our discussion, so we will confine ourselves to the mixed linear model specification. See de Leeuw and Meijer [15] for a more extensive discussion of the model and the various ways to interpret it.

The parameters of the model described by (11.1a) can be divided into a set of *fixed* parameters, the elements of $\boldsymbol{\beta}$, and a set of *random* parameters, i.e., σ^2, the variance of the level-1 residuals, and the elements of $\boldsymbol{\Omega}$, the variances and covariances of the level-2 residuals. The random parameters are commonly referred to as *variance components*. Under the given normality assumptions for the residuals, `FIML` estimates are obtained by maximizing the loglikelihood function with respect to all model parameters. The asymptotic covariance matrix of the estimators is the inverse of the information matrix, which is derived from standard `ML` theory. Standard errors for both fixed and random parameters are obtained by taking the square roots of the diagonal elements of an estimate of this matrix.

The assumptions underlying `ML` estimation need closer examination. Two general assumptions were briefly mentioned above. The first assumption is a sufficiently large sample size. Hierarchical data structures, however, make sample size a more complicated issue: besides the total sample size, the sample size at each level of the hierarchy has to be considered. The second assumption is (multivariate) normality of the residuals. In multilevel models, each level in the data generates its own residuals. Therefore, several distributions are involved in the estimation procedure for which the normality assumption must be met. One further assumption, often made more implicitly, is that the estimated model is correct. In other words, it is assumed that the conditional expectation $E(\underline{\boldsymbol{y}}_j \mid \boldsymbol{X}_j, \boldsymbol{Z}_j) = \boldsymbol{X}_j \boldsymbol{\beta}$ and the covariance matrix $\boldsymbol{V}_j$ of $\underline{\boldsymbol{y}}_j$ conditional on $\boldsymbol{X}_j$ and $\boldsymbol{Z}_j$, equal to $\boldsymbol{V}_j = \boldsymbol{Z}_j \boldsymbol{\Omega} \boldsymbol{Z}_j' + \sigma^2 \boldsymbol{I}_{n_j}$, are specified correctly.

Obviously, in practical research, the aforementioned assumptions may easily be violated, especially the assumption of a large sample size. The effects that these violations have on the quality of the estimators and their standard errors in multilevel (or similar) models have been discussed by several authors. It is well understood, theoretically, that FIML estimators of the variance components are (negatively) biased [e.g., 56, p. 240]. Moreover, several simulation studies show that this bias can be substantial. Particularly when the sample size is small, FIML may fail [7, 31, 67]. Magnus [40] and Breusch [6], however, proved that the ML estimators of the fixed parameters are unbiased if the random component $\underline{\boldsymbol{r}}_j = \boldsymbol{Z}_j \underline{\boldsymbol{\delta}}_j + \underline{\boldsymbol{\epsilon}}_j$ is symmetrically distributed. Even if the components of $\underline{\boldsymbol{\delta}}_j$ and $\underline{\boldsymbol{\epsilon}}_j$ are skewed, this requirement will be approximately satisfied due to a central limit theorem argument. Furthermore, theory for ML estimation under distributional misspecification [69, 71] ensures that even in such cases, the ML estimators of the fixed parameters will be virtually unbiased. Asymptotic calculations of Breslow and Lin [5] for a particular class of mixed linear models confirm that bias in the fixed parameter estimators is negligible, whereas variance component estimators may be seriously biased.

Standard errors are based on large sample theory as well. The idea is that as the sample size goes to infinity, the distribution of the estimators converges to a (multivariate) normal distribution with covariance matrix equal to the inverse of the information matrix. The standard errors of the ML estimators, as reported by the various multilevel analysis programs, are the square roots of the diagonal elements of an estimate of this matrix. In finite samples, the covariance matrix of the estimators may not be approximated well by the asymptotic covariance matrix. Moreover, if the distributional assumptions are incorrect, the asymptotic covariance matrix differs from the inverse of the information matrix as assumed by the ML method. As a result, asymptotically correct standard errors based on a so-called sandwich estimator of the asymptotic covariance matrix may be quite different from standard errors based on the inverse of the information matrix. This difference appears to be small for the fixed parameters, but may be large for the variance components [70]. Furthermore, convergence to normality may not be satisfactory for the distributions of the estimators. Busing [7] shows that they can be severely skewed for the variance components.

11.3 General Theory of Bootstrap and Jackknife

In this section, we discuss theory and general principles of the bootstrap and the jackknife. Ordinary regression models and multilevel models share similar characteristics. Therefore, to prepare the stage for bootstrapping multilevel models, we illustrate the principles of the bootstrap with respect to regression models. The jackknife is introduced in its classic form.

11.3.1 The Bootstrap

The bootstrap is a general approach to estimate the bias and the variance (and consequently the standard error) of an estimator under minimal assumptions [14, 20, 28]. Let $\underline{z}$ be a random variable with distribution function F, and let $\{\underline{z}_1, \underline{z}_2, \ldots, \underline{z}_n\}$ be a random sample of size n from F. The underlying idea of the bootstrap is that the empirical distribution function $\underline{\hat{F}}_n$, generated by this sample, is a consistent estimator of the distribution function F in the population [e.g., 44, p. 507].

Let θ_0 be the true value of a parameter θ associated with the distribution F, $\theta_0 = \theta(F)$, and let $\underline{\hat{\theta}}$ be an estimator of θ from the sample, $\underline{\hat{\theta}} = \theta(\underline{z}_1, \underline{z}_2, \ldots, \underline{z}_n) = \theta(\underline{\hat{F}}_n)$. The bootstrap simulates the sampling and estimation process by drawing samples with replacement from $\hat{F}_n$, which is completely known once the original sample is obtained. In this simulation, the distribution $\hat{F}_n$ plays the role of F, and $\hat{\theta}$ plays the role of θ_0. Simulation samples, referred to as *bootstrap samples*, are drawn from $\hat{F}_n$ and $\hat{\theta}$ is estimated by $\underline{\theta}^*$ in the same way θ_0 was estimated by $\underline{\hat{\theta}}$.

Because $\underline{\hat{F}}_n \xrightarrow{\mathcal{P}} F$, it is assumed that the properties of the estimator $\underline{\theta}^*$ based on the distribution $\hat{F}_n$ give information about the properties of $\underline{\hat{\theta}}$ based on the distribution F. For example, the bias of $\underline{\theta}^*$ based on the distribution $\hat{F}_n$ is taken as an estimator of the bias of $\underline{\hat{\theta}}$ based on the distribution F. It has been proven by many authors that this approach works in many cases: It leads to consistent estimators of the properties of $\underline{\hat{\theta}}$ [e.g., 47].

The bootstrap is implemented as follows: B bootstrap samples $\{\underline{z}^*_{b1}, \underline{z}^*_{b2}, \ldots, \underline{z}^*_{bn}\}$, $b = 1, \ldots, B$, are drawn from $\hat{F}_n$ (i.e., drawn with replacement from $\{z_1, z_2, \ldots, z_n\}$). From each of the B samples, the parameter θ is estimated, thereby obtaining B estimates θ^*_b, $b = 1, \ldots, B$. The expectation of $\underline{\theta}^*$ (given $\hat{F}_n$) is estimated by the mean of the estimates θ^*_b: $\theta^*_{(.)} = \sum_{b=1}^{B} \theta^*_b / B$. The variance of $\underline{\theta}^*$ (given $\hat{F}_n$) is estimated by the variance of the estimates θ^*_b: $\widehat{\mathrm{Var}}(\underline{\theta}^*) = \sum_{b=1}^{B} (\theta^*_b - \theta^*_{(.)})^2/(B-1)$.

The bias of $\underline{\hat{\theta}}$ is estimated by the (estimated) bias of $\underline{\theta}^*$:

$$\widehat{\mathrm{Bias}}_B = \widehat{\mathrm{Bias}}(\underline{\theta}^*) = \theta^*_{(.)} - \hat{\theta}, \tag{11.3}$$

and the bias-corrected estimator of θ is therefore [see, e.g., 28, pp. 8–9]

$$\hat{\underline{\theta}}_B = \hat{\underline{\theta}} - \widehat{\underline{\text{Bias}}}_B = 2\hat{\underline{\theta}} - \underline{\theta}^*_{(.)}. \tag{11.4}$$

The standard error of $\hat{\underline{\theta}}$ is estimated by the square root of the estimated variance of $\underline{\theta}^*$ [see, e.g., 20, p. 47]:

$$\widehat{\underline{\text{se}}}_B = \sqrt{\widehat{\underline{\text{Var}}}(\underline{\theta}^*)} = \sqrt{\frac{1}{B-1}\sum_{b=1}^{B}\left(\underline{\theta}^*_b - \underline{\theta}^*_{(.)}\right)^2}. \tag{11.5}$$

The bootstrap described above is called the *nonparametric bootstrap*, because the bootstrap samples are drawn from the nonparametric empirical distribution function $\hat{F}_n$. Frequently, however, F is assumed to be a specific distribution $F(\phi)$, only depending on a parameter (or parameter vector of fixed dimension) ϕ, which may or may not be the same parameter as θ. Then, if ϕ is estimated by $\hat{\underline{\phi}}$, F can also be estimated by $\tilde{\underline{F}}_n = F(\hat{\underline{\phi}})$, instead of $\hat{\underline{F}}_n$, and a *parametric bootstrap* can be defined [see, e.g., 20, Section 6.5]. If the distributional assumption about F is correct, one benefits from the fact that this *parametric* empirical distribution function $\tilde{\underline{F}}_n$ will generally be a more efficient estimator of F.

The parametric bootstrap is defined exactly analogous to the nonparametric bootstrap, except that bootstrap samples are drawn from $\tilde{F}_n$ instead of $\hat{F}_n$. For example, if it is assumed that F is a normal distribution function with mean μ and variance σ^2, then bootstrap samples are drawn from a normal distribution with mean $\bar{x}$ and variance s^2, where $\bar{x}$ and s^2 are the mean and variance of the observed, original sample. A consequence of the parametric bootstrap is that samples are drawn from a generally smoother distribution than the nonparametric empirical distribution, which is a step function. Therefore, in contrast to the nonparametric bootstrap, the values z^*_{bi} encountered in the bootstrap samples will usually not be present in the original sample.

Bootstrapping Regression Models

Consider the linear regression model

$$\underline{y}_i = \alpha + \beta x_i + \underline{\epsilon}_i,$$

where $\underline{\epsilon}_i$ is a residual with zero mean and variance σ^2. Suppose that we have observed the sample $\{(y_1, x_1), \ldots, (y_n, x_n)\}$. Then parameter estimates $\hat{\alpha}$, $\hat{\beta}$, and $\hat{\sigma}^2$ can be obtained by using the familiar ordinary least squares method. Depending on the assumptions made, several different bootstrap methods can be used.

If the explanatory variable x is considered a *random* variable, therefore now denoted as $\underline{x}$, nonparametric bootstrap samples can easily be obtained from a straightforward generalization of the basic method discussed above to vector-valued variables. In this case, bootstrap samples are drawn from the bivariate empirical distribution function of $(\underline{y}, \underline{x})$. This means drawing complete *cases* with replacement: Bootstrap samples $\{(y_1^*, x_1^*), \ldots, (y_n^*, x_n^*)\}$ consist of pairs (y_i^*, x_i^*) that are also elements of the original sample: for each $i = 1, \ldots, n$, there exists a j, $1 \leq j \leq n$, such that $(y_i^*, x_i^*) = (y_j, x_j)$. Next, the regression parameters can be estimated from each bootstrap sample and bias-corrected estimates can be obtained, as well as standard errors of the estimators, using formulas (11.4) and (11.5).

The situation is different if the exogenous variable x is a fixed design variable, determined by the research problem or chosen by the experimenter. This happens, for instance, if x is the dose of some drug administered to a group of experimental subjects. Now, each bootstrap sample should have exactly the same x values, that is, $x_i^* = x_i$ for each i in each bootstrap sample. The distribution function from which bootstrap samples should be drawn is the empirical conditional distribution of $\underline{y}$ given x. Sampling from this distribution amounts to resampling the residuals instead of complete cases. First, the residuals are estimated from the original sample by

$$\hat{\epsilon}_i = y_i - \hat{\alpha} - \hat{\beta} x_i, \tag{11.6}$$

with $\hat{\alpha}$ and $\hat{\beta}$ as above. Then, bootstrap samples $\{\underline{\epsilon}_1^*, \ldots, \underline{\epsilon}_n^*\}$ are drawn from $\{\hat{\epsilon}_1, \ldots, \hat{\epsilon}_n\}$, and bootstrap samples of $\underline{y}$ are obtained from

$$y_i^* = \hat{\alpha} + \hat{\beta} x_i + \epsilon_i^*. \tag{11.7}$$

When B bootstrap samples have been obtained, bias-corrected estimates of the parameters and bootstrap estimates of the covariance matrix of the parameters may be computed in the usual way, although for the simple linear regression model, resampling is not necessary and the bootstrap results can be computed analytically [e.g., 20, p. 112]. This will not be the case for multilevel models, however.

If x is a fixed design variable and $\underline{\epsilon}$ is assumed to be normally distributed, the parametric estimator of the conditional distribution of $\underline{y}_i$ given x_i, is a normal distribution with mean $\hat{\alpha} + \hat{\beta} x_i$ and variance $\hat{\sigma}^2$. Therefore, parametric bootstrap samples can be obtained by drawing samples $\{\underline{\epsilon}_1^*, \ldots, \underline{\epsilon}_n^*\}$ from a normal distribution with mean zero and variance $\hat{\sigma}^2$ and then adding these to the estimated mean $\hat{\alpha} + \hat{\beta} x_i$. Then, the parametric bootstrap method proceeds in the same way as the nonparametric bootstrap method with fixed x. Similar parametric bootstrap methods can be designed for random $\underline{x}$.

The bootstrap methods discussed here for regression models are the conventional implementations as, for instance, discussed by Freedman [22], Efron [18], Hinkley [30], Hall [28, pp. 170–171], and Efron and Tibshirani [20, Chapter 9]. These methods have some drawbacks and, therefore, several alternative resampling methods have been proposed that are, for example, robust to heteroskedasticity. A thorough discussion is given by Wu [73].

11.3.2 The Jackknife

The jackknife was originally introduced by Quenouille [48, 49] to estimate the bias of an estimator and to correct for it. Tukey [65] proposed an accompanying estimator for the variance of the estimator, and hence for its standard error.

The idea of the jackknife is as follows. Consider an independently and identically distributed sample of size n from some distribution and an estimator $\underline{\hat{\theta}}_n$ of a parameter θ obtained from this sample. Furthermore, consider removing a group of m observations from the sample, and let $\underline{\hat{\theta}}_{n-m}$ be the estimator of the same parameter θ based on this sample of size $n-m$. The difference between $\hat{\theta}_n$ and $\hat{\theta}_{n-m}$ can then be used to estimate the bias of $\underline{\hat{\theta}}_n$ and this estimate can be used to obtain the bias-corrected jackknife estimator $\underline{\hat{\theta}}_J$. It is known that the bias of $\underline{\hat{\theta}}_J$ is generally of order n^{-2} if m is relatively small compared to n. This is typically much smaller than the bias of $\underline{\hat{\theta}}_n$, which is generally of order n^{-1}.

Obviously, there are many possibilities for selecting a group of observations of size m from the sample. If m is equal for each group, the simplest case is obtained for $m = 1$. Now, the sample is divided into n "groups" of size one, i.e., the n observations. In all other cases with $m > 1$ and n a multiple of m, the sample is divided into g mutually exclusive groups of size m, with $g = n/m$. In the remainder of this section, we will give the details of the standard jackknife procedures for these situations. Justifications can be found in the standard jackknife literature [e.g., 57] or as special cases of the discussion in Section 11.5 below.

Delete-1 Jackknife

Suppose $\underline{\hat{\theta}}_n$ is an estimator of θ based on a sample of size n. Now, remove the i-th observation from the sample, and let $\underline{\hat{\theta}}_{(i)}$ be the estimator of θ based on a sample size of $n-1$. The delete-1 jackknife estimator of θ is now given by

$$\underline{\hat{\theta}}_{J(1)} = n\underline{\hat{\theta}}_n - (n-1)\underline{\bar{\theta}}_{(1)}, \tag{11.8}$$

where $\underline{\bar{\theta}}_{(1)} = n^{-1}\sum_{i=1}^{n}\underline{\hat{\theta}}_{(i)}$.

The delete-1 jackknife variance estimator [65], based on the *pseudo-values*

$$\underline{\tilde{\theta}}_{(i)} = n\underline{\hat{\theta}}_n - (n-1)\underline{\hat{\theta}}_{(i)}, \qquad i = 1, \ldots, n,$$

is given by

$$\begin{aligned}\hat{\underline{\sigma}}^2_{J(1)} &= \frac{1}{n}\sum_{i=1}^{n}\frac{1}{n-1}\left(\tilde{\underline{\theta}}_{(i)} - \frac{1}{n}\sum_{k=1}^{n}\tilde{\underline{\theta}}_{(k)}\right)^2 \\ &= \frac{n-1}{n}\sum_{i=1}^{n}\left(\hat{\underline{\theta}}_{(i)} - \bar{\underline{\theta}}_{(1)}\right)^2.\end{aligned} \tag{11.9}$$

As mentioned above and discussed in more detail in Section 11.5, the bias of $\hat{\underline{\theta}}_{J(1)}$ is typically $O(n^{-2})$, whereas the bias of $\hat{\underline{\theta}}_n$ is typically $O(n^{-1})$. Furthermore, $\hat{\underline{\sigma}}^2_{J(1)}$ is a consistent estimator of the asymptotic variance of both $\hat{\underline{\theta}}_n$ and $\hat{\underline{\theta}}_{J(1)}$.

Delete-m Jackknife

Suppose the sample is divided into g mutually exclusive and independent groups of (equal) size m ($m > 1$), where $m = n/g$. Now remove the m observations of group j from the sample, and let $\hat{\underline{\theta}}_{(j)}$ be the estimator of θ based on the corresponding reduced sample of size $n - m$. The delete-m jackknife (or grouped jackknife) estimator of θ is now given by

$$\hat{\underline{\theta}}_{J(m)} = g\hat{\underline{\theta}}_n - (g-1)\bar{\underline{\theta}}_{(m)}, \tag{11.10}$$

with $\bar{\underline{\theta}}_{(m)} = g^{-1}\sum_{j=1}^{g}\hat{\underline{\theta}}_{(j)}$. Hence, $\hat{\underline{\theta}}_{J(m)}$ is based on g estimators $\hat{\underline{\theta}}_{(j)}$ of θ, each based on a subsample of size $n - m$. Clearly, for $m = 1$, (11.10) reduces to (11.8).

The delete-m jackknife variance estimator is defined similarly to (11.9). It is based on the pseudo-values

$$\tilde{\underline{\theta}}_{(j)} = g\hat{\underline{\theta}}_n - (g-1)\hat{\underline{\theta}}_{(j)}, \qquad j = 1, \ldots, g,$$

and given by

$$\begin{aligned}\hat{\underline{\sigma}}^2_{J(m)} &= \frac{1}{g}\sum_{j=1}^{g}\frac{1}{g-1}\left(\tilde{\underline{\theta}}_{(j)} - \frac{1}{g}\sum_{k=1}^{g}\tilde{\underline{\theta}}_{(k)}\right)^2 \\ &= \frac{g-1}{g}\sum_{j=1}^{g}\left(\hat{\underline{\theta}}_{(j)} - \bar{\underline{\theta}}_{(m)}\right)^2.\end{aligned} \tag{11.11}$$

The mathematics leading to (11.8)–(11.11) can be found in the standard jackknife literature. For example, Shao and Tu [57] provide a systematic introduction to the theory of the jackknife, including a discussion of its theoretical properties.

11.4 Bootstrapping Two-Level Models

In order to make the bootstrap succeed, the simulation must reflect the properties of the stochastic model that is assumed to have generated the data. Therefore, a resampling scheme for multilevel models must, first of all, take into account the hierarchical data structure, that is, the fact that observations are subject to intraclass dependency. Multilevel models can be viewed as linear regression models with a complex structure for the residuals. Hence, we may consider the methods for bootstrapping regression models to implement the bootstrap for multilevel models. In order to deal properly with the intraclass dependency, we have to make several adaptations, which we will discuss below. In exactly the same manner as with regression models, it is useful to distinguish between two different kinds of models: incorporating fixed or random explanatory variables.

Analogous to the discussion in Section 11.3, we discuss three approaches to apply the bootstrap to two-level models: (1) the parametric bootstrap; (2) the residual bootstrap, in which the residuals are resampled; and (3) the cases bootstrap, in which entire cases are resampled. The three bootstrap methods discussed here are based on different assumptions. The parametric bootstrap requires the strongest assumptions: The explanatory variables are considered fixed, and both the model (specification) and the distribution(s) are assumed to be correct. The residual bootstrap requires weaker assumptions: Apart from considering the explanatory variables as fixed, only the model (specification) is assumed to be correct. This implies, for example, that the residuals are assumed to be homoskedastic. The cases bootstrap, finally, requires minimal assumptions: Only the hierarchical dependency in the data is assumed to be specified correctly.

11.4.1 Parametric Bootstrap

The parametric bootstrap uses the parametrically estimated distribution function of the data to generate bootstrap samples. In the two-level model discussed here, two of these distribution functions are involved. For the level-1 residuals $\underline{\epsilon}_j$, we use the $\mathcal{N}(\emptyset, \hat{\sigma}^2 \boldsymbol{I}_{n_j})$ distribution function, and for the level-2 residuals, contained in the vectors $\underline{\boldsymbol{\delta}}_j$, we use the $\mathcal{N}(\emptyset, \hat{\boldsymbol{\Omega}})$ distribution function. Compared to the other two bootstrap approaches, we could say that the parametric bootstrap is "closest" to `FIML`.

Let $\hat{\boldsymbol{\beta}}$ be the `FIML` estimate of $\boldsymbol{\beta}$. The (re)sampling procedure is now as follows:

1. Draw J vectors $\underline{\boldsymbol{\delta}}_j^*$, $j = 1, \ldots, J$, of level-2 residuals from a (multivariate) normal distribution with mean zero and covariance matrix $\hat{\boldsymbol{\Omega}}$.

2. Draw J vectors $\boldsymbol{\epsilon}_j^*$ of sizes n_j, $j = 1, \ldots, J$, containing level-1 residuals from a normal distribution with means zero and covariance matrices $\hat{\sigma}^2 \boldsymbol{I}_{n_j}$.
3. Generate the bootstrap samples $\boldsymbol{y}_j^*$, $j = 1, \ldots, J$, from $\boldsymbol{y}_j^* = \boldsymbol{X}_j\hat{\boldsymbol{\beta}} + \boldsymbol{Z}_j\boldsymbol{\delta}_j^* + \boldsymbol{\epsilon}_j^*$.
4. Compute estimates for all parameters of the two-level model.
5. Repeat steps 1–4 B times and compute bias-corrected estimates and bootstrap standard errors using formulas (11.4) and (11.5).

In this procedure the explanatory variables are assumed to be fixed. Note that the values on the outcome variable encountered in the bootstrap samples will generally not be present in the observed, original sample.

11.4.2 Residual Bootstrap

If the variables contained in $\boldsymbol{X}_j$ and $\boldsymbol{Z}_j$ are considered to be fixed explanatory (design) variables, bootstrap samples can be obtained by resampling the residuals. To implement this strategy, called *residual bootstrap*, the residuals at each level need to be estimated first. We can study (at least) two approaches: (1) estimation by the method of *shrinkage*, which gives $\hat{\boldsymbol{\delta}}_j = \hat{\boldsymbol{\Omega}}\boldsymbol{Z}_j'\hat{\boldsymbol{V}}_j^{-1}(\boldsymbol{y}_j - \boldsymbol{X}_j\hat{\boldsymbol{\beta}})$ and $\hat{\boldsymbol{\epsilon}}_j = \boldsymbol{y}_j - \boldsymbol{X}_j\hat{\boldsymbol{\beta}} - \boldsymbol{Z}_j\hat{\boldsymbol{\delta}}_j$, and (2) estimation of *raw* residuals, using `OLS` decomposition of the total residuals, which gives $\hat{\boldsymbol{\delta}}_j = (\boldsymbol{Z}_j'\boldsymbol{Z}_j)^{-1}\boldsymbol{Z}_j'(\boldsymbol{y}_j - \boldsymbol{X}_j\hat{\boldsymbol{\beta}})$ and $\hat{\boldsymbol{\epsilon}}_j = \boldsymbol{y}_j - \boldsymbol{X}_j\hat{\boldsymbol{\beta}} - \boldsymbol{Z}_j\hat{\boldsymbol{\delta}}_j$. (See Snijders and Berkhof [60] for an extensive and complementary discussion of residuals in a multilevel model.) In both cases, $\hat{\boldsymbol{\beta}}$ is the `FIML` estimate of $\boldsymbol{\beta}$. Resampling of both types of residuals may be considered for the following reason. Raw residuals are unbiased, but inefficient, estimators. Shrunken residuals are (asymptotically) more efficient than their raw counterparts, but are biased toward zero and may therefore not adequately reflect the true variation in the residuals, which could have undesirable effects upon bootstrap results. Note that the estimation of these residuals is mathematically equivalent to the estimation of factor scores in factor analysis. Given this similarity, it follows that one could also estimate the residuals by a covariance preserving method [e.g., 62], so that the (sample) covariance matrix of the estimated transformed residuals is equal to the estimate of the covariance matrix of the corresponding random variables. This idea has been elaborated by Carpenter et al. [13], although their two-step method does not coincide with one of the two optimal solutions given by Ten Berge et al. [62].

Unlike in regression analysis, the estimated residuals in multilevel analysis do not necessarily have an average of zero. Therefore, the residuals must be centered first. Otherwise, the possibly nonzero average of the residuals would lead to biased bootstrap estimators [cf. 22].

Let $\{\hat{\epsilon}_{ij}\}$ and $\{\hat{\boldsymbol{\delta}}_j\}$, $j = 1, \ldots, J$, $i = 1, \ldots, n_j$, be the sets of (centered) estimates of the level-1 and level-2 residuals, respectively. Further, let $\hat{\boldsymbol{\beta}}$ be the `FIML` estimate of $\boldsymbol{\beta}$. Now, bootstrap samples are obtained by the following procedure:

1. Draw a sample $\{\underline{\boldsymbol{\delta}}_j^*\}$ of size J with replacement from the set $\{\hat{\boldsymbol{\delta}}_j\}$ of estimated level-2 residuals.
2. Draw J samples $\{\underline{\epsilon}_{ij}^*\}$ of sizes n_j, $j = 1, \ldots, J$, with replacement from the elements of $\{\hat{\epsilon}_{ij}\}$.
3. Generate the bootstrap samples $\boldsymbol{y}_j^*$, $j = 1, \ldots, J$, from $\boldsymbol{y}_j^* = \boldsymbol{X}_j\hat{\boldsymbol{\beta}} + \boldsymbol{Z}_j\boldsymbol{\delta}_j^* + \boldsymbol{\epsilon}_j^*$.
4. Compute estimates for all parameters of the two-level model.
5. Repeat steps 1–4 B times and compute bias-corrected estimates and bootstrap standard errors using formulas (11.4) and (11.5).

The level-1 and level-2 residuals are assumed to be independently distributed and therefore, in the above procedure, they are also independently resampled. As a result, the level-1 and level-2 residuals corresponding to the same individual observation are not kept together during resampling. If it is suspected that $\underline{\boldsymbol{\delta}}_j$ and $\underline{\boldsymbol{\epsilon}}_j$ are not independent, a bootstrap method that is robust to this kind of dependence is obtained by drawing the level-1 residuals $\underline{\boldsymbol{\epsilon}}_j^*$ from the estimated level-1 residuals for the same original level-2 unit the drawing $\boldsymbol{\delta}_j^*$ happens to come from: If $\boldsymbol{\delta}_j^* = \hat{\boldsymbol{\delta}}_k$, then $\{\underline{\epsilon}_{ij}^*, i = 1, \ldots, n_j\}$ are drawn with replacement from $\{\hat{\epsilon}_{hk}, h = 1, \ldots, n_k\}$. This is called the *linked* residual bootstrap [25]. A theoretical discussion of the virtues of various forms of linking and shrinkage in balanced two-level models without covariates is given by Davison and Hinkley [14, p. 102].

A variant on the residual bootstrap is obtained if we consider the multilevel model as a regression model $\underline{\boldsymbol{y}} = \boldsymbol{X}\boldsymbol{\beta} + \underline{\boldsymbol{r}}$, with $\underline{\boldsymbol{y}} = (\underline{\boldsymbol{y}}_1', \ldots, \underline{\boldsymbol{y}}_J')'$, $\boldsymbol{X} = (\boldsymbol{X}_1', \ldots, \boldsymbol{X}_J')'$, and where $\underline{\boldsymbol{r}}$ is the resulting vector of non-i.i.d. residuals. The covariance matrix of $\underline{\boldsymbol{r}}$ is $\boldsymbol{V} = \bigoplus_{j=1}^J \boldsymbol{V}_j$. Let $\boldsymbol{L}$ be a matrix such that $\boldsymbol{L}\boldsymbol{L}' = \boldsymbol{V}$. Then we can define $\underline{\boldsymbol{\zeta}} = \boldsymbol{L}^{-1}\underline{\boldsymbol{r}}$, so that $E(\underline{\boldsymbol{\zeta}}\,\underline{\boldsymbol{\zeta}}') = \boldsymbol{I}_n$ and $\underline{\boldsymbol{r}} = \boldsymbol{L}\underline{\boldsymbol{\zeta}}$. Hence, given the `FIML` estimates, we can compute $\hat{\boldsymbol{L}}$ from $\hat{\boldsymbol{V}}$, $\hat{\boldsymbol{r}} = \boldsymbol{y} - \boldsymbol{X}\hat{\boldsymbol{\beta}}$, and $\hat{\boldsymbol{\zeta}} = \hat{\boldsymbol{L}}^{-1}\hat{\boldsymbol{r}}$. After centering $\hat{\boldsymbol{\zeta}}$, residuals $\underline{\zeta}_{ij}^*$ can be drawn with replacement from $\{\hat{\zeta}_{hk}, k = 1, \ldots, J; h = 1, \ldots, n_k\}$, and $\boldsymbol{y}^* = \boldsymbol{X}\hat{\boldsymbol{\beta}} + \hat{\boldsymbol{L}}\boldsymbol{\zeta}^*$ can be used in the bootstrap procedure. This method is statistically correct under a wider range of assumptions than the residual bootstrap discussed thus far and performed very well in a simulation study using error component models for panel (longitudinal) data [1]. However, it uses the multilevel structure in the data only implicitly through the covariance matrix $\boldsymbol{V}$. It is therefore less intuitively appealing, which explains why the method has (to our knowledge) not been used in multilevel analysis yet.

11.4.3 Cases Bootstrap

If the explanatory variables contained in $\boldsymbol{X}_j$ and $\boldsymbol{Z}_j$ are considered to be realizations of random variables, bootstrap samples can be obtained by resampling entire cases. Therefore, this method is called the *cases bootstrap.* The resampling procedure is as follows [2, 8]:

1. Draw a sample of size J with replacement from the *level-2 units*; that is, draw a sample $\{\underline{j}_k^*, k = 1, \ldots, J\}$ (with replacement) of level-2 unit numbers.
2. For each k, draw a sample of entire cases, with replacement, from (the original) level-2 unit $j = j_k^*$. This sample has the same size $n_k^* = n_{j_k^*} = n_j$ as the original unit from which the cases are drawn. (Note that this implies that the total sample size of the bootstrap samples may not be n.) Then, for each k, we have a set of data $\{(y_{ik}^*, \boldsymbol{X}_{ik}^*, \boldsymbol{Z}_{ik}^*), i = 1, \ldots, n_k^*\}$.
3. Compute estimates for all parameters of the two-level model.
4. Repeat steps 1–3 B times and compute bias-corrected estimates and bootstrap standard errors using formulas (11.4) and (11.5).

An alternative formulation is: (1) draw one entire level-2 unit $(\boldsymbol{y}_j, \boldsymbol{X}_j, \boldsymbol{Z}_j)$, containing n_j level-1 cases, with replacement; (2) from this level-2 unit, draw a bootstrap sample $(\underline{\boldsymbol{y}}_j^*, \underline{\boldsymbol{X}}_j^*, \underline{\boldsymbol{Z}}_j^*)$ of size n_j with replacement; (3) repeat steps 1 and 2 J times; (4) compute all parameter estimates for the two-level model; (5) repeat steps 1–4 B times and compute bias-corrected estimates and bootstrap standard errors using formulas (11.4) and (11.5).

The above procedure shows that for the cases bootstrap each observed response y_{ij} keeps joined together with the observed scores on the explanatory variables in $\boldsymbol{X}_{ij}$ and $\boldsymbol{Z}_{ij}$.

The cases bootstrap must be handled with some care. It depends on the nature of the data whether it makes sense to resample units from both levels, or only from level 2 or level 1. Two examples may give insight into this problem.

1. If the level-2 units are individuals and the level-1 units are repeated measures of some variables for these individuals [58, 59, 66], it makes sense to resample only the individuals and keep the values of the $\boldsymbol{y}$, $\boldsymbol{X}$, and $\boldsymbol{Z}$ variables constant for each individual. Thus, only level-2 units are resampled and once a level-2 unit enters the bootstrap sample, all level-1 units within this level-2 unit are collected from the original sample and are not resampled.
2. If the level-2 units are countries and the level-1 units are individuals from these countries, it makes sense to resample only the individuals and keep the countries and the country-specific (level-2) variables constant. Now only level-1 units are resampled within each level-2 unit. The level-2

units and their variables are taken from the original sample and are not resampled.

Many more examples can be given in which only the level-2 units, or only the level-1 units, or both level-1 and level-2 units should be resampled. Which of these three possibilities is most appropriate depends mainly on two factors: (1) the degree of randomness of the sampling at both levels, and (2) the (average) sample size at both levels. If, for example, students (level 1) from all Dutch universities (level 2) are compared, it is clear that the sample of universities is nonrandom, and that only level-1 units should be resampled. The reverse may be true for a sample of families, with all family members from each family present in the sample.

If the sample size at one level (usually level 1) is very small [see 52, for examples], the sample size at this level may be too small to give accurate bootstrap results. Furthermore, resampling the units at this level may lead to numerical problems, because it can easily happen that in the bootstrap sample only one original unit is present, repeated J or n_j times.

If the most appropriate resampling scheme is selected in this way, the cases bootstrap leads to consistent estimators. Goldstein [26, p. 82] suggests that this would not be the case if both levels are resampled, because the within-group dependence would be lost, but this is incorrect. The independent resampling at level 1 is conditional on the level-2 unit; observations from different original level-2 units cannot be assigned to the same level-2 unit in the resampling. Conditional on the level-2 unit, the observations at level 1 are assumed independent in the model. This is exactly how the resampling is done and hence it gives consistent estimators. After resampling at level 1, the bootstrap observations within the same level-2 unit share the same (unconditional) dependence as in the original sample. However, cases bootstrap estimators are typically less efficient than parametric and residuals bootstrap, but this is because they use considerable weaker assumptions. For example, cases bootstrap is still consistent under heteroskedasticity. Thus, as is often the case, there is a trade-off between robustness and efficiency.

11.4.4 Bootstrap Confidence Intervals

Up till now, we have used the bootstrap only for bias correction and computation of standard errors. However, an important and nontrivial application of the bootstrap is the computation of confidence intervals. We will now discuss a number of different types of bootstrap confidence intervals for a typical parameter θ with true value θ_0. We will only discuss two-sided intervals; one-sided intervals are defined analogously. The intended nominal coverage of the confidence interval will be denoted by $1 - \alpha$, so that the probability that the interval contains the true parameter value should be approximately $1 - \alpha$.

Notation

Before we introduce the different bootstrap confidence intervals, we will introduce some useful notation. Let $\Phi(z)$ be the standard normal distribution function. Then z_α is the α-th quantile of the standard normal distribution, $z_\alpha = \Phi^{-1}(\alpha)$. Let the distribution function of the estimator $\underline{\hat{\theta}}$ be $H(\theta)$, that is, $H(\theta) = \Pr(\underline{\hat{\theta}} \leq \theta)$. A consistent estimator of this distribution function is obtained from the B bootstrap replications $\underline{\theta}_b^*$, $b = 1, \ldots, B$, of $\underline{\hat{\theta}}$:

$$\underline{\hat{H}}(\theta) = \frac{\#\{b : \underline{\theta}_b^* \leq \theta\}}{B} . \tag{11.12}$$

Note that $\hat{H}$ is invariant under monotonic transformation, in the sense that if $g(\theta)$ is a monotonically increasing function of θ, then the estimate of its distribution function is

$$\tilde{H}(g(\theta)) = \frac{\#\{b : g(\theta_b^*) \leq g(\theta)\}}{B} = \hat{H}(\theta) .$$

This property has been used in the derivations of some of the confidence intervals described below.

Bootstrap Normal Confidence Interval

If the assumptions of the model, including the normality assumptions, hold, then the estimators are asymptotically normally distributed with a certain covariance matrix, derived from the likelihood function. Hence, for our typical parameter θ, we have

$$\sqrt{n}(\underline{\hat{\theta}} - \theta_0) \xrightarrow{\mathcal{L}} \mathcal{N}(0, \psi), \tag{11.13}$$

say. The distribution of $\underline{\hat{\theta}} - \theta_0$ can be approximated by the normal distribution with mean zero and variance $\underline{\hat{\psi}}/n$, where $\underline{\hat{\psi}}$ is a consistent estimator of ψ derived from the likelihood function. The usual confidence intervals for θ_0 are therefore

$$\left[\underline{\hat{\theta}} + z_{\frac{1}{2}\alpha} \underline{\widehat{\text{se}}}_{\mathcal{N}}(\underline{\hat{\theta}});\ \underline{\hat{\theta}} + z_{1-\frac{1}{2}\alpha} \underline{\widehat{\text{se}}}_{\mathcal{N}}(\underline{\hat{\theta}}) \right], \tag{11.14}$$

where $\underline{\widehat{\text{se}}}_{\mathcal{N}}(\underline{\hat{\theta}}) = \sqrt{\underline{\hat{\psi}}/n}$ is the estimator of the asymptotic standard deviation of $\underline{\hat{\theta}}$. Under mild regularity conditions, the estimators are asymptotically normally distributed, even if the random terms in the model are not. In that case, $\underline{\widehat{\text{se}}}_{\mathcal{N}}$ may not be a consistent estimator of the standard deviation of the estimators of the variance components, although it is still consistent for the fixed parameters. This suggests replacing $\underline{\widehat{\text{se}}}_{\mathcal{N}}$ in (11.14) by a bootstrap estimator. This gives the *bootstrap normal* confidence interval

$$\left[\underline{\hat{\theta}} + z_{\frac{1}{2}\alpha} \underline{\widehat{\text{se}}}_B(\underline{\hat{\theta}});\ \underline{\hat{\theta}} + z_{1-\frac{1}{2}\alpha} \underline{\widehat{\text{se}}}_B(\underline{\hat{\theta}}) \right], \tag{11.15}$$

in which $\widehat{\text{se}}_B$ is the bootstrap estimator of the standard deviation of $\hat{\underline{\theta}}$. Alternatively, one might use

$$\left[\hat{\underline{\theta}}_B + z_{\frac{1}{2}\alpha}\widehat{\underline{\text{se}}}_B(\hat{\underline{\theta}});\ \hat{\underline{\theta}}_B + z_{1-\frac{1}{2}\alpha}\widehat{\underline{\text{se}}}_B(\hat{\underline{\theta}})\right], \tag{11.16}$$

where $\hat{\underline{\theta}}_B$ is the bootstrap bias-corrected estimator of θ.

The bootstrap normal confidence interval relaxes the assumption of normality of the data, but it still heavily relies on the asymptotic normality of the estimators. In finite samples, however, the estimators may not be approximately normally distributed [7].

Hall's Percentile Interval

Hall's percentile interval [28, p. 12] takes the bootstrap normal interval (11.15) as its starting point. That interval is based on the idea that

$$\Pr\left(\hat{\underline{\theta}} + z_{\frac{1}{2}\alpha}\widehat{\underline{\text{se}}}_B(\hat{\underline{\theta}}) \leq \theta_0 \leq \hat{\underline{\theta}} + z_{1-\frac{1}{2}\alpha}\widehat{\underline{\text{se}}}_B(\hat{\theta})\right) \longrightarrow 1-\alpha, \tag{11.17}$$

because $\hat{\underline{\theta}}$ is asymptotically normally distributed and $\widehat{\underline{\text{se}}}_B(\hat{\underline{\theta}})$ is a consistent estimator of its standard deviation. In finite samples, however, the distribution of $\hat{\underline{\theta}}$ may not be approximately normal [7]. Therefore, instead of using quantiles of the normal distribution, using bootstrap quantiles may give more accurate results.

To derive the necessary bootstrap quantiles, let us rewrite (11.17) into the following form:

$$\Pr\left(z_{\frac{1}{2}\alpha}\widehat{\underline{\text{se}}}_B(\hat{\underline{\theta}}) \leq \theta_0 - \hat{\underline{\theta}} \leq z_{1-\frac{1}{2}\alpha}\widehat{\underline{\text{se}}}_B(\hat{\underline{\theta}})\right) \longrightarrow 1-\alpha.$$

The estimated quantiles $q_{\frac{1}{2}\alpha} = z_{\frac{1}{2}\alpha}\widehat{\text{se}}_B(\hat{\underline{\theta}})$ and $q_{1-\frac{1}{2}\alpha} = z_{1-\frac{1}{2}\alpha}\widehat{\text{se}}_B(\hat{\underline{\theta}})$ of the normal distribution have to be replaced by corresponding quantiles of the distribution of $\theta_0 - \hat{\underline{\theta}}$. These are estimated by quantiles $\hat{q}_{\frac{1}{2}\alpha}$ and $\hat{q}_{1-\frac{1}{2}\alpha}$ of the bootstrap distribution of $\hat{\theta} - \underline{\theta}^*$. From the definition $\Pr(\hat{\theta} - \underline{\theta}^* \leq \hat{q}_{\frac{1}{2}\alpha}) = \frac{1}{2}\alpha$, it follows that $\hat{q}_{\frac{1}{2}\alpha} = \hat{\theta} - \hat{H}^{-1}(1-\frac{1}{2}\alpha)$ and, consequently, the confidence interval for θ_0 becomes the interval $[\hat{\underline{\theta}} + \hat{\underline{q}}_{\frac{1}{2}\alpha}; \hat{\underline{\theta}} + \hat{\underline{q}}_{1-\frac{1}{2}\alpha}]$, which reduces to

$$\left[2\hat{\underline{\theta}} - \underline{\hat{H}}^{-1}(1-\tfrac{1}{2}\alpha);\ 2\hat{\underline{\theta}} - \underline{\hat{H}}^{-1}(\tfrac{1}{2}\alpha)\right]. \tag{11.18}$$

Note that the upper quantile of $\hat{H}$ ends up (in reverse) in the lower confidence point and vice versa. This tends to give a small bias and skewness correction.

Percentile-*t*

The percentile-t (also called bootstrap-t) is a combination of the ideas of the bootstrap normal and Hall's percentile intervals. It is derived by rewriting (11.17) into the following form:

$$\Pr\left(z_{\frac{1}{2}\alpha} \leq \frac{\theta_0 - \hat{\underline{\theta}}}{\widehat{\mathrm{se}}_B(\hat{\underline{\theta}})} \leq z_{1-\frac{1}{2}\alpha} \right) \longrightarrow 1 - \alpha. \tag{11.19}$$

The quantiles of the normal distribution are now replaced by quantiles of the distribution of $\hat{\underline{t}} = (\theta_0 - \hat{\underline{\theta}})/\widehat{\mathrm{se}}_B(\hat{\underline{\theta}})$. These are estimated by quantiles of the bootstrap distribution of $\underline{t}^* = (\hat{\theta} - \underline{\theta}^*)/\mathrm{se}^*_B(\underline{\theta}^*)$. Let $\hat{G}(t)$ be the bootstrap-estimated distribution function of this quantity, i.e.,

$$\hat{G}(t) = \frac{\#\left\{ b : \dfrac{\hat{\theta} - \theta^*_b}{\mathrm{se}^*_{B,b}(\underline{\theta}^*)} \leq t \right\}}{B},$$

and let $\hat{t}_{\frac{1}{2}\alpha}$ and $\hat{t}_{1-\frac{1}{2}\alpha}$ be the $\frac{1}{2}\alpha$-th and $(1 - \frac{1}{2}\alpha)$-th quantiles of $\hat{G}$, respectively; that is, $\hat{t}_{\frac{1}{2}\alpha} = \hat{G}^{-1}(\frac{1}{2}\alpha)$ and $\hat{t}_{1-\frac{1}{2}\alpha} = \hat{G}^{-1}(1 - \frac{1}{2}\alpha)$. The percentile-$t$ interval is obtained by replacing $z_{\frac{1}{2}\alpha}$ by $\hat{\underline{t}}_{\frac{1}{2}\alpha}$ and $z_{1-\frac{1}{2}\alpha}$ by $\hat{\underline{t}}_{1-\frac{1}{2}\alpha}$ in (11.15) and is thus

$$\left[\hat{\underline{\theta}} + \hat{\underline{t}}_{\frac{1}{2}\alpha}\widehat{\mathrm{se}}_B(\hat{\underline{\theta}});\ \hat{\underline{\theta}} + \hat{\underline{t}}_{1-\frac{1}{2}\alpha}\widehat{\mathrm{se}}_B(\hat{\underline{\theta}}) \right]. \tag{11.20}$$

This confidence interval requires an estimate $\mathrm{se}^*_{B,b}(\underline{\theta}^*)$ of the standard deviation of $\underline{\theta}^*$ for each bootstrap resample b. This is usually obtained by performing a small bootstrap within each bootstrap resample. Thus, for example, $B = 1000$ bootstrap samples are drawn with replacement from the original sample and within each sample $b = 1, \ldots, B$, $B_2 = 25$ samples are drawn with replacement from the bootstrap sample. From the B_2 samples, $\mathrm{se}^*_{B,b}(\underline{\theta}^*)$ is obtained. This means that $B \cdot B_2$ bootstrap samples have to be drawn and $B \cdot B_2$ times the estimator of θ has to be computed. In the example, this amounts to $1000 \cdot 25 = 25{,}000$ bootstrap samples and 25,000 times computing the estimator.

The percentile-t interval tends to perform better than the bootstrap normal and Hall's percentile interval, because it uses the nonnormality of the distribution of the estimator (as opposed to the former) and $\hat{\underline{t}}$ is more nearly *pivotal* than $\theta_0 - \hat{\underline{\theta}}$ in a number of important cases, which means that its distribution depends less on the parameters that are being estimated. The quantity $\hat{\underline{t}}$ is not always nearly pivotal, however, and in those cases in which it is not, the percentile-t confidence interval performs less well. A complicated extension that aims at transforming the parameter to a near-pivotal quantity is the *variance stabilized percentile-t* interval, see, e.g., Efron and Tibshirani [20, Section 12.6].

Efron's Percentile Interval

The idea behind this interval is quite different from the ideas behind the bootstrap normal interval and its extensions. It was stated above that $\hat{\underline{H}}(\theta)$ is a consistent estimator of the distribution function of $\hat{\underline{\theta}}$. Therefore, an asymptotic $1-\alpha$ confidence interval can be obtained by taking the relevant quantiles from $\hat{\underline{H}}$, which leads to the interval

$$\left[\hat{\underline{H}}^{-1}(\tfrac{1}{2}\alpha);\ \hat{\underline{H}}^{-1}(1-\tfrac{1}{2}\alpha)\right]. \tag{11.21}$$

Efron's percentile interval does not rely on the asymptotic normality of $\hat{\underline{\theta}}$. Its coverage performance in finite samples is, however, frequently not very well, because the end points of the interval tend to be a little biased. Note the difference with Hall's percentile interval. Here, percentiles of the distribution of $\hat{\underline{\theta}}$ are approximated by percentiles of the distribution of $\underline{\theta}^*$, whereas in Hall's percentile interval, percentiles of the distribution of $\theta_0 - \hat{\underline{\theta}}$ are approximated by percentiles of the distribution of $\hat{\theta} - \underline{\theta}^*$.

Bias-Corrected (BC) and Bias-Corrected and Accelerated (BC_a) Percentile Intervals

The BC and BC_a intervals have been introduced to correct for some bias in the endpoints of Efron's percentile interval (11.21). Assume that there exists a monotonically increasing function $g(\theta)$ such that

$$\frac{g(\hat{\underline{\theta}}) - g(\theta_0)}{1 + a g(\theta_0)} \sim \mathcal{N}(-z_0, 1). \tag{11.22}$$

The constant z_0 allows for some bias in the estimator $g(\hat{\underline{\theta}})$ of $g(\theta_0)$ and the constant a, called the *acceleration constant*, expresses the speed at which the standard deviation of the estimator increases with the parameter being estimated. In typical estimation problems, $a = O(n^{-1/2})$ and $z_0 = O(n^{-1/2})$.

From the likelihood based on (11.22), it can now be derived that the exact confidence interval for θ_0 is equal, up to order $O(n^{-1})$, to the BC_a interval given by

$$\left[\hat{\underline{H}}^{-1}\left(\Phi(z[\tfrac{1}{2}\alpha])\right); \hat{\underline{H}}^{-1}\left(\Phi(z[1-\tfrac{1}{2}\alpha])\right)\right], \tag{11.23}$$

where

$$z[\tfrac{1}{2}\alpha] = z_0 + \frac{z_0 + z_{\frac{1}{2}\alpha}}{1 - a(z_0 + z_{\frac{1}{2}\alpha})}$$

and $z[1-\frac{1}{2}\alpha]$ is similarly defined. Note that this interval does not depend on the specific transformation $g(\cdot)$, which follows from the invariance property of $\hat{H}$ discussed earlier. In practice, the constants z_0 and a have to be estimated, but this does not alter the results up to order $O(n^{-1})$. Moreover, even if (11.22) does not hold, the BC_a endpoints are correct up to order $O(n^{-1})$, whereas in many cases the endpoints of the intervals discussed previously are only correct up to order $O(n^{-1/2})$.

A simple consistent estimator of z_0 is $\hat{\underline{z}}_0 = \Phi^{-1}\big(\hat{\underline{H}}(\hat{\underline{\theta}})\big)$. The estimation of a is the most important problem with the BC_a method. If it is assumed that $a = 0$, we obtain the BC interval, which is discussed, e.g., in Efron [18]. Usually, however, the BC interval is only correct up to order $O(n^{-1/2})$ and is therefore typically worse than the BC_a interval.

Efron [19] provided several formulas for a. In a one-parameter parametric model where $\hat{\underline{\theta}}$ is the ML estimator, a good approximation for a is $a \approx \frac{1}{6}\mathrm{Skew}_{\theta=\hat{\theta}}(\dot{\underline{l}}_\theta)$, where "Skew" denotes the skewness of a random variable and $\dot{\underline{l}}_\theta$ is the score function (derivative of the loglikelihood with respect to θ).

When more parameters are to be estimated, which is the case in multilevel analysis, these results are no longer valid. Efron [19] gave a formula for a based on reducing the multiparameter problem to a one-parameter problem defined by the least favorable direction. This is defined by the following formulas: Let $f_{\boldsymbol{\eta}}(\boldsymbol{y})$ be the density function of the data dependent on a parameter vector $\boldsymbol{\eta}$, and let $\ddot{\boldsymbol{l}}_{\hat{\boldsymbol{\eta}}} = \partial^2 \log f_{\boldsymbol{\eta}}(\boldsymbol{y})/\partial\boldsymbol{\eta}\,\partial\boldsymbol{\eta}'$, evaluated in the value $\hat{\boldsymbol{\eta}}$ of the estimator. Let $\theta = t(\boldsymbol{\eta})$ be the parameter of interest and let $\hat{\boldsymbol{\nabla}} = \partial t(\boldsymbol{\eta})/\partial\boldsymbol{\eta}$, evaluated in $\hat{\boldsymbol{\eta}}$. The least favorable direction at $\boldsymbol{\eta} = \hat{\boldsymbol{\eta}}$ is now defined as $\hat{\boldsymbol{\mu}} = (\ddot{\boldsymbol{l}}_{\hat{\boldsymbol{\eta}}})^{-1}\hat{\boldsymbol{\nabla}}$. The multiparameter problem is now reduced to a one-parameter problem by considering only parameter values of the form $\hat{\boldsymbol{\eta}} + \lambda\hat{\boldsymbol{\mu}}$. Now, $a \approx \frac{1}{6}\mathrm{Skew}_{\lambda=0}[\partial \log f_{\hat{\boldsymbol{\eta}}+\lambda\hat{\boldsymbol{\mu}}}(\underline{\boldsymbol{y}}^*)/\partial\lambda]$. Note that the parameter vector is called $\boldsymbol{\eta}$ in this definition. This formula may be used for the parametric bootstrap in multilevel analysis, although the formulas are quite complicated. This was done for a simple two-level variance components model by LeBlond [36, Section 7.3.4]. Usually, confidence intervals are required for each parameter separately, so that $t(\boldsymbol{\eta}) = \boldsymbol{e}_i'\boldsymbol{\eta}$ and $\hat{\boldsymbol{\nabla}} = \boldsymbol{e}_i$, where $\boldsymbol{e}_i$ is the i-th unit vector.

For the nonparametric bootstrap, the standard formula for a is based on the empirical influence function of $\hat{\underline{\theta}}$. This is, however, not well defined for multilevel data, so that this formula cannot be used. Tu and Zhang [64] proposed to estimate a by the jackknife according to the formula

$$\hat{\underline{a}}_J = \frac{(n-1)^3}{6n^3(\hat{\underline{\sigma}}^2_{J(1)})^{3/2}} \sum_{i=1}^{n} \left(\hat{\underline{\theta}}_{(i)} - \bar{\underline{\theta}}_{(1)}\right)^3 . \tag{11.24}$$

A similar formula for a was used by Frangos and Schucany [21], who also studied a corresponding method using the *positive* jackknife, which extends the data set by duplicating a data point, so that we get a sample size of $n+1$ and estimators $\hat{\underline{\theta}}_{(+i)}$ instead of $\hat{\underline{\theta}}_{(i)}$ for the standard jackknife, which they call *negative* jackknife. For multilevel data, we have to replace these jackknife formulas with grouped jackknife methods for unequal group sizes (see Section 11.5). It is, however, doubtful whether the jackknife for multilevel models will give a reasonable estimate of a third-order moment. A bootstrap analog of (11.24) would be

$$\underline{\hat{a}}_B = \frac{1}{6} \frac{\frac{1}{B}\sum_{b=1}^{B}\left(\underline{\theta}_b^* - \underline{\theta}_{(.)}^*\right)^3}{\left[\frac{1}{B}\sum_{b=1}^{B}\left(\underline{\theta}_b^* - \underline{\theta}_{(.)}^*\right)^2\right]^{3/2}}.$$

It is still an open question whether this gives reasonable results.

11.5 Jackknifing Two-Level Models

In Section 11.3, we discussed the classic jackknife approach to estimating the bias of an estimator and to obtain a bias-corrected version of this estimator. Using the pseudo-values, an accompanying estimator for the variance of the (original or bias-corrected) estimator, and hence for the standard error, can be obtained as well.

The jacknife version we discussed is based on subsamples obtained from the original sample by successively removing mutually exclusive groups of observations of size m. Furthermore, it relies on the assumption of independently and identically distributed observations. Both features influence the formulation of a jackknife resampling scheme for multilevel data and models.

The independence assumption restricts the application of the jackknife to the highest level in the data. In the two-level case, independence can only be assumed for the groups. Within the groups, data are dependent. Consequently, a multilevel jackknife approach must be based on subsamples obtained by removing complete level-2 units. In fact, Wolter [72, Section 4.6] already stated that the delete-m jackknife can be used in cluster sampling, when the data within clusters are dependent. In multilevel data, however, groups are usually not of *equal* size m. Therefore, to make the jackknife suitable for multilevel data and models, the delete-m jackknife needs to be generalized to a grouped jackknife for unequal group sizes, called the delete-m_j jackknife by Busing et al. [10].

11.5.1 Delete-m_j Jackknife

To apply the delete-m jackknife (with $m > 1$, and n a multiple of m), the sample is divided into g mutually exclusive groups of size m, with $g = n/m$. In multilevel analysis, the sample is divided into J groups of (usually) varying size n_j; that is, n_j is not equal for each group and n/n_j will not necessarily be equal to J. As a result, the formulas discussed earlier have to be adapted slightly. Let $\hat{\underline{\theta}}_{(j^*)}$ be an estimator of θ based on a sample from which group j with size n_j is removed. The delete-m_j jackknife estimator of θ is now given by

$$\hat{\underline{\theta}}_{J(m_j)} = J\hat{\underline{\theta}}_n - \sum_{j=1}^{J}\left(1 - \frac{n_j}{n}\right)\hat{\underline{\theta}}_{(j^*)}. \tag{11.25}$$

The estimator $\hat{\underline{\theta}}_{J(m_j)}$ can be justified as follows. Consider an estimator $\hat{\underline{\theta}}_n$ of a parameter θ obtained from a sample of size n from some distribution. In general, the expected value of such estimators can be written as the true value θ_0 plus a power series expansion in $1/n$; that is,

$$E(\hat{\underline{\theta}}_n) = \theta_0 + \frac{b_1}{n} + \frac{b_2}{n^2} + \frac{b_3}{n^3} + \cdots, \tag{11.26}$$

where $b_1, b_2, \ldots$ are unknown constants, independent of sample size, and frequently not equal to zero [see, e.g., 49, 55]. If $b_1 \neq 0$, the bias in (11.26) is clearly of order n^{-1}. Let $h_j = n/n_j$. Then, the total sample size can be written as $n = n_j h_j$. Hence,

$$E(h_j\hat{\underline{\theta}}_n) = h_j\theta_0 + \frac{b_1}{n_j} + \frac{b_2}{h_j n_j^2} + \frac{b_3}{h_j^2 n_j^3} + \cdots \tag{11.27}$$

and

$$E(\hat{\underline{\theta}}_{(j^*)}) = \theta_0 + \frac{b_1}{(h_j-1)n_j} + \frac{b_2}{(h_j-1)^2 n_j^2} + \frac{b_3}{(h_j-1)^3 n_j^3} + \cdots. \tag{11.28}$$

Combining (11.27) and (11.28) gives

$$\begin{aligned} E&\left[h_j\hat{\underline{\theta}}_n - (h_j-1)\hat{\underline{\theta}}_{(j^*)}\right] \\ &= \theta_0 + \frac{b_2}{n_j^2}\left(\frac{1}{h_j} - \frac{1}{h_j-1}\right) + \frac{b_3}{n_j^3}\left(\frac{1}{h_j^2} - \frac{1}{(h_j-1)^2}\right) + \cdots \\ &= \theta_0 - \frac{b_2}{n^2}\frac{h_j}{h_j-1} - \frac{b_3}{n^3}\frac{h_j(2h_j-1)}{(h_j-1)^2} + \cdots. \end{aligned} \tag{11.29}$$

Finally, to prevent loss of efficiency, the weighted average of the J possible estimators is used [cf. 49]. This gives

$$\begin{aligned} \hat{\underline{\theta}}_{J(m_j)} &= \sum_{j=1}^{J}\frac{n_j}{n}\left(h_j\hat{\underline{\theta}}_n - (h_j-1)\hat{\underline{\theta}}_{(j^*)}\right) \\ &= J\hat{\underline{\theta}}_n - \sum_{j=1}^{J}\left(1 - \frac{n_j}{n}\right)\hat{\underline{\theta}}_{(j^*)}. \end{aligned} \tag{11.30}$$

The expectation of (11.30) is

$$E\left[\sum_{j=1}^{J}\frac{n_j}{n}\left(h_j\hat{\underline{\theta}}_n-(h_j-1)\hat{\underline{\theta}}_{(j^*)}\right)\right]$$

$$=\sum_{j=1}^{J}\frac{n_j}{n}\theta_0-\sum_{j=1}^{J}\frac{1}{h_j}\left(\frac{b_2}{n^2}\frac{h_j}{h_j-1}\right)-\sum_{j=1}^{J}\frac{1}{h_j}\left(\frac{b_3}{n^3}\frac{h_j(2h_j-1)}{(h_j-1)^2}\right)+\cdots$$

$$=\theta_0-\frac{b_2}{n^2}\sum_{j=1}^{J}\frac{1}{h_j-1}-\frac{b_3}{n^3}\sum_{j=1}^{J}\frac{2h_j-1}{(h_j-1)^2}+\cdots,$$

so that the bias is of order n^{-2} if $b_2 \neq 0$ and if n_j is relatively small compared to the total sample size n.

The corresponding estimator of the variance of $\hat{\underline{\theta}}_{J(m_j)}$, based on the pseudo-values

$$\tilde{\underline{\theta}}_{(j^*)}=h_j\hat{\underline{\theta}}_n-(h_j-1)\hat{\underline{\theta}}_{(j^*)},\qquad j^*=1,\ldots,J,$$

is given by

$$\hat{\underline{\sigma}}^2_{J(m_j)}=\frac{1}{J}\sum_{j=1}^{J}\frac{1}{h_j-1}\left(\tilde{\underline{\theta}}_{(j^*)}-\hat{\underline{\theta}}_{J(m_j)}\right)^2$$

$$=\frac{1}{J}\sum_{j=1}^{J}\frac{1}{h_j-1}\left(h_j\hat{\underline{\theta}}_n-(h_j-1)\hat{\underline{\theta}}_{(j^*)}-J\hat{\underline{\theta}}_n+\sum_{k=1}^{J}\left(1-\frac{n_k}{n}\right)\hat{\underline{\theta}}_{(k^*)}\right)^2. \tag{11.31}$$

Note that when all groups are of equal size, (11.10) follows from (11.30), that is, the delete-m_j jackknife estimator reduces to the delete-m jackknife estimator. Analogously, (11.31) reduces to the expression for the delete-m jackknife variance estimator (11.11).

11.5.2 Jackknife Confidence Intervals

The delete-m_j jackknife estimator and the delete-m_j jackknife variance estimator can be used to construct the jackknife normal confidence interval

$$\left[\hat{\underline{\theta}}_{J(m_j)}+z_{\frac{1}{2}\alpha}\hat{\underline{\sigma}}_{J(m_j)};\ \hat{\underline{\theta}}_{J(m_j)}+z_{1-\frac{1}{2}\alpha}\hat{\underline{\sigma}}_{J(m_j)}\right]. \tag{11.32}$$

The jackknife normal confidence interval relaxes the normality assumption for the data. However, the interval relies on the asymptotic normality of the estimators, which may in finite samples not be approximately satisfied [7]. Other jackknife confidence intervals are not applicable or are probably worse, due to the limited use of the pseudo-values.

11.6 Software

In this section, we briefly discuss resampling options within the available multilevel software packages. Basically, there are two programs containing built-in options for resampling: `MLwiN` [51] and `MLA` [11]. The other major programs for multilevel analysis, `HLM` [53] and `VARCL` [38], do not contain resampling options. In principle, bootstrap and jackknife methods as discussed in the preceding sections could be implemented within general-purpose packages such as `SAS`, `SPSS`, `R`, and `S-Plus`. However, these procedures need to be entirely developed and programmed by the user.

11.6.1 MLwiN

Goldstein [24] proposed a parametric bootstrap procedure with iterative bias correction, based on the results of Kuk [32]. This procedure has been implemented in the `MLwiN` program (versions 1.1 and higher), under the name of iterated bootstrap. In a series of steps, bootstrap simulation and bias correction are performed alternatingly. The process starts by using an estimated parameter as the "true value" from which a set of (parametric) bootstrap replicates is obtained. From these bootstrap replicates, the bias-corrected estimate is computed. In the next step, the corrected estimate serves as the "true value" for a new set of bootstrap replicates. From this set, an update for the bias-corrected estimate is computed, and so on. Steps are repeated until the successive corrected estimates converge. Although Rasbash et al. [51] present a promising illustration of this iterated bootstrap, they still suggest using the procedure with care. This phrasing is also used by Goldstein [26, p. 126], who states that a certain correction to the bootstrap estimates must be done to obtain "approximately correct standard errors and quantile estimates." He also states (on p. 84) an important problem with the iterated bootstrap, namely that the procedure may not converge.

A second bootstrap method that is implemented in `MLwiN` is the residual ("nonparametric") bootstrap with covariance-preserving residuals as proposed by Carpenter et al. [13] and mentioned earlier.

11.6.2 MLA

The `MLA` program has been developed primarily for research on resampling methods in two-level models. It is extensively documented in Busing et al. [8, 9, 11]. Langer [35] discusses in detail many examples of analyses with `MLA`. The latest version of `MLA` is version 4.1. Among several options mainly reflecting the research interests of the authors, the program provides `FIML` estimates of parameters and standard errors, and their counterparts obtained with the parametric bootstrap, the residual bootstrap with raw and shrunken residuals with and without linking, and the cases bootstrap. Furthermore, a number of bootstrap confidence intervals have been implemented, viz. the bootstrap normal confidence interval, the percentile-t interval, Efron's percentile interval, and the bias-corrected (BC) percentile interval. A final option of interest for our discussion here is the delete-m_j jackknife.

11.7 Empirical Evidence

Bootstrap and jackknife estimation applied to multilevel models has not yet been studied by many authors, and for some specific models or situations only. Information about the performance of the different approaches comes from a relatively limited number of Monte Carlo simulation studies. In this section, we summarize the results that are most relevant for our discussion.

11.7.1 Bootstrap Bias Correction and Standard Errors

Van der Leeden et al. [68] evaluated the parametric bootstrap, the two versions of the residual bootstrap, and the cases bootstrap. In an extensive simulation study they addressed the question whether these bootstrap estimators of model parameters and standard errors are less biased and have smaller mean squared error (MSE) than their `FIML` counterparts. Data were generated for a two-level model containing one predictor variable at each level. The (conditional) intraclass correlation was set to 0.2, and the intercept-slope correlation to 0.5. Assumptions were violated by using moderately small sample sizes (especially at level 2), and severely skewed distributions for the residuals.

The main conclusion of this study was that the shrunken residual bootstrap works for variance component estimation; that is, it works for cases like the one simulated (small sample size at level 2 and heavily skewed distribution of the residuals). In such cases, this type of bootstrap provides nearly unbiased estimators of the variance components at both levels, with relatively small MSE. It may be considered a valuable alternative to `FIML` estimation, especially when the interest is in estimating the "true value" of a variance component. Regarding bias, the other three bootstrap methods do not produce useful results for this case, although the cases bootstrap has MSEs similar to those of the shrunken residual bootstrap.

Results confirmed the finding that bias in the `FIML` fixed parameter estimators is negligible [cf. 5, 40]. The application of the bootstrap to estimate these parameters has clearly no surplus value. However, bootstrap confidence intervals may be useful for testing the fixed parameters, since their `FIML` standard errors can be substantially downward biased, making the commonly used t-ratios suspicious.

The biases of the standard errors are about equally bad for FIML and the different bootstrap methods. Compared to the other methods, the cases bootstrap seemed to perform best. This holds for the MSEs of the standard errors as well. The cases bootstrap yielded standard errors with the relatively most satisfactory MSEs.

Although the results of Van der Leeden et al. [68] show that the shrunken bootstrap method works for the case they had simulated, the study merely provides empirical evidence in the same tradition as the bootstrap method itself: using raw computing power. The four bootstrap methods differ in their assumptions. The extent to which these are violated in the simulation study gives a little grip on the explanation of the findings.

Compared to the other methods, the parametric bootstrap requires the strongest assumptions. In particular, it leans heavily upon the assumption of a (multivariate) normal distribution for the residuals. In the simulation, this is exactly the assumption that is severely violated. This explains the relatively bad performance here. In cases where data are less skewed (and the model is specified correctly), we expect this method to yield more satisfactory results.

For the other methods, matters are more complicated. In the simulation study, we know that there is homoskedasticity and that the model is correct. Under these conditions, we should expect good results for the cases bootstrap, as well as for the residual bootstrap. However, this is not corroborated for the cases bootstrap. Still, from a theoretical perspective, this method seems to be the most attractive since it comes closest to the ideal of an assumption-free method. To explain the bad performance of the cases bootstrap in this study, we could hypothesize that this method is possibly more sensitive to the distorting effects of small sample size than the residual bootstrap. In the study, this method was implemented in its most simple, conventional form. Future research may include several ways of refining, for instance by balancing, to improve the performance of this method.

The residual bootstrap treats the regression design as fixed, but "at the cost" (in contrast to the cases bootstrap, which can deal with heteroskedasticity) of assuming homoskedasticity, that is, a single empirical error distribution for both levels. In the study, this method works very satisfactory, but only as far as the resampling of shrunken residuals is concerned. Apparently, these residuals adequately reflect the true variation of the residuals in the population, whereas the raw residuals do not.

The simulation results affirm the theoretical insight that a z-test for the null hypothesis that a variance component is zero is unreliable and not well founded. Distributions of the variance component estimators are far from normal. Therefore, chi-square tests and likelihood ratio tests have been recommended instead. However, likelihood ratio tests are still founded upon the assumption of normality, whereas chi-square tests rely on asymptotic properties (a sufficiently large sample size at level 2). Bootstrap methods may provide nonparametric confidence intervals that could replace (or could be used in addition to) these approaches.

The discussion of resampling methods has so far been limited to `FIML` estimation. As stated in Section 11.1, in some cases the bias of `FIML` is well understood. An accepted alternative procedure for the estimation of the variance components is the method of restricted maximum likelihood (`REML`), which is claimed to provide less biased estimators [e.g., 56, Chapter 6]. There are, however, two drawbacks associated with this approach. First, `REML` still relies on the assumption of normality, whereas the bootstrap does not. Second, compared to `FIML`, the use of `REML` limits the application of likelihood ratio tests. `REML` optimizes a transformed likelihood function that does not contain the fixed parameters, that is, with respect to the variance components only. Hence, changes in model specification can only be tested as far as the random part is concerned. Nevertheless, it is useful and necessary to compare `REML` with bootstrap estimation, in particular by the method of shrunken residuals. Bootstrapping the `REML` estimators is yet another option.

11.7.2 Bootstrap Confidence Intervals

Meijer et al. [42] studied the performance of bootstrap confidence intervals for multilevel models by means of a simulation study. They used essentially the same design as Van der Leeden et al. [68] in their study of bootstrap bias correction and standard errors as discussed above. The only exception is that Meijer et al. [42] only studied the cases bootstrap. They compared the bootstrap normal, percentile-t, Efron's percentile, and BC intervals to standard intervals for `FIML` and `REML` estimators.

Their results showed that the different bootstrap confidence intervals of the fixed parameters were all satisfactory. The `FIML` and `REML` confidence intervals of the fixed parameters were also satisfactory, except the confidence intervals for the intercept, which showed some undercoverage. The `FIML` and `REML` confidence intervals for the variance components were dramatically bad, with a coverage of about 40% for a nominal value of 95%. The bootstrap confidence intervals studied were a great improvement, but with coverage percentages around 70%, they were still far from satisfactory. Hence, further research is needed to find improvements. Possibly, the BC_a interval or the double bootstrap (see below) may give satisfactory results, given their higher-order accuracies.

11.7.3 Jackknife

Busing et al. [10] studied the performance of the delete-m_j jackknife in a simulation study. Data for a (two-level) random effects ANOVA model were generated, that is, an intercept-only model including one variance component at each level. Residuals were drawn from a skewed distribution and the intraclass correlation was set to 0.2. In the simulation design, the number of groups J, group sizes n_j, and the skewness of the residual distributions were varied in turn.

The results showed that the delete-m_j jackknife estimator offers a minor reduction in bias compared to FIML and REML estimators, in exchange for a minor decrease in efficiency. A distinct reduction in bias was found for the delete-m_j standard error. When sample size is moderate, this standard error even improves in efficiency. It can be concluded that the delete-m_j jackknife standard errors are to be preferred to FIML and REML standard errors in the situations studied.

Due to the assumption of independence, the application of the jackknife for multilevel models is restricted to the highest level in the hierarchy. However, this limitation makes the implementation of the jackknife relatively simple. Researchers may easily use this jackknife method within their own general-purpose packages (e.g., SAS, S-Plus, R).

By using a very simple multilevel model in the simulation study, the merits and drawbacks of the presented jackknife approach have not been fully explored yet. More research is needed to reveal the full potential of the jackknife for multilevel models.

11.8 Extensions

Up till now, we have limited our discussion of bootstrap and jackknife procedures for multilevel models to the simple two-level mixed linear model, estimated with FIML. However, the discussion generalizes straightforwardly to other types of estimation methods, such as REML or two-step OLS (although in such cases, shrunken residuals may not be appropriate) and to more levels, in which case the delete-m_j jackknife must be applied to the highest level. Also, the methods, especially cases bootstrap, apply with little or no adaptation to estimators that handle missing data, such as multiple imputation and the EM algorithm [39].

Moreover, from the treatment in this chapter, resampling methods for other types of models can be easily derived. If a parametric distribution for the random terms is assumed, for example with generalized linear mixed models [54], the parametric bootstrap can be applied [4, 32]. If all variables are assumed to be random, as for example in many multilevel structural equation models [37, 46], the cases bootstrap or delete-m_j jackknife can be applied. (The multilevel SEM model of du Toit and du Toit [17] includes nonrandom covariates, though.) If (some of) the covariates are assumed nonrandom and residuals can be estimated meaningfully, as in nonlinear regression models, the residual bootstrap can be applied.

Note, however, that some care is needed in the application of the residual bootstrap, because the residuals can in some cases not be estimated satisfactorily or are not independent of the covariates. A typical example is a multilevel logistic regression (or mixed logit) model [e.g., 29], where the observable dependent variable $\underline{y}_i$ is binary. We may postulate an underlying continuous random variable that satisfies a regression model with continuous residuals, but these cannot be estimated. On the other hand, we could compute the residuals as $\hat{r}_i = y_i - E(\underline{y}_i \mid \boldsymbol{X}_i)$, the values of the dependent variable minus their (conditional) expectations according to the model. If we would perform a residual bootstrap using these residuals, the bootstrap-generated values $y_i^* = E(\underline{y}_i \mid \boldsymbol{X}_i) + r_i^*$ of the dependent variable would not be binary anymore. Moreover, the distribution of the residuals $\underline{r}_i$ depends on the value of the conditional expectation, $E(\underline{y}_i \mid \boldsymbol{X}_i)$.

With non-hierarchical models [50], parametric and residual bootstrap techniques can generally be used, but cases bootstrap sampling and jackknife are problematic due to the dependency structure (the data cannot be decomposed in disjoint independent subsets). It may be possible to define versions of cases bootstrap analogous to the moving blocks bootstrap for time series that give satisfactory results, but this is not straightforward.

In the discussion in this chapter, the various characteristics of the distribution $\hat{F}_n$ were obtained by simple random sampling from this distribution. In many cases, this may be computationally inefficient and with too small values of B lead to unstable estimators. The literature on Monte Carlo methods contains a number of strategies to improve the computational and statistical efficiency, such as antithetic sampling, importance sampling, and control variates. These, as well as a number of bootstrap-specific issues such as balanced resampling, are discussed by, e.g., Hall [28, Appendix II], Efron and Tibshirani [20, Chapter 23], and Davison and Hinkley [14, Chapter 9].

As we have stated, the bootstrap confidence intervals studied by Meijer et al. [42] do not perform satisfactorily, although they have better coverage rates than the standard `FIML` and `REML` confidence intervals. It is expected that the BC_a interval has better coverage, but the estimation of the acceleration constant is a little problematic in multilevel data. A promising alternative is the *double bootstrap*, which is a computationally highly intensive method, but this may nevertheless be feasible with today's (and tomorrow's) computers. The double bootstrap is obtained by performing a bootstrap within each bootstrap sample. The coverage rates of the bootstrap confidence intervals at the lower level of simulation are used to determine the quantiles of the bootstrap-estimated distribution function $\hat{H}(\theta)$ at the higher level that must be used in order to obtain a suitable confidence interval. The theoretical properties of the double bootstrap are comparable to those of the BC_a method, but the computations are easier, although they are more extensive as well.

See, e.g., McCullough and Vinod [41] for the details of the implementation and the statistical properties of the double bootstrap.

Finally, note that, although the most obvious parameters θ that may be subjected to bootstrap and jackknife procedures are the fixed parameters and variance components of the multilevel model, we can let θ be almost any characteristic of the model or the data, as illustrated by LeBlond [36].

References

1. M. K. Andersson and S. Karlsson. Bootstrapping error component models. *Computational Statistics*, 16:221–231, 2001.
2. J. G. Bagaka's. *Two Level Nested Hierarchical Linear Model with Random Intercepts via the Bootstrap*. PhD thesis, Michigan State University, East Lansing, MI, 1992. (Available from University Microfilms International, publication number AAT 9302972, `http://www.umi.com`)
3. L. Bellmann, J. Breitung, and J. Wagner. Bias correction and bootstrapping of error component models for panel data: Theory and applications. *Empirical Economics*, 14:329–342, 1989.
4. J. Booth. Bootstrap methods for generalized linear mixed models with applications to small area estimation. In G. U. H. Seeber, B. J. Francis, R. Hatzinger, and G. Steckel-Berger, editors, *Statistical Modelling: Proceedings of the 10th International Workshop on Statistical Modelling, Innsbruck, Austria, 10–14 July, 1995*, pages 43–51. Springer, New York, 1995.
5. N. E. Breslow and X. Lin. Bias correction in generalised linear mixed models with a single component of dispersion. *Biometrika*, 82:81–91, 1995.
6. T. S. Breusch. Useful invariance results for generalized regression models. *Journal of Econometrics*, 13:327–340, 1980.
7. F. M. T. A. Busing. Distribution characteristics of variance estimates in two-level models: A Monte Carlo study. Technical Report PRM 93-04, Leiden University, Department of Psychology, Leiden, 1993.
8. F. M. T. A. Busing, E. Meijer, and R. Van der Leeden. MLA: Software for multilevel analysis of data with two levels. User's guide for version 1.0b. Technical Report PRM 94-01, Leiden University, Department of Psychology, Leiden, 1994.
9. F. M. T. A. Busing, E. Meijer, and R. Van der Leeden. The MLA program for two-level analysis with resampling options. In T. A. B. Snijders, B. Engel, J. C. Van Houwelingen, A. Keen, G. J. Stemerdink, and M. Verbeek, editors, *SSS'95: Toeval zit Overal*, pages 37–58. iec ProGAMMA, Groningen, 1995.
10. F. M. T. A. Busing, E. Meijer, and R. Van der Leeden. Delete-m jackknife for unequal m. *Statistics and Computing*, 9:3–8, 1999.
11. F. M. T. A. Busing, E. Meijer, and R. Van der Leeden. *MLA: Software for MultiLevel Analysis of Data with Two Levels. User's Guide for Version 4.1.* Leiden University, Department of Psychology, Leiden, 2005.
12. F. M. T. A. Busing, R. Van der Leeden, and E. Meijer. MLA: Software for two-level analysis with resampling options. *Multilevel Modelling Newsletter*, 7 (3):11–13, 1995.

13. J. Carpenter, H. Goldstein, and J. Rasbash. A non-parametric bootstrap for multilevel models. *Multilevel Modelling Newsletter*, 11(1):2–5, 1999.
14. A. C. Davison and D. V. Hinkley. *Bootstrap Methods and their Application.* Cambridge University Press, Cambridge, UK, 1997.
15. J. de Leeuw and E. Meijer. Introduction to multilevel analysis. In J. de Leeuw and E. Meijer, editors, *Handbook of Multilevel Analysis*, Chapter 1. Springer, New York, 2008. (this volume)
16. D. Draper. Bayesian multilevel analysis and MCMC. In J. de Leeuw and E. Meijer, editors, *Handbook of Multilevel Analysis*, Chapter 2. Springer, New York, 2008. (this volume)
17. S. H. C. du Toit and M. du Toit. Multilevel structural equation modeling. In J. de Leeuw and E. Meijer, editors, *Handbook of Multilevel Analysis*, Chapter 12. Springer, New York, 2008. (this volume)
18. B. Efron. *The Jackknife, the Bootstrap and Other Resampling Plans.* SIAM, Philadelphia, 1982.
19. B. Efron. Better bootstrap confidence intervals. *Journal of the American Statistical Association*, 82:171–200, 1987. (with discussion)
20. B. Efron and R. J. Tibshirani. *An Introduction to the Bootstrap.* Chapman & Hall, New York, 1993.
21. C. C. Frangos and W. R. Schucany. Jackknife estimation of the bootstrap acceleration constant. *Computational Statistics & Data Analysis*, 9:271–282, 1990.
22. D. A. Freedman. Bootstrapping regression models. *The Annals of Statistics*, 9: 1218–1228, 1981.
23. W. R. Gilks, S. Richardson, and D. J. Spiegelhalter, editors. *Markov Chain Monte Carlo in Practice.* Chapman & Hall, London, 1996.
24. H. Goldstein. Consistent estimators for multilevel generalised linear models using an iterated bootstrap. *Multilevel Modelling Newsletter*, 8(1):3–6, 1996.
25. H. Goldstein. Bootstrapping for multilevel models. Working paper, Multilevel Models Project, London, 1998.
26. H. Goldstein. *Multilevel Statistical Models*, 3rd edition. Edward Arnold, London, 2003.
27. C. Gouriéroux and A. Monfort. *Simulation-Based Econometric Methods.* Oxford University Press, Oxford, UK, 1996.
28. P. Hall. *The Bootstrap and Edgeworth Expansion.* Springer, New York, 1992.
29. D. Hedeker. Multilevel models for ordinal and nominal variables. In J. de Leeuw and E. Meijer, editors, *Handbook of Multilevel Analysis*, Chapter 6. Springer, New York, 2008. (this volume)
30. D. V. Hinkley. Bootstrap methods. *Journal of the Royal Statistical Society, Series B*, 50:321–337, 1988.
31. I. G. G. Kreft, J. de Leeuw, and R. Van der Leeden. Review of five multilevel analysis programs: BMDP-5V, GENMOD, HLM, ML3, VARCL. *The American Statistician*, 48:324–335, 1994.
32. A. Y. C. Kuk. Asymptotically unbiased estimation in generalized linear models with random effects. *Journal of the Royal Statistical Society, Series B*, 57: 395–407, 1995.

33. N. M. Laird and T. A. Louis. Empirical Bayes confidence intervals based on bootstrap samples. *Journal of the American Statistical Association*, 82:739–757, 1987. (with discussion)
34. N. M. Laird and T. A. Louis. Empirical Bayes confidence intervals for a series of related experiments. *Biometrics*, 45:481–495, 1989.
35. W. Langer. *Mehrebenenanalyse: eine Einführung für Forschung und Praxis.* VS Verlag für Sozialwissenschaften, Wiesbaden, Germany, 2004.
36. D. J. LeBlond. *Methodology for Predicting Batch Manufacturing Risk.* Master's thesis, Colorado State University, Fort Collins, 2005.
37. S.-Y. Lee. Multilevel analysis of structural equation models. *Biometrika*, 77: 763–772, 1990.
38. N. T. Longford. *VARCL. Software for Variance Component Analysis of Data with Nested Random Effects (Maximum Likelihood).* Educational Testing Service, Princeton, NJ, 1990.
39. N. T. Longford. Missing data. In J. de Leeuw and E. Meijer, editors, *Handbook of Multilevel Analysis*, Chapter 10. Springer, New York, 2008. (this volume)
40. J. R. Magnus. Maximum likelihood estimation of the GLS model with unknown parameters in the disturbance covariance matrix. *Journal of Econometrics*, 7: 281–312, 1978.
41. B. D. McCullough and H. D. Vinod. Implementing the double bootstrap. *Computational Economics*, 12:79–95, 1998.
42. E. Meijer, F. M. T. A. Busing, and R. Van der Leeden. Estimating bootstrap confidence intervals for two-level models. In J. J. Hox and E. D. de Leeuw, editors, *Assumptions, Robustness, and Estimation Methods in Multivariate Modeling*, pages 35–47. TT Publicaties, Amsterdam, 1998.
43. E. Meijer, R. Van der Leeden, and F. M. T. A. Busing. Implementing the bootstrap for multilevel models. *Multilevel Modelling Newsletter*, 7(2):7–11, 1995.
44. A. M. Mood, F. A. Graybill, and D. C. Boes. *Introduction to the Theory of Statistics*, 3rd edition. McGraw-Hill, Singapore, 1974.
45. L. H. Moulton and S. L. Zeger. Analyzing repeated measures on generalized linear models via the bootstrap. *Biometrics*, 45:381–394, 1989.
46. B. O. Muthén. Multilevel covariance structure analysis. *Sociological Methods & Research*, 22:376–398, 1994.
47. H. Putter. *Consistency of Resampling Methods.* PhD thesis, Leiden University, Leiden, 1994.
48. M. H. Quenouille. Approximate tests of correlation in time series. *Journal of the Royal Statistical Society, Series B*, 11:68–84, 1949.
49. M. H. Quenouille. Notes on bias in estimation. *Biometrika*, 43:353–360, 1956.
50. J. Rasbash and W. Browne. Non-hierarchical multilevel models. In J. de Leeuw and E. Meijer, editors, *Handbook of Multilevel Analysis*, Chapter 8. Springer, New York, 2008. (this volume)
51. J. Rasbash, F. Steele, W. J. Browne, and B. Prosser. *A User's Guide to MLwiN. Version 2.0.* Centre for Multilevel Modelling, University of Bristol, Bristol, UK, 2005.

52. S. W. Raudenbush. Many small groups. In J. de Leeuw and E. Meijer, editors, *Handbook of Multilevel Analysis*, Chapter 5. Springer, New York, 2008. (this volume)
53. S. W. Raudenbush, A. S. Bryk, Y. F. Cheong, and R. Congdon. *HLM 6: Hierarchical Linear and Nonlinear Modeling.* Scientific Software International, Chicago, 2004.
54. G. Rodríguez. Multilevel generalized linear models. In J. de Leeuw and E. Meijer, editors, *Handbook of Multilevel Analysis*, Chapter 9. Springer, New York, 2008. (this volume)
55. W. R. Schucany, H. L. Gray, and D. B. Owen. On bias reduction in estimation. *Journal of the American Statistical Association*, 66:524–533, 1971.
56. S. R. Searle, G. Casella, and C. E. McCulloch. *Variance Components.* Wiley, New York, 1992.
57. J. Shao and D. Tu. *The Jackknife and Bootstrap.* Springer, New York, 1995.
58. A. Skrondal and S. Rabe-Hesketh. Multilevel and related models for longitudinal data. In J. de Leeuw and E. Meijer, editors, *Handbook of Multilevel Analysis*, Chapter 7. Springer, New York, 2008. (this volume)
59. T. A. B. Snijders. Analysis of longitudinal data using the hierarchical linear model. *Quality & Quantity*, 30:405–426, 1996.
60. T. A. B. Snijders and J. Berkhof. Diagnostic checks for multilevel models. In J. de Leeuw and E. Meijer, editors, *Handbook of Multilevel Analysis*, Chapter 3. Springer, New York, 2008. (this volume)
61. S. Stern. Simulation-based estimation. *Journal of Economic Literature*, 35: 2006–2039, 1997.
62. J. M. F. Ten Berge, W. P. Krijnen, T. Wansbeek, and A. Shapiro. Some new results on correlation-preserving factor scores prediction methods. *Linear Algebra and its Applications*, 289:311–318, 1999.
63. K. E. Train. *Discrete Choice Methods with Simulation.* Cambridge University Press, Cambridge, UK, 2003.
64. D. Tu and L. Zhang. Jackknife approximations for some nonparametric confidence intervals of functional parameters based on normalizing transformations. *Computational Statistics*, 7:3–15, 1992.
65. J. W. Tukey. Bias and confidence in not-quite large samples. *The Annals of Mathematical Statistics*, 29:614, 1958.
66. R. Van der Leeden. Multilevel analysis of repeated measures data. *Quality & Quantity*, 32:15–29, 1998.
67. R. Van der Leeden and F. M. T. A. Busing. First iteration versus final IGLS/RIGLS estimates in two-level models: A Monte Carlo study with ML3. Technical Report PRM 02-94, Leiden University, Department of Psychology, Leiden, 1994.
68. R. Van der Leeden, F. M. T. A. Busing, and E. Meijer. Bootstrap methods for two-level models. Technical Report PRM 97-04, Leiden University, Department of Psychology, Leiden, 1997.
69. G. Verbeke and E. Lesaffre. Large sample properties of the maximum likelihood estimators in linear mixed models with misspecified random-effects distribu-

tions. Technical Report 1996.1, Catholic University of Leuven, Biostatistical Centre for Clinical Trials, Leuven, 1996.

70. G. Verbeke and E. Lesaffre. The effect of misspecifying the random-effects distribution in linear mixed models for longitudinal data. *Computational Statistics & Data Analysis*, 23:541–556, 1997.
71. H. White. Maximum likelihood estimation of misspecified models. *Econometrica*, 50:1–25, 1982.
72. K. M. Wolter. *Introduction to Variance Estimation.* Springer, New York, 1985.
73. C. F. J. Wu. Jackknife, bootstrap and other resampling methods in regression analysis. *The Annals of Statistics*, 14:1261–1350, 1986. (with discussion)

12

Multilevel Structural Equation Modeling

Stephen H. C. du Toit and Mathilda du Toit

Scientific Software International

12.1 Introduction

Multilevel analysis allows characteristics of different groups to be included in models of individual behavior. Most analyses of social data entail the analysis of data with built-in hierarchies, usually obtained as a consequence of complex sampling methods. At each level of the hierarchy, a different set of variables may be defined.

Random regression models have been developed to model continuous data [6], and also dichotomous repeated measures data [16] where certain characteristics of the data preclude the use of traditional ANOVA models. Random regression models, however, do not allow for the possibility of including higher-level variables. It has been shown by Aitkin and Longford [2] that the aggregation of variables over individual observations may lead to misleading results. Both the aggregation of individual variables to a higher level of observations and the disaggregation of higher-order variables to an individual level in the analysis of multilevel data have been shown to be inadequate [11, 20]. Thus, the need for statistical models that take account of the sampling scheme is well recognized.

The use of multilevel models was initially hampered by the fact that closed-form mathematical expressions to estimate the variance and covariance components have only been available for perfectly balanced designs. Iterative numerical procedures must be used to obtain efficient estimates for unbalanced designs. Among the procedures suggested are full maximum likelihood [18, 31] and restricted maximum likelihood as proposed by Mason et al. [32] and Raudenbush and Bryk [42]. Another approach is the procedure of Bayes estimation [13]. Fitting the Mason et al. model, using the method of scoring, was illustrated by de Leeuw and Kreft [12].

J. de Leeuw, E. Meijer (eds.), *Handbook of Multilevel Analysis*,
© Springer 2008

At the same time, interest in latent variables, i.e., variables that can not be directly observed or alternatively only imperfectly observed, led to theory providing for the definition, fitting, and testing of general models for linear structural relations with latent variables for data from simple random sample(s). General applications based on this theory followed important contributions by Jöreskog and Sörbom [24] and McArdle and McDonald [33].

A more general model for multilevel structural relations, accommodating latent variables and the possibility of missing data at any level of the hierarchy and providing the combination of developments in these two fields, was a logical next step. In papers by Goldstein and McDonald [19], McDonald and Goldstein [36], Lee [28], and McDonald [34, 35], such a model was proposed. Attention was also given to the problem of estimation in the case of both balanced and unbalanced designs for linear structural relations in two-level data. Muthén [38, 39, 40] proposed a partial maximum likelihood solution as simplification in the case of an unbalanced design, entailing the computation of a single between-groups covariance matrix and an ad hoc estimator/scaling parameter. An overview of the latter can be found in Hox [21]. Raudenbush [41], Lee and Poon [29], and Liang and Bentler [30] developed full maximum likelihood estimators using the EM algorithm.

Liang and Bentler [30] discussed the similarities and differences between the various formulations of two-level structural equation models and presented a computationally efficient EM algorithm for obtaining ML estimates for unbalanced designs with cases missing at random.

In this chapter we describe a general two-level structural model that is similar to Liang and Bentler [30], the main difference being the estimation procedure. In our approach, we use the Fisher scoring [see 10] algorithm to obtain ML estimates. An advantage of this method is that it uses the expected values of the second-order derivatives and hence standard errors of the estimated parameters are readily available. We also make use of the special structure of the population covariance matrix to derive computationally efficient expressions (cf. Appendix 12.B) for the log-likelihood function and derivatives. An algorithm for full maximum likelihood estimation of the model is proposed, and a likelihood-based discrepancy function and test for goodness of fit is derived. Two examples, illustrating the implementation of the results for unbalanced designs with missing data at both levels of the hierarchy, are given.

12.2 A General Two-Level Structural Equation Model

We use McDonald's [34] formulation of a multilevel structural equations model (cf. (12.1)). In his paper, he derived minimal sufficient statistics for a balanced sampling design, i.e., $n_j = n$, $j = 1, 2, \ldots, J$, no missing values, and unrestricted means.

We have concentrated on the full information normal maximum likelihood procedure for these types of models when data are missing at random and the samples lead to an unbalanced design. Particular attention is paid to the derivation of results that can be directly used by researchers who would like to write their own multilevel SEM programs.

Suppose we have measures y_{ijk} on $k = 1, 2, \ldots, p$ variables from $i = 1, 2, \ldots, n_j$ level-1 units (for example students) from $j = 1, 2, \ldots, J$ randomly sampled level-2 units (for example schools). It is further supposed that we have x_{jl}, $l = 1, 2, \ldots, q$, variables characterizing the level-2 units.

For the j-th level-2 unit, we write the observed data as

$$\boldsymbol{y}'_j = (\boldsymbol{y}'_{1j}, \boldsymbol{y}'_{2j}, \ldots, \boldsymbol{y}'_{n_j j}, \boldsymbol{x}'_j),$$

where

$$\boldsymbol{y}'_{ij} = (y_{ij1}, y_{ij2}, \ldots, y_{ijp})$$

and

$$\boldsymbol{x}'_j = (x_{j1}, x_{j2}, \ldots, x_{jq}).$$

We assume that $\boldsymbol{y}_{ij}$ and $\boldsymbol{x}_j$ can be written as

$$\underline{\boldsymbol{y}}_{ij} = \boldsymbol{X}_{(y)ij}\boldsymbol{\beta}_y + \boldsymbol{S}_{ij}\underline{\boldsymbol{v}}_j + \boldsymbol{S}_{ij}\underline{\boldsymbol{u}}_{ij} \,, \tag{12.1}$$

$$\underline{\boldsymbol{x}}_j = \boldsymbol{X}_{(x)j}\boldsymbol{\beta}_x + \boldsymbol{R}_j\underline{\boldsymbol{w}}_j \,. \tag{12.2}$$

It is assumed that $\underline{\boldsymbol{v}}_1, \underline{\boldsymbol{v}}_2, \ldots, \underline{\boldsymbol{v}}_N$ are i.i.d. $\mathcal{N}(\boldsymbol{\emptyset}, \boldsymbol{\Sigma}_B)$ and that $\underline{\boldsymbol{u}}_{1j}$, $\underline{\boldsymbol{u}}_{2j}$, $\ldots$, $\underline{\boldsymbol{u}}_{n_j j}$ are i.i.d. $\mathcal{N}(\boldsymbol{\emptyset}, \boldsymbol{\Sigma}_W)$. It is additionally assumed that $\mathrm{Cov}(\underline{\boldsymbol{v}}_j, \underline{\boldsymbol{u}}'_{ij}) = \boldsymbol{\emptyset}$ for $j = 1, 2, \ldots, J$; $i = 1, 2, \ldots, n_j$.

Note that the matrices $\boldsymbol{X}_{(y)}$ and $\boldsymbol{S}_{ij}$ defined by (12.1) allow for the handling of incomplete data. For example, suppose $p = 4$ and that for a specific (level-2, level-1) combination only two measurements (say y_1 and y_3) are available, then

$$\boldsymbol{S} = \begin{pmatrix} 1\,0\,0\,0 \\ 0\,0\,1\,0 \end{pmatrix}, \quad \text{so that} \quad \boldsymbol{S}\boldsymbol{v} = \begin{pmatrix} v_1 \\ v_3 \end{pmatrix}.$$

In general, $\boldsymbol{S}_{ij}$ can be regarded as a selection matrix [14] consisting of a subset p_{ij} of the rows of the $p \times p$ identity matrix $\boldsymbol{I}_p$, where the rows of $\boldsymbol{S}_{ij}$ correspond to the response measurements available for the (i, j)-th unit. Likewise, $\boldsymbol{R}_j$ can be regarded as a subset q_j of the rows of $\boldsymbol{I}_q$.

Additional distributional assumptions are

$$\begin{aligned} \text{Cov}(\underline{\boldsymbol{w}}_j) &= \boldsymbol{\Sigma}_{xx}\,, && j = 1, 2, \ldots, J, \\ \text{Cov}(\underline{\boldsymbol{y}}_{ij}, \underline{\boldsymbol{w}}'_j) &= \boldsymbol{\Sigma}_{yx}\,, && j = 1, 2, \ldots, J;\ i = 1, 2, \ldots, n_j, \\ \text{Cov}(\underline{\boldsymbol{u}}_{ij}, \underline{\boldsymbol{w}}'_j) &= \boldsymbol{\emptyset}. \end{aligned} \tag{12.3}$$

From (12.1) and (12.2) it follows that

$$\underline{\boldsymbol{y}}_j = \begin{pmatrix} \boldsymbol{X}_{(y)j}\boldsymbol{\beta}_y + \boldsymbol{S}_j\underline{\boldsymbol{v}}_j + \sum_{j=1}^{n_j} \boldsymbol{Z}_{ij}\underline{\boldsymbol{u}}_{ij} \\ \boldsymbol{X}_{(x)j}\boldsymbol{\beta}_x + \boldsymbol{R}_j\underline{\boldsymbol{w}}_j \end{pmatrix}, \tag{12.4}$$

where

$$\boldsymbol{X}_{(y)j} = \begin{pmatrix} \boldsymbol{X}_{(y)1j} \\ \vdots \\ \boldsymbol{X}_{(y)n_jj} \end{pmatrix}, \quad \boldsymbol{S}_j = \begin{pmatrix} \boldsymbol{S}_{1j} \\ \boldsymbol{S}_{2j} \\ \vdots \\ \boldsymbol{S}_{n_jj} \end{pmatrix},$$

$$\boldsymbol{R}_j = \begin{pmatrix} \boldsymbol{R}_{1j} \\ \boldsymbol{R}_{2j} \\ \vdots \\ \boldsymbol{R}_{n_jj} \end{pmatrix}, \text{ and } \boldsymbol{Z}_{ij} = \begin{pmatrix} \boldsymbol{\emptyset} \\ \vdots \\ \boldsymbol{\emptyset} \\ \boldsymbol{S}_{ij} \\ \vdots \\ \boldsymbol{\emptyset} \end{pmatrix}. \tag{12.5}$$

From the distributional assumptions given above, it follows that

$$\underline{\boldsymbol{y}}_j \sim \mathcal{N}(\boldsymbol{\mu}_j, \boldsymbol{\Sigma}_j),$$

where

$$\boldsymbol{\mu}_j = \begin{pmatrix} \boldsymbol{X}_{(y)j} & \boldsymbol{\emptyset} \\ \boldsymbol{\emptyset} & \boldsymbol{X}_{(x)j} \end{pmatrix} \begin{pmatrix} \boldsymbol{\beta}_y \\ \boldsymbol{\beta}_x \end{pmatrix} = \boldsymbol{X}_j\boldsymbol{\beta}, \tag{12.6}$$

$$\boldsymbol{\Sigma}_j = \begin{pmatrix} \boldsymbol{V}_j & \boldsymbol{S}_j\boldsymbol{\Sigma}_{yx}\boldsymbol{R}'_j \\ \boldsymbol{R}_j\boldsymbol{\Sigma}_{xy}\boldsymbol{S}'_j & \boldsymbol{R}_j\boldsymbol{\Sigma}_{xx}\boldsymbol{R}'_j \end{pmatrix}, \tag{12.7}$$

and

$$\boldsymbol{V}_j = \text{Cov}\begin{pmatrix} \underline{\boldsymbol{y}}_{1j} \\ \vdots \\ \underline{\boldsymbol{y}}_{n_jj} \end{pmatrix} = \boldsymbol{S}_j\boldsymbol{\Sigma}_B\boldsymbol{S}'_j + \sum_{i=1}^{n_j} \boldsymbol{Z}_{ij}\boldsymbol{\Sigma}_W\boldsymbol{Z}'_{ij}.$$

Remark

If $\boldsymbol{R}_j = \boldsymbol{I}_q$ and $\boldsymbol{S}_{ij} = \boldsymbol{I}_p$, corresponding to the case of no missing y or x variables, then $\boldsymbol{S}_j\boldsymbol{\Sigma}_{yx}\boldsymbol{R}'_j = \boldsymbol{1}_{n_j} \otimes \boldsymbol{\Sigma}_{yx}$, where $\boldsymbol{1}_{n_j}$ is an $n_j \times 1$ column vector $(1, 1, \ldots, 1)'$.

Furthermore, for $\boldsymbol{S}_{ij} = \boldsymbol{I}_p$, $i = 1, \ldots, n_j$,

$$\boldsymbol{V}_j = \boldsymbol{I}_{n_j} \otimes \boldsymbol{\Sigma}_W + \boldsymbol{1}_{n_j}\boldsymbol{1}'_{n_j} \otimes \boldsymbol{\Sigma}_B$$

[see, e.g., 36]. The unknown parameters in (12.6) and (12.7) are $\boldsymbol{\beta}$, vecs $\boldsymbol{\Sigma}_B$, vecs $\boldsymbol{\Sigma}_W$, vec $\boldsymbol{\Sigma}_{xy}$, and vecs $\boldsymbol{\Sigma}_{xx}$, where vecs $\boldsymbol{A}$ denotes the $\frac{1}{2}p(p+1) \times 1$ vector of nonduplicated elements of the $p \times p$ symmetric matrix $\boldsymbol{A}$. The unknown parameters are contained in a $k^* \times 1$ vector $\boldsymbol{\pi}$.

Structural models for the type of data described above may be defined by restricting the elements of $\boldsymbol{\beta}$, $\boldsymbol{\Sigma}_B$, $\boldsymbol{\Sigma}_W$, $\boldsymbol{\Sigma}_{xy}$, and $\boldsymbol{\Sigma}_{xx}$ to be functions of some basic set of parameters $\boldsymbol{\gamma}' = (\gamma_1, \gamma_2, \ldots, \gamma_k)$, $k < k^*$.

For example, assume the following pattern for the matrices $\boldsymbol{\Sigma}_W$ and $\boldsymbol{\Sigma}_B$, where $\boldsymbol{\Sigma}_W$ refers to the within (level-1) covariance matrix and $\boldsymbol{\Sigma}_B$ to the between (level-2) covariance matrix:

$$\begin{aligned} \boldsymbol{\Sigma}_W &= \boldsymbol{\Lambda}_W \boldsymbol{\Psi}_W \boldsymbol{\Lambda}'_W + \boldsymbol{D}_W, \\ \boldsymbol{\Sigma}_B &= \boldsymbol{\Lambda}_B \boldsymbol{\Psi}_B \boldsymbol{\Lambda}'_B + \boldsymbol{D}_B. \end{aligned} \tag{12.8}$$

Factor analysis models typically have the covariance structures defined by (12.8). Consider a confirmatory factor analysis model with 2 factors and assume $p = 6$.

$$\boldsymbol{\Lambda}_W = \begin{pmatrix} \lambda_{11} & 0 \\ \lambda_{21} & 0 \\ \lambda_{31} & 0 \\ 0 & \lambda_{42} \\ 0 & \lambda_{52} \\ 0 & \lambda_{62} \end{pmatrix}, \quad \boldsymbol{\Psi}_W = \begin{pmatrix} \psi_{11} & \psi_{12} \\ \psi_{21} & \psi_{22} \end{pmatrix}, \quad \text{and} \quad \boldsymbol{D}_W = \begin{pmatrix} \theta_{11} & \cdots & \\ \vdots & \ddots & \\ & & \theta_{66} \end{pmatrix}.$$

If we restrict all the parameters across the level-1 and level-2 units to be equal, then

$$\boldsymbol{\gamma}' = (\lambda_{11}, \lambda_{21}, \ldots, \lambda_{62}, \psi_{11}, \psi_{21}, \psi_{22}, \theta_{11}, \ldots, \theta_{66})$$

is the vector of unknown parameters.

12.3 Maximum Likelihood for General Means and Covariance Structures

In this section we give a general framework for normal maximum likelihood estimation of the unknown parameters. In practice, the number of variables $(p + q)$ and the number of level-1 units within a specific level-2 unit may be quite large, which leads to $\boldsymbol{\Sigma}_j$ matrices of very high order. It is therefore apparent that further simplification of the likelihood function derivatives and

Hessian is required if the goal is to implement the theoretical results in a computer program. These aspects are addressed in Appendix 12.B.

Denote the expected value and covariance matrix of $\underline{\boldsymbol{y}}_j$ by $\boldsymbol{\mu}_j$ and $\boldsymbol{\Sigma}_j$, respectively (see (12.6) and (12.7)). The log-likelihood function of $\underline{\boldsymbol{y}}_1, \underline{\boldsymbol{y}}_2, \dots, \underline{\boldsymbol{y}}_J$ may then be expressed as

$$\ln L = -\tfrac{1}{2}\sum_{j=1}^{J}\{n_j \ln 2\pi + \ln|\boldsymbol{\Sigma}_j| + \operatorname{tr}\boldsymbol{\Sigma}_j^{-1}(\boldsymbol{y}_j - \boldsymbol{\mu}_j)(\boldsymbol{y}_j - \boldsymbol{\mu}_j)'\}. \qquad (12.9)$$

Instead of maximizing $\ln L$, maximum normal likelihood estimates of the unknown parameters are obtained by minimizing $-\ln L$ with the constant term omitted, i.e., by minimizing the function

$$F(\boldsymbol{\gamma}) = \tfrac{1}{2}\sum_{j=1}^{J}\{\ln|\boldsymbol{\Sigma}_j| + \operatorname{tr}\boldsymbol{\Sigma}_j^{-1}\boldsymbol{G}_{y_j}\}, \qquad (12.10)$$

where

$$\boldsymbol{G}_{y_j} = (\boldsymbol{y}_j - \boldsymbol{\mu}_j)(\boldsymbol{y}_j - \boldsymbol{\mu}_j)'. \qquad (12.11)$$

The first-order condition $\partial F(\boldsymbol{\gamma})/\partial\boldsymbol{\gamma} = \boldsymbol{\emptyset}$ yields the normal maximum likelihood estimator $\underline{\hat{\boldsymbol{\gamma}}}$ of the unknown vector of parameters $\boldsymbol{\gamma}$.

Unless the model yields maximum likelihood estimators in closed form, it will be necessary to make use of an iterative procedure to minimize the discrepancy function. The optimization procedure described next [see 10] is based on the so-called Fisher scoring algorithm, which in the case of structured means and covariances may be regarded as a sequence of Gauss-Newton steps with quantities to be fitted as well as the weight matrix changing at each step. Fisher scoring algorithms require the gradient vector and an approximation to the Hessian matrix. Elements of the gradient vector $\boldsymbol{g}(\boldsymbol{\gamma})$ and approximate Hessian matrix $\boldsymbol{H}(\boldsymbol{\gamma})$ of $F(\boldsymbol{\gamma})$ are given by

$$\frac{\partial F}{\partial \gamma_r} = [\boldsymbol{g}(\boldsymbol{\gamma})]_r = -\sum_{j=1}^{J}\left\{\operatorname{tr}\boldsymbol{Q}_j\frac{\partial\boldsymbol{\mu}_j}{\partial\gamma_r} + \tfrac{1}{2}\operatorname{tr}\boldsymbol{P}_j\frac{\partial\boldsymbol{\Sigma}_j}{\partial\gamma_r}\right\}, \qquad (12.12)$$

where

$$\boldsymbol{Q}_j = (\boldsymbol{y}_j - \boldsymbol{\mu}_j)'\boldsymbol{\Sigma}_j^{-1} \qquad (12.13)$$

and

$$\boldsymbol{P}_j = \boldsymbol{\Sigma}_j^{-1}(\boldsymbol{G}_{y_j} - \boldsymbol{\Sigma}_j)\boldsymbol{\Sigma}_j^{-1}. \qquad (12.14)$$

Let

$$[\boldsymbol{H}(\boldsymbol{\gamma})]_{r,s} = -E\left(\frac{\partial^2 \ln \underline{L}}{\partial \gamma_r\, \partial \gamma_s}\right).$$

In the remainder of this chapter, $\boldsymbol{H}(\boldsymbol{\gamma})$ will be referred to as the Hessian. Hence,

$$\begin{aligned}\frac{\partial^2 F}{\partial \gamma_r\, \partial \gamma_s} &\approx [\boldsymbol{H}(\boldsymbol{\gamma})]_{r,s} \\ &= \sum_{j=1}^{J}\left\{\mathrm{tr}\left(\frac{\partial \boldsymbol{\mu}_j'}{\partial \gamma_r}\boldsymbol{\Sigma}_j^{-1}\frac{\partial \boldsymbol{\mu}_j}{\partial \gamma_s}\right) + \tfrac{1}{2}\left(\boldsymbol{\Sigma}_j^{-1}\frac{\partial \boldsymbol{\Sigma}_j}{\partial \gamma_r}\boldsymbol{\Sigma}_j^{-1}\frac{\partial \boldsymbol{\Sigma}_j}{\partial \gamma_s}\right)\right\}. \quad (12.15)\end{aligned}$$

Suppose that $\boldsymbol{\gamma}_k$ is the k-th approximation to the $\hat{\gamma}$ that minimizes $F(\boldsymbol{\gamma})$. Let $\boldsymbol{g}_k = \boldsymbol{g}(\boldsymbol{\gamma}_k)$, $\boldsymbol{H}_k = \boldsymbol{H}(\boldsymbol{\gamma}_k)$, and $F_k = F(\boldsymbol{\gamma}_k)$. The next approximation is obtained from

$$\boldsymbol{\gamma}_{k+1} = \boldsymbol{\gamma}_k + \alpha_k \boldsymbol{\delta}_k\,, \quad (12.16)$$

where

$$\boldsymbol{\delta}_k = -\boldsymbol{H}_k^{-1}\boldsymbol{g}_k \quad (12.17)$$

and α_k is a step size parameter chosen initially as 1 and then successively halved until $F_{k+1} \leq F_k$.

Agresti [1] pointed out that the Fisher scoring method resembles the Newton-Raphson method, the distinction being that the Fisher scoring (cf. (12.15)) uses the expected value of the second derivative matrix.

A convenient feature of the Fisher scoring algorithm is that an estimate $\{\boldsymbol{H}(\hat{\boldsymbol{\gamma}})\}^{-1}$ of the asymptotic covariance matrix of estimators $\boldsymbol{\gamma}$ is available on convergence as a by-product of the calculations.

It may be necessary to minimize $F(\boldsymbol{\gamma})$ subject to r nonlinear constraints of the form

$$\boldsymbol{c}(\boldsymbol{\gamma}) = \boldsymbol{\emptyset}, \quad (12.18)$$

where $\boldsymbol{c}(\boldsymbol{\gamma})$ is a continuously differentiable $r \times 1$ vector-valued function of $\boldsymbol{\gamma}$. Let $\boldsymbol{c}_k = \boldsymbol{c}(\boldsymbol{\gamma}_k)$ and $\boldsymbol{L}_k = \boldsymbol{L}(\boldsymbol{\gamma}_k)$. Then the linear Taylor approximation for the constraint function is

$$\boldsymbol{c}(\boldsymbol{\gamma}) \approx \boldsymbol{c}_k + \boldsymbol{L}_k \boldsymbol{\delta}, \quad (12.19)$$

where $\boldsymbol{\delta} = \boldsymbol{\gamma} - \boldsymbol{\gamma}_k$. A typical element of the Jacobian matrix $\boldsymbol{L}_k$ is given by

$$[\boldsymbol{L}_k]_{ij} = \left.\frac{\partial c_i}{\partial \gamma_j}\right|_{\boldsymbol{\gamma}=\boldsymbol{\gamma}_k}, \quad (12.20)$$

where $\boldsymbol{c}_i = [\boldsymbol{c}(\boldsymbol{\gamma})]_i$. Consequently, the nonlinear constraints (12.18) may be approximated by the linear constraints

$$\boldsymbol{L}_k \boldsymbol{\delta} = -\boldsymbol{c}_k \,. \tag{12.21}$$

The increment vector $\boldsymbol{\delta}_k$ is obtained [10] as the solution of

$$\begin{pmatrix} \boldsymbol{\delta}_k \\ \boldsymbol{\lambda}_k \end{pmatrix} = \begin{pmatrix} \boldsymbol{H}_k + \boldsymbol{L}_k' \boldsymbol{D}_k \boldsymbol{L}_k & \boldsymbol{L}_k' \\ \boldsymbol{L}_k & \boldsymbol{\emptyset} \end{pmatrix}^{-1} \begin{pmatrix} -(\boldsymbol{g}_k + \boldsymbol{L}_k' \boldsymbol{D}_k \boldsymbol{c}_k) \\ -\boldsymbol{c}_k \end{pmatrix}, \tag{12.22}$$

where $\boldsymbol{\lambda}_k$ is an $r \times 1$ vector of Lagrange multipliers and $\boldsymbol{D}_k$ is an arbitrary nonnegative definite matrix. The scaling matrix $\boldsymbol{D}_k$ does not affect the solution and is often chosen to be the null matrix [17]. The next approximation γ_{k+1} for $\hat{\gamma}$ is obtained from

$$\gamma_{k+1} = \gamma_k + \alpha_k \boldsymbol{\delta}_k \,,$$

where α_k is chosen initially as 1 and is halved successively until

$$F_b + 2\sum_{i=1}^{r} \left| [\boldsymbol{\lambda}_k]_j [\boldsymbol{c}_{k+1}]_j \right| < F_\alpha + 2\sum_{i=1}^{r} \left| [\boldsymbol{\lambda}_k]_j [\boldsymbol{c}_k]_j \right|, \tag{12.23}$$

where (cf. (12.10)) $F_b = F(\gamma_{k+1})$ and $F_\alpha = F(\gamma_k)$. If no constraints are imposed, all terms involving $\boldsymbol{c}_k$ and $\boldsymbol{L}_k$ are omitted.

It can happen that the matrix to be inverted in (12.22) is singular or near singular. An adaptation of the Jennrich and Sampson [22] stepwise regression procedure may be used to obtain an appropriate conditional inverse. Their procedure for imposing bounds on the estimates may also be employed.

Let $\boldsymbol{\pi}$ denote a $k^* \times 1$ vector containing the elements of the $m \times 1$ parameter vector $\boldsymbol{\beta}$, and the nonduplicated elements of $\boldsymbol{\Sigma}_B$, $\boldsymbol{\Sigma}_W$, $\boldsymbol{\Sigma}_{xy}$, and $\boldsymbol{\Sigma}_{xx}$. It follows that

$$k^* = m + 2\left(\tfrac{1}{2}p(p+1)\right) + pq + \tfrac{1}{2}q(q+1)\,.$$

In Appendix 12.B, results are derived for the gradient vector $\boldsymbol{g} = \boldsymbol{g}(\boldsymbol{\pi})$ and Hessian $\boldsymbol{H} = \boldsymbol{H}(\boldsymbol{\pi})$ in terms of the parameters of a two-level model when no restrictions are imposed on the elements of $\boldsymbol{\beta}$ and the parameter matrices $\boldsymbol{\Sigma}_B$ to $\boldsymbol{\Sigma}_{xx}$.

Two-level structural equation models impose restrictions on the between (level-2) and within (level-1) variance components. Formally, suppose that $\boldsymbol{\beta} = \boldsymbol{\beta}(\boldsymbol{\gamma})$, $\boldsymbol{\Sigma}_B = \boldsymbol{\Sigma}_B(\boldsymbol{\gamma})$, $\boldsymbol{\Sigma}_W = \boldsymbol{\Sigma}_W(\boldsymbol{\gamma})$, $\boldsymbol{\Sigma}_{xy} = \boldsymbol{\Sigma}_{xy}(\boldsymbol{\gamma})$, and $\boldsymbol{\Sigma}_{xx} = \boldsymbol{\Sigma}_{xx}(\boldsymbol{\gamma})$, where $\boldsymbol{\gamma}$ is a $k \times 1$ vector of unknown parameters, $k < k^*$. Derivatives of the form $\partial \boldsymbol{\Sigma}_B / \partial \gamma_r$, ..., $\partial \boldsymbol{\Sigma}_{xx} / \partial \gamma_r$ form an inherent part of the estimation procedure in structural equation models and are relatively straightforward to compute. For example (see (12.8)),

$$\frac{\partial \boldsymbol{\Sigma}_W}{\partial [\boldsymbol{\Lambda}_W]_{r,s}} = \boldsymbol{J}_{rs} \boldsymbol{\Psi}_W \boldsymbol{\Lambda}_W' + \boldsymbol{\Lambda}_W \boldsymbol{\Psi}_W \boldsymbol{J}_{rs}' \,,$$

where $[\boldsymbol{\Lambda}_W]_{r,s}$ denotes an element of the parameter vector $\boldsymbol{\pi}$ and $\boldsymbol{J}_{rs}$ is a null matrix except for the (r, s)-th element, which equals 1.

In general, let $\boldsymbol{C}$: $(k^* \times k)$ denote the matrix of derivatives

$$\boldsymbol{C} = \frac{\partial \boldsymbol{\pi}}{\partial \boldsymbol{\gamma}'} .$$

Using the chain rule for matrix differentiation, it follows that

$$\frac{\partial \ln L}{\partial \boldsymbol{\gamma}'} = \frac{\partial \ln L}{\partial \boldsymbol{\pi}'} \frac{\partial \boldsymbol{\pi}}{\partial \boldsymbol{\gamma}'} ,$$

and hence

$$\frac{\partial \ln L}{\partial \boldsymbol{\gamma}} = \boldsymbol{C}' \boldsymbol{g}(\boldsymbol{\pi}). \tag{12.24}$$

Similarly,

$$\boldsymbol{H}(\boldsymbol{\gamma}) = -E\left(\frac{\partial^2 \ln \underline{L}}{\partial \boldsymbol{\gamma}\, \partial \boldsymbol{\gamma}'}\right) = \boldsymbol{C}' \boldsymbol{H}(\boldsymbol{\pi}) \boldsymbol{C}. \tag{12.25}$$

The matrix $\boldsymbol{C}$ and expressions (12.24) and (12.25) are instrumental in the analysis of multilevel structural equation models and the derivation of a χ^2 goodness of fit statistic. Note that $\boldsymbol{C} = \boldsymbol{I}$ when $\boldsymbol{\gamma} = \boldsymbol{\pi}$ and hence no restrictions are imposed on the multilevel variance components. Standard errors of the estimated parameters are obtained as the square roots of the diagonal elements of $[\boldsymbol{H}(\underline{\hat{\boldsymbol{\gamma}}})]^{-1}$, where $\underline{\hat{\boldsymbol{\gamma}}}$ is the maximum likelihood estimator of $\boldsymbol{\gamma}$. In Appendix 12.B, detailed computational formulas for $\partial \ln L / \partial \boldsymbol{\gamma}$ and $\boldsymbol{H}(\boldsymbol{\gamma})$ are derived.

Starting Values and Convergence Issues

In fitting a structural equation model to a hierarchical data set, one may encounter convergence problems unless good starting values are provided.

We have implemented the following procedure in LISREL [15]. As a first step, estimates of the fixed components $\boldsymbol{\beta}$ and the variance components $\boldsymbol{\Sigma}_B$, $\boldsymbol{\Sigma}_{xy}$, $\boldsymbol{\Sigma}_{xx}$, and $\boldsymbol{\Sigma}_W$ are obtained. This is accomplished by setting $\boldsymbol{C} = \partial \boldsymbol{\pi} / \partial \boldsymbol{\gamma}' = \boldsymbol{I}$ (see (12.23) and (12.25)), where $\boldsymbol{\pi}'$ is the vector of parameters $\big(\boldsymbol{\mu}', (\text{vecs}\, \boldsymbol{\Sigma}_B)', \ldots, (\text{vecs}\, \boldsymbol{\Sigma}_{xx})'\big)$ and $\boldsymbol{\gamma}$ the set of parameters when restrictions are imposed on $\boldsymbol{\mu}$, vecs $\boldsymbol{\Sigma}_B$, ..., vecs $\boldsymbol{\Sigma}_{xx}$.

Our experience with the Gauss-Newton algorithm described above is that convergence is usually obtained in less than 10 iterations, where initially $\boldsymbol{\beta} = \boldsymbol{\emptyset}$, $\boldsymbol{\Sigma}_B = \boldsymbol{I}_p$, $\boldsymbol{\Sigma}_{xy} = \boldsymbol{\emptyset}$, $\boldsymbol{\Sigma}_{xx} = \boldsymbol{I}_q$, and $\boldsymbol{\Sigma}_W = \boldsymbol{I}_p$. At convergence, the value of $-2 \ln L$ is computed.

Next, we treat

$$\boldsymbol{S}_B = \begin{pmatrix} \hat{\boldsymbol{\Sigma}}_B & \hat{\boldsymbol{\Sigma}}_{yx} \\ \hat{\boldsymbol{\Sigma}}_{xy} & \hat{\boldsymbol{\Sigma}}_{xx} \end{pmatrix} \quad \text{and} \quad \boldsymbol{S}_W = \begin{pmatrix} \hat{\boldsymbol{\Sigma}}_W & \boldsymbol{\emptyset} \\ \boldsymbol{\emptyset} & \boldsymbol{\emptyset} \end{pmatrix}$$

as sample covariance matrices and fit a two-group structural equation model to the between and within groups. Parameter estimates obtained in this manner are used as the elements of the initial parameter vector $\boldsymbol{\gamma}_0$. The estimators obtained from this step are consistent and are therefore typically already close to the final ML estimators.

In the third step, the iterative procedure is restarted and $\boldsymbol{\gamma}_k$ updated from $\boldsymbol{\gamma}_{k-1}$, $k = 1, 2, \ldots$, until convergence is obtained. This value is denoted by $\hat{\underline{\boldsymbol{\gamma}}}$, the maximum likelihood estimator of $\boldsymbol{\gamma}$. Standard errors of the elements of $\hat{\underline{\boldsymbol{\gamma}}}$ are calculated as the square roots of the diagonal elements of $[\boldsymbol{H}(\hat{\underline{\boldsymbol{\gamma}}})]^{-1}$.

12.4 Fit Statistics and Hypothesis Testing

The multilevel structural equation model, $M(\boldsymbol{\gamma})$, and its assumptions imply a covariance structure $\boldsymbol{\Sigma}_B(\boldsymbol{\gamma})$, $\boldsymbol{\Sigma}_W(\boldsymbol{\gamma})$, $\boldsymbol{\Sigma}_{xy}(\boldsymbol{\gamma})$, $\boldsymbol{\Sigma}_{xx}(\boldsymbol{\gamma})$ and mean structure $\boldsymbol{\mu}(\boldsymbol{\gamma})$ for the observable random variables, where $\boldsymbol{\gamma}$ is a $k \times 1$ vector of parameters in the statistical model. It is assumed that the empirical data is a random sample of J level-2 units and $n = \sum_{j=1}^{J} n_j$ level-1 units, where n_j denotes the number of level-1 units within the j-th level-2 unit. From this data, we can compute estimates of $\boldsymbol{\mu}$, $\boldsymbol{\Sigma}_B$, $\ldots$, $\boldsymbol{\Sigma}_{xx}$ if no restrictions are imposed on their elements. The number of parameters for the unrestricted model (see Section 12.3) is $k^* = m + 2\left[\frac{1}{2}p(p+1)\right] + pq + \frac{1}{2}q(q+1)$ and is summarized in the $k^* \times 1$ vector $\boldsymbol{\pi}$. To test the model $M(\boldsymbol{\gamma})$, we use the likelihood ratio test statistic

$$\underline{c} = -2 \ln L(\hat{\underline{\boldsymbol{\gamma}}}) + 2 \ln L(\hat{\underline{\boldsymbol{\pi}}}). \tag{12.26}$$

If the unrestricted model $M(\boldsymbol{\pi})$ holds, $\underline{c}$ has a χ^2 distribution with $d = k^* - k$ degrees of freedom.

If the model does not hold, $\underline{c}$ has a noncentral χ^2 distribution with d degrees of freedom and noncentrality parameter λ that may be estimated as [see 9]

$$\hat{\underline{\lambda}} = \max(\underline{c} - d,\ 0). \tag{12.27}$$

Browne and Cudeck [9] also show how to set up a confidence interval for λ.

It is possible that the researcher has specified a number of competing models $M_1(\boldsymbol{\gamma}_1)$, $M_2(\boldsymbol{\gamma}_2)$, $\ldots$, $M_K(\boldsymbol{\gamma}_K)$. If the models are nested in the sense that $\boldsymbol{\gamma}_j$: $k_j \times 1$ is a subset of $\boldsymbol{\gamma}_i$: $k_i \times 1$, one may use the likelihood ratio test $\underline{c}^* = -2 \ln L(\hat{\underline{\boldsymbol{\gamma}}}_j) + 2 \ln L(\hat{\underline{\boldsymbol{\gamma}}}_i)$ with degrees of freedom $k_i - k_j$ to test $M_j(\boldsymbol{\gamma}_j)$ against $M_i(\boldsymbol{\gamma}_i)$.

Another approach is to compare models on the basis of some criteria that take parsimony as well as fit into account. This approach can be used regardless of whether or not the models can be ordered in a nested sequence. Two strongly related criteria are the AIC measure of Akaike [3] and the CAIC of Bozdogan [7]:

$$\text{AIC} = \underline{c} + 2k, \tag{12.28}$$

$$\text{CAIC} = \underline{c} + (1 + \ln n)k. \tag{12.29}$$

The use of $\underline{c}$ as a central χ^2 statistic is based on the assumption that the model holds exactly in the population. A consequence of this assumption is that models that hold approximately in the population will be rejected in large samples.

Steiger [43] proposed the root mean square error of approximation (RMSEA) statistic that takes particular account of the error of approximation in the population,

$$\text{RMSEA} = \sqrt{\hat{\underline{F}}_0/d}\,, \tag{12.30}$$

where $\hat{\underline{F}}_0$ is a function of the sample size, degrees of freedom, and the fit function. To use the RMSEA as a fit measure in multilevel SEM, we propose

$$\hat{\underline{F}}_0 = \max\left\{\frac{\underline{c} - d}{n},\ 0\right\}. \tag{12.31}$$

Browne and Cudeck [9] suggest that an RMSEA value of 0.05 indicates a close fit and that values of up to 0.08 represent reasonable errors of approximation in the population.

12.5 A Simple Illustration

The following example illustrates the steps outlined above. The data set used in this section forms part of the data library of the Multilevel Project at the University of London, and comes from the Junior School Project [37]. Mathematics and language tests were administered in three consecutive years to more than 1000 students from 49 primary schools that were randomly selected from primary schools maintained by the Inner London Education Authority.

The following variables were selected from the data file:

School School code (1–49)
Math1 Score on mathematics test in year 1 (score 1–40)
Math2 Score on mathematics test in year 2 (score 1–40)
Math3 Score on mathematics test in year 3 (score 1–40).

The school number (*School*) is used as the level-2 identification.

A simple confirmatory factor analysis model (see Fig. 12.1) is fitted to the data:

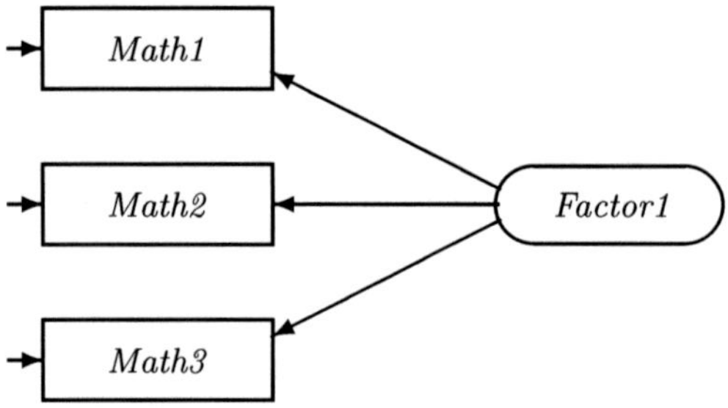

Fig. 12.1 Confirmatory factor analysis model.

$$\boldsymbol{\Sigma}_B = \boldsymbol{\lambda}\psi\boldsymbol{\lambda}' + \boldsymbol{D}_B \ ,$$
$$\boldsymbol{\Sigma}_W = \boldsymbol{\lambda}\psi\boldsymbol{\lambda}' + \boldsymbol{D}_W \ ,$$

where $\boldsymbol{\lambda}' = (1, \lambda_{21}, \lambda_{31})$ and $\boldsymbol{D}_B$ and $\boldsymbol{D}_W$ are diagonal matrices with diagonal elements equal to the unique (error) variances of *Math1*, *Math2*, and *Math3*. The variance of the factor is denoted by ψ. Note that we assume equal factor loadings and factor variance across the between and within groups, leading to a model with 3 degrees of freedom. The SIMPLIS [see 25] syntax to fit the factor analysis model is shown below. Note that the between- and within-groups covariance matrices are the estimated $\boldsymbol{\Sigma}_B$ and $\boldsymbol{\Sigma}_W$ obtained in the first step by fitting the unrestricted model.

```
Group 1: Between Schools JSP data (Level 2)
Observed Variables: Math1 Math2 Math3
Covariance matrix
   3.38885
   2.29824    5.19791
   2.31881    3.00273    4.69663
Sample Size=24 ! Taken as (n1 + n2 + ... + nN)/N
               ! and rounded to nearest integer
Latent Variables: Factor1
Relationships
Math1=1*Factor1
Math2-Math3=Factor1

Group 2: Within Schools JSP data (Level 1)
Covariance matrix
  47.04658
  38.56798   55.37006
  30.81049   36.04099   40.71862
Sample Size=1192 ! Total number of pupils
! Uncomment the following line to free the parameter
! Set the Variance of Factor1 Free
```

```
Set the Error Variance of Math1 Free
Set the Error Variance of Math2 Free
Set the Error Variance of Math3 Free
Path Diagram
LISREL OUTPUT ND=3
End of Problem
```

Table 12.1 shows the parameter estimates, standard errors, and χ^2 statistic from the SIMPLIS output and the corresponding values from the multilevel SEM output.

Remarks

1. The between-groups sample size of 24 used in the SIMPLIS syntax was computed as $J^{-1}\sum_{j=1}^{J} n_j$, where J is the number of schools and n_j the number of children within school j. Since this value is only used to obtain starting values, it is not really crucial how the between-group sample size is computed. See, for example, Muthén [38, 39] for an alternative formula.
2. The within-group sample size of 1192 used in the SIMPLIS syntax is equal to the total number of school children.

Table 12.1 Parameter estimates and standard errors for factor analysis model.

	SIMPLIS		Multilevel SEM	
	Estimate	Standard error	Estimate	Standard error
Factor loadings				
Math1	1.000	—	1.000	—
Math2	1.173	0.031	1.177	0.032
Math3	0.939	0.026	0.947	0.028
Factor variance				
ψ	32.109	1.821	31.235	1.808
Error variances (between)				
Math1	1.640	0.787	1.656	0.741
Math2	2.123	1.059	2.035	0.942
Math3	1.868	0.779	1.840	0.734
Error variances (within)				
Math1	14.114	0.810	14.209	0.890
Math2	10.274	0.884	10.256	0.993
Math3	11.910	0.699	11.837	0.806
Chi-square	36.233		46.560	
Degrees of freedom	3		3	

3. The number of missing values per variable is as follows:
   Math1: 38
   Math2: 63
   Math3: 239.
   The large percentage missing for the *Math3* variable may partially explain the relatively large difference in χ^2 values from the SIMPLIS and multilevel SEM outputs.
4. If one allows for the factor variance parameter to be free over groups, the χ^2 fit statistic becomes 1.087 at 2 degrees of freedom. The total number of multilevel SEM iterations required to obtain convergence equals 8.

In conclusion, a small number of variables and a single factor SEM model were used to illustrate the starting values procedure that we adopted. The next section contains two additional examples, also based on a schools data set. It should be noted that in the applications to follow, we focus on the parameters of the latent variable submodel and do not present (although they may be important in their own right) the regression coefficients of the exogenous variables.

12.6 Practical Applications

The two examples discussed in this section are based on school data that were collected during a 1994 survey in South Africa.[1]

A brief description of the `SASchools94.dat` data set is as follows: $J = 136$ schools were selected and the total number of children within schools $n = \sum_{j=1}^{J} n_j = 6047$, where n_j varies from 20 to 60. The data set contains 20 variables as shown in Table 12.2.

The variables *Language* and *Socio* are school-level variables and their values do not vary within schools. Listwise deletion of missing cases results in a data set containing only 2691 of the original 6047 cases.

12.6.1 Example 1: Confirmatory Factor Analysis

For this example we use the variables *Classif*, *Compar*, *Verbal*, *Figure*, *Pattcomp*, and *Numserie* from the schools data set discussed in the previous section. Two common factors are hypothesized: verbal and numeric ability. The first three variables are assumed to measure *Verbfac* and the last three to measure *Numfac*. A path diagram of the assumed factor model is shown in Fig. 12.2. Appropriate LISREL syntax is given in the appendix (see p. 458) of this chapter.

[1] This data set is available on request from the authors.

Table 12.2 Description of variables in `SASchool94.dat`.

Var.	Name	Description	Number missing
1	*Student*	Level-1 identification	0
2	*School*	Level-2 identification	0
3	*Constant*	All values equal to 1	0
4	*Grade*	0 = Grade 2, 1 = Grade 3, 2 = Grade 4	0
5	*Language*	0 = White[a], 1 = Black[a]	0
6	*Gender*	1 = Male, 2 = Female	1
7	*Mothedu*	Mother's level of education on a scale from 1 to 7	783
8	*Fathede*	Father's level of education on a scale from 1 to 7	851
9	*Read*	Teacher's evaluation on a scale from 1 to 5[b]	482
10	*Speech*	Teacher's evaluation on a scale from 1 to 5[b]	470
11	*Write*	Teacher's evaluation on a scale from 1 to 5[b]	467
12	*Arithm*	Teacher's evaluation on a scale from 1 to 5[b]	451
13	*Socio*	Socio-economic status indicator, scale 0 to 5 on school level	0
14	*Classif*	Classification: total correct out of 30 items	23
15	*Compar*	Comparison: total correct out of 23 items	27
16	*Verbal*	Verbal Instructions: total correct out of 50 items	20
17	*Figure*	Figure Series: total correct out of 24 items	118
18	*Pattcomp*	Pattern Completion: total correct out of 24 items	109
19	*Knowled*	Knowledge: total correct out of 32 items	112
20	*Numserie*	Number Series: total correct out of 15 items	2305

[a] "White" = Afrikaans or English;
"Black" = One of the 11 official black languages.
[b] 1 = Poor, ..., 5 = Excellent.

The between- and within-school structural equation models are

$$\boldsymbol{\Sigma}_B = \boldsymbol{\Lambda}\boldsymbol{\Psi}_B\boldsymbol{\Lambda}' + \boldsymbol{D}_B, \tag{12.32}$$

$$\boldsymbol{\Sigma}_W = \boldsymbol{\Lambda}\boldsymbol{\Psi}_W\boldsymbol{\Lambda}' + \boldsymbol{D}_W, \tag{12.33}$$

respectively, where

$$\boldsymbol{\Lambda}\colon (6 \times 2) = \begin{pmatrix} 1 & 0 \\ \lambda_{21} & 0 \\ \lambda_{31} & 0 \\ 0 & 1 \\ 0 & \lambda_{52} \\ 0 & \lambda_{62} \end{pmatrix},$$

and where factor loadings are assumed to be equal over the between (schools) and within (children) levels. The 2×2 matrices $\boldsymbol{\Psi}_B$ and $\boldsymbol{\Psi}_W$ denote unconstrained factor covariance matrices. Diagonal elements of $\boldsymbol{D}_B$ and $\boldsymbol{D}_W$ are the unique (error) variances.

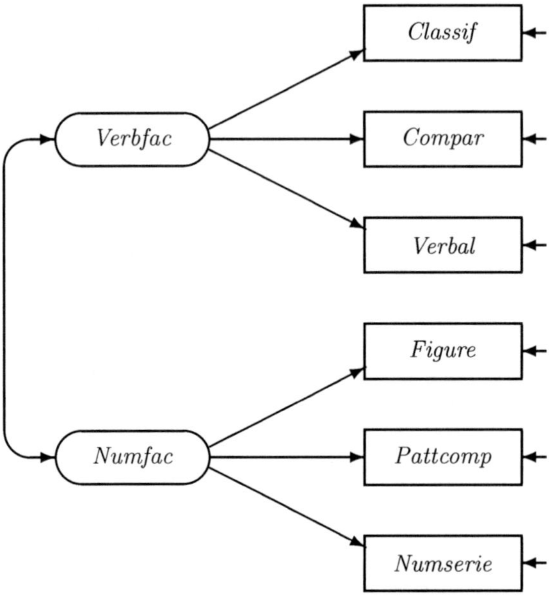

Fig. 12.2 Confirmatory factor analysis model for 6 variables.

Gender and *Grade* differences were accounted for in the means part of the model,

$$E(\underline{y}_{ijk}) = \beta_{k0} + \beta_{k1} Gender + \beta_{k2} Grade,$$

where the subscripts i, j, and k denote students, schools, and variables $k = 1, 2, \ldots, 6$, respectively.

From the description of the school data set, we note that the variable *Numserie* has 2505 missing values. An inspection of the data set reveals that the pattern of missingness can hardly be described as missing at random. To establish how well the proposed algorithm performs in terms of the handling of missing cases, we have nevertheless decided to retain this variable in both examples.

Table 12.3 shows the estimated between-schools covariance matrix $\hat{\boldsymbol{\Sigma}}_B$ when no restrictions are imposed on its elements, and the fitted covariance matrix $\boldsymbol{\Sigma}_B(\hat{\gamma})$ with γ the vector of parameters of the CFA models (12.32) and (12.33). Likewise, Table 12.4 shows $\hat{\boldsymbol{\Sigma}}_W$ for the unrestricted model and $\boldsymbol{\Sigma}_W(\hat{\gamma})$ for the CFA model.

The goodness of fit statistics for the CFA model (with 6047 students in 136 schools) are: $\chi^2 = 159.87$, RMSEA $= 0.061$, df $= 20$. Parameter estimates and standard errors are given in Table 12.5.

It is typical of structural equation models to produce large χ^2 values when sample sizes are large, as in the present case. The RMSEA may be a more meaningful measure of goodness of fit and the value of 0.061 indicates that the assumption of equal factor loadings between and within schools is reasonable.

Table 12.3 Estimated between-schools covariance matrix, $\boldsymbol{\Sigma}_B$.

(i) $\hat{\boldsymbol{\Sigma}}_B$ *unrestricted*

	Classif	*Compar*	*Verbal*	*Figure*	*Pattcomp*	*Numserie*
Classif	1.29					
Compar	1.27	2.66				
Verbal	2.83	3.54	10.42			
Figure	2.06	2.70	6.89	5.53		
Pattcomp	2.17	2.60	6.58	5.09	5.34	
Numserie	1.46	1.85	4.93	3.85	3.81	3.16

(ii) $\boldsymbol{\Sigma}_B(\hat{\gamma})$ *for the CFA model*

	Classif	*Compar*	*Verbal*	*Figure*	*Pattcomp*	*Numserie*
Classif	1.61					
Compar	1.74	3.86				
Verbal	2.21	3.22	6.76			
Figure	2.52	3.67	4.66	5.82		
Pattcomp	2.29	3.33	4.22	5.03	4.93	
Numserie	1.79	2.60	3.30	3.93	3.56	3.11

Table 12.4 Estimated within-schools covariance matrix, $\boldsymbol{\Sigma}_W$.

(i) $\hat{\boldsymbol{\Sigma}}_W$ *unrestricted*

	Classif	*Compar*	*Verbal*	*Figure*	*Pattcomp*	*Numserie*
Classif	8.49					
Compar	4.59	18.77				
Verbal	5.52	7.64	17.26			
Figure	4.45	7.21	8.49	16.27		
Pattcomp	4.30	7.21	8.45	9.55	16.19	
Numserie	2.69	4.05	5.28	7.31	5.80	7.31

(ii) $\boldsymbol{\Sigma}_W(\hat{\gamma})$ *for the CFA model*

	Classif	*Compar*	*Verbal*	*Figure*	*Pattcomp*	*Numserie*
Classif	7.81					
Compar	3.31	17.12				
Verbal	4.19	6.11	15.07			
Figure	3.72	5.42	6.87	14.58		
Pattcomp	3.37	4.91	6.23	8.13	14.61	
Numserie	2.64	3.84	4.87	6.36	5.76	7.21

Table 12.5 Parameter estimates and standard errors for the CFA model.

	Estimate	Standard error
Factor loadings		
λ_{11}	1.000	—
λ_{21}	1.456	0.048
λ_{31}	1.846	0.054
λ_{42}	1.000	—
λ_{52}	0.906	0.017
λ_{62}	0.708	0.014
Factor covariances (between schools)		
Ψ_{11}	1.196	0.185
Ψ_{21}	2.524	0.342
Ψ_{22}	5.546	0.729
Error variances (between schools)		
Classif	0.413	0.081
Compar	1.327	0.223
Verbal	2.673	0.388
Figure	0.279	0.090
Pattcomp	0.377	0.092
Numserie	0.325	0.069
Factor covariances (within schools)		
Ψ_{11}	2.271	0.114
Ψ_{21}	3.722	0.128
Ψ_{22}	8.976	0.272
Error variances (within schools)		
Classif	5.538	0.119
Compar	12.305	0.262
Verbal	7.328	0.222
Figure	5.606	0.169
Pattcomp	7.239	0.178
Numserie	2.710	0.098

12.6.2 Example 2: Structural Equation Model

We now consider a more elaborate model and use the following variables: *Gender*, *Grade*, *Classif*, *Compar*, *Verbal*, *Knowled*, *Figure*, *Pattcomp*, *Numserie*, *Read*, *Speech*, *Write*, *Arithm*, *Mothedu*, *Fathedu*, *Language*, and *Socio*. The variables *Language* and *Socio* are so-called school variables in the sense that their values vary across, but not within, schools. A path diagram for the structural equation model is given in Fig. 12.3.

Using Jöreskog's [23] LISREL notation, the latent variable model is written as

$$\underline{\eta} = \boldsymbol{B}\underline{\eta} + \boldsymbol{\Gamma}\underline{\xi} + \underline{\zeta}. \tag{12.34}$$

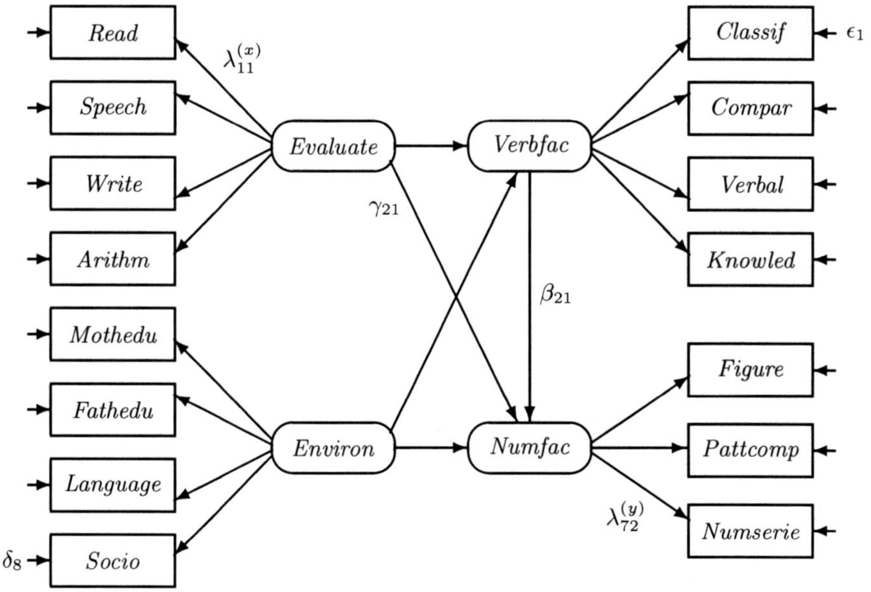

Fig. 12.3 Path diagram for the between-schools model.

The $\underline{\boldsymbol{\eta}}$: (2×1) vector contains the latent endogeneous variables *Verbfac* and *Numfac*. The coefficient matrix $\boldsymbol{B}$: (2×2) gives the effect of the $\underline{\boldsymbol{\eta}}$'s on each other. It is usually assumed that $\boldsymbol{B}$ is non-singular and has all diagonal elements equal to zero. The $\underline{\boldsymbol{\xi}}$: (2×1) vector contains the latent exogeneous variables *Evaluate* and *Environ*. The coefficient matrix $\boldsymbol{\Gamma}$: (2×2) contains the coefficients for the impact of $\underline{\boldsymbol{\xi}}$ on $\underline{\boldsymbol{\eta}}$. Disturbances for each latent endogeneous variable are contained in the 2×1 vector $\underline{\boldsymbol{\zeta}}$, and $\text{Cov}(\underline{\boldsymbol{\zeta}}) = \boldsymbol{\Psi}$.

The measurement part of the model is

$$\underline{\boldsymbol{y}} = \boldsymbol{\Lambda}_y \underline{\boldsymbol{\eta}} + \underline{\boldsymbol{\epsilon}},$$
$$\underline{\boldsymbol{x}} = \boldsymbol{\Lambda}_x \underline{\boldsymbol{\xi}} + \underline{\boldsymbol{\delta}}.$$

Elements of the 7×1 vector $\boldsymbol{y}$ are the 7 endogeneous variables *Classif* to *Numserie* that are the indicators of *Verbfac* and *Numfac*. The coefficient matrix $\boldsymbol{\Lambda}_y$: (7×2) (the factor loadings) gives the impact of *Verbfac* and *Numfac* on the variables *Classif*, *Compar*, ..., *Numserie*. The unique variables or "errors" are in the vector $\underline{\boldsymbol{\epsilon}}$: (7×1). It is assumed that $E(\underline{\boldsymbol{\epsilon}}) = \boldsymbol{\emptyset}$, $\text{Cov}(\underline{\boldsymbol{\epsilon}}) = \boldsymbol{\Theta}_\epsilon$ (usually a diagonal matrix) and that $\text{Cov}(\underline{\boldsymbol{\eta}}, \underline{\boldsymbol{\epsilon}}') = \boldsymbol{\emptyset}$.

Analogous definitions and assumptions hold for the 8×1 vector $\boldsymbol{x}$ representing the eight exogeneous variables *Read*, *Speech*, ..., *Socio*. Hence, $E(\underline{\boldsymbol{\delta}}) = \boldsymbol{\emptyset}$, $\text{Cov}(\underline{\boldsymbol{\delta}}) = \boldsymbol{\Theta}_\delta$, and $\text{Cov}(\underline{\boldsymbol{\xi}}, \underline{\boldsymbol{\delta}}') = \boldsymbol{\emptyset}$.

It is customary to scale each latent variable by selecting one of its indicators and setting its factor loading to 1.

From the above assumptions, it follows for the between-schools group that (see also Jöreskog and Sörbom [26, equation (1.4)])

$$\begin{pmatrix} \boldsymbol{\Sigma}_{B11} & & \\ \boldsymbol{\Sigma}_{B21} & \boldsymbol{\Sigma}_{B22} & \\ \boldsymbol{\Sigma}_{B1xy} & \boldsymbol{\Sigma}_{B2xy} & \boldsymbol{\Sigma}_{xx} \end{pmatrix} = \begin{pmatrix} \boldsymbol{\Lambda}_y \boldsymbol{A}(\boldsymbol{\Gamma}\boldsymbol{\Phi}\boldsymbol{\Gamma}' + \boldsymbol{\Psi})\boldsymbol{A}'\boldsymbol{\Lambda}_y' + \boldsymbol{\Theta}_\epsilon & \boldsymbol{\Lambda}_y \boldsymbol{A}\boldsymbol{\Gamma}\boldsymbol{\Phi}\boldsymbol{\Lambda}_x' \\ \boldsymbol{\Lambda}_x \boldsymbol{\Phi}\boldsymbol{\Gamma}'\boldsymbol{A}'\boldsymbol{\Lambda}_y' & \boldsymbol{\Lambda}_x \boldsymbol{\Phi}\boldsymbol{\Lambda}_x' + \boldsymbol{\Theta}_\delta \end{pmatrix}, \quad (12.35)$$

where $\boldsymbol{A} = (\boldsymbol{I} - \boldsymbol{B})^{-1}$,

$$\boldsymbol{\Sigma}_B\colon (13 \times 13) = \begin{pmatrix} \boldsymbol{\Sigma}_{B11}\colon (7 \times 7) & \boldsymbol{\Sigma}_{B12} \\ \boldsymbol{\Sigma}_{B21}\colon (6 \times 7) & \boldsymbol{\Sigma}_{B22}\colon (6 \times 6) \end{pmatrix}, \quad \text{and} \quad \boldsymbol{\Sigma}_{xx}\colon (2 \times 2).$$

Note that the seven variables *Classif*, *Compar*, ..., *Numserie* are endogeneous variables in the LISREL model ($\boldsymbol{\Sigma}_{B11}$ part), while the next six variables *Read*, *Speech*, ..., *Fathedu* are exogeneous variables ($\boldsymbol{\Sigma}_{B22}$ part). In the theoretical framework, these thirteen variables are considered y-variables with between-schools covariance matrix $\boldsymbol{\Sigma}_B$ and within-schools matrix $\boldsymbol{\Sigma}_W$. The two school-level variables *Language* and *Socio* have covariance matrix $\boldsymbol{\Sigma}_{xx}$.

As described in the previous example, we controlled for *Gender* and *Grade* effects through the inclusion of these variables in the means part of the model.

Table 12.6 shows the estimated between-schools covariance matrix

$$\begin{pmatrix} \boldsymbol{\Sigma}_B & \boldsymbol{\Sigma}_{yx} \\ \boldsymbol{\Sigma}_{xy} & \boldsymbol{\Sigma}_{xx} \end{pmatrix}$$

when no restrictions are imposed on its elements and also the fitted covariance matrix for the structural equations model (12.34).

In both between- and within-school models, the residual covariance matrix of the latent variables *Verbfac* and *Numfac* was assumed to be diagonal. Furthermore, in the between-schools model it was assumed that there is no effect of endogeneous latent variables on each other, and hence $\boldsymbol{B} = \boldsymbol{\emptyset}$. For the within-schools model,

$$\boldsymbol{B} = \begin{pmatrix} 0 & 0 \\ \beta_{21} & 0 \end{pmatrix}.$$

Since the estimated error variance for the exogeneous variable *Fathedu* was negative (but close to zero), this error variance was fixed at zero. The appropriate LISREL syntax is given in the appendix (see p. 458) of this chapter.

Table 12.7 shows the estimated within-school covariance matrix for the unrestricted and restricted cases.

Estimates and standard errors of the unknown parameters in the structural equation models for the between- and within-school models are given in Table 12.8.

Table 12.6 Estimated between-school covariance matrices for $\boldsymbol{\Sigma}_B$ unrestricted and $\boldsymbol{\Sigma}_B$ restricted according to (12.34).

(i) $\hat{\boldsymbol{\Sigma}}_B$ *with no restrictions imposed on the elements*

Classif	1.29														
Compar	1.26	2.63													
Verbal	2.83	3.52	10.40												
Knowled	1.76	2.64	5.67	4.57											
Figure	2.06	2.70	6.87	3.69	5.51										
Pattcomp	2.16	2.58	6.54	3.87	5.06	5.31									
Numserie	1.46	1.84	4.89	2.75	3.81	3.77	3.11								
Read	0.18	0.17	0.59	0.46	0.37	0.34	0.27	0.14							
Speech	0.12	0.13	0.42	0.36	0.24	0.25	0.20	0.12	0.12						
Write	0.14	0.14	0.48	0.42	0.25	0.25	0.18	0.12	0.11	0.12					
Arithm	0.14	0.16	0.55	0.40	0.34	0.34	0.23	0.12	0.10	0.11	0.13				
Mothedu	0.49	0.53	1.26	0.58	1.18	1.15	0.85	0.06	0.04	0.02	0.04	0.70			
Fathedu	0.50	0.55	1.33	0.57	1.28	1.21	0.92	0.06	0.02	0.02	0.04	0.72	0.78		
Language	0.26	0.22	0.92	0.36	0.77	0.76	0.61	0.03	0.01	0.00	0.02	0.20	0.23	0.21	
Socio	0.86	0.97	2.39	1.26	1.96	2.01	1.44	0.12	0.10	0.09	0.13	0.61	0.67	0.32	1.24

(ii) $\boldsymbol{\Sigma}_B(\hat{\gamma})$ *with restrictions imposed according to the model (12.34)*

Classif	1.31														
Compar	1.21	2.69													
Verbal	2.78	3.66	10.16												
Knowled	1.80	2.36	5.44	4.39											
Figure	1.33	1.75	4.04	2.61	5.39										
Pattcomp	1.31	1.72	3.96	2.56	4.99	5.28									
Numserie	1.01	1.32	3.05	1.97	3.84	3.77	3.17								
Read	0.22	0.28	0.65	0.42	0.41	0.40	0.31	0.14							
Speech	0.19	0.24	0.56	0.36	0.35	0.35	0.27	0.11	0.12						
Write	0.20	0.26	0.61	0.39	0.38	0.37	0.29	0.12	0.10	0.12					
Arithm	0.20	0.26	0.60	0.39	0.37	0.37	0.28	0.12	0.10	0.11	0.13				
Mothedu	0.43	0.57	1.30	0.84	1.27	1.25	0.96	0.04	0.03	0.04	0.04	0.68			
Fathedu	0.47	0.61	1.41	0.91	1.38	1.35	1.04	0.04	0.04	0.04	0.04	0.70	0.76		
Language	0.15	0.20	0.46	0.30	0.45	0.44	0.34	0.01	0.01	0.01	0.01	0.23	0.25	0.21	
Socio	0.43	0.56	1.29	0.83	1.26	1.24	0.95	0.04	0.03	0.04	0.04	0.64	0.70	0.23	1.24

Finally, the goodness of fit measures for the multilevel structural equation model fitted to the school data set are: $\chi^2 = 691.198$, RMSEA $= 0.029$, df $= 145$. The χ^2 statistic is for testing the null hypothesis that model (12.34) holds for the between- and within-schools covariance structures against the alternative hypotheses that no restrictions are imposed on the covariance matrices. According to this measure of fit, one should reject the null hypothesis. The RMSEA value of 0.029 and inspection of Tables 12.6 and 12.7 shows that the fitted model may be quite acceptable.

12.7 Conclusion and Discussion

A Fisher Scoring algorithm is employed to obtain full maximum likelihood estimation of a general two-level structural equations model. An explicit feature of our approach is that the likelihood function and derivatives (see

Table 12.7 Estimated within-school covariance matrices for Σ_W unrestricted and Σ_W restricted according to (12.34).

(i) $\hat{\Sigma}_W$ *with no restrictions imposed on the elements*

Classif	8.49												
Compar	4.59	18.78											
Verbal	5.52	7.65	17.27										
Knowled	3.49	5.40	7.10	16.34									
Figure	4.46	7.22	8.49	5.59	16.29								
Pattcomp	4.31	7.22	8.46	5.56	9.56	16.20							
Numserie	2.71	4.08	5.28	3.65	7.31	5.81	7.31						
Read	0.66	0.80	1.21	0.88	1.02	0.87	0.78	1.21					
Speech	0.51	0.66	1.01	0.67	0.88	0.78	0.66	0.75	1.04				
Write	0.63	0.85	1.23	0.84	1.02	0.94	0.81	0.83	0.73	1.08			
Arithm	0.64	0.82	1.21	0.81	1.04	0.92	0.82	0.73	0.64	0.75	1.11		
Mothedu	0.23	0.28	0.43	0.20	0.24	0.10	0.14	0.08	0.05	0.05	0.06	1.44	
Fathedu	0.20	0.20	0.38	0.15	0.25	0.20	0.16	0.08	0.04	0.03	0.04	0.87	2.00

(ii) $\Sigma_W(\hat{\gamma})$ *with restrictions imposed according to the model (12.34)*

Classif	7.88												
Compar	3.39	17.32											
Verbal	4.23	6.17	14.86										
Knowled	2.71	3.95	4.94	14.75									
Figure	3.76	5.48	6.85	4.38	14.72								
Pattcomp	3.34	4.87	6.08	3.89	8.11	14.49							
Numserie	2.64	3.85	4.81	3.08	6.42	5.70	7.18						
Read	0.63	0.92	1.15	0.74	1.08	0.96	0.76	1.18					
Speech	0.56	0.82	1.02	0.65	0.96	0.85	0.67	0.71	1.02				
Write	0.63	0.93	1.16	0.74	1.08	0.96	0.76	0.81	0.72	1.06			
Arithm	0.57	0.84	1.05	0.67	0.98	0.87	0.69	0.73	0.65	0.73	1.11		
Mothedu	0.20	0.30	0.37	0.24	0.19	0.17	0.14	0.06	0.05	0.06	0.05	1.45	
Fathedu	0.16	0.24	0.30	0.19	0.15	0.14	0.11	0.05	0.04	0.05	0.04	0.88	2.01

Appendix 12.B) are expressed in terms of matrix operations of order less or equal to $p + q$ leading to a very significant reduction in computational workload. In the case of most EM algorithms, estimates of the standard errors of the estimated population parameters are not readily available as is the case in our method.

Results are given for an unbalanced design with responses possibly missing at random. The model allows for regression on fixed explanatory variables and structured residual covariance matrices on both levels of the hierarchy. A number of fit statistics are discussed and practical examples are given to demonstrate the feasibility of the derived procedures. Additional examples, including examples dealing with structured mean vectors, are included with LISREL [15].

In structural equation modeling (SEM), it is typically assumed that the data to be analyzed is obtained from a simple random sample (SRS). In many research studies, however, data has a hierarchical structure. For example, students nested within schools or patients nested within hospitals. By ignoring the hierarchical structure of the data, incorrect parameter estimates, standard errors, and inappropriate fit statistics may be obtained.

Table 12.8 Parameter estimates and standard errors for the unknown parameters in model (12.34).

		Between schools		Within schools	
		Estimate	Standard error	Estimate	Standard error
$\boldsymbol{\Lambda}_y$	λ_{21}	1.313	0.144	1.459	0.049
	λ_{31}	3.027	0.255	1.822	0.053
	λ_{41}	1.954	0.172	1.167	0.044
	λ_{62}	0.981	0.040	0.888	0.019
	λ_{72}	0.756	0.033	0.703	0.015
$\boldsymbol{\Lambda}_x$	λ_{21}	0.863	0.058	0.886	0.014
	λ_{31}	0.934	0.052	1.003	0.014
	λ_{41}	0.918	0.062	0.909	0.014
	λ_{62}	1.084	0.032	0.804	0.153
	λ_{72}	0.354	0.040	—	—
	λ_{82}	0.990	0.090	—	—
$\boldsymbol{B}$	β_{21}	2.186	0.195	1.605	0.055
$\boldsymbol{\Gamma}$	γ_{11}	1.488	0.220	0.776	0.033
	γ_{12}	0.571	0.088	0.143	0.035
	γ_{21}	2.587	0.415	0.088	0.058
	γ_{22}	1.801	0.176	−0.126	0.049
$\boldsymbol{\Phi}$	ϕ_{11}	0.130	0.021	0.804	0.022
	ϕ_{21}	0.041	0.029	0.060	0.016
	ϕ_{22}	0.649	0.087	1.090	0.208
$\boldsymbol{\Psi}$	ψ_{11}	0.353	0.076	1.803	0.095
	ψ_{22}	1.735	0.284	3.032	0.179
$\boldsymbol{\Theta}_\epsilon$	*Classif*	0.391	0.076	5.555	0.117
	Compar	1.100	0.189	12.380	0.259
	Verbal	1.736	0.388	7.153	0.200
	Knowled	0.876	0.195	11.590	0.234
	Figure	0.302	0.094	5.583	0.171
	Pattcomp	0.391	0.102	7.293	0.178
	Numserie	0.267	0.066	2.673	0.098
$\boldsymbol{\Theta}_\delta$	*Read*	0.015	0.004	0.375	0.010
	Speech	0.020	0.004	0.394	0.009
	Write	0.010	0.003	0.248	0.008
	Arithm	0.022	0.005	0.445	0.010
	Mothedu	0.030	0.008	0.358	0.207
	Fathedu	0.000	—	1.302	0.136
	Language	0.126	0.016	—	—
	Socio	0.606	0.078	—	—

To examine the effect of clustering, one could fit the level-1 SEM models described in the previous section by treating the data as an SRS. In general, if the parameter estimates and estimated standard errors are close to those obtained using the multilevel SEM approach, it can be assumed that there is a negligible cluster effect. For example, student characteristics under study do not vary significantly across schools.

Acknowledgements This research was supported by Scientific Software International, Inc.

The authors would like to thank Roderick P. McDonald who provided us with copies of published papers on multilevel models and whose related research served as our inspiration. Thanks are also due to Leo Stam, who assisted us with typesetting issues in the original manuscript and to Gerhard Mels, who worked through the algebra. A special word of thanks is due to Erik Meijer who carefully worked through the manuscript and suggested many changes and improvements and also took care of the LaTeX revision.

Appendix

12.A LISREL Programs

LISREL Syntax for Example 1 (CFA Model)

```
Group1: Between Schools HSRC School Project
DA NI=7 NO=0 NG=2 MA=CM MI=-9.0
LA
Classif Compar Verbal Figure Pattcomp Numserie
RA = SASchools94.dat
$CLuster School
SE
1 2 3 4 5 6 /
MO NY=6 NE=2 LY=FU,FI PS=SY,FR TE=DI,FR
LE
Verbfac Numfac
FR LY(2,1) LY(3,1) LY(5,2) LY(6,2)
VA 1.00 LY(1,1) LY(4,2)
PD
OU ME=ML

Group2: Within Schools HSRC School Project
LA
Classif Compar Verbal Figure Pattcomp Numserie
```

```
DA NI=7 NO=0 NG=2 MA=CM MI=-9.0
RA = SASchools94.dat
SE
1 2 3 4 5 6 /
MO NY=6 NE=2 LY=IN PS=IN TE=IN
LE
Verbfac Numfac
FR PS(1,1) PS(2,1) PS(2,2) TE(1,1) TE(2,2)
FR TE(3,3) TE(4,4) TE(5,5) TE(6,6)
! FR LY(2,1) LY(3,1) LY(5,2) LY(6,2)
OU
```

LISREL Syntax for Example 2 (Structural Equation Model)

```
Group1: Between Schools Data, HSRC project
DA NI=16 NO=0 NG=2 MA=CM MI=-9.0
LA
Classif Compar Verbal Knowled Figure Pattcomp Numserie
Read Speech Write Arithm Mothedu Fathedu Language Socio
RA=SA_Schools94.dat
$CLuster School
SE
1 2 3 4 5 6 7 8 9 10 11 12 13 14 15/
MO NX=8 NY=7 NK=2 NE=2 LY=FU,FI LX=FU,FI BE=FU,FI c
   GA=FU,FI PH=SY,FR PS=DI,FR TE=DI,FR TD=DI,FR
LE
Verbfac Numfac
LK
Evaluate Environ
FI TD(6,6)
FR LY(2,1) LY(3,1) LY(4,1) LY(6,2) LY(7,2) BE(2,1)
FR LX(2,1) LX(3,1) LX(4,1)
FR LX(6,2) LX(7,2) LX(8,2)
FR GA(1,1) GA(1,2) GA(2,1) GA(2,2)
VA 1.000 LY(1,1) LY(5,2) LX(1,1) LX(5,2)
VA 0.001 TD(6,6)
PD
OU ME=ML ND=2
Group2: Within Schools Data, HSRC project
DA NI=16 NO=0 NG=2 MA=CM MI=-9.0
LA
Classif Compar Verbal Knowled Figure Pattcomp Numserie
Read Speech Write Arithm Mothedu Fathedu Language Socio
```

```
RA=SA_Schools.dat
MO NX=8 NY=7 NK=2 NE=2 LY=FU,FI LX=FU,FI BE=FU,FI c
   GA=FU,FI PH=SY,FR PS=DI,FR TE=DI,FR TD=DI,FR
LE
Verbfac Numfac
LK
Evaluate Environ
FR LY(2,1) LY(3,1) LY(4,1) LY(6,2) LY(7,2) BE(2,1)
FR LX(2,1) LX(3,1) LX(4,1)
FR LX(6,2)
FR GA(1,1) GA(1,2) GA(2,1) GA(2,2)
VA 1.000 LY(1,1) LY(5,2) LX(1,1) LX(5,2)
VA 0.0 LX(7,2) LX(8,2)
!EQ PS(2,2) PS(1,2,2)
!EQ PS(1,1) PS(1,1,1)
!EQ PH(2,2) PH(1,2,2)
!EQ PH(2,1) PH(1,2,1)
!EQ PH(1,1) PH(1,1,1)
!EQ BE(2,1) BE(1,2,1)
OU
```

12.B Computational Details

12.B.1 Expressions for the Inverse and Determinant of $\boldsymbol{\Sigma}$

The estimation procedures outlined in the previous section can be implemented in a computer program for a general two-level structural equation model. A major problem, however, is how to deal efficiently with the calculation of the high-order matrix products, determinants, and inverses that are part and parcel of multilevel models. Due to the particular structure of $\boldsymbol{\Sigma}_j$ defined by (12.7), it is shown that storage space and execution time considerations can be eliminated to a large extent. We show that the likelihood function, derivatives, and Hessian can be expressed in terms of matrix operations of order less or equal to $p+q$.

From (12.5) and (12.7) it follows that the covariance matrix of $(\underline{\boldsymbol{y}}_{1j}, \underline{\boldsymbol{y}}_{2j}, \ldots, \underline{\boldsymbol{y}}_{n_j j})'$ can be written as

$$\boldsymbol{\Sigma}_{jyy} = \boldsymbol{V}_j = \boldsymbol{S}_j \boldsymbol{\Sigma}_B \boldsymbol{S}_j' + \boldsymbol{\Lambda}_j \,, \tag{12.36}$$

where $\boldsymbol{\Lambda}_j$ is a block-diagonal matrix

$$\boldsymbol{\Lambda}_j = \begin{pmatrix} \boldsymbol{\Lambda}_{1j} & & & \\ & \boldsymbol{\Lambda}_{2j} & & \\ & & \ddots & \\ & & & \boldsymbol{\Lambda}_{n_j j} \end{pmatrix}$$

and where

$$\boldsymbol{\Lambda}_{ij}\colon (p_{ij} \times p_{ij}) = \boldsymbol{S}_{ij}\boldsymbol{\Sigma}_W\boldsymbol{S}'_{ij}, \quad p_{ij} \le p. \tag{12.37}$$

Let $\boldsymbol{A}_j\colon (p \times p) = \boldsymbol{S}'_j\boldsymbol{\Lambda}_j^{-1}\boldsymbol{S}_j$. From (12.5) and (12.6) it follows that $\boldsymbol{A}_j = \sum_{i=1}^{n_j} \boldsymbol{A}_{ij}$, where

$$\boldsymbol{A}_{ij} = \boldsymbol{S}'_{ij}\boldsymbol{\Lambda}_{ij}^{-1}\boldsymbol{S}_{ij}\,. \tag{12.38}$$

Also, define

$$\boldsymbol{B}_j\colon (p \times p) = (\boldsymbol{\Sigma}_B^{-1} + \boldsymbol{S}'_j\boldsymbol{\Lambda}_j^{-1}\boldsymbol{S}_j)^{-1} = (\boldsymbol{\Sigma}_B^{-1} + \boldsymbol{A}_j)^{-1}. \tag{12.39}$$

The following matrix expressions are defined in terms of (12.38) and (12.39):

$$\boldsymbol{C}_j\colon (p \times p) = (\boldsymbol{I}_p - \boldsymbol{A}_j\boldsymbol{B}_j), \tag{12.40}$$

$$\boldsymbol{D}_j\colon (p \times p) = (\boldsymbol{I}_p - \boldsymbol{A}_j\boldsymbol{B}_j)\boldsymbol{A}_j = \boldsymbol{C}_j\boldsymbol{A}_j\,. \tag{12.41}$$

In order to obtain expressions for the inverse of the patterned covariance matrix $\boldsymbol{\Sigma}_j$ defined by (12.7), define the $q_j \times q_j$ matrix $\boldsymbol{\Sigma}_{22.1}$ as

$$\boldsymbol{\Sigma}_{22.1} = (\boldsymbol{\Sigma}_{jxx} - \boldsymbol{\Sigma}_{jxy}\boldsymbol{V}_j^{-1}\boldsymbol{\Sigma}_{jyx}), \quad q_j \le q \tag{12.42}$$

where (see (12.7)) $\boldsymbol{\Sigma}_{jxx} = \boldsymbol{R}_j\boldsymbol{\Sigma}_{xx}\boldsymbol{R}'_j$ and $\boldsymbol{\Sigma}_{jyx} = \boldsymbol{S}_j\boldsymbol{\Sigma}_{yx}\boldsymbol{R}'_j$.

Using a well-known matrix identity [see, e.g., 27], it follows from (12.36) that

$$\begin{aligned} \boldsymbol{V}_j^{-1} &= \boldsymbol{\Lambda}_j^{-1} - \boldsymbol{\Lambda}_j^{-1}\boldsymbol{S}_j\left(\boldsymbol{\Sigma}_B^{-1} + \boldsymbol{S}'_j\boldsymbol{\Lambda}_j^{-1}\boldsymbol{S}_j\right)^{-1}\boldsymbol{S}'_j\boldsymbol{\Lambda}_j^{-1} \\ &= \boldsymbol{\Lambda}_j^{-1} - \boldsymbol{\Lambda}_j^{-1}\boldsymbol{S}_j\boldsymbol{B}_j\boldsymbol{S}'_j\boldsymbol{\Lambda}_j^{-1}. \end{aligned} \tag{12.43}$$

Hence, using (12.39), it follows that

$$\boldsymbol{\Sigma}_{22.1} = \boldsymbol{R}_j\left(\boldsymbol{\Sigma}_{xx} - \boldsymbol{\Sigma}_{xy}\boldsymbol{D}_j\boldsymbol{\Sigma}_{yx}\right)\boldsymbol{R}'_j.$$

Using another well-known result for the inverse of a partitioned matrix [see, e.g., 4],

$$\boldsymbol{\Sigma}_j^{-1} = \begin{pmatrix} \boldsymbol{\Sigma}_j^{11} & \boldsymbol{\Sigma}_j^{12} \\ \boldsymbol{\Sigma}_j^{21} & \boldsymbol{\Sigma}_j^{22} \end{pmatrix}, \tag{12.44}$$

it follows that

$$\begin{aligned} \boldsymbol{\Sigma}_j^{11} &= \boldsymbol{V}_j^{-1}\boldsymbol{\Sigma}_{jyx}\boldsymbol{\Sigma}_{22.1}^{-1}\boldsymbol{\Sigma}_{jxy}\boldsymbol{V}_j^{-1} + \boldsymbol{V}_j^{-1}, \\ \boldsymbol{\Sigma}_j^{21} &= -\boldsymbol{\Sigma}_{22.1}^{-1}\boldsymbol{\Sigma}_{jxy}\boldsymbol{V}_j^{-1}, \\ \boldsymbol{\Sigma}_j^{12} &= -\boldsymbol{V}_j^{-1}\boldsymbol{\Sigma}_{jyx}\boldsymbol{\Sigma}_{22.1}^{-1}, \\ \boldsymbol{\Sigma}_j^{22} &= \boldsymbol{\Sigma}_{22.1}^{-1}\,. \end{aligned}$$

Let

$$\boldsymbol{E}_j\colon (q \times q) = \boldsymbol{R}_j'\boldsymbol{\Sigma}_j^{22}\boldsymbol{R}_j = \boldsymbol{R}_j'\boldsymbol{\Sigma}_{22.1}^{-1}\boldsymbol{R}_j\,. \tag{12.45}$$

Using (12.43) and (12.45), it follows after simplification that

$$\begin{aligned}\boldsymbol{\Sigma}_j^{11} &= \boldsymbol{V}_j^{-1} + \boldsymbol{\Lambda}_j^{-1}\boldsymbol{S}_j\boldsymbol{C}_j\boldsymbol{\Sigma}_{yx}\boldsymbol{R}_j'\boldsymbol{\Sigma}_{22.1}^{-1}\boldsymbol{R}_j\boldsymbol{\Sigma}_{xy}\boldsymbol{C}_j'\boldsymbol{S}_j'\boldsymbol{\Lambda}_j^{-1}\\ &= \boldsymbol{V}_j^{-1} + \boldsymbol{\Lambda}_j^{-1}\boldsymbol{S}_j\boldsymbol{C}_j\boldsymbol{\Sigma}_{yx}\boldsymbol{E}_j\boldsymbol{\Sigma}_{xy}\boldsymbol{C}_j'\boldsymbol{S}_j'\boldsymbol{\Lambda}_j^{-1}.\end{aligned}$$

A more compact expression for $\boldsymbol{\Sigma}_j^{11}$ is obtained by defining the $p \times p$ matrix $\boldsymbol{F}_j$ as

$$\boldsymbol{F}_j = \boldsymbol{C}_j\boldsymbol{\Sigma}_{yx}\boldsymbol{R}_j'\boldsymbol{\Sigma}_{22.1}^{-1}\boldsymbol{R}_j\boldsymbol{\Sigma}_{xy}\boldsymbol{C}_j' = \boldsymbol{C}_j\boldsymbol{\Sigma}_{yx}\boldsymbol{E}_j\boldsymbol{\Sigma}_{xy}\boldsymbol{C}_j'\,. \tag{12.46}$$

Finally, define $\boldsymbol{H}_j$ as

$$\boldsymbol{H}_j = \boldsymbol{F}_j - \boldsymbol{B}_j\,. \tag{12.47}$$

Then from (12.43) it follows that $\boldsymbol{\Sigma}_j^{11}$ can be written as

$$\boldsymbol{\Sigma}_j^{11} = \boldsymbol{V}_j^{-1} + \boldsymbol{\Lambda}_j^{-1}\boldsymbol{S}_j\boldsymbol{F}_j\boldsymbol{S}_j'\boldsymbol{\Lambda}_j^{-1} = \boldsymbol{\Lambda}_j^{-1} + \boldsymbol{\Lambda}_j^{-1}\boldsymbol{S}_j\boldsymbol{H}_j\boldsymbol{S}_j'\boldsymbol{\Lambda}_j^{-1}. \tag{12.48}$$

It can also be verified (see (12.38) and (12.44)) that

$$\boldsymbol{\Sigma}_j^{12} = -\boldsymbol{\Lambda}_j^{-1}\boldsymbol{S}_j\boldsymbol{C}_j'\boldsymbol{\Sigma}_{yx}\boldsymbol{R}_j'\boldsymbol{\Sigma}_{22.1}^{-1}\,. \tag{12.49}$$

From (12.7) and applying well-known results [see, e.g., 4], for partitioned matrices, it follows that $|\boldsymbol{\Sigma}_j| = |\boldsymbol{V}_j|\,|\boldsymbol{\Sigma}_{22.1}|$, where

$$|\boldsymbol{V}_j| = |\boldsymbol{S}_j\boldsymbol{\Sigma}_B\boldsymbol{S}_j' + \boldsymbol{\Lambda}_j| = |\boldsymbol{\Lambda}_j|\,|\boldsymbol{\Sigma}_B|\,|\boldsymbol{\Sigma}_B^{-1} + \boldsymbol{A}_j| \tag{12.50}$$

with $\boldsymbol{A}_j$ as defined in (12.38). Hence,

$$|\boldsymbol{\Sigma}_j| = \left\{\prod_{i=1}^{n_j}|\boldsymbol{\Lambda}_{ij}|\right\}|\boldsymbol{\Sigma}_B|\,|\boldsymbol{\Sigma}_B^{-1} + \boldsymbol{A}_j|\,|\boldsymbol{\Sigma}_{22.1}|.$$

12.B.2 Likelihood Function

From the equations (12.1)–(12.7) it follows that

$$\ln L_j = -\tfrac{1}{2}\left\{\sum_{i=1}^{n_j} p_{ij}\ln 2\pi + \ln|\boldsymbol{\Sigma}_j| + \boldsymbol{e}_j'\boldsymbol{\Sigma}_j^{-1}\boldsymbol{e}_j\right\},$$

where $\boldsymbol{e}_j = \boldsymbol{y}_j - \boldsymbol{\mu}_j$ and $p_{ij} = \text{rank}(\boldsymbol{S}_{ij})$. If no y_{ij} values are missing, $p_{ij} = p$ and $\boldsymbol{S}_{ij} = \boldsymbol{I}_p$. From (12.10) it follows that the function to be minimized is

$$F(\gamma) = \tfrac{1}{2}\sum_{j=1}^{J}(\ln|\boldsymbol{\Sigma}_j| + \boldsymbol{e}_j'\boldsymbol{\Sigma}_j^{-1}\boldsymbol{e}_j). \tag{12.51}$$

Partition $\boldsymbol{e}_j$ as

$$\boldsymbol{e}_j = \begin{pmatrix} \boldsymbol{e}_{(1)j} \\ \boldsymbol{e}_{(2)j} \end{pmatrix}, \tag{12.52}$$

where

$$\boldsymbol{e}'_{(1)j} = (\boldsymbol{e}'_{1j}, \boldsymbol{e}'_{2j}, \ldots, \boldsymbol{e}'_{n_j j}) = (\boldsymbol{y}'_{1j} - \boldsymbol{\mu}'_{1j}, \boldsymbol{y}'_{2j} - \boldsymbol{\mu}'_{2j}, \ldots, \boldsymbol{y}'_{n_j j} - \boldsymbol{\mu}'_{n_j j})$$

and where

$$\boldsymbol{e}_{(2)j} = \boldsymbol{x}_j - \boldsymbol{\mu}_{(x)j} \,. \tag{12.53}$$

In order to simplify the terms in (12.51), we define the following vectors in terms of $\boldsymbol{e}_j$. Recall from (12.1) that $\boldsymbol{\mu}_{ij} = \boldsymbol{X}_{(y)ij}\boldsymbol{\beta}_y$ and $\boldsymbol{\mu}_{(x)j} = \boldsymbol{X}_{(x)j}\boldsymbol{\beta}_x$. Let $\boldsymbol{p}_j$ be a $p \times 1$ vector defined as

$$\boldsymbol{p}_j = \boldsymbol{S}'_j \boldsymbol{\Lambda}_j^{-1} \boldsymbol{e}_{(1)j} \,; \tag{12.54}$$

then

$$\boldsymbol{p}_j = \sum_{i=1}^{n_j} \boldsymbol{S}'_{ij} \boldsymbol{\Lambda}_{ij}^{-1} \boldsymbol{e}_{ij} = \sum_{i=1}^{n_j} \boldsymbol{p}_{ij} \,.$$

Also, let

$$\boldsymbol{q}_{ij} = \boldsymbol{Z}'_{ij} \boldsymbol{\Sigma}_j^{11} \boldsymbol{e}_{(1)j} + \boldsymbol{Z}'_{ij} \boldsymbol{\Sigma}_j^{12} \boldsymbol{e}_{(2)j} \,, \tag{12.55}$$

$$\boldsymbol{r}_j = \boldsymbol{S}'_j \boldsymbol{\Sigma}_j^{11} \boldsymbol{e}_{(1)j} + \boldsymbol{S}'_j \boldsymbol{\Sigma}_j^{12} \boldsymbol{e}_{(2)j} \,. \tag{12.56}$$

From (12.38), (12.48), and (12.54) it follows that

$$\boldsymbol{q}_{ij} = \boldsymbol{p}_{ij} + \boldsymbol{A}_{ij} \boldsymbol{H}_j \boldsymbol{p}_j - \boldsymbol{A}_{ij} \boldsymbol{C}'_j \boldsymbol{\Sigma}_{yx} \boldsymbol{R}'_j \boldsymbol{\Sigma}_{22.1}^{-1} \boldsymbol{e}_{(2)j} \,,$$

$$\boldsymbol{r}_j = (\boldsymbol{I}_p + \boldsymbol{A}_j \boldsymbol{H}_j) \boldsymbol{p}_j - \boldsymbol{A}_j \boldsymbol{C}'_j \boldsymbol{\Sigma}_{yx} \boldsymbol{R}'_j \boldsymbol{\Sigma}_{22.1}^{-1} \boldsymbol{e}_{(2)j} \,,$$

and

$$\begin{aligned} \boldsymbol{s}_j &= \boldsymbol{R}'_j \boldsymbol{\Sigma}_j^{21} \boldsymbol{e}_{(1)j} + \boldsymbol{R}'_j \boldsymbol{\Sigma}_j^{22} \boldsymbol{e}_{(2)j} \\ &= \boldsymbol{R}'_j \boldsymbol{\Sigma}_{22.1}^{-1} \boldsymbol{R}_j \boldsymbol{\Sigma}_{xy} \boldsymbol{C}_j \boldsymbol{p}_j + \boldsymbol{R}'_j \boldsymbol{\Sigma}_{22.1}^{-1} \boldsymbol{e}_{(2)j} \,. \end{aligned} \tag{12.57}$$

Calculation of $\boldsymbol{e}'_j \boldsymbol{\Sigma}_j^{-1} \boldsymbol{e}_j$

From (12.52) it follows that $\boldsymbol{e}'_j \boldsymbol{\Sigma}_j^{-1} \boldsymbol{e}_j = t_{11} + 2t_{12} + t_{22}$, where (see (12.43)–(12.47) and (12.54))

$$t_{11} = \boldsymbol{e}'_{(1)j} \boldsymbol{\Sigma}_j^{11} \boldsymbol{e}_{(1)j} = \sum_{i=1}^{n_j} \boldsymbol{e}'_{ij} \boldsymbol{\Lambda}_{ij}^{-1} \boldsymbol{e}_{ij} + \boldsymbol{p}'_j \boldsymbol{H}_j \boldsymbol{p}_j \,,$$

$$t_{12} = \boldsymbol{e}'_{(1)j} \boldsymbol{\Sigma}_j^{12} \boldsymbol{e}_{(2)j} = -\boldsymbol{p}'_j \boldsymbol{C}_j \boldsymbol{\Sigma}_{yx} \boldsymbol{R}'_j \boldsymbol{\Sigma}_{22.1}^{-1} \boldsymbol{e}_{(2)j} \,,$$

and

$$t_{22} = \boldsymbol{e}'_{(2)j} \boldsymbol{\Sigma}_{22.1}^{-1} \boldsymbol{e}_{(2)j} \,. \tag{12.58}$$

12.B.3 Gradient Vector

From (12.12), (12.14), and (12.53) it follows that

$$\frac{\partial F}{\partial[\boldsymbol{\pi}_1]_r} = -\sum_{j=1}^{J} \operatorname{tr}\left\{(\boldsymbol{\Sigma}_j^{-1}\boldsymbol{e}_j\boldsymbol{e}_j'\boldsymbol{\Sigma}_j^{-1} - \boldsymbol{\Sigma}_j^{-1})\frac{\partial \boldsymbol{\Sigma}_j}{\partial[\boldsymbol{\pi}_1]_r}\right\}, \tag{12.59}$$

$$\frac{\partial F}{\partial[\boldsymbol{\pi}_2]_r} = -2\sum_{j=1}^{J} \operatorname{tr}\left\{\boldsymbol{e}_j'\boldsymbol{\Sigma}_j^{-1}\frac{\partial \boldsymbol{\mu}_j}{\partial[\boldsymbol{\pi}_2]_r}\right\}, \tag{12.60}$$

where the $k^* \times 1$ vector $\boldsymbol{\pi}$ is partitioned as $\boldsymbol{\pi}' = (\boldsymbol{\pi}_1', \boldsymbol{\pi}_2')$, with $\boldsymbol{\pi}_1' = (\boldsymbol{\beta}_y', \boldsymbol{\beta}_x')$ and $\boldsymbol{\pi}_2' = \big((\text{vecs}\,\boldsymbol{\Sigma}_B)', (\text{vecs}\,\boldsymbol{\Sigma}_W)', (\text{vec}\,\boldsymbol{\Sigma}_{xy})', (\text{vecs}\,\boldsymbol{\Sigma}_{xx})'\big)$.

Calculation of $\partial F/\partial[\boldsymbol{\Sigma}_B]_{r,s}$

From (12.7) it follows that

$$\frac{\partial \boldsymbol{\Sigma}_j}{\partial[\boldsymbol{\Sigma}_B]_{r,s}} = \begin{pmatrix} \dfrac{\partial \boldsymbol{V}_j}{\partial[\boldsymbol{\Sigma}_B]_{r,s}} & \boldsymbol{\emptyset} \\ \boldsymbol{\emptyset} & \boldsymbol{\emptyset} \end{pmatrix},$$

where

$$\frac{\partial \boldsymbol{V}_j}{\partial[\boldsymbol{\Sigma}_B]_{r,s}} = \boldsymbol{S}_j\boldsymbol{G}_{rs}\boldsymbol{S}_j',$$

$$\boldsymbol{G}_{rs} = \boldsymbol{J}_{rs} + (1-\delta_{rs})\boldsymbol{J}_{sr}, \tag{12.61}$$

and δ_{rs} is Kronecker's delta, i.e., $\delta_{rs} = 1$ if $r = s$, and 0 otherwise.

Therefore, after some simplification using the partitioning (12.44) of $\boldsymbol{\Sigma}_j^{-1}$, it follows that

$$\begin{aligned}\frac{\partial F}{\partial[\boldsymbol{\Sigma}_B]_{r,s}} &= -\sum_{j=1}^{J} \operatorname{tr}\{(\boldsymbol{\Sigma}_j^{11}\boldsymbol{e}_{(1)j} + \boldsymbol{\Sigma}_j^{12}\boldsymbol{e}_{(2)j})(\boldsymbol{e}_{(1)j}'\boldsymbol{\Sigma}_j^{11} + \boldsymbol{e}_{(2)j}'\boldsymbol{\Sigma}_j^{21})(\boldsymbol{S}_j\boldsymbol{G}_{rs}\boldsymbol{S}_j')\} \\ &\quad + \sum_{j=1}^{J} \operatorname{tr}\boldsymbol{\Sigma}_j^{11}\boldsymbol{S}_j\boldsymbol{G}_{rs}\boldsymbol{S}_j' \\ &= \sum_{j=1}^{J} \operatorname{tr}\boldsymbol{S}_j'\boldsymbol{\Sigma}_j^{11}\boldsymbol{S}_j\boldsymbol{G}_{rs} - \sum_{j=1}^{J} \operatorname{tr}\boldsymbol{r}_j\boldsymbol{r}_j'\boldsymbol{G}_{rs},\end{aligned}$$

with $\boldsymbol{r}_j$ defined in (12.56). Equivalently,

$$\frac{\partial F}{\partial[\boldsymbol{\Sigma}_B]_{r,s}} = (2-\delta_{rs})\sum_{j=1}^{J} \left[\boldsymbol{S}_j'\boldsymbol{\Sigma}_j^{11}\boldsymbol{S}_j - \boldsymbol{r}_j\boldsymbol{r}_j'\right]_{r,s}. \tag{12.62}$$

Calculation of $S_j'\Sigma_j^{11}S_j$

From (12.48) it follows that

$$S_j'\Sigma_j^{11}S_j = S_j'\Lambda_j^{-1}S_j + S_j'\Lambda_j^{-1}S_jH_jS_j'\Lambda_j^{-1}S_j\,.$$

Therefore, using (12.38) and (12.66),

$$S_j'\Sigma_j^{11}S_j = A_j(I_p + H_jA_j) = K_j\,. \tag{12.63}$$

Calculation of r_j

From the expressions (12.48) and (12.56), it follows that $S_j'\Sigma_j^{11}e_{(1)j} = S_j'\Lambda_j^{-1}e_{(1)j} + S_j'\Lambda_j^{-1}S_jH_jS_j'\Lambda_j^{-1}e_{(1)j}$. Therefore, using (12.54),

$$S_j'\Sigma_j^{11}e_{(1)j} = p_j + A_jH_jp_j = (I_p + A_jH_j)p_j\,. \tag{12.64}$$

Similarly, from (12.40) and (12.49),

$$\begin{aligned} S_j'\Sigma_j^{12}e_{(2)j} &= -S_j'\Lambda_j^{-1}S_jC_j'\Sigma_{yx}R_j'\Sigma_{22.1}^{-1}e_{(2)j} \\ &= -A_jC_j'\Sigma_{yx}R_j'\Sigma_{22.1}^{-1}e_{(2)j}\,. \end{aligned} \tag{12.65}$$

Calculation of $\partial F/\partial[\Sigma_W]_{r,s}$

From (12.5), (12.7), and (12.36) it follows that

$$\frac{\partial V_j}{\partial[\Sigma_W]_{r,s}} = \sum_{i=1}^{n_j} Z_{ij}G_{rs}Z_{ij}'\,,$$

with G_{rs} defined by (12.61), and Z_{ij} by (12.5). Therefore,

$$\begin{aligned} \frac{\partial F}{\partial[\Sigma_W]_{r,s}} &= -\sum_{j=1}^{J}\sum_{i=1}^{n_j} \operatorname{tr}\{(\Sigma_j^{11}e_{(1)j} + \Sigma_j^{12}e_{(2)j})(e_{(1)j}'\Sigma_j^{11} + e_{(2)j}'\Sigma_j^{21}) \\ &\qquad \times Z_{ij}G_{rs}Z_{ij}'\} + \sum_{j=1}^{J}\operatorname{tr}\Sigma_j^{-1}\frac{\partial\Sigma_j}{\partial[\Sigma_W]_{r,s}} \\ &= \sum_{j=1}^{J}\operatorname{tr}\Sigma_j^{-1}\frac{\partial\Sigma_j}{\partial[\Sigma_W]_{r,s}} - \sum_{j=1}^{J}\sum_{i=1}^{n_j}\operatorname{tr} q_{ij}q_{ij}'G_{rs} \\ &= \sum_{j=1}^{J}\operatorname{tr}\Sigma_j^{-1}\frac{\partial\Sigma_j}{\partial[\Sigma_W]_{r,s}} - (2-\delta_{rs})\sum_{j=1}^{J}\sum_{i=1}^{n_j}[q_{ij}q_{ij}']_{r,s}\,, \end{aligned}$$

where q_{ij} is defined in (12.55). Since $Z_{ij}' = (\emptyset, \ldots, S_{ij}', \ldots, \emptyset)$, it follows that (see also (12.54))

$$
\begin{aligned}
Z'_{ij}\Sigma_j^{11}e_{(1)j} &= Z'_{ij}\left[\Lambda_j^{-1} + \Lambda_j^{-1}S_j(F_j - B_j)S'_j\Lambda_j^{-1}\right]e_{(1)j} \\
&= S_{ij}\Lambda_{ij}^{-1}e_{ij} + A_{ij}H_jp_j \\
&= p_{ij} + A_{ij}H_jp_j \,.
\end{aligned}
$$

Furthermore, $Z'_{ij}\Sigma_j^{12}e_{(2)j} = -A_{ij}C'_j\Sigma_{yx}R'_j\Sigma_{22.1}^{-1}e_{(2)j}$. Finally (see (12.38) and (12.47)),

$$
\begin{aligned}
&\operatorname{tr}\Sigma_j^{-1}\frac{\partial\Sigma}{\partial[\Sigma_W]_{r,s}} \\
&\quad = \operatorname{tr}\left\{\sum_{i=1}^{n_j} Z'_{ij}\Sigma_j^{11}Z_{ij}G_{rs}\right\} \\
&\quad = \sum_{i=1}^{n_j}\operatorname{tr}(Z'_{ij}\Lambda_j^{-1}Z_{ij}G_{rs} + Z'_{ij}\Lambda_j^{-1}S_jH_jS'_j\Lambda_j^{-1}Z_{ij}G_{rs}) \\
&\quad = (2-\delta_{rs})\sum_{i=1}^{n_j}\left[A_{ij}(I_p + H_jA_{ij})\right]_{r,s} \,.
\end{aligned}
$$

Hence,

$$
\frac{\partial F}{\partial[\Sigma_W]_{r,s}} = (2-\delta_{rs})\sum_{j=1}^{J}\sum_{i=1}^{n_j}\left[A_{ij}(I_p + H_j)A_{ij} - q_{ij}q'_{ij}\right]_{r,s} \,.
$$

Calculation of $\partial F/\partial[\Sigma_{xx}]_{r,s}$

$$
\begin{aligned}
\frac{\partial F}{\partial[\Sigma_{xx}]_{r,s}} &= -\sum_{j=1}^{J}\operatorname{tr}\{(\Sigma_j^{21}e_{(1)j} + \Sigma_j^{22}e_{(2)j})\,(e'_{(1)j}\Sigma_j^{12} + e'_{(2)j}\Sigma_j^{22})R_jG_{rs}R'_j\} \\
&\quad + \sum_{j=1}^{J}\operatorname{tr}\left\{\Sigma_j^{-1}\begin{pmatrix}\emptyset & \emptyset \\ \emptyset & R_jG_{rs}R'_j\end{pmatrix}\right\} \\
&= \sum_{j=1}^{J}\operatorname{tr}\Sigma_j^{22}R_jG_{rs}R'_j - \sum_{j=1}^{J}\operatorname{tr}s_js'_jG_{rs} \\
&= (2-\delta_{rs})\sum_{j=1}^{J}\left[R'_j\Sigma_j^{22}R_j - \sum_{j=1}^{J}s_js'_j\right]_{r,s} \,,
\end{aligned}
$$

where s_j is defined in (12.57).

12.B.4 Hessian Matrix

The following matrix expressions are defined in terms of (12.38)–(12.41) and (12.45)–(12.47) in order to simplify elements of the Hessian:

$$\boldsymbol{K}_j = \boldsymbol{S}_j'\boldsymbol{\Sigma}_j^{11}\boldsymbol{S}_j = \boldsymbol{A}_j(\boldsymbol{I}_p + \boldsymbol{H}_j\boldsymbol{A}_j) \tag{12.66}$$

and

$$\boldsymbol{L}_j\colon (p \times q) = \boldsymbol{S}_j'\boldsymbol{\Sigma}_j^{12}\boldsymbol{R}_j = -\boldsymbol{A}_j\boldsymbol{C}_j'\boldsymbol{\Sigma}_{yx}\boldsymbol{R}_j'\boldsymbol{\Sigma}_{22.1}^{-1}\boldsymbol{R}_j = -\boldsymbol{D}_j'\boldsymbol{\Sigma}_{yx}\boldsymbol{E}_j\,. \tag{12.67}$$

Let

$$\boldsymbol{M}_{jrs} = \sum_{i=1}^{n_j}\boldsymbol{A}_{ij}\boldsymbol{G}_{rs}\boldsymbol{A}_{ij} = \sum_{i=1}^{n_j}\{\boldsymbol{a}_{ijr}\boldsymbol{a}_{ijs}' + (1-\delta_{rs})\boldsymbol{a}_{ijs}\boldsymbol{a}_{ijr}'\}, \tag{12.68}$$

with $\boldsymbol{a}_{ijr}$ the r-th column of $\boldsymbol{A}_{ij} = \boldsymbol{S}_{ij}'\boldsymbol{\Lambda}_{ij}^{-1}\boldsymbol{S}_{ij}$. Also, let

$$\boldsymbol{N}_{ij} = \boldsymbol{A}_{ij}\boldsymbol{H}_j\boldsymbol{A}_{ij}\,. \tag{12.69}$$

Note

To compute $\boldsymbol{M}_{jrs} = \sum_{i=1}^{n_j}\boldsymbol{A}_{ij}\boldsymbol{G}_{rs}\boldsymbol{A}_{ij}$ let $\boldsymbol{j}_r$ be a column vector with all elements equal to zero except for the r-th element, which equals 1. Hence (see (12.61)),

$$\boldsymbol{G}_{rs} = \boldsymbol{J}_{rs} + (1-\delta_{rs})\boldsymbol{J}_{sr} = \boldsymbol{j}_r\boldsymbol{j}_s' + (1-\delta_{rs})\boldsymbol{j}_s\boldsymbol{j}_r'\,.$$

Note that $\boldsymbol{a}_{ijr} = \boldsymbol{A}_{ij}\boldsymbol{j}_r$ is the r-th column of the symmetric matrix $\boldsymbol{A}_{ij}$. Therefore,

$$\sum_{i=1}^{n_j}\boldsymbol{A}_{ij}\boldsymbol{G}_{rs}\boldsymbol{A}_{ij} = \sum_{i=1}^{n_j}\{\boldsymbol{a}_{ijr}\boldsymbol{a}_{ijs}' + (1-\delta_{rs})\boldsymbol{a}_{ijs}\boldsymbol{a}_{ijr}'\}.$$

Finally, define the block-diagonal matrix $\boldsymbol{D}_{jrs}$ as

$$\boldsymbol{D}_{jrs} = \bigoplus_{i=1}^{n_j}\boldsymbol{S}_{ij}\boldsymbol{G}_{rs}\boldsymbol{S}_{ij}'\,. \tag{12.70}$$

Let $\boldsymbol{\sigma}_B = \operatorname{vecs}\boldsymbol{\Sigma}_B$, $\boldsymbol{\sigma}_W = \operatorname{vecs}\boldsymbol{\Sigma}_W$, $\boldsymbol{\sigma}_{xx} = \operatorname{vecs}\boldsymbol{\Sigma}_{xx}$, $\boldsymbol{\sigma}_{xy} = \operatorname{vec}\boldsymbol{\Sigma}_{xy}$, and $\boldsymbol{\pi}_2' = (\boldsymbol{\sigma}_B', \boldsymbol{\sigma}_W', \boldsymbol{\sigma}_{xy}', \boldsymbol{\sigma}_{xx}')$. The Hessian with respect to $\boldsymbol{\pi}_2$: $\left[p(p+1) + pq + \frac{1}{2}q(q+1)\right] \times 1$ is

$$\boldsymbol{H}(\boldsymbol{\pi}_2) = E\begin{pmatrix} \dfrac{\partial^2\underline{F}}{\partial\boldsymbol{\sigma}_B\,\partial\boldsymbol{\sigma}_B'} & & & \text{[sym.]} \\ \dfrac{\partial^2\underline{F}}{\partial\boldsymbol{\sigma}_W\,\partial\boldsymbol{\sigma}_B'} & \dfrac{\partial^2\underline{F}}{\partial\boldsymbol{\sigma}_W\,\partial\boldsymbol{\sigma}_W'} & & \\ \dfrac{\partial^2\underline{F}}{\partial\boldsymbol{\sigma}_{xy}\,\partial\boldsymbol{\sigma}_B'} & \dfrac{\partial^2\underline{F}}{\partial\boldsymbol{\sigma}_{xy}\,\partial\boldsymbol{\sigma}_W'} & \dfrac{\partial^2\underline{F}}{\partial\boldsymbol{\sigma}_{xy}\,\partial\boldsymbol{\sigma}_{xy}'} & \\ \dfrac{\partial^2\underline{F}}{\partial\boldsymbol{\sigma}_{xx}\,\partial\boldsymbol{\sigma}_B'} & \dfrac{\partial^2\underline{F}}{\partial\boldsymbol{\sigma}_{xx}\,\partial\boldsymbol{\sigma}_W'} & \dfrac{\partial^2\underline{F}}{\partial\boldsymbol{\sigma}_{xx}\,\partial\boldsymbol{\sigma}_{xy}'} & \dfrac{\partial^2\underline{F}}{\partial\boldsymbol{\sigma}_{xx}\,\partial\boldsymbol{\sigma}_{xx}'} \end{pmatrix}. \tag{12.71}$$

The simplification of the terms for the between and within components, $\partial^2 F/\partial[\boldsymbol{\Sigma}_B]_{u,v}\,\partial[\boldsymbol{\Sigma}_B]_{r,s}$ and $\partial^2 F/\partial[\boldsymbol{\Sigma}_W]_{u,v}\,\partial[\boldsymbol{\Sigma}_W]_{r,s}$ are given in the form of propositions, followed by a listing of the simplified results for the remaining elements of the Hessian.

Proposition 12.1

$$E\,\frac{\partial^2 \underline{F}}{\partial[\boldsymbol{\Sigma}_B]_{u,v}\,\partial[\boldsymbol{\Sigma}_B]_{r,s}} \tag{12.72}$$

$$= \tfrac{1}{2}\sum_{j=1}^{J} \operatorname{tr} \boldsymbol{K}_j \boldsymbol{G}_{rs} \boldsymbol{K}_j \boldsymbol{G}_{uv}$$

$$= \frac{(2-\delta_{rs})(2-\delta_{uv})}{4}\sum_{j=1}^{J}([\boldsymbol{K}_j]_{r,u}[\boldsymbol{K}_j]_{s,v} + [\boldsymbol{K}_j]_{r,v}[\boldsymbol{K}_j]_{s,u}) \tag{12.73}$$

with $\boldsymbol{K}_j$ defined in (12.66) and F by (12.10).

Proof. From (12.61),

$$\frac{\partial \boldsymbol{\Sigma}_j}{\partial[\boldsymbol{\Sigma}_B]_{r,s}} = \begin{pmatrix} \boldsymbol{S}_j \boldsymbol{G}_{rs} \boldsymbol{S}_j' & \emptyset \\ \emptyset & \emptyset \end{pmatrix}.$$

Using this result and (12.44), $[\boldsymbol{H}(\boldsymbol{\Sigma}_B)]_{rs,uv}$ can be written as

$$[\boldsymbol{H}(\boldsymbol{\Sigma}_B)]_{rs,uv} = \tfrac{1}{2}\sum_{j=1}^{J} \operatorname{tr}\left\{\boldsymbol{\Sigma}_j^{-1}\frac{\partial \boldsymbol{\Sigma}_j}{\partial[\boldsymbol{\Sigma}_B]_{r,s}}\boldsymbol{\Sigma}_j^{-1}\frac{\partial \boldsymbol{\Sigma}_j}{\partial[\boldsymbol{\Sigma}_B]_{u,v}}\right\}$$

$$= \tfrac{1}{2}\sum_{j=1}^{J} \operatorname{tr}\{\boldsymbol{\Sigma}_j^{11}\boldsymbol{S}_j\boldsymbol{G}_{rs}\boldsymbol{S}_j'\boldsymbol{\Sigma}_j^{11}\boldsymbol{S}_j\boldsymbol{G}_{uv}\boldsymbol{S}_j'\}.$$

Since $\operatorname{tr}\boldsymbol{AB} = \operatorname{tr}\boldsymbol{BA}$ with $\boldsymbol{A} = \boldsymbol{S}_j'$ and $\boldsymbol{B} = \boldsymbol{\Sigma}_j^{11}\boldsymbol{S}_j\boldsymbol{G}_{rs}\boldsymbol{S}_j'\boldsymbol{\Sigma}_j^{11}\boldsymbol{S}_j\boldsymbol{G}_{uv}$,

$$[\boldsymbol{H}(\boldsymbol{\Sigma}_B)]_{rs,uv} = \tfrac{1}{2}\sum_{j=1}^{J} \operatorname{tr}\{\boldsymbol{S}_j'\boldsymbol{\Sigma}_j^{11}\boldsymbol{S}_j\boldsymbol{G}_{rs}\boldsymbol{S}_j'\boldsymbol{\Sigma}_j^{11}\boldsymbol{S}_j\boldsymbol{G}_{uv}\}.$$

From (12.66), $\boldsymbol{K}_j = \boldsymbol{S}_j'\boldsymbol{\Sigma}_j^{11}\boldsymbol{S}_j$, so that

$$[\boldsymbol{H}(\boldsymbol{\Sigma}_B)]_{rs,uv} = \tfrac{1}{2}\sum_{j=1}^{J} \operatorname{tr} \boldsymbol{K}_j\boldsymbol{G}_{rs}\boldsymbol{K}_j\boldsymbol{G}_{uv}\,.$$

From Bargmann [5] the following property holds: $\operatorname{tr}\boldsymbol{A}\boldsymbol{J}_{ij}\boldsymbol{B}\boldsymbol{J}_{rs} = [\boldsymbol{A}]_{si}[\boldsymbol{B}]_{jr}$. Using (12.63) and the definition of $\boldsymbol{G}_{rs}$ (see (12.61)), it follows that the formula for $[\boldsymbol{H}(\boldsymbol{\Sigma}_B)]_{rs,uv}$ can be simplified to the form (12.73). □

Proposition 12.2

$$E\,\frac{\partial^2 \underline{F}}{\partial[\boldsymbol{\Sigma}_W]_{u,v}\,\partial[\boldsymbol{\Sigma}_W]_{r,s}} = \tfrac{1}{2}\sum_{j=1}^{J}\operatorname{tr}\left\{\boldsymbol{\Sigma}_j^{11}\frac{\partial \boldsymbol{V}_j}{\partial[\boldsymbol{\Sigma}_W]_{r,s}}\boldsymbol{\Sigma}_j^{11}\frac{\partial \boldsymbol{V}_j}{\partial[\boldsymbol{\Sigma}_W]_{r,s}}\right\}$$

$$= \tfrac{1}{2}\sum_{j=1}^{J}\operatorname{tr}\boldsymbol{\Sigma}_j^{11}\boldsymbol{D}_{jrs}\boldsymbol{\Sigma}_j^{11}\boldsymbol{D}_{uv}$$

$$= \tfrac{1}{2}\sum_{j=1}^{J}\sum_{i=1}^{n_j}\operatorname{tr}\boldsymbol{A}_{ij}\boldsymbol{G}_{rs}\boldsymbol{A}_{ij}\boldsymbol{G}_{uv}$$

$$+ \tfrac{1}{2}\sum_{j=1}^{J}\operatorname{tr}\boldsymbol{M}_{jrs}\boldsymbol{H}_j\boldsymbol{M}_{juv}\boldsymbol{H}_j$$

$$+ \sum_{j=1}^{J}\sum_{i=1}^{n_j}\operatorname{tr}\boldsymbol{N}_{ij}\boldsymbol{G}_{rs}\boldsymbol{A}_{ij}\boldsymbol{G}_{uv}\,,$$

where

$$\frac{\partial \boldsymbol{V}_j}{\partial[\boldsymbol{\Sigma}_W]_{r,s}} = \sum_{i=1}^{n_j}\boldsymbol{Z}_{ij}\boldsymbol{G}_{rs}\boldsymbol{Z}'_{ij}$$

and where $\boldsymbol{G}_{rs}$ and $\boldsymbol{D}_{rs}$ have been defined in (12.61) and (12.70), respectively.

Proof. From (12.48), $\boldsymbol{\Sigma}_j^{11} = \boldsymbol{\Lambda}_j^{-1} + \boldsymbol{\Lambda}_j^{-1}\boldsymbol{S}_j\boldsymbol{H}_j\boldsymbol{S}'_j\boldsymbol{\Lambda}_j^{-1}$, so that

$$\tfrac{1}{2}\sum_{j=1}^{J}\operatorname{tr}\boldsymbol{\Sigma}_j^{11}\boldsymbol{D}_{jrs}\boldsymbol{\Sigma}_j^{11}\boldsymbol{D}_{juv}$$

$$= \tfrac{1}{2}\sum_{j=1}^{J}\operatorname{tr}\{(\boldsymbol{\Lambda}_j^{-1} + \boldsymbol{\Lambda}_j^{-1}\boldsymbol{S}_j\boldsymbol{H}_j\boldsymbol{S}'_j\boldsymbol{\Lambda}_j^{-1})\boldsymbol{D}_{jrs}(\boldsymbol{\Lambda}_j^{-1} + \boldsymbol{\Lambda}_j^{-1}\boldsymbol{S}_j\boldsymbol{H}_j\boldsymbol{S}'_j\boldsymbol{\Lambda}_j^{-1})\boldsymbol{D}_{juv}\}$$

$$= \tfrac{1}{2}\sum_{j=1}^{J}\operatorname{tr}\boldsymbol{\Lambda}_j^{-1}\boldsymbol{D}_{jrs}\boldsymbol{\Lambda}_j^{-1}\boldsymbol{D}_{juv}$$

$$+ \tfrac{1}{2}\sum_{j=1}^{J}\operatorname{tr}(\boldsymbol{\Lambda}_j^{-1}\boldsymbol{S}_j\boldsymbol{H}_j\boldsymbol{S}'_j\boldsymbol{\Lambda}_j^{-1}\boldsymbol{D}_{jrs}\boldsymbol{\Lambda}_j^{-1}\boldsymbol{S}_j\boldsymbol{H}_j\boldsymbol{S}'_j\boldsymbol{\Lambda}_j^{-1}\boldsymbol{D}_{juv})$$

$$+ \sum_{j=1}^{J}\operatorname{tr}(\boldsymbol{S}'_j\boldsymbol{\Lambda}_j^{-1}\boldsymbol{D}_{jrs}\boldsymbol{\Lambda}_j^{-1}\boldsymbol{D}_{juv}\boldsymbol{\Lambda}_j^{-1}\boldsymbol{S}_j\boldsymbol{H}_j). \tag{12.74}$$

The three terms can be simplified as follows.

Term 1: Using (12.37) and (12.38),

$$\tfrac{1}{2}\sum_{j=1}^{J} \operatorname{tr} \boldsymbol{\Lambda}_j^{-1}\boldsymbol{D}_{jrs}\boldsymbol{\Lambda}_j^{-1}\boldsymbol{D}_{juv}$$

$$= \tfrac{1}{2}\sum_{j=1}^{J}\sum_{i=1}^{n_j} \operatorname{tr}\{(\boldsymbol{S}'_{ij}\boldsymbol{\Lambda}_{ij}^{-1}\boldsymbol{S}_{ij})\boldsymbol{G}_{rs}(\boldsymbol{S}'_{ij}\boldsymbol{\Lambda}_{ij}^{-1}\boldsymbol{S}_{ij})\boldsymbol{G}_{uv}\}$$

$$= \tfrac{1}{2}\sum_{j=1}^{J}\sum_{i=1}^{n_j} \operatorname{tr} \boldsymbol{A}_{ij}\boldsymbol{G}_{rs}\boldsymbol{A}_{ij}\boldsymbol{G}_{uv}\,.$$

Term 2: Using the well-known property $\operatorname{tr}\boldsymbol{AB} = \operatorname{tr}\boldsymbol{BA}$ with $\boldsymbol{A} = \boldsymbol{\Lambda}_j^{-1}\boldsymbol{S}_j\boldsymbol{H}_j$,

$$\tfrac{1}{2}\sum_{j=1}^{J} \operatorname{tr}(\boldsymbol{\Lambda}_j^{-1}\boldsymbol{S}_j\boldsymbol{H}_j\boldsymbol{S}'_j\boldsymbol{\Lambda}_j^{-1}\boldsymbol{D}_{jrs}\boldsymbol{\Lambda}_j^{-1}\boldsymbol{S}_j\boldsymbol{H}_j\boldsymbol{S}'_j\boldsymbol{\Lambda}_j^{-1}\boldsymbol{D}_{juv})$$

$$= \tfrac{1}{2}\sum_{j=1}^{J} \operatorname{tr}(\boldsymbol{S}'_j\boldsymbol{\Lambda}_j^{-1}\boldsymbol{D}_{jrs}\boldsymbol{\Lambda}_j^{-1}\boldsymbol{S}_j\boldsymbol{H}_j\boldsymbol{S}'_j\boldsymbol{\Lambda}_j^{-1}\boldsymbol{D}_{juv}\boldsymbol{\Lambda}_j^{-1}\boldsymbol{S}_j\boldsymbol{H}_j).$$

Using (12.38) and (12.70), $\boldsymbol{S}'_j\boldsymbol{\Lambda}_j^{-1}\boldsymbol{D}_{jrs}\boldsymbol{\Lambda}_j^{-1}\boldsymbol{S}_j$ can be rewritten as

$$\begin{aligned}\boldsymbol{S}'_j\boldsymbol{\Lambda}_j^{-1}\boldsymbol{D}_{jrs}\boldsymbol{\Lambda}_j^{-1}\boldsymbol{S}_j &= \sum_{i=1}^{n_j} \boldsymbol{S}'_{ij}\boldsymbol{\Lambda}_{ij}^{-1}\boldsymbol{S}_{ij}\boldsymbol{G}_{rs}\boldsymbol{S}'_{ij}\boldsymbol{\Lambda}_{ij}^{-1}\boldsymbol{S}_{ij}\\ &= \sum_{i=1}^{n_j} \boldsymbol{A}_{ij}\boldsymbol{G}_{rs}\boldsymbol{A}_{ij}\\ &= \boldsymbol{M}_{jrs}\,,\end{aligned}$$

with $\boldsymbol{M}_{jrs}$ defined in (12.68). Substitution of this result allows the simplification of the second term to

$$\tfrac{1}{2}\sum_{j=1}^{J} \operatorname{tr}(\boldsymbol{\Lambda}_j^{-1}\boldsymbol{S}_j\boldsymbol{H}_j\boldsymbol{S}'_j\boldsymbol{\Lambda}_j^{-1}\boldsymbol{D}_{jrs}\boldsymbol{\Lambda}_j^{-1}\boldsymbol{S}_j\boldsymbol{H}_j\boldsymbol{S}'_j\boldsymbol{\Lambda}_j^{-1}\boldsymbol{D}_{juv})$$

$$= \tfrac{1}{2}\sum_{j=1}^{J} \operatorname{tr} \boldsymbol{M}_{jrs}\boldsymbol{H}_j\boldsymbol{M}_{juv}\boldsymbol{H}_j\,.$$

Term 3: The final term can be simplified using the definitions of $\boldsymbol{A}_{ij}$, $\boldsymbol{D}_{jrs}$, and $\boldsymbol{G}_{rs}$ (see (12.38), (12.70), and (12.61)) to write

$$\boldsymbol{S}'_j\boldsymbol{\Lambda}_j^{-1}\boldsymbol{D}_{jrs}\boldsymbol{\Lambda}_j^{-1}\boldsymbol{D}_{juv}\boldsymbol{\Lambda}_j^{-1}\boldsymbol{S}_j = \sum_{i=1}^{n_j} \boldsymbol{A}_{ij}\boldsymbol{G}_{rs}\boldsymbol{A}_{ij}\boldsymbol{G}_{uv}\boldsymbol{A}_{ij}\,.$$

Thus (see (12.69)),

$$\sum_{j=1}^{J} \operatorname{tr}(\boldsymbol{S}_j' \boldsymbol{\Lambda}_j^{-1} \boldsymbol{D}_{jrs} \boldsymbol{\Lambda}_j^{-1} \boldsymbol{D}_{juv} \boldsymbol{\Lambda}_j^{-1} \boldsymbol{S}_j \boldsymbol{H}_j) = \sum_{j=1}^{J} \sum_{i=1}^{n_j} \operatorname{tr} \boldsymbol{N}_{ij} \boldsymbol{G}_{rs} \boldsymbol{A}_{ij} \boldsymbol{G}_{uv} .$$

When the results obtained for the three terms are substituted in (12.74), the proposition follows. □

The remaining components of the Hessian are obtained in a similar way. Expressions for each remaining component of (12.71) are given below.

$\boldsymbol{H}(1,2)$:

$$\begin{aligned}
E \frac{\partial^2 \underline{F}}{\partial[\boldsymbol{\Sigma}_B]_{r,s}\, \partial[\boldsymbol{\Sigma}_W]_{u,v}}
&= \tfrac{1}{2} \sum_{j=1}^{J} \operatorname{tr}(\boldsymbol{S}_j' \boldsymbol{\Sigma}_j^{11} \boldsymbol{D}_{juv} \boldsymbol{\Sigma}_j^{11} \boldsymbol{S}_j \boldsymbol{G}_{rs}) \\
&= \tfrac{1}{2} \sum_{j=1}^{J} \operatorname{tr}\{(\boldsymbol{S}_j' \boldsymbol{\Lambda}_j^{-1} + \boldsymbol{A}_j \boldsymbol{H}_j \boldsymbol{S}_j' \boldsymbol{\Lambda}_j^{-1}) \boldsymbol{D}_{juv} (\boldsymbol{\Lambda}_j^{-1} \boldsymbol{S}_j + \boldsymbol{\Lambda}_j^{-1} \boldsymbol{S}_j \boldsymbol{H}_j \boldsymbol{A}_j) \boldsymbol{G}_{rs}\} \\
&= \tfrac{1}{2} \sum_{j=1}^{J} \operatorname{tr}\{(\boldsymbol{I}_p + \boldsymbol{A}_j \boldsymbol{H}_j) \boldsymbol{S}_j' \boldsymbol{\Lambda}_j^{-1} \boldsymbol{D}_{juv} \boldsymbol{\Lambda}_j^{-1} \boldsymbol{S}_j (\boldsymbol{I}_p + \boldsymbol{H}_j \boldsymbol{A}_j) \boldsymbol{G}_{rs}\} \\
&= \tfrac{1}{2} \sum_{j=1}^{J} \operatorname{tr}\{(\boldsymbol{I}_p + \boldsymbol{A}_j \boldsymbol{H}_j) \boldsymbol{M}_{juv} (\boldsymbol{I}_p + \boldsymbol{H}_j \boldsymbol{A}_j) \boldsymbol{G}_{rs}\}.
\end{aligned}$$

$\boldsymbol{H}(1,3)$:

$$\begin{aligned}
E \frac{\partial^2 \underline{F}}{\partial[\boldsymbol{\Sigma}_B]_{r,s}\, \partial[\boldsymbol{\Sigma}_{xy}]_{u,v}}
&= \tfrac{1}{2} \sum_{j=1}^{J} \operatorname{tr}(\boldsymbol{S}_j' \boldsymbol{\Sigma}_j^{11} \boldsymbol{S}_j \boldsymbol{G}_{rs} \boldsymbol{S}_j' \boldsymbol{\Sigma}_j^{12} \boldsymbol{R}_j \boldsymbol{J}_{uv} + \boldsymbol{R}_j \boldsymbol{J}_{uv} \boldsymbol{S}_j' \boldsymbol{\Sigma}_j^{11} \boldsymbol{S}_j \boldsymbol{G}_{rs} \boldsymbol{S}_j' \boldsymbol{\Sigma}_j^{12}) \\
&= \sum_{j=1}^{J} \operatorname{tr}(\boldsymbol{S}_j' \boldsymbol{\Sigma}_j^{11} \boldsymbol{S}_j \boldsymbol{G}_{rs} \boldsymbol{S}_j' \boldsymbol{\Sigma}_j^{12} \boldsymbol{R}_j \boldsymbol{J}_{uv}) \\
&= \sum_{j=1}^{J} \operatorname{tr} \boldsymbol{K}_j \boldsymbol{G}_{rs} \boldsymbol{L}_j \boldsymbol{J}_{uv} .
\end{aligned}$$

$\boldsymbol{H}(2,3)$: Using (12.45) and (12.49),

$$
\begin{aligned}
E\,&\frac{\partial^2 \underline{F}}{\partial[\boldsymbol{\Sigma}_W]_{r,s}\,\partial[\boldsymbol{\Sigma}_{xy}]_{u,v}} \\
&= \tfrac{1}{2}\sum_{j=1}^{J} \operatorname{tr}(\boldsymbol{\Sigma}_j^{11}\boldsymbol{D}_{jrs}\boldsymbol{\Sigma}_j^{12}\boldsymbol{R}_j\boldsymbol{J}_{uv}\boldsymbol{S}_j' + \boldsymbol{\Sigma}_j^{21}\boldsymbol{D}_{jrs}\boldsymbol{\Sigma}_j^{11}\boldsymbol{S}_j\boldsymbol{J}_{uv}'\boldsymbol{R}_j') \\
&= \sum_{j=1}^{J} \operatorname{tr}(\boldsymbol{S}_j'\boldsymbol{\Sigma}_j^{11}\boldsymbol{D}_{jrs}\boldsymbol{\Sigma}_j^{12}\boldsymbol{R}_j\boldsymbol{J}_{uv}) \\
&= -\sum_{j=1}^{J} \operatorname{tr}\{(\boldsymbol{I}_p + \boldsymbol{A}_j\boldsymbol{H}_j)\boldsymbol{S}_j'\boldsymbol{\Lambda}_j^{-1}\boldsymbol{D}_{jrs}\boldsymbol{\Lambda}_j^{-1}\boldsymbol{S}_j\boldsymbol{C}_j'\boldsymbol{\Sigma}_{yx}\boldsymbol{E}_j\boldsymbol{J}_{uv}\}.
\end{aligned}
$$

But (see (12.68))

$$
\boldsymbol{S}_j'\boldsymbol{\Lambda}_j^{-1}\boldsymbol{D}_{jrs}\boldsymbol{\Lambda}_j^{-1}\boldsymbol{S}_j = \sum_{i=1}^{n_j} \boldsymbol{A}_{ij}\boldsymbol{G}_{rs}\boldsymbol{A}_{ij} = \boldsymbol{M}_{jrs}
$$

and $\boldsymbol{K}_j = \boldsymbol{S}_j'\boldsymbol{\Sigma}_j^{11}\boldsymbol{S}_j$, so that

$$
\begin{aligned}
E\,&\frac{\partial^2 \underline{F}}{\partial[\boldsymbol{\Sigma}_W]_{r,s}\,\partial[\boldsymbol{\Sigma}_{xy}]_{u,v}} \\
&= -\sum_{j=1}^{J} \operatorname{tr}\{(\boldsymbol{I}_p + \boldsymbol{A}_j\boldsymbol{H}_j)\boldsymbol{S}_j'\boldsymbol{\Lambda}_j^{-1}\boldsymbol{D}_{jrs}\boldsymbol{\Lambda}_j^{-1}\boldsymbol{S}_j\boldsymbol{C}_j'\boldsymbol{\Sigma}_{yx}\boldsymbol{E}_j\boldsymbol{J}_{uv}\} \\
&= -\sum_{j=1}^{J} \operatorname{tr}(\boldsymbol{K}_j\boldsymbol{M}_{jrs}\boldsymbol{C}_j'\boldsymbol{\Sigma}_{yx}\boldsymbol{E}_j\boldsymbol{J}_{uv}).
\end{aligned}
$$

$\boldsymbol{H}(3,3)$:

$$
\begin{aligned}
E\,&\frac{\partial^2 \underline{F}}{\partial[\boldsymbol{\Sigma}_{xy}]_{r,s}\,\partial[\boldsymbol{\Sigma}_{xy}]_{u,v}} \\
&= \sum_{j=1}^{J} \operatorname{tr}(\boldsymbol{S}_j'\boldsymbol{\Sigma}_j^{12}\boldsymbol{R}_j\boldsymbol{J}_{rs}\boldsymbol{S}_j'\boldsymbol{\Sigma}_j^{12}\boldsymbol{R}_j\boldsymbol{J}_{uv} + \boldsymbol{S}_j'\boldsymbol{\Sigma}_j^{11}\boldsymbol{S}_j\boldsymbol{J}_{rs}'\boldsymbol{R}_j'\boldsymbol{\Sigma}_j^{22}\boldsymbol{R}_j\boldsymbol{J}_{uv}) \\
&= \sum_{j=1}^{J} \operatorname{tr}(\boldsymbol{L}_j\boldsymbol{J}_{rs}\boldsymbol{L}_j\boldsymbol{J}_{uv} + \boldsymbol{K}_j\boldsymbol{J}_{rs}'\boldsymbol{E}_j\boldsymbol{J}_{uv}).
\end{aligned}
$$

$\boldsymbol{H}(1,4)$:

$$
\begin{aligned}
E\,\frac{\partial^2 \underline{F}}{\partial[\boldsymbol{\Sigma}_B]_{r,s}\,\partial[\boldsymbol{\Sigma}_{xx}]_{u,v}} &= \tfrac{1}{2}\sum_{j=1}^{J} \operatorname{tr}(\boldsymbol{S}_j'\boldsymbol{\Sigma}_j^{12}\boldsymbol{R}_j\boldsymbol{G}_{uv}\boldsymbol{R}_j'\boldsymbol{\Sigma}_j^{21}\boldsymbol{S}_j\boldsymbol{G}_{rs}) \\
&= \tfrac{1}{2}\sum_{j=1}^{J} \operatorname{tr}\boldsymbol{L}_j'\boldsymbol{G}_{rs}\boldsymbol{L}_j\boldsymbol{G}_{uv}\,,
\end{aligned}
$$

with $\boldsymbol{G}_{rs}$ defined in (12.61).

$\boldsymbol{H}(2,4)$:

$$E\frac{\partial^2 \underline{F}}{\partial[\boldsymbol{\Sigma}_W]_{r,s}\,\partial[\boldsymbol{\Sigma}_{xx}]_{u,v}}$$
$$= \tfrac{1}{2}\sum_{j=1}^{J}\mathrm{tr}(\boldsymbol{R}_j'\boldsymbol{\Sigma}_j^{21}\boldsymbol{D}_{jrs}\boldsymbol{\Sigma}_j^{12}\boldsymbol{R}_j\boldsymbol{G}_{uv})$$
$$= \tfrac{1}{2}\sum_{j=1}^{J}\mathrm{tr}(\boldsymbol{R}_j'\boldsymbol{\Sigma}_{22.1}^{-1}\boldsymbol{R}_j\boldsymbol{\Sigma}_{xy}\boldsymbol{C}_j\boldsymbol{S}_j'\boldsymbol{\Lambda}_j^{-1}\boldsymbol{D}_{jrs}\boldsymbol{\Lambda}_j^{-1}\boldsymbol{S}_j\boldsymbol{C}_j'\boldsymbol{\Sigma}_{yx}\boldsymbol{R}_j'\boldsymbol{\Sigma}_{22.1}^{-1}\boldsymbol{R}_j\boldsymbol{G}_{uv})$$
$$= \tfrac{1}{2}\sum_{j=1}^{J}\mathrm{tr}(\boldsymbol{E}_j\boldsymbol{\Sigma}_{xy}\boldsymbol{C}_j\boldsymbol{M}_{jrs}\boldsymbol{C}_j'\boldsymbol{\Sigma}_{yx}\boldsymbol{E}_j\boldsymbol{G}_{uv}).$$

$\boldsymbol{H}(3,4)$: From (12.45) and (12.67),

$$E\frac{\partial^2 \underline{F}}{\partial[\boldsymbol{\Sigma}_{xx}]_{r,s}\,\partial[\boldsymbol{\Sigma}_{xy}]_{u,v}}$$
$$= \tfrac{1}{2}\sum_{j=1}^{J}\mathrm{tr}(\boldsymbol{S}_j'\boldsymbol{\Sigma}_j^{12}\boldsymbol{R}_j\boldsymbol{G}_{rs}\boldsymbol{R}_j'\boldsymbol{\Sigma}_j^{22}\boldsymbol{R}_j\boldsymbol{J}_{uv} + \boldsymbol{R}_j\boldsymbol{J}_{uv}\boldsymbol{S}_j'\boldsymbol{\Sigma}_j^{12}\boldsymbol{R}_j\boldsymbol{G}_{rs}\boldsymbol{R}_j'\boldsymbol{\Sigma}_j^{22})$$
$$= \tfrac{1}{2}\sum_{j=1}^{J}\mathrm{tr}(\boldsymbol{S}_j'\boldsymbol{\Sigma}_j^{12}\boldsymbol{R}_j\boldsymbol{G}_{rs}\boldsymbol{R}_j'\boldsymbol{\Sigma}_j^{22}\boldsymbol{R}_j\boldsymbol{J}_{uv} + \boldsymbol{S}_j'\boldsymbol{\Sigma}_j^{12}\boldsymbol{R}_j\boldsymbol{G}_{rs}\boldsymbol{R}_j'\boldsymbol{\Sigma}_j^{22}\boldsymbol{R}_j\boldsymbol{J}_{uv})$$
$$= \sum_{j=1}^{J}\mathrm{tr}(\boldsymbol{S}_j'\boldsymbol{\Sigma}_j^{12}\boldsymbol{R}_j\boldsymbol{G}_{rs}\boldsymbol{R}_j'\boldsymbol{\Sigma}_j^{22}\boldsymbol{R}_j\boldsymbol{J}_{uv})$$
$$= \sum_{j=1}^{J}\mathrm{tr}\,\boldsymbol{L}_j\boldsymbol{G}_{rs}\boldsymbol{E}_j\boldsymbol{J}_{uv}\,.$$

$\boldsymbol{H}(4,4)$:

$$E\frac{\partial^2 \underline{F}}{\partial[\boldsymbol{\Sigma}_{xx}]_{r,s}\,\partial[\boldsymbol{\Sigma}_{xx}]_{u,v}} = \tfrac{1}{2}\sum_{j=1}^{J}\mathrm{tr}(\boldsymbol{R}_j'\boldsymbol{\Sigma}_j^{22}\boldsymbol{R}_j\boldsymbol{G}_{rs}\boldsymbol{R}_j'\boldsymbol{\Sigma}_j^{22}\boldsymbol{R}_j\boldsymbol{G}_{uv})$$
$$= \tfrac{1}{2}\sum_{j=1}^{J}\mathrm{tr}\,\boldsymbol{E}_j\boldsymbol{G}_{rs}\boldsymbol{E}_j\boldsymbol{G}_{uv}\,.$$

12.B.5 Gradient and Hessian for the Fixed Part of the Model

For the fixed part of the model, the gradient and Hessian are derived in a similar way. Partition $\underline{\boldsymbol{y}}_j$ as $\underline{\boldsymbol{y}}_j' = (\underline{\boldsymbol{y}}_{(1)j}', \underline{\boldsymbol{x}}_j')$, with $E(\underline{\boldsymbol{y}}_{(1)j}) = \boldsymbol{X}_{(y)j}\boldsymbol{\beta}_y = \boldsymbol{\mu}_{(1)j}$ and $E(\underline{\boldsymbol{x}}_j) = \boldsymbol{X}_{(x)j}\boldsymbol{\beta}_x = \boldsymbol{\mu}_{(x)j}$. Then

$$\frac{\partial F}{\partial[\boldsymbol{\beta}_y]_r} = -\sum_{j=1}^{J} \mathrm{tr} \begin{pmatrix} \boldsymbol{y}_{(1)j} - \boldsymbol{\mu}_{(1)j} \\ \boldsymbol{x}_j - \boldsymbol{\mu}_{(x)j} \end{pmatrix}' \begin{pmatrix} \boldsymbol{\Sigma}_j^{11} & \boldsymbol{\Sigma}_j^{12} \\ \boldsymbol{\Sigma}_j^{21} & \boldsymbol{\Sigma}_j^{22} \end{pmatrix} \begin{pmatrix} \boldsymbol{X}_{(y)j}\boldsymbol{j}_r \\ \boldsymbol{\emptyset} \end{pmatrix}$$

$$= -\sum_{j=1}^{J} \mathrm{tr}\{(\boldsymbol{y}_{(1)j} - \boldsymbol{\mu}_{(1)j})'\boldsymbol{\Sigma}_j^{11}\boldsymbol{X}_{(y)j}\boldsymbol{j}_r + (\boldsymbol{x}_j - \boldsymbol{\mu}_{(x)j})'\boldsymbol{\Sigma}_j^{21}\boldsymbol{X}_{(y)j}\boldsymbol{j}_r\}$$

$$= -\sum_{j=1}^{J} (\boldsymbol{e}'_{(1)j}\boldsymbol{\Sigma}_j^{11}\boldsymbol{X}_{(y)j}\boldsymbol{j}_r + \boldsymbol{e}'_{(2)j}\boldsymbol{\Sigma}_j^{21}\boldsymbol{X}_{(y)j}\boldsymbol{j}_r),$$

where $\boldsymbol{e}_{(1)j} = \boldsymbol{y}_{(1)j} - \boldsymbol{\mu}_{(1)j}$ and $\boldsymbol{e}_{(2)j} = \boldsymbol{x}_j - \boldsymbol{\mu}_{(x)j}$. Similarly,

$$\frac{\partial F}{\partial[\boldsymbol{\beta}_x]_r} = -\sum_{j=1}^{J} \mathrm{tr} \begin{pmatrix} \boldsymbol{y}_{(1)j} - \boldsymbol{\mu}_{(1)j} \\ \boldsymbol{x}_j - \boldsymbol{\mu}_{(x)j} \end{pmatrix}' \begin{pmatrix} \boldsymbol{\Sigma}_j^{11} & \boldsymbol{\Sigma}_j^{12} \\ \boldsymbol{\Sigma}_j^{21} & \boldsymbol{\Sigma}_j^{22} \end{pmatrix} \begin{pmatrix} \boldsymbol{\emptyset} \\ \boldsymbol{X}_{(x)j}\boldsymbol{j}_r \end{pmatrix}$$

$$= -\sum_{j=1}^{J} \mathrm{tr}\{(\boldsymbol{y}_{(1)j} - \boldsymbol{\mu}_{(1)j})'\boldsymbol{\Sigma}_j^{12}\boldsymbol{X}_{(x)j}\boldsymbol{j}_r + (\boldsymbol{x}_j - \boldsymbol{\mu}_{(x)j})'\boldsymbol{\Sigma}_j^{22}\boldsymbol{X}_{(x)j}\boldsymbol{j}_r\}$$

$$= -\sum_{j=1}^{J} (\boldsymbol{e}'_{(1)j}\boldsymbol{\Sigma}_j^{12}\boldsymbol{X}_{(x)j}\boldsymbol{j}_r + \boldsymbol{e}'_{(2)j}\boldsymbol{\Sigma}_j^{22}\boldsymbol{X}_{(x)j}\boldsymbol{j}_r).$$

Turning to the elements of the Hessian, we have

$$[\boldsymbol{H}(\boldsymbol{\beta})]_{r,s} = \sum_{j=1}^{J} \left\{ \frac{\partial \boldsymbol{\mu}'_j}{\partial[\boldsymbol{\beta}]_r} \boldsymbol{\Sigma}_j^{-1} \frac{\partial \boldsymbol{\mu}_j}{\partial[\boldsymbol{\beta}]_s} \right\}.$$

In terms of $\boldsymbol{\beta}_y$, we find that

$$E \frac{\partial^2 \underline{F}}{\partial[\boldsymbol{\beta}_y]_r \, \partial[\boldsymbol{\beta}_y]_s} = \sum_{j=1}^{J} \mathrm{tr}(\boldsymbol{j}'_r \boldsymbol{X}'_{(y)j} \boldsymbol{\Sigma}_j^{11} \boldsymbol{X}_{(y)j} \boldsymbol{j}_s).$$

Also,

$$E \frac{\partial^2 \underline{F}}{\partial[\boldsymbol{\beta}_x]_r \, \partial[\boldsymbol{\beta}_x]_s} = \sum_{j=1}^{J} \mathrm{tr}(\boldsymbol{j}'_r \boldsymbol{X}'_{(x)j} \boldsymbol{\Sigma}_j^{22} \boldsymbol{X}_{(x)j} \boldsymbol{j}_s).$$

Finally,

$$E \frac{\partial^2 \underline{F}}{\partial[\boldsymbol{\beta}_y]_r \, \partial[\boldsymbol{\beta}_x]_s} = \sum_{j=1}^{J} \mathrm{tr}(\boldsymbol{j}'_r \boldsymbol{X}'_{(y)j} \boldsymbol{\Sigma}_j^{12} \boldsymbol{X}_{(x)j} \boldsymbol{j}_s).$$

12.B.6 Simplifications for Special Cases

When there are no missing data, and no x-variables on level 2, the reader may verify that expressions for the gradient vector and Hessian matrix simplify considerably [see 14]. For example,

$$\boldsymbol{H}(\boldsymbol{\sigma}_B, \boldsymbol{\sigma}_B') = \tfrac{1}{2}\boldsymbol{G}_p' \left\{ \sum_{j=1}^{J} \boldsymbol{K}_j \otimes \boldsymbol{K}_j \right\} \boldsymbol{G}_p ,$$

where $\boldsymbol{K}_j = \boldsymbol{A}_j(\boldsymbol{I}_p - \boldsymbol{B}_j\boldsymbol{A}_j)$, with $\boldsymbol{A}_j = n_j\boldsymbol{\Sigma}_W^{-1}$ and $\boldsymbol{B}_j = (\boldsymbol{\Sigma}_B^{-1} + n_j\boldsymbol{\Sigma}_W^{-1})^{-1}$. Also,

$$\boldsymbol{H}(\boldsymbol{\sigma}_B, \boldsymbol{\sigma}_W') = \tfrac{1}{2}\boldsymbol{G}_p' \left\{ \sum_{j=1}^{J} \boldsymbol{\Sigma}_W^{-1}(\boldsymbol{I}_p - \boldsymbol{B}_j\boldsymbol{A}_j) \otimes \boldsymbol{\Sigma}_W^{-1}(\boldsymbol{I}_p - \boldsymbol{B}_j\boldsymbol{A}_j) \right\} \boldsymbol{G}_p ,$$

$$\begin{aligned}\boldsymbol{H}(\boldsymbol{\sigma}_W, \boldsymbol{\sigma}_W') = \tfrac{1}{2}\boldsymbol{G}_p' \Bigg\{ \sum_{j=1}^{J} & n_j(\boldsymbol{\Sigma}_W^{-1} - \boldsymbol{\Sigma}_W^{-1}\boldsymbol{B}_j\boldsymbol{\Sigma}_W^{-1}) \otimes (\boldsymbol{\Sigma}_W^{-1} - \boldsymbol{\Sigma}_W^{-1}\boldsymbol{B}_j\boldsymbol{\Sigma}_W^{-1}) \\ & + n_j(n_j - 1)(\boldsymbol{\Sigma}_W^{-1}\boldsymbol{B}_j\boldsymbol{\Sigma}_W^{-1}) \otimes (\boldsymbol{\Sigma}_W^{-1}\boldsymbol{B}_j\boldsymbol{\Sigma}_W^{-1}) \Bigg\} \boldsymbol{G}_p .\end{aligned}$$

These results follow, since $\boldsymbol{S}_{ij} = \boldsymbol{I}_p$ for $j = 1, 2, \ldots, J$, $i = 1, 2, \ldots, n_j$. Use is also made of the result that $\operatorname{tr} \boldsymbol{A}\boldsymbol{G}_{rs}\boldsymbol{A}\boldsymbol{G}_{uv}$ is a typical element of $\boldsymbol{G}_p'(\boldsymbol{A} \otimes \boldsymbol{A})\boldsymbol{G}_p$, where [see, e.g., 8] $\boldsymbol{G}_p$ is a unique $p^2 \times \frac{1}{2}p(p+1)$ matrix such that $\operatorname{vec} \boldsymbol{S} = \boldsymbol{G}_p \operatorname{vecs} \boldsymbol{S}$, with $\boldsymbol{S}$ a symmetric $p \times p$ matrix, and $\boldsymbol{G}_{rs}$ was defined in (12.61).

Note that we preferred to obtain expressions for the gradient and Hessian in terms of individual elements and not in terms of the resulting Kronecker products. Suppose, for example, that $p = 10$ and $\boldsymbol{\Sigma}_W$ is constrained to be equal to a diagonal matrix. The corresponding Hessian will only contain the 55 nonduplicated elements $[\boldsymbol{H}(\boldsymbol{\Sigma}_W)]_{rr,ss}$ instead of the $\frac{1}{2}(55 \times 56)$ elements of the unconstrained case.

In general, if allowance is to be made for some elements of $\boldsymbol{\Sigma}_W$ and $\boldsymbol{\Sigma}_B$ to be fixed, it is more efficient to have elementwise expressions for the corresponding gradient and Hessian.

One could also compute the patterns of missingness within each level-2 unit, so that, for example,

$$\boldsymbol{A}_j = \sum_{i=1}^{n_j} \boldsymbol{S}_{ij}'\boldsymbol{A}_{ij}\boldsymbol{S}_{ij} = n_j^*\boldsymbol{\Sigma}_W^{-1} + \sum_{k=1}^{m} n_k\boldsymbol{S}_{ik}'\boldsymbol{A}_{ik}\boldsymbol{S}_{ik} ,$$

where n_j^* equals the number of complete patterns, m equals the number of patterns for missing value cases, and n_k denotes the number of cases belonging to pattern k, $k = 1, 2, \ldots, m$. Note that $\sum_{k=1}^{m} n_k = n_j - n_j^*$.

References

1. A. Agresti. *Categorical Data Analysis.* Wiley, New York, 1990.
2. M. Aitkin and N. Longford. Statistical modelling issues in school effectiveness studies. *Journal of the Royal Statistical Society, Series A*, 149:1–43, 1986. (with discussion)
3. H. Akaike. Factor analysis and AIC. *Psychometrika*, 52:317–332, 1987.
4. T. W. Anderson. *An Introduction to Multivariate Statistical Analysis*, 2nd edition. Wiley, New York, 1984.
5. R. E. Bargmann. Matrices and determinants. In R. C. Weast and S. M. Selby, editors, *C.R.C. Handbook of Tables in Mathematics*, 4th edition, pages 507–534. Chemical Rubber, Cleveland, OH, 1975.
6. R. D. Bock. Within-subject experimentation in psychiatric research. In R. D. Gibbons and M. W. Dysken, editors, *Statistical and Methodological Advances in Psychiatric Research*, pages 59–90. Spectrum, New York, 1983.
7. H. Bozdogan. Model selection and Akaike's Information Criterion (AIC): The general theory and its analytical extensions. *Psychometrika*, 52:345–370, 1987.
8. M. W. Browne. Generalized least squares estimators in the analysis of covariance structures. *South African Statistical Journal*, 8:1–24, 1974. (Reprinted in D. J. Aigner and A. S. Goldberger, editors, *Latent Variables in Socio-Economic Models*, pages 205–226, North-Holland, Amsterdam, 1977.)
9. M. W. Browne and R. Cudeck. Alternative ways of assessing model fit. In K. A. Bollen and J. S. Long, editors, *Testing Structural Equation Models*, pages 136–162. Sage, Thousand Oaks, CA, 1993.
10. M. W. Browne and S. H. C. du Toit. Automated fitting of nonstandard models for mean vectors and covariance matrices. *Multivariate Behavioral Research*, 27: 269–300, 1992.
11. A. S. Bryk and S. W. Raudenbush. *Hierarchical Linear Models: Applications and Data Analysis Methods.* Sage, Newbury Park, CA, 1992.
12. J. de Leeuw and I. G. G. Kreft. Random coefficient models for multilevel analysis. *Journal of Educational Statistics*, 11:57–85, 1986.
13. A. P. Dempster, D. B. Rubin, and R. K. Tsutakawa. Estimation in covariance components models. *Journal of the American Statistical Association*, 76:341–353, 1981.
14. M. du Toit. *The Analysis of Hierarchical and Unbalanced Complex Survey Data using Multilevel Models.* PhD thesis, University of Pretoria, Pretoria, South Africa, 1995.
15. M. du Toit and S. H. C. du Toit. *Interactive LISREL: User's Guide.* Scientific Software International, Chicago, 2002.
16. R. D. Gibbons and R. D. Bock. Trend in correlated proportions. *Psychometrika*, 52:113–124, 1987.
17. P. E. Gill, W. Murray, and M. H. Wright. *Practical Optimization.* Academic Press, London, 1981.
18. H. Goldstein. Multilevel mixed linear model analysis using iterative generalized least squares. *Biometrika*, 73:43–56, 1986.
19. H. Goldstein and R. P. McDonald. A general model for the analysis of multilevel data. *Psychometrika*, 53:455–467, 1988.

20. D. Holt, A. J. Scott, and P. D. Ewings. Chi-squared tests with survey data. *Journal of the Royal Statistical Society, Series A*, 143:303–320, 1980.
21. J. J. Hox. Factor analysis of multilevel data: Gauging the Muthén model. In J. H. L. Oud and R. A. W. van Blokland-Vogelesang, editors, *Advances in Longitudinal and Multivariate Analysis in the Behavioral Sciences*, pages 141–156. ITS, Nijmegen, The Netherlands, 1993.
22. R. I. Jennrich and P. F. Sampson. Application of stepwise regression to nonlinear estimation. *Technometrics*, 10:63–72, 1968.
23. K. G. Jöreskog. Structural equation models in the social sciences: Specification, estimation and testing. In P. R. Krishnaiah, editor, *Applications of Statistics*, pages 265–287. North-Holland, Amsterdam, 1977.
24. K. G. Jöreskog and D. Sörbom. *Advances in Factor Analysis and Structural Equation Models.* Abt Books, Cambridge, MA, 1979.
25. K. G. Jöreskog and D. Sörbom. *LISREL 8: Structural Equation Modeling with the SIMPLIS Command Language.* Scientific Software International, Chicago, 1993.
26. K. G. Jöreskog and D. Sörbom. *LISREL 8 User's Reference Guide.* Scientific Software International, Chicago, 1996.
27. C. G. Khatri. A note on a MANOVA model applied to problems in growth curves. *Annals of the Institute of Statistical Mathematics*, 181:75–86, 1966.
28. S.-Y. Lee. Multilevel analysis of structural equation models. *Biometrika*, 77: 763–772, 1990.
29. S.-Y. Lee and W.-Y. Poon. Analysis of two-level structural equation models via EM type algorithms. *Statistica Sinica*, 8:749–766, 1998.
30. J. Liang and P. M. Bentler. An EM algorithm for fitting two-level structural equation models. *Psychometrika*, 69:101–122, 2004.
31. N. T. Longford. A fast scoring algorithm for maximum likelihood estimation in unbalanced mixed models with nested random effects. *Biometrika*, 74:817–827, 1987.
32. W. M. Mason, G. Y. Wong, and B. Entwisle. Contextual analysis through the multilevel linear model. *Sociological Methodology*, 14:72–103, 1983.
33. J. J. McArdle and R. P. McDonald. Some algebraic properties of the Reticular Action Model for moment structures. *British Journal of Mathematical and Statistical Psychology*, 37:234–251, 1984.
34. R. P. McDonald. A general model for two-level data with responses missing at random. *Psychometrika*, 58:575–585, 1993.
35. R. P. McDonald. The Bilevel Reticular Action Model for path analysis with latent variables. *Sociological Methods & Research*, 22:399–413, 1994.
36. R. P. McDonald and H. Goldstein. Balanced versus unbalanced designs for linear structural relations in two-level data. *British Journal of Mathematical and Statistical Psychology*, 42:215–232, 1989.
37. P. Mortimore, P. Sammons, L. Stoll, D. Lewis, and R. Ecob. *School Matters, the Junior Years.* Open Books, Wells, UK, 1988.
38. B. O. Muthén. Mean and covariance structure analysis of hierarchical data. UCLA Statistics Preprint 62, UCLA, Department of Statistics, Los Angeles, 1990.

39. B. O. Muthén. Multilevel factor analysis of class and student achievement components. *Journal of Educational Measurement*, 28:338–354, 1991.
40. B. O. Muthén. Multilevel covariance structure analysis. *Sociological Methods & Research*, 22:376–398, 1994.
41. S. W. Raudenbush. Maximum likelihood estimation for unbalanced multilevel covariance structure models via the EM algorithm. *British Journal of Mathematical and Statistical Psychology*, 48:359–370, 1995.
42. S. W. Raudenbush and A. S. Bryk. A hierarchical model for studying school effects. *Sociology of Education*, 59:1–17, 1986.
43. J. H. Steiger. Structural model evaluation and modification: An interval estimation approach. *Multivariate Behavioral Research*, 25:173–180, 1990.

Author Index

Subject Index